Applied Case Studies and Solutions in Molecular Docking-Based Drug Design

Siavoush Dastmalchi
Tabriz University of Medical Sciences, Iran

Maryam Hamzeh-Mivehroud
Tabriz University of Medical Sciences, Iran

Babak Sokouti
Tabriz University of Medical Sciences, Iran

A volume in the Advances in Medical Technologies and Clinical Practice (AMTCP) Book Series

Medical Information Science Reference
An Imprint of IGI Global

Published in the United States of America by
Medical Information Science Reference (an imprint of IGI Global)
701 E. Chocolate Avenue
Hershey PA, USA 17033
Tel: 717-533-8845
Fax: 717-533-8661
E-mail: cust@igi-global.com
Web site: http://www.igi-global.com

Library of Congress Cataloging-in-Publication Data

Names: Dastmalchi, Siavoush, 1966- editor. | Hamzeh-Mivehroud, Maryam, 1981- editor. | Sokouti, Babak, 1978- editor.
Title: Applied case studies and solutions in molecular docking-based drug design / Siavoush Dastmalchi, Maryam Hamzeh-Mivehroud, and Babak Sokouti, editors.
Description: Hersey, PA : Medical Information Science Reference, [2016] | Includes bibliographical references and index.
Identifiers: LCCN 2016006034| ISBN 9781522503620 (hardcover) | ISBN 9781522503637 (ebook)
Subjects: LCSH: Pharmaceutical technology. | Drugs--Design--Computer simulation.
Classification: LCC RM301.25 A673 2016 | DDC 615.1/9--dc23 LC record available at https://lccn.loc.gov/2016006034

This book is published in the IGI Global book series Advances in Medical Technologies and Clinical Practice (AMTCP) (ISSN: 2327-9354; eISSN: 2327-9370)

British Cataloguing in Publication Data
A Cataloguing in Publication record for this book is available from the British Library.

All work contributed to this book is new, previously-unpublished material. The views expressed in this book are those of the authors, but not necessarily of the publisher.

For electronic access to this publication, please contact: eresources@igi-global.com.

Advances in Medical Technologies and Clinical Practice (AMTCP) Book Series

Srikanta Patnaik
SOA University, India
Priti Das
S.C.B. Medical College, India

ISSN: 2327-9354
EISSN: 2327-9370

Mission

Medical technological innovation continues to provide avenues of research for faster and safer diagnosis and treatments for patients. Practitioners must stay up to date with these latest advancements to provide the best care for nursing and clinical practices.

The **Advances in Medical Technologies and Clinical Practice (AMTCP) Book Series** brings together the most recent research on the latest technology used in areas of nursing informatics, clinical technology, biomedicine, diagnostic technologies, and more. Researchers, students, and practitioners in this field will benefit from this fundamental coverage on the use of technology in clinical practices.

Coverage

- Biomedical Applications
- Nursing Informatics
- Neural Engineering
- Clinical High-Performance Computing
- Medical Informatics
- Clinical Nutrition
- E-Health
- Biomechanics
- Clinical Data Mining
- Patient-Centered Care

IGI Global is currently accepting manuscripts for publication within this series. To submit a proposal for a volume in this series, please contact our Acquisition Editors at Acquisitions@igi-global.com or visit: http://www.igi-global.com/publish/.

The Advances in Medical Technologies and Clinical Practice (AMTCP) Book Series (ISSN 2327-9354) is published by IGI Global, 701 E. Chocolate Avenue, Hershey, PA 17033-1240, USA, www.igi-global.com. This series is composed of titles available for purchase individually; each title is edited to be contextually exclusive from any other title within the series. For pricing and ordering information please visit http://www.igi-global.com/book-series/advances-medical-technologies-clinical-practice/73682. Postmaster: Send all address changes to above address.

Titles in this Series

For a list of additional titles in this series, please visit: www.igi-global.com

Advancing Pharmaceutical Processes and Tools for Improved Health Outcomes
Tagelsir Mohamed Gasmelseid (International University of Africa, Sudan)
Medical Information Science Reference • copyright 2016 • 388pp • H/C (ISBN: 9781522502487) • US $185.00 (our price)

Classification and Clustering in Biomedical Signal Processing
Nilanjan Dey (Techno India College of Technology, India) and Amira Ashour (Tanta University, Egypt)
Medical Information Science Reference • copyright 2016 • 463pp • H/C (ISBN: 9781522501404) • US $225.00 (our price)

Virtual Reality Enhanced Robotic Systems for Disability Rehabilitation
Fei Hu (University of Alabama, USA) Jiang Lu (University of Houston -Clear Lake, USA) and Ting Zhang (University of Alabama, USA)
Medical Information Science Reference • copyright 2016 • 383pp • H/C (ISBN: 9781466697409) • US $210.00 (our price)

Software Innovations in Clinical Drug Development and Safety
Partha Chakraborty (Cognizant Technology Solutions, India) and Amit Nagal (GVK Bioscience, India)
Medical Information Science Reference • copyright 2016 • 305pp • H/C (ISBN: 9781466687264) • US $225.00 (our price)

Handbook of Research on Holistic Perspectives in Gamification for Clinical Practice
Daniel Novák (Czech Technical University in Prague, Czech Republic) Bengisu Tulu (Worcester Polytechnic Institute, USA) and Håvar Brendryen (Norwegian Center for Addiction Research, University of Oslo, Norway)
Medical Information Science Reference • copyright 2016 • 636pp • H/C (ISBN: 9781466695221) • US $340.00 (our price)

Optimizing Assistive Technologies for Aging Populations
Yosry S. Morsi (Swinburne University of Technology, Australia) Anupam Shukla (ABV - Indian Institute of Information Technology and Management Gwalior, India) and Chandra Prakash Rathore (Oracle India Private Limited, India)
Medical Information Science Reference • copyright 2016 • 494pp • H/C (ISBN: 9781466695306) • US $220.00 (our price)

IGI GLOBAL
DISSEMINATOR OF KNOWLEDGE
www.igi-global.com

701 E. Chocolate Ave., Hershey, PA 17033
Order online at www.igi-global.com or call 717-533-8845 x100
To place a standing order for titles released in this series, contact: cust@igi-global.com
Mon-Fri 8:00 am - 5:00 pm (est) or fax 24 hours a day 717-533-8661

Table of Contents

Detailed Table of Contents

Molecular Docking is widely used in CADD (Computer-Aided Drug Designing), SBDD (Structure-Based Drug Designing) and LBDD (Ligand-Based Drug Designing). It is a method used to predict the binding orientation of one molecule with the other and used for any kind of molecule based on the interaction like, small drug molecule with its protein target, protein – protein binding or a DNA – protein binding. Docking is very much popular technique due to its reliable prediction properties. This book chapter will provide an overview of diverse docking methodologies present that are used in drug design and development. There will be discussion on several case studies, pertaining to each method, followed by advantages and disadvantages of the discussed methodology. It will typically aim professionals in the field of cheminformatics and bioinformatics, both in academia and in industry and aspiring scientists and students who want to take up this as a profession in the near future. We will conclude with our opinion on the effectiveness of this technology in the future of pharmaceutical industry.

The computational strategies permeate all aspects of drug discovery such as virtual screening techniques. Virtual screening can be classified into ligand based and structure based methods. The ligand based method such as Quantitative Structure Activity Relationship (QSAR) is used when a set of active ligand compounds is recognized and slight or no structural information is available for the receptors. In structure based drug design, the most widespread method is molecular docking. It is widely accepted that drug activity is obtained through the molecular binding of one ligand to receptor. In their binding conformations, the molecules exhibit geometric and chemical complementarity, both of which are essential for successful drug activity. The molecular docking approach can be used to model the interaction between a small drug molecule and a protein, which allow us to characterize the performance of small molecules in the binding site of target proteins as well as to clarify fundamental biochemical processes.

Today, the development of new drugs is a challenging task of science. Researchers already applied molecular docking in the drug design field to simulate ligand- receptor interactions. Docking is a term used for computational schemes that attempt to find the "best" matching between two molecules in a complex formed from constituent molecules. It has a wide range of uses and applications in drug discovery. However, some defects still exist; the accuracy and speed of docking calculation is a challenge to explore and these methods can be enhanced as a solution to docking problem. The molecular docking problem can be defined as follows: Given the atomic coordinates of two molecules, predict their "correct" bound association. The chapter discusses common challenges critical aspects of docking method such as ligand- and receptor- conformation, flexibility and cavity detection, etc. It emphasis to the challenges and inadequacies with the theories behind as well as the examples.

Enzymes play an important role in many biologically relevant processes and are some attractive targets in the therapy and pharmaceutical research. The interaction between drugs and enzymes in vitro might account for a variety of biological processes and has attracted scientists' great interest for several decades. Investigation of the interaction can explore their mechanism of biological activities and provide useful knowledge for optimizing molecular structure of drug, prescriptions and route of administration and it can also provide the information for their bioavailability and bioactivity. In this chapter, the bindings of natural products (including flavionoids and coumarins) with three enzymes, including pepsin, hyaluronidase and acetylcholinesterase, were investigated by fluorescence spectroscopy and molecular docking. The present studies provide direct evidence at a molecular level to understand the mechanism of inhibitory effect of natural products against enzymes.

Molecular docking of ligands to DNA-targets is of great importance for the design of new anticancer drugs. Unfortunately, most docking programs were developed for protein-ligand docking which raises a

entry via membrane fusion. The present study models, optimizes and demonstrates the 3D monomer of the responsible human protein. The essential residue (studied from wet-laboratory research) was observed to be functionally conserved from multiple-sequence alignment. Thus, after performing point-mutation, the mutant protein was satisfactorily re-modelled; keeping its functionality preserved. Comparable residual participation in GP2 and each of the proteins was examined, individually. Stability of the proteins and protein-GP2 complexes on mutation; were discerned via energy calculations, solvent-accessibility area and conformational switching, with supportive statistical significances. Therefore, this probe paves a pathway to examine the weaker interaction of the stable mutated human protein with Ebola GP2 protein, thereby defending the Ebola viral entry.

Chapter 9
Molecular-Docking-Based Anti-Allergic Drug Design 232
Anamika Basu, Gurudas College, India
Piyali Basak, Jadavpur University, India
Anasua Sarkar, Jadavpur University, India

Allergens are foreign proteins that when come in contact of part(s) of human body stimulate the production of immunoglobulin types of proteins (antibodies). These allergens react with antibodies (immunoglobulin type E or IgE) produces allergic reactions, also known as immediate-type hypersensitivity reactions. As much as 20% of the general population may be affected by grass pollen as a major cause of allergic disease. EXPB class of proteins are known in the immunological literature as group-1 grass pollen allergens Molecular docking method can be used to identify the predicated the interaction of pollen allergen EXPB1 (Zea m 1), a beta-expansin and group-1 pollen allergen from maize with IgE molecules of human. The World Health Organization recognised allergen immunotherapy, as therapeutics for allergic diseases. RNA Interference (RNAi) is a biological process in which RNA molecules e.g. Small Interfering RNAs (siRNAs) inhibit gene expression, by cleavage and destruction of specific mRNA molecules. Use of Small Interfering RNA (siRNA) is a novel method in the induction of RNA Interference (RNAi), which is a potent method for therapeutics of allergic reactions. Due to various effects of STAT 6 proteins during hypersensitivity reactions caused by pollen allergens, mRNA of STAT6 gene is selected as target gene for allergy therapeutics via Post-Transcriptional Gene Silencing (PTGS). Using molecular docking study a specific sense siRNA is identified as anti allergic drug to treat allergic asthma during immediate type of hypersensitivity reaction, caused by Zea m 1 pollen allergen.

Chapter 10
Protein-Protein Interactions (PPIs) as an Alternative to Targeting the ATP Binding Site of Kinase: In Silico Approach to Identify PPI Inhibitors 249
Sailu Sarvagalla, Pondicherry University, India
Mohane Selvaraj Coumar, Pondicherry University, India

Most of the developed kinase inhibitor drugs are ATP competitive and suffer from drawbacks such as off-target kinase activity, development of resistance due to mutation in the ATP binding pocket and unfavorable intellectual property situations. Besides the ATP binding pocket, protein kinases have binding sites that are involved in Protein-Protein Interactions (PPIs); these PPIs directly or indirectly regulate the protein kinase activity. Of recent, small molecule inhibitors of PPIs are emerging as an alternative to ATP competitive agents. Rational design of inhibitors for kinase PPIs could be carried out using molecular modeling techniques. In silico tools available for the prediction of hot spot residues and

Foreword

Here is my stamp of approval for this book. I give it unreservedly.

This book has expertly assembled a diverse set of researchers to cover all aspects of Molecular Docking. The book will find favour with all comers, starting with those new to the area, but the specifics also ensure important lessons for all. Part of the secret of the approach is to have significant coverage of both methodologies as well as application and case studies.

Dastmalchi, who most credibly leads the editors of this book, is with the Medicinal Chemistry Department at the School of Pharmacy, at the Tabriz University of Medical Sciences, Tabriz and also has been Director of the Biotechnology Research Centre at the University. He is co-located with his co-editors Hamzeh-Mivehroud and Sokouti. The Centre encompasses Biochemistry, Genetics, Molecular Biology, Structural Biology and Molecular Design. The Molecular Design remit of the Centre has spawned an active and well balanced interest in all aspects of pharmaceutical biotechnology and computational biology including Molecular Docking and bioinformatics, just to mention these specific aspects. Also, at least some of Dastmalchi and Hamzeh-Mivehroud's interests have been supported by interludes at the University of Sydney.

The analogies of the discovery of new drug activities with the approaches taken to searches for valuable land by the explorers and navigators of the western world of the 17 and 18th Century I attribute to Richard (Dick) Cramer. Cramer, one of the doyens of the area, started working on the relationship between shape and biological activity in 1975, on a programme that became Comparative Molecular Field Analysis, the technique more often going by the name "CoMFA" alone. Mostly the "finds" of the explorers and navigators of 3 and 4 centuries ago were to be of small islands which had great import, although there are also some sizeable islands out there. So the mission in drug discovery becomes something akin to seeking and developing the "islands of activity" in "chemistry space". Of course land mass is valuable for different reasons, as there are also different drug activities.

The game changers in computation are now fast becoming algorithms that are termed artificial intelligence or machine learning. So, an important component is to accumulate the data from which the algorithms can learn, but the underlying assumption is fast computing. In principle all the required processing power exists…in the last 10 years supercomputers have broken the petaflop (10 to the power 15 floating operations per second) barrier, but now over 80 supercomputers have a performance over 1 petaflop. Biological applications can increasingly take advantage not only of clusters of multi-core processors, but also GPGPUs (general-purpose graphical processing units) and highly parallel coprocessors. Research teams can use local servers, shared high performance computer (HPC) centres, or turn to private cloud providers to create their own virtual computing environment. This volume has within it the clues about the algorithms that will be providing information on drug activity.

In the context of this book, there is no better way to point out recognition of the role of closely related areas than to point to the award of the 2013 Nobel Prize for Chemistry to Martin Karplus, Michael Levitt, and Arieh Warshel "for the development of multiscale models for complex chemical systems". They have contributed to computational chemistry at many levels for many years, and for those unfamiliar with all their work, be assured that many of their contributions are captured in much existing software. By all means seek the individual published works of researchers, but also be reassured that the output of researchers the calibre of these Nobel Prize winners has undergone much scrutiny. Remember that in science, there should be no obligation to fully trust any dogma. Helping one's faith can be important, and this is often achieved by great communicators and their willingness to communicate. I have crossed paths with Michael Levitt, none of the times more notable than in 1984, at the start of my career in research, when he was briefly in-house at the University of Sydney as a guest of the Inorganic Chemistry Department. He gave a series of lectures- speaking about *Protein Structure and Dynamics* - to chemists, albeit primarily to structural chemists and biochemists and protein chemists. A willingness to communicate can easily kick off a career, but communication takes the form of both the spoken and written word. The written word is amply on display and accessible in this volume.

Computational fields have generally blossomed due to the advances in computer architecture, but so too have areas of endeavour using three-dimensional structure. While I might ask you to assume for the moment that methods in small molecule structure determination have improved, it is a simple matter to cite the growth of the protein database, itself a protein structure database, and therefore representing a significant source of information on target structures of interest… after its establishment in 1972 it hit 114,697 entries at the end of 2015. It is a more simple matter to cite the protein structure numbers as there is really only the one database for protein structure, and hopefully there is nothing like a simple statistic to be convincing. Be aware that the depositions can represent structures with a different ligand or mutant proteins, rather than solely a new protein structure, but they can also merely represent a protein solved in different conditions. Unsurprisingly over the period of the accrual of these database entries all methodological aspects of protein structure determination have developed, and as these entries themselves are models, this is also an important consideration relevant to achieving a true comprehension of our biological world.

This book is a "has it all" volume for those wanting entry into the field of Structure-based Drug Design, with coverage from the theoretical to practical, and also from the here-and-now to the possible in the near future. The pedigree is right for these editors to produce a volume on Molecular Docking-Based Drug Design and Discovery. While the techniques discussed in the book are crucial, so too is the fundamental knowledge of the Drug Discovery field, including the interplay with biology, medicinal and pharmaceutics. The exquisite requirements we want to design into a therapeutic dictate that multi-disciplinary approaches are a necessity. Let there be no mistake: let's have strong computational input in all the disciplines that feed into our studies, but clearly positive clinical studies are the ultimate- the clinic is the endpoint which ever way you look at it, but there can and should be a leading role for computational effort and algorithms at every point along the path of discovery.

Let me end by ensuring I clearly commend this book to you. The editors are most ably led by my colleague and friend Siavoush Dastmalchi, and the result is a wonderfully comprehensive achievement. The principles and learnedness of Siavoush are very worthy- a not so veiled reference to the Siavoush, another Prince of Persia!

W. Bret Church
University of Sydney, Australia

Preface

The ultimate goal of medicinal chemistry is designing new drugs effective in alleviating disease states using most rational approaches. It seems that drug design process without application of computational techniques is unimaginable, and *in silico* approaches help developing potent, bioavailable, safe and well tolerated therapeutic agents. In this regard, molecular docking has been known as a valuable technique in modern structure-based drug design and become a first line technique in early phase of hit and lead identification. Location, conformation, and orientation of a molecule at the binding site of a target bio-macromolecule is predicted through molecular docking techniques. This volume continues to discuss methodologies and importance of Molecular Docking in drug design, but with more emphasis on different applications of the technique presented in the following chapters aiming to familiarize the reader with different aspects of the molecular docking technique.

The first chapter, authored by Subhabrata Sen, Rahul Agarwal, and Ashutosh Singh, has overviewed most of docking methodologies along with some case studies related to the corresponding methods. This will provide the readers a comprehensive application-based chapter entitled "Role of Molecular Docking in Computer-Aided Drug Design and Development".

In Chapter 2 titled "Application of Docking Methodologies in QSAR-Based Studies", Omar Deeb, Heidy Martinez-Pachecho, Guillermo Ramirez-Galicia, and Ramon Garduno-Juarez have discussed the important role of docking methodologies in the QSAR studies in which molecular descriptors are important properties of the target compounds. Furthermore, the relationships between 3D-QSAR studies carried out by molecular docking are presented which have also been included by a case study for providing the readers more comprehensive content.

Although the molecular docking has passed the infancy period and tremendous advances have been achieved, however, accuracy and speed of docking calculation are still challenging issues. Jahan B. Ghasemi, Azizeh Abdolmaleki, and Fereshteh Shiri contributed a chapter titled "Molecular Docking Challenges and Limitations" (Chapter 3). In this chapter, different aspects of molecular docking processes such as receptor flexibility, ligand conformation and sampling, entropy in biomolecular interaction, and role of structural water molecules in docking are reviewed. At the end solutions and recommendations are suggested.

Natural products constitute a significant portion of pharmaceuticals. Moreover, they are in most cases the main source of lead identification for drug design and discovery. Flavonoids and coumarins are a group of plant secondary metabolites found in many dietary sources of foods and beverages. These compounds exhibit broad pharmaceutical activities and wide variety of the enzymes are influenced by these nutraceuticals. Chapter 4 titled "Application of Molecular Docking in Studies on the Binding Mechanism of Three Enzymes with Natural Products" contributed by Hua-jin Zeng, Ran Yang and

Ling-bo Qu, reveals the inhibitory activity and binding of flavonoids and coumarins to the enzymes like pepsin, hyaluronidase and acetylcholinesterase using molecular docking and fluorescence spectroscopy.

Nucleic acids can be a suitable target for many therapeutics available in the market. Understanding the molecular interactions between biologically active compounds and nucleic acid is of great importance especially in designing anticancer drugs. Kateryna V. Miroshnychenko and Anna V. Shestopalova applied different docking programs such as Vina, and different versions of AutoDock for the prediction of the structures of 50 DNA-ligand complexes taken from the Nucleic Acid Database. The results were analyzed using RMSD and demonstrated in Chapter 5.

G protein-Coupled Receptors (GPCRs) are very important receptors which are widely distributed in cell surface. These receptors are the target for approximately 30% of marketed drugs. Lack of specificity observed for GPCR targeted drugs can be a major obstacle leading to undesirable side effect. Chapter 6 titled "Molecular Docking-Based Drug Design and Discovery: Rational Drug Design for the Subtype Selective GPCR Ligands" authored by Soo-Kyung Kim and William A Goddard III, focuses on the molecular docking based design of subtype-selective GPCR ligands. The interactions of the ligands with 5-HT2B (5-Hydroxytryptamine, 5-HT) serotonin receptor (HT2BR), H3 histamine receptor (H3HR) and A3 adenosine receptor (A3AR) at the atomic level using molecular docking studies are discussed in this chapter. Understanding the mode of interactions between these subtype selective ligands and GPCR subtypes would be helpful in designing of novel classes of compounds with minimum level of undesirable side effects.

Chapter 7 with the title of "Molecular Modelling, Dynamics, and Docking of Membrane Proteins: Still a Challenge" contributed by Nanda Kumar Yellapu addresses the challenging issues regarding molecular modeling, dynamics and docking of membrane proteins. Since there are difficulties associated with membrane proteins purification and crystallization, there is a great tendency toward the use of molecular modeling techniques for determining the three dimensional structures of these proteins required for docking studies and investigation of the interactions between membrane proteins and related ligands.

Ebola Virus Disease (EVD) is caused by a fatal viral hemorrhagic fever in human, which is recognized by Toll-Like Receptor-4 (TLR4) in the body. Chapter 8 contributed by Sujay Ray and Arundhati Banerjee, presents the interactions between Ebola glycoprotein and human TLR4 using molecular docking studies. The chapter starts with an introduction on Ebola virus and TLR4. Then the homology based modeling of the TLR4 along with its mutated form is presented. Moreover, the result of docking studies of Ebola glycoprotein into the modeled TLR4 and mutated form are described.

Allergens can stimulate allergic reactions mediated by different cytokines leading to production of antibodies. Antiallergic drugs reduce the signs and symptoms of the allergic reactions. The major class of the allergens belongs to grass pollens. Chapter 9, "Molecular-Docking-Based Antiallergic Drug Design," contributed by Anamika Basu, Piyali Basak, and Anasua Sarkar, explains the design of antiallergic agent against the Zea m 1 pollen allergen using molecular docking study. In this chapter, a sense siRNA is introduced as antiallergen drug against homo sapiens STAT6 mRNA as target.

As Protein-Protein Interactions (PPIs) are the source of many cellular functions and are important in wide array of signal transduction processes. This makes them a suitable target for intervention in many pathological conditions such as cancer. Among which protein kinases are important targets for pharmaceuticals such as kinase inhibitors. In Chapter 10 authored by Sailu Sarvagalla and Mohane Selvaraj Coumar, an explanation is given to highlight the important proteins involved in PPIs. Then, authors provide some practical examples related to structure-based drug design of PPI inhibitors using in silico studies such as molecular docking.

Apart from drug design and discovery, molecular docking has been applied in other fields of science. Chapter 11 titled "Applications of Molecular Docking: Its Impact and Importance Outside the Purview of Drug Discovery" contributed by Josephine Anthony, Vijaya Raghavan Rangamaran, Kumar T Sivashankarasubbiah, Dharani Gopal and Kirubagaran Ramalingam, presents the other application of the molecular docking. The authors first give a brief explanation about structure based drug design and peptidomimetic designing, then other applications of molecular docking studies in bioremediation process, fatty acid biosynthesis, nutraceuticals and nanomaterial interactions are described by providing case studies.

Taken together, we hope that the chapters in this volume are useful for newcomers as well as those in the field and provide practice-oriented applications for the molecular docking and provide thought-provoking ensemble for the readers with recipes for an appropriate application of the algorithms.

Siavoush Dastmalchi
Tabriz University of Medical Sciences, Iran

Maryam Hamzeh-Mivehroud
Tabriz University of Medical Sciences, Iran

Babak Sokouti
Tabriz University of Medical Sciences, Iran
December 30, 2015

Chapter 1
Role of Molecular Docking in Computer-Aided Drug Design and Development

Rahul Agarwal
Shiv Nadar University, India

Ashutosh Singh
Shiv Nadar University, India

Subhabrata Sen
Shiv Nadar University, India

ABSTRACT

Molecular Docking is widely used in CADD (Computer-Aided Drug Designing), SBDD (Structure-Based Drug Designing) and LBDD (Ligand-Based Drug Designing). It is a method used to predict the binding orientation of one molecule with the other and used for any kind of molecule based on the interaction like, small drug molecule with its protein target, protein – protein binding or a DNA – protein binding. Docking is very much popular technique due to its reliable prediction properties. This book chapter will provide an overview of diverse docking methodologies present that are used in drug design and development. There will be discussion on several case studies, pertaining to each method, followed by advantages and disadvantages of the discussed methodology. It will typically aim professionals in the field of cheminformatics and bioinformatics, both in academia and in industry and aspiring scientists and students who want to take up this as a profession in the near future. We will conclude with our opinion on the effectiveness of this technology in the future of pharmaceutical industry.

1. INTRODUCTION

Computer aided drug designing or CADD, is a strategy that harnesses state of the art technology to expedite the drug development process. Traditionally the drug development /discovery (that includes random screening, serendipitous discovery and process optimization) takes nearly a decade to complete with an average expense of ~300 million dollar. CADD tends to curtail this expenditure and timeline by providing

DOI: 10.4018/978-1-5225-0362-0.ch001

a holistic view of a drug discovery project and enables a SWOT (strength-weakness-opportunity-threat) analysis to evaluate the viability of the program.

Molecular docking is one of the fundamental pillars of CADD. It analyzes the binding interaction between a target protein and small molecules (Lengauer & Rarey, 1996). These small molecules also referred as ligands are the potential drug candidates that are being developed against the phenotypes and the therapeutic model for which the protein in the CADD is the target (Sliwoski, Kothiwale, Meiler, & Lowe, 2014).

2. HISTORY OF CADD

The evolution of CADD began in 1900s when Emiel Fischer (in 1894) and later Paul Ehlrich (in 1909) propagated the concept of receptors and lock and key mechanism (Fischer, 1894). Lock and key concept explains the interaction of a drug and its receptor. Just as there are innumerable pairs of lock and key, where each key inserts into a specific lock to open it, drugs are designed to be tailor made for a particular receptor. Depending on the perfection of the design, the drugs its binding capacity of the drugs with the receptor excels. But just as duplicate keys can be made for a single lock, which makes the lock vulnerable, there is a chance that multiple drugs can work on single receptor thereby giving rise to undue side effects. Additionally one drug can also interact with multiple receptors again giving rise to undue side effects. It was Paul Ehlrich, who proposed the concept of "magic bullet" the elusive drug that only bind to the chosen receptor thereby exhibiting no side effects (Elrich P, 1909,1957).

Nearly 70 years later with the advent of quantitative structure activity relationship (QSAR), CADD took its next leap towards advancement. At that time only 2-dimentional medchem properties were considered and it was much later that the 3-dimentional properties came into the picture. Gradually, with the evolution of the concepts of molecular biology, protein X-ray crystallography (refer Figure 1) and multidimensional nuclear magnetic resonance spectroscopy (NMR), CADD too evolved into a more reliable and viable strategy in drug discovery.

Advancements in high performance computing, availability of 3-dimensional structure of important pharmaceutical targets, has opened new possibilities for computational drug design. An article in Fortune magazine entitled "Next Industrial Revolution: Designing Drugs by Computer at Merck" (Sliwoski et al., 2014; Van Drie, 2007) clearly postulated the beginning of a new era for drug discovery methods involving a bevy of computational approaches.

3. COMPUTER AIDED DRUG DESIGN (CADD)

Computer Aided Drug Designing (CADD) broadly focuses on two major verticals; 1) Ligand Based Drug Designing (LBDD) and 2) Structure Based Drug Designing (SBDD). These two methods depend on the information available about the protein structure and the ligands binding to them. Ligand based drug designing involved structure activity relationship studies where structure of the receptor (mostly proteins) is unknown. Ligands were tested for their activity and pharmacophore generation. Ligand based drug discovery propagates that similar ligands are assumed to bind to similar proteins and thus to have similar biological activities.

In general LBDD involved four major steps:

Figure 1. Flowsheet of the operations in CADD

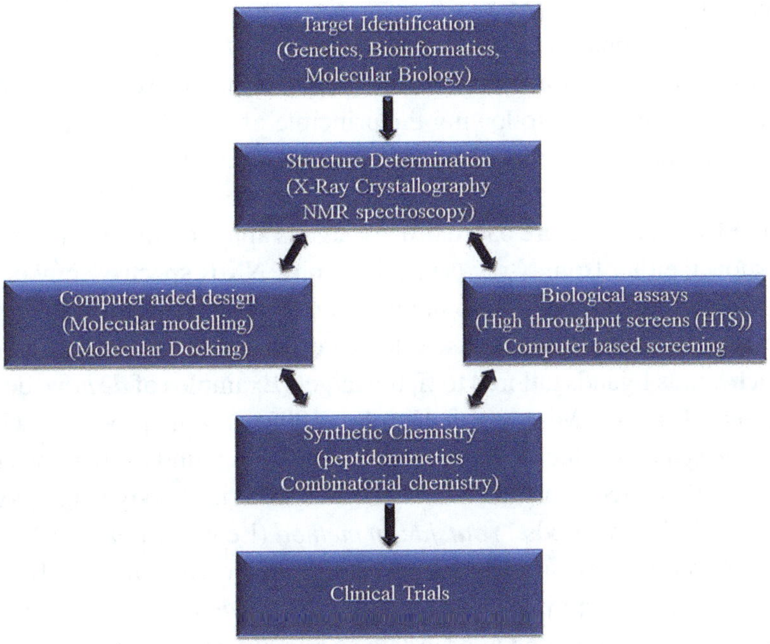

Figure 2. PDB protein structure growth per annum graph that displays the number of searchable protein structures per year

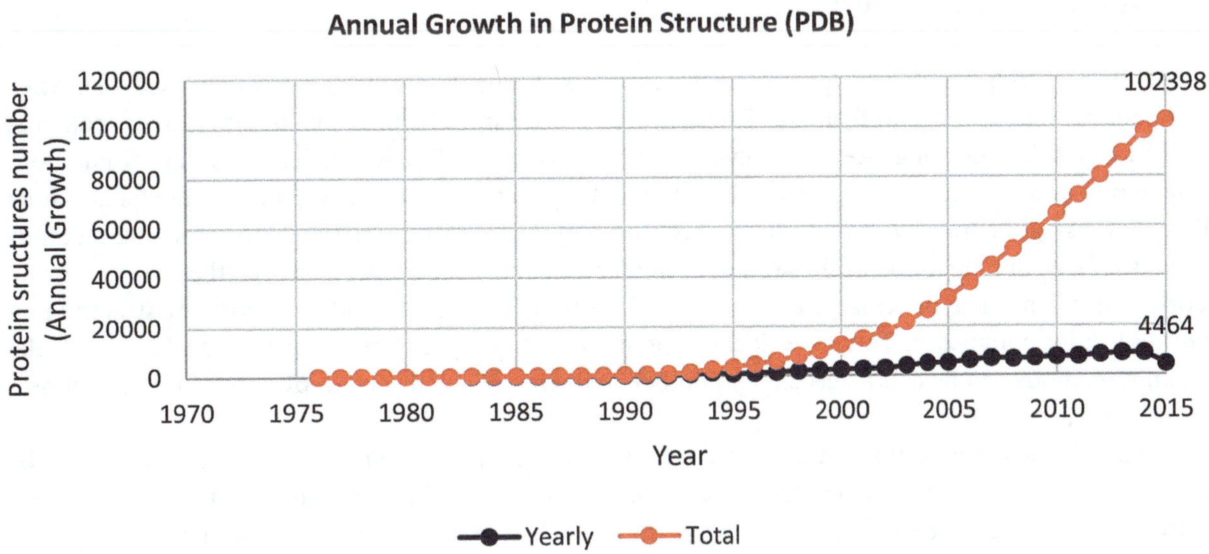

1. Prefiltering in virtual screening: it involves Lipinski's Rule of Five (Lipinski, 2004; Lipinski, Lombardo, Dominy, & Feeney, 2001) to find out compounds with drug likeliness features
    a. A molecular weight of less than 500 g/mol
    b. A calculated lipophilicity (log P) of less than 5

   c. Fewer than five H-bond donors
   d. Fewer than 10 H-bond acceptors
   e. Number of rotable bonds is less than 10 or one of the four rules can be violated.
2. *In silico* Absorption, Distribution, Metabolism, Elimination, and Toxicity (ADMET)
3. Similarity searches: achieved by following the principle of "similar structure- similar property"
4. Development of pharmacophore model by 3D structure alignment.

On the other hand SBDD or Structure based drug design is applied only when the 3D structure of the target of interest is known, either from X-ray crystallography, NMR spectroscopy or homology modeling. Here the known or predicted shape of the binding site is used to optimize the ligand to best fit the receptor. Existing algorithms for structure-based design can be categorized into two classes: *de novo design methods*, which builds ligands tailored to fit the target (Examples of *de novo* design include Lewis (Lewis et al., 1992) and Miranker (Miranker & Karplus, 1995) and the program LUDI (Böhm, 1992a, 1992b), and the *docking methods*, which searches for existing compounds with good complementarity to the target. In both the methods receptors are usually treated as a rigid body single conformation. There are two types of *de novo* design methods 1) *outside-in method* (Lewis & Leach, 1994), the binding sites is first analyzed to determine where specific functional groups might bind tightly. These groups were then connected together to transform into real molecules. 2) *inside-out approach* where molecules are allowed to grow inside binding site using appropriate search algorithm evaluating the each suggestion by energy functions.

4. MOLECULAR DOCKING

Molecular Docking is the central part of the structure based drug designing and is widely used for CADD applications. Fitting a ligand from a 3-D structure database into the binding site of a target protein is called docking. The docking process involves identification of "true" pose (conformation and orientation) of ligand inside receptor binding site and assessment of the binding affinity. Docking methods can be illustrated using a combined approach of search strategy and scoring function. Most docking methods generated large number of possible structures and hence require a parameter to rank the structure using scoring functions. The docking problem is linked with the generation and evaluation of structures of intermolecular complexes. To tackle docking problem, many algorithms has been developed. All the docking methods can be characterized by number of degree of freedom. An overview of docking programs is given in Table 1.

Before ligands can be docked against an enzyme or receptor, usually the binding site has to be discovered first. To minimize the search space on the receptor surface and thus minimize the degrees of freedom that has to be searched. This information can be obtained from crystal structures of ligand-bound receptors or can be predicted using binding site detection programs or online servers, e.g. GRID (Goodford, 1985; Kastenholz, Pastor, Cruciani, Haaksma, & Fox, 2000), POCKET (Levitt & Banaszak, 1992), SurfNet (Laskowski, 1995), PASS (Brady & Stouten, 2000) and MMC (Mezei, 2003). Binding site information significantly increases the docking efficiency. Generally, the largest cavity on a protein surface is considered as an active site, but this is not a universal fact.

Table 1. List of docking programs with web availability, the key references and search algorithms

Method	Reference(s)	Website	Search Algorithm
Autodock 4.0	(Morris et al., 2009; Österberg, Morris, Sanner, Olson, & Goodsell, 2002)	autodock.scripps.edu/	GA
CDOCKER	(Wu, Robertson, Brooks, & Vieth, 2003)	N/A	MD-SA
CHARMM	(Vieth, Hirst, Kolinski, & Brooks, 1998)	www.charmm.org/_	GA and MC
CLIX	(Lawrence & Davis, 1992)	N/A	RBD
DARWIN	(Taylor & Burnett, 2000)	N/A	GA
DIVALI	(Clark & Ajay, 1995)	N/A	GA
DOCK 6	(Kang, Shafer, & Kuntz, 2004; Knegtel, Kuntz, & Oshiro, 1997; Moustakas et al., 2006; Oshiro, Kuntz, & Dixon, 1995)	dock.compbio.ucsf.edu/	IC
eHiTs	(Zsoldos, Reid, Simon, Sadjad, & Peter Johnson, 2006)	www.simbiosys.ca/ehits/index.html_	RBD of fragments followed by reconstruction
FITTED	(Corbeil, Englebienne, & Moitessier, 2007)	www.fitted.ca_	GA
FlexX 2.2	(Rarey, Kramer, Lengauer, & Klebe, 1996)	www.biosolveit.de/_	IC
FlipDock (formerly pyDock)	(Zhao & Sanner, 2007)	www.scripps.edu/Byongzhao/FLIPDock/_	GA
FRED	(McGann, Almond, Nicholls, Grant, & Brown, 2003)	www.eyesopen.com/products/applications/fred.html_	RBD
FTDock	(Gabb, Jackson, & Sternberg, 1997)	www.bmm.icnet.uk/docking/ftdock.html_	RBD
Glide 4.0	(Richard A Friesner et al., 2004)	www.schrodinger.com/_ProductDescription.php? mID ¼ 6&sID ¼ 6&cID ¼ 0	Hierarchical filters and MC
GOLD 3.1	(Verdonk et al., 2005)	www.ccdc.cam.ac.uk/products/life_sciences/gold/_	GA
HADDOCK	(Dominguez, Boelens, & Bonvin, 2003)	www.nmr.chem.uu.nl/haddock/_	SA
MacDOCK	(Fradera, Kaur, & Mestres, 2004)	N/A	IC
MolDock	(Thomsen & Christensen, 2006)	N/A	DE
PatchDOCK	(Schneidman-Duhovny, Inbar, Nussinov, & Wolfson, 2005)	N/A	Shape complementarity
PAS-Dock	(Tøndel, Anderssen, & Drabløs, 2006)	N/A	TS
PhDOCK	(Joseph-McCarthy, Thomas, Belmarsh, Moustakas, & Alvarez, 2003)	N/A	MA
PIPER	(Kozakov, Brenke, Comeau, & Vajda, 2006)	N/A	Fast Fourier transform
ProPose	(Seifert, 2005)	N/A	IC
PSI_DOCK	(Pei et al., 2006)	N/A	GA with TS
RiboDock	(Morley & Afshar, 2004)	N/A	MC
ROSETTALIGAND	(Meiler & Baker, 2006)	N/A	MC
SDOCKER	(Wu & Vieth, 2004)	N/A	Random walk
SLIDE	(Schnecke & Kuhn, 2000; Zavodszky & Kuhn, 2005)	N/A	IC
SODOCK	(Chen, Liu, Huang, Hwang, & Ho, 2007)	N/A	Swarm Optimisation
SG-DOCK/SP-DOCK	(Fradera, Knegtel, & Mestres, 2000)	N/A	IC
Surflex 2.1	(Jain, 2003, 2007)	N/A	IC with MA

Table directly adapted from (Moitessier et al., 2008).

Abbreviation: N/A, not available.

Search Algorithms notations: DE, differential evolution; EA, evolutionary algorithm; GA, genetic algorithm; IC, incremental construction; MA, matching algorithm; MD, molecular dynamics; RBD, rigid-body docking; SA, simulated annealing; TS, Tabu search.

5. TYPES OF MOLECULAR DOCKING

Based on conformational flexibility of the ligand, molecular docking can be either rigid, flexible can be divided into two sub categories rigid and flexible docking.

Rigid Docking is the simplest algorithm that considers both the receptor and ligand as rigid entities. It explores only six degrees of translational and rotational freedom. Rigid docking approaches were being used in the initial docking algorithms for docking small molecules into the binding sites of proteins and DNA. DOCK is an excellent example of rigid docking methods (Kuntz, Blaney, Oatley, Langridge, & Ferrin, 1982). It works on identifying the small molecules with high degree of shape complementarity to the receptor binding site. The DOCK algorithm uses a geometric matching algorithm to superimpose the ligand onto a negative image of the binding pocket. The receptor spheres generated inside the binding sites of the ligand atoms is viewed in a negative image of the active site, ligand shape can be identified by looking at the negative image of the active site. Rigid docking involved docking with rigid ligands and mostly used pre-generated ligand conformers. This reduced the computational burden however was not the best docking solution provider.

Flexible Docking can be done where either ligand or receptor is flexible or both are flexible. Generally the active site or the surrounding residues of the receptor active site are made flexible rather than the whole receptor. This is done because the docking takes place in the active site or nearby active site residues. Flexible docking strategies do not require any prior knowledge of the conformation of the ligand. Hence the search space consists of 6+N translational, rotational and conformational variables. Flexible docking is useful if there is no conformational information about the ligand. Four different strategies are applied, they are a) Monte Carlo; b) in-site combinatorial search; c) Ligand building; d) site mapping and fragment assembly.

Distance geometry can also be used to perform molecular docking, however incremental construction of the ligand within binding sites is used by a number of programs.in this method a series of conformations would be generated by systematically varying dihedral angles between 0° and 360°.This involves the grid search, where all base fragments were identified within the ligand. Base fragments will be docked into the binding site and clustered to remove duplicate orientations. Each docked orientation of the base fragment then represents the starting point for the conformational analysis of the rest of the ligand.

Genetic algorithms, molecular dynamics approaches can also be used to perform molecular docking. The ideal docking method would allow both ligand and receptor to explore their conformational degrees of freedom. However such calculations are computation intensive and are more useful for refinement of docked structures.

6. SCORING FUNCTIONS

The binding of ligand and receptor are quantum mechanical in nature, but due to computational limitations and biological complexities quantum theories becomes impractical and are not always accurate. The strength of ligand receptor interaction can be measured experimentally and is often reported as the dissociation constant, K_d or by the concentration of ligand inhibiting activity by 50% (IC 50). Most of the methods attempt to estimate binding free energy of the ligand binding to receptor, which can be given by:

$$\Delta G_{bind} = \Delta G_{complex} - (\Delta G_{ligand} + \Delta G_{receptor}) \qquad (1)$$

The relationship between the binding free energy ΔG and the experimentally determined K_d or IC_{50} is defined in Equation 2

$$\Delta G_{\text{bind}} = -RT \ln K_{\text{eq}} = -RT \ln K_d = -RT \ln 1/\text{IC50} \quad (2)$$

With Equation (2) it is possible to calculate the binding free energy where K_d is defined with respect to standard state (Atkins, De Paula, & Walters, 2006). The pressure,1 atm, and the activity of the solutions, namely 1M, define standard state.

Molecular docking have played an increasing role in the functional study of proteins and structure-based drug design with the increase in experimentally determined three-dimensional protein structures. However, one important problem is the development of scoring function that can heuristically and precisely describe the protein and ligand interaction. Scoring function plays major role in ascertaining three major applications. Firstly, defining the binding mode and ligand site on a protein. For a target molecule like protein, molecular docking generates numerous putative ligand binding orientations/conformations at the active site around the protein. A scoring function is used to rank these ligand orientations/conformations by evaluating the binding tightness of each of the putative complexes. Gold standard for scoring is to rank the experimentally determined binding mode most highly. Another application is linked to first application and it involves prediction of absolute binding affinity between protein and ligand, which is very much important for lead optimization. Lead optimization is used to improvise the binding of low-affinity hits or lead compounds. An idle scoring function can significantly increase the optimization efficiency and prevent costs of ligand synthesis and experimental testing. Structure-based drug design is another important application of scoring functions and is used to screen potential drug hits (leads) for a receptor (protein target) by searching a large ligand database this is also known as virtual screening. Scoring function is commonly used for ranking known binders most highly according to their binding scores during virtual screening (Huang, Grinter, & Zou, 2010).

Large number of poses are generated by docking algorithms, few got rejected because of high-energy clash with the protein. Scoring functions will be used further to assess the left out poses. Scoring function is required to rank the ligand molecules (if using database of molecules for screening against a receptor) relative to each other. Moreover, scoring functions are essential to identify the docked orientation that could be the real representation of 'true' structure of the intermolecular complex or docking mode.

Molecular docking is used to predict how a ligand will interact with the receptor binding site. The ligand is mostly a small molecule but can be a protein also. Likewise, receptor is mostly a protein of pharmacological significance however it could be a DNA or RNA molecule as well. The next step in docking is consists of scoring functions which will distinguish among the generated binding modes and the optimal solution that is fairly matches to actual binding. The most promising predicted protein-ligand complex is considered to be the biologically relevant one. A large number of different scoring functions have been developed to model protein-small molecule docking which can be divided into three types or classes: 1) Force-field based 2) empirical and 3) knowledge-based scoring functions.

Force-field Scoring Functions: Force field (FF) based scoring tries to quantify internal ligand energy as well as ligand-receptor binding energy. Molecular mechanics force fields based scoring functions are based on different parameter sets, like AutoDock is based on the AMBER force field (Kitchen, Decornez, Furr, & Bajorath, 2004).

The ligand-receptor binding interactions can be best described by using van der Waals (VDW), electrostatic energy terms, bond stretching/bending/torsional energies, etc. Intermolecular forces are calculated using "6-12" Lennard-Jones potential functions representing van der Waals interactions however the interactions between atomic charges can be calculated using Coulomb's equation. Examples of force field based scoring functions include D-Score, G-Score, GOLD, AUTODOCK, DOCK. Following equations shows the use of distance-dependent dielectric constant ε (*rij*) force field scoring function in DOCK:

$$\Delta G_{bind} = \sum_{i}^{lig}\sum_{j}^{rec}\left[\frac{A_{ij}}{r_{ij}^{12}} - \frac{B_{ij}}{r_{ij}^{6}} + \frac{q_i q_j}{\varepsilon(r_{ij})r_{ij}}\right] \tag{3}$$

where *rij* stands for the distance between protein atom *i* and ligand atom *j*, *Aij* and *Bij* are the VDW parameters, and *qi* and *qj* are the atomic charges. *ε* (*rij*) is usually set to *4rij*, reflecting the screening effect of water on electrostatic interactions.

Force field scoring functions is widely used but faces some challenges which involves the explanation of solvent effect and entropy effect.

Empirical Scoring Functions: Emperical scoring functions uses weighted energy terms to estimate the binding affinity of a complex. Ludi empirical scoring functions conceptualized that binding energies can be estimated by addition of individual unrelated terms.

These unrelated terms can be obtained from receptor-ligand crystal structure or training the binding constants of thousands of protein-ligand complexes. In empirical scoring function simple scalable contributions like hydrogen bonds, metal ligations, hydrophobic effect rotable bonds evaluated for estimating the energetics of ligand- receptor binding.

$$\Delta G_{bind} = \Delta G^0 + \Delta G_{HB}\sum_{HB} f(\Delta r) f(\Delta 1\alpha) + \Delta G_{met}\sum_{met} f(\Delta r) + \Delta G_{lipo}\sum_{lipo} f(\Delta r) + \Delta G_{tot} N'_{rot} \tag{4}$$

Or it can also be given as,

$$\Delta G = \sum_{i} Wi.\Delta G_{i} \tag{5}$$

In Equation 5 (Huang et al., 2010), ΔG_i represents different energy terms such as Van Der Waal energy, electrostatic energy, hydrogen bond, desolvation, entropy, hydrophobicity, etc. Here, *Wi* are determined by fitting the binding affinity data of a training set of protein–ligand complexes with known three-dimensional structures.

Because of their simple energy terms, empirical scoring functions are much more faster in binding score calculations compared to force field scoring functions.

LUDI, F-score, Chemscore, Score, x-score are some examples of empirical based scoring function.

Knowledge based scoring functions: Knowledge based scoring functions (also referred to as statistical-potential based scoring functions). Knowledge based scoring functions recruits energy potentials directly derived from structural information of experimentally determined atomic structures. Knowledge based scoring functions for protein- ligand complex are modelled using simple atomic interaction-pair potentials. This scoring functions is based on the potential of mean force, which is defined by the inverse

Boltzmann relation. The GOLD fitness function is a measure of Chemscore or GOLD score, represented by $\sigma_{0,}$ which is measure of water contribution $\sigma_{1,}$

Is loss in the entropy of the water molecule and σ_w is the intrinsic binding affinity of the water.

$$\text{Fitness} = \sigma_{0+} \sum_{w} O(w)(\sigma p + \sigma i(w)) \quad (6)$$

Drugscore, ITScore, DFIRE, GOLD/ASP,Kscore, PMF and SMoG, BLEEP, MScore are examples of knowledge based scoring functions.

Compared to the force field and empirical scoring functions, the knowledge-based scoring functions provide more balanced option between precision and heuristics. Because the potentials in Equation (6) are extracted from the structures rather than from attempting to reproduce the known affinities by fitting, and because the training structural database can be large and diverse, the knowledge-based scoring functions are quite robust and relatively insensitive to the training set (Huang et al., 2010).

7. APPLICATIONS OF MOLECULAR DOCKING

Molecular docking has been one of the most extensively used technique in CADD (Computer Aided Drug Designing) and is the central dogma of Structure-Based Drug Designing (SBDD). SBDD is applied when the structure of target macromolecules generally, protein is known. The main aim of the SBDD is to identify the molecules which favorably interact with the target protein at its binding site. Molecular docking fulfills this objective by predicting the binding orientation of one molecule with the other in 3D space thereby identifying the lowest free energy complex structures. Another aim of the SB-CADD is the lead identification in the drug discovery process. In general, lead identification is done by two methods: High Throughput Screening (HTS) and Virtual High Throughput Screening (vHTS). HTS is the traditional technique which is slow and expensive compared to vHTS. In vHTS, a database containing large number of small molecules are screened against a protein target using molecular docking techniques and then ranked hits for lead generation. Previously, molecular docking was also used in SBDD pipeline for screening large compounds database and to identify the binding orientation with the target. With the advancement of the docking methodologies, presently molecular docking is used for various other applications viz. identification of differential binding of a ligand to different target biomacromolecules, de novo design for lead generation, adverse drug reaction prediction in the drug development and in some cases molecular docking is also used for non drug related projects like identification of putative pollutants for biodegradative enzymes. This exhibits the diversity of application of molecular docking.

In the following section we will discuss major applications of molecular docking along with case studies.

7.1 Virtual Screening

Virtual screening is a tool to screen large compound libraries against a drug target that can be a protein or an enzyme or any other biomacromolecules. It is used to find the compounds from a large library which exhibited good binding interaction with the drug target. It is one of the most useful technique in the drug development process.

Case Study (Virtual Screening): Identification of BCR-ABL Tyrosine Kinase Hits Using Molecular Operating Environment (MOE) Software

In this case study (Rahul & Ashutosh, unpublished) we discuss how virtual screening is performed using MOE (Molecular Operating Environment (MOE), 2015). We screened an in-house database of ~2000 compounds against BCR-ABL tyrosine kinase, durg target for chronic myeloid leukemia (CML) treatment. The crystallographic co-ordinates of BCR-ABL is available on Protein Data bank (Berman et al., 2000) (PDB ID: 3OXZ, (Zhou et al., 2011)). The coordinates pdb file was downloaded and prepared using MOE Strucuture preparation wizard (Molecular Operating Environment (MOE), 2015). During struture prepration the hydrogens was added to the protein structure (Note: at times the pdb file has some missing side chains or missing residue which could be fixed by various tools like MOE (Molecular Operating Environment (MOE), 2015) or chimera (Pettersen et al., 2004)). Appropriate charges were added and water molecules which were more than 4.5 A^0 apart from the receptor were removed. Finally the energy was minimized using MMFF94x forcefield (Halgren, 1996) (There are various other force field are available like Amber (Cornell et al., 1995), OPLS (Jorgensen & Tirado-Rives, 1988), CHARMM (MacKerell et al., 1998) etc. with different versions here we are using the default).

In order to prepare the ligands the conformations of the 2000 compounds structure database were generated using MMFF94x forcefield (Halgren, 1996). A limit of 4.5 kcal/mol strain energy was imposed with a 500 conformation limit for each molecule. Duplicate conformations were removed with a heavy atom RMSD tolerance of 0.6 A^0 (0.75 A^0 for conformations with strain greater than 3.5 kcal/mol). The fact that a ligand was already bound at the active site of the protein, enabled us to define the active site. Next step involved generation of possible poses from the various ligand conformations. Diverse placement methods are available viz. Alpha Triangle, Alpha PMI, Flex, GOLD, Proxy Triangle, Template Forcing, Template Searching, Triangle Matcher. Here we used the default triangle matcher for the placement step. Next rescoring, refinement and again rescoring took place using London DG methods (other methods such as ASE, Affinity dG, Alpha HB, Electron Density, GBVI/WSA dG are also available). Further refinement is done using forcefield method and again rescoring was done using GBVI/WSA dG method. Finally 30 poses were kept and total 46,129 conformations were generated which were docked into the active site of BCR-ABL tyrosine kinase. MOE (Molecular Operating Environment (MOE), 2015) gave an output in form of a table having 10 fields which are depicted in Table 2 and the output of top 10 compounds on the basis of final score S was given in Table 3. These 10 compounds having score in order of (-9.2595) to (-8.8398) were obtained. All the top 10 compounds were docked into active site, the results are shown in Figure 3. These compounds could be used further for invitro assays.

Case Study (Virtual Screening): Identification of BRD4(1) Inhibitors Using Protein-Ligand Docking and Structure-Guided Design.

Bromodomains (BRDs) are recently discovered as a potential new drug targets for cancer, inflammation, diabetes and cardiovascular therapeutics. BRD-containing proteins choses epigenetics route for binding against acetylated lysines (KAc) on chromatin structures. Duffy et al., in 2015 employed a virtual screening & X-ray crystallography along with other cheminformatics techniques to identify several scaffold

Table 2. Fields with description of MOE output (Molecular Operating Environment (MOE), 2015)

Field(s)	Description
Mol	An output pose.
receptor	The receptor after the refinement stage.
Rseq	The receptor sequence number.
Mseq	The molecule sequence number.
S	The final score, which is the score of the last stage that was not set to None.
Rmsd	The root mean square deviation of the pose, in Å, from the original ligand. This field is present if the site definition was identical to the ligand definition.
rmsd_refine	The root mean square deviation between the pose before refinement and the pose after refinement.
E_conf	The energy of the conformer. If there is a refinement stage, this is the energy calculated at the end of the refinement. Note that for Forcefield refinement, by default, this energy is calculated with the solvation option set to Born.
E_place	Score from the placement stage.
E_score1 E_score2	Score from rescoring stages 1 and 2.
E_refine	Score from the refinement stage, calculated to be the sum of the van der Waals electrostatics and solvation energies, under the Generalized Born solvation model (GB/VI).

Figure 3. 3D view of bound ligands with the BCR-ABL protein. A) Full view of the BCR-ABL protein with ligand. B) zoomed view of the ligands present into the active site of BCR-ABL protein. Red is the ligand already present in crystal structure while green are the top 10 compounds which are screened using docking studies.

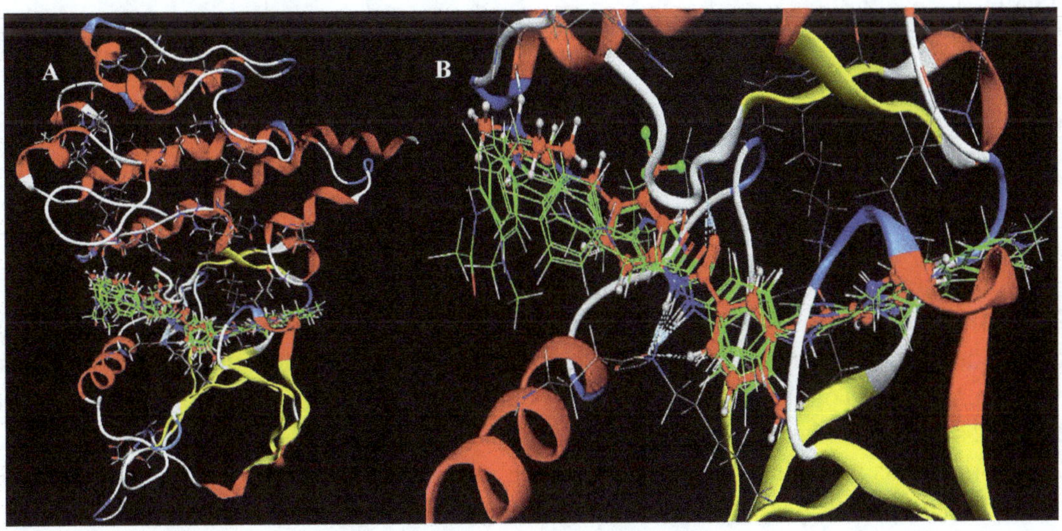

clustershits binding to BRD4(1) (Duffy et al., 2015). Database of approx 25 million compounds from various publicly available sources were filtered on the basis of common drug-like property. Approximately 230,000 filtered compounds were selected through multiple rounds of random selections followed by diversity confirmation *via* 2-D atom-pair similarity calculations. These were docked against two X-ray crystal structures of BRD4(1) with different co-crystallized ligand using Glide (Richard A. Friesner

Table 3. Top 10 compounds docked on the basis of MOE dock final score S (Molecular Operating Environment (MOE), 2015)

Mol	Rseq	Mseq	S	rmsd_ refine	E_conf	E_place	E_score1	E_refine	E_ score2
O=C(NCC[NH+](C)C)c1cc(ccc1)-c1nc(N2CCN(CC2)c2cc[nH+]cc2)cnc1	1	1299	-9.25949	2.229406	85.04276	-52.6258	-12.1489	-11.9827	-9.25949
O=C(Nc1ccc(Nc2ncnc(c2)-c2cc3c(cc2)cccc3)cc1)c1ccc(N)cc1	1	505	-9.15959	1.383814	-6.63474	-86.3349	-12.6617	-21.7742	-9.15959
O(C(C)(C)C)C(=O)Nc1ccc(Nc2ncnc(c2)-c2ccc(N(C)C)cc2)cc1	1	550	-9.13703	1.297704	-55.6844	-108.026	-11.3118	-14.6355	-9.13703
O=C(Nc1ccc(Nc2ncnc(c2)-c2cc(N(C)C)ccc2)cc1)c1ccc(N)cc1	1	510	-9.02742	1.506958	-27.5484	-102.505	-11.7205	-30.7333	-9.02742
o1c2c(cc1-c1ncnc(Nc3ccc(NC(=O)c4ccc(N)cc4)cc3)c1)cccc2	1	507	-8.93877	0.817132	-22.9512	-130.471	-13.7038	-17.7379	-8.93877
S(C)c1ccc(cc1)-c1ncnc(Nc2ccc(NC(=O)c3ccc(N)cc3)cc2)c1	1	508	-8.92662	1.648885	-18.9101	-88.5567	-11.9491	-23.8481	-8.92662
O=C(Nc1ccc(Nc2ncnc(c2)-c2ccc(NC(=O)C)cc2)cc1)c1ccc(N)cc1	1	515	-8.91988	2.205891	-49.7061	-64.7249	-11.8485	-31.4016	-8.91988
O=C(Nc1cc(ccc1)-c1nc(N2CCN(CC2)c2cc[nH+]cc2)cnc1)C	1	1294	-8.8837	1.633233	105.6261	-59.2689	-12.4821	-11.4932	-8.8837
O=C(Nc1ccc(Nc2ncnc(c2)-c2cc3c(cc2)cccc3)cc1)c1ccc(N)cc1	1	505	-8.88196	2.717045	-7.07057	-77.8603	-12.1259	-23.2881	-8.88196
O=C(Nc1cc(ccc1)-c1nc(N2CCN(CC2)CCN(C)C)cnc1)C	1	1441	-8.83977	1.08584	150.4833	-60.5674	-10.335	-11.6034	-8.83977

et al., 2004). The binding pose was considered if potential hydrogen bond is present with the Asn140 sidechain amide, that mimics the native histone acetylated lysine binding location. Finally, 153 available compounds were selected on the basis of hydrogen bond requirements & Glide score. These compounds represented 82 scaffolds and were purchased from commercial vendors and evaluated in vitro screening. These 153 compounds were tested by HTRF binding inhibition to BRD4(1) and BRD2(2) at 50 μM concentrations. Compounds having 50% or more binding to BRD4(1) were submitted for dose-response testing and IC_{50} calculations, while hits in the 30–49% binding range at 50 μM were grouped by scaffold. The compound showing highest activity was retest for IC50 calculation and also co-crystallized in BRD4(1) to identify the actual binding pose. A small series of compounds analogs of the highest active compound were prepared and tested. The most active compound in this series, shows greater activity. The molecular docking with other experimental method process allowed to identify BRD4(1) inhibitors also helps in improving the activity of a compound in BRD4(1) inhibition.

7.2 Identifying the Binding Orientation ('True' Pose) of Ligand with the Drug Target

Molecular docking can be used as a tool for virtual screening if screening against ligand database or if single or few ligands are searched for identification of true pose inside receptor binding site. This information is extremely advantageous to identify the binding sites or groups which are involved in binding. This will also help in designing the new potent compounds by knowing the binding pattern of the existing compounds. There are two steps involved in identifying the binding orientation of ligand withn the drug target. They are preparation of the protein and the liagnd and then the final docking. Preparation of the protien involves the searching the protein structure in PDB database for the availability of their crystal structure. If the protein crystal structure is not available then the protein structrue is modelled using bioinformatics methods. While the preparation of ligand involves the generation of energy minimized 3D structure of the lignad.

The docking depends on individual cases which is depicted below.

Case Study: True Pose Identification of Acarbose with α-Glucosidase Enzyme

This study which was conducted in our lab (unpublished results), involved the step wise procedure to identify the binding orientation of Acarbose with α-Glucosidase enzyme. In this example α-Glucosidase protein model was built using I-Tasser (Yang et al., 2015) while docking analysis was done using Autodock (Morris et al., 2009). α-Glucosidase is a proven drug target for Diabetes. α-Glucosidase inhibitors decrease the absorption of carbohydrates from the digestive tract, thereby lowering the aftermeal glucose levels. Crystal structure of α-Glucosidase enzyme is not available on PDB (Berman et al., 2000). So, α-Glucosidase protein structure was modelled using bioinformatics methods. Modelling of the protein structure involved few steps and primarily depends on the identity of sequence similarity with the template sequence. Here we had 44% sequence identity with two human maltase-glucoamylase templates (PDBID: 2QLY and 3L4T) with query coverage 90% each. On this sequence identity we used threading method to model the protein. I-Tasser (Yang et al., 2015) was applied to model α-Glucosidase protein structure. The modelled protein structure was validated through ramachandran plot. Since there was no ligand bound to the protein we identified the active site residues in the protein structure *via* centroid co-ordinates of the complex ligands in the crystal structure (PDB ID: 3L4T). Finally docking was performed using Autodock program (Morris et al., 2009). It is one of the widely used freely available docking tool. The docking procedure for this molecule is shown in Figure 4.

After successful execution of autodock program docked log file (dlg) was generated in which the docked conformation co-ordinates of ligand with their binding energy were present. Total 10 docked conformation were obtained and the best conformation is having -4.45 Kcal/mol binding energy. This docked acarbose with protein target can be visulized using pymol (The PyMOL Molecular Graphics System) as shown in Figure 5 and save as complex pdb file. This complex pdb file is used to analyze the 2D and 3D binding orientation of Acarbose with protein using Ligplot as shown in Figure 6 & 7 resp.

The major limitation of this case study is the quality of the alpha-glucosidase protein structure because the whole docking is depends on the receptor structure. While performing docking studies using predicted structure one have to ensure that the modelled protein structure quality is upto the mark or not.

Figure 4. The procedure followed to perform this study

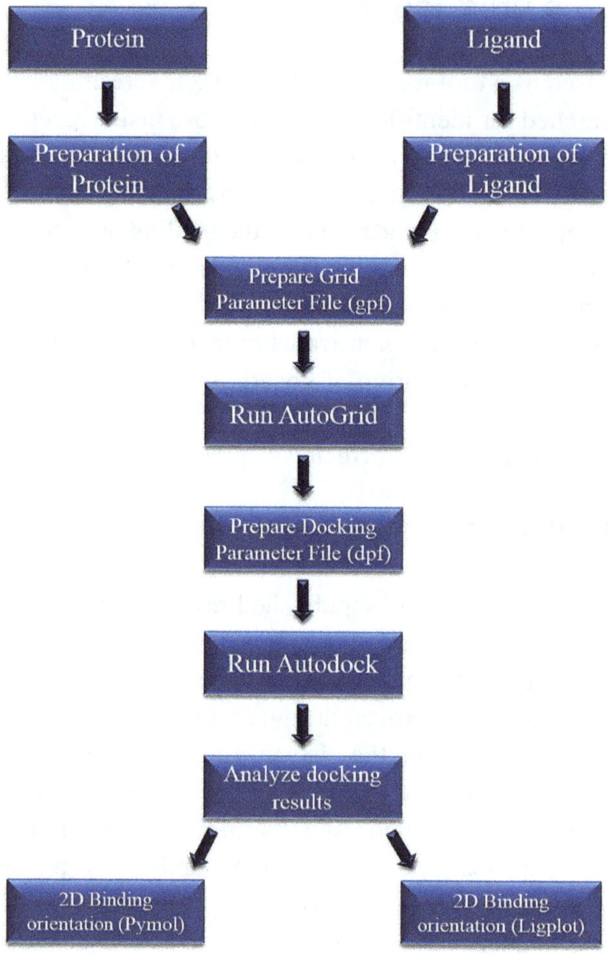

7.3 Hydrated Docking

Hydrated docking involved docking of protein-ligand in presence of water molecules. Water molecules play an important role during protein-ligand interaction. Forli & Olson in 2012 developed a force field with discrete displaceable water and desolvation entropy for studying hydrated ligand docking (Forli & Olson, 2012). Autodock tool provided a script for implementing hydrated docking.

Case Study: Identification of Noncamptothecin Topoisomerase I (Top1) Inhibitors Using Hydrated Docking

Taliani et al. in 2013 used hydrated docking to identify the binding orientation of their noncamptothecin Topoisomerase I (Top1) inhibitors (Taliani et al., 2013). To begin with, the binding pose of HTOP1 (human topoisomerase 1) with inhibitor Topotecan (inform the readers about Topotecan) was observed from X-ray diffraction technique and Autodock docking method (Morris et al., 2009) (as shown in Figure 8 below). Here, hydrated docking between HTop1 and Topotecan provided useful information about the

Figure 5. Full view of the Acarbose docked in the modelled alpha-glucosidase active site. Acarbose are shown as sticks model while the protein are shown as cartoons.
This image is viewed using Pymol (The PyMOL Molecular Graphics System).

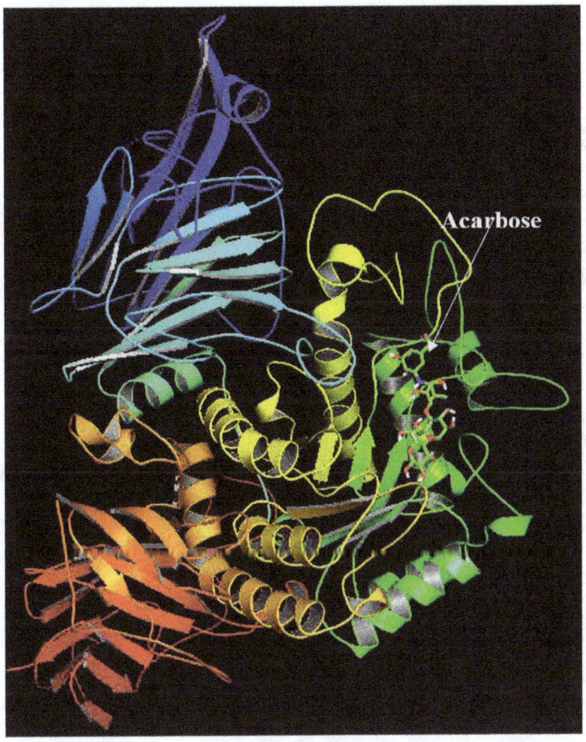

binding pattern of the compound in presence of water molecules. This information was used to design novel compounds with diverse architectures that resulted in the improvement of Top1 inhibitory activity.

The problem with hydrated docking is the difficulty in comparing the results with the crystal complex having water interactions which is not able to distiguish whether water is really involved or it is apart of crystal effect.

7.4 Reverse Docking

Reverse docking is a technique that identified the target proteins of known biological active compounds. In this method small compounds or ligands are docked into the active sites of multiple drug targets, their binding orientation were analysed and ranked on the basis of their binding interaction strength. It is very useful when target information is unknown.

Case Study: Identification of Targets which are binding to Dioscin

Recently Yen et al. described a workflow in which they coupled reverse docking with other bioinformatics methods to predict drug targets, biological activities, signal pathway and regulating networks of Dioscin that could be implemented on other available biological active compounds (Yin et al., 2015). Dioscin is a natural steroidal saponin present in many Chinese medical herbs (like, Dioscorea nipponica Makino,

Figure 6. 2D interaction of modelled alpha glucosidase with the Acarbose inhibitor. Acarbose is majorilly interacted with Asp194, Ser435, Asn436, Phe437, Ala467, Gly563, Leu589, Leu590, Ser591.
Image is created by Ligplot (Wallace, Laskowski, & Thornton, 1995).

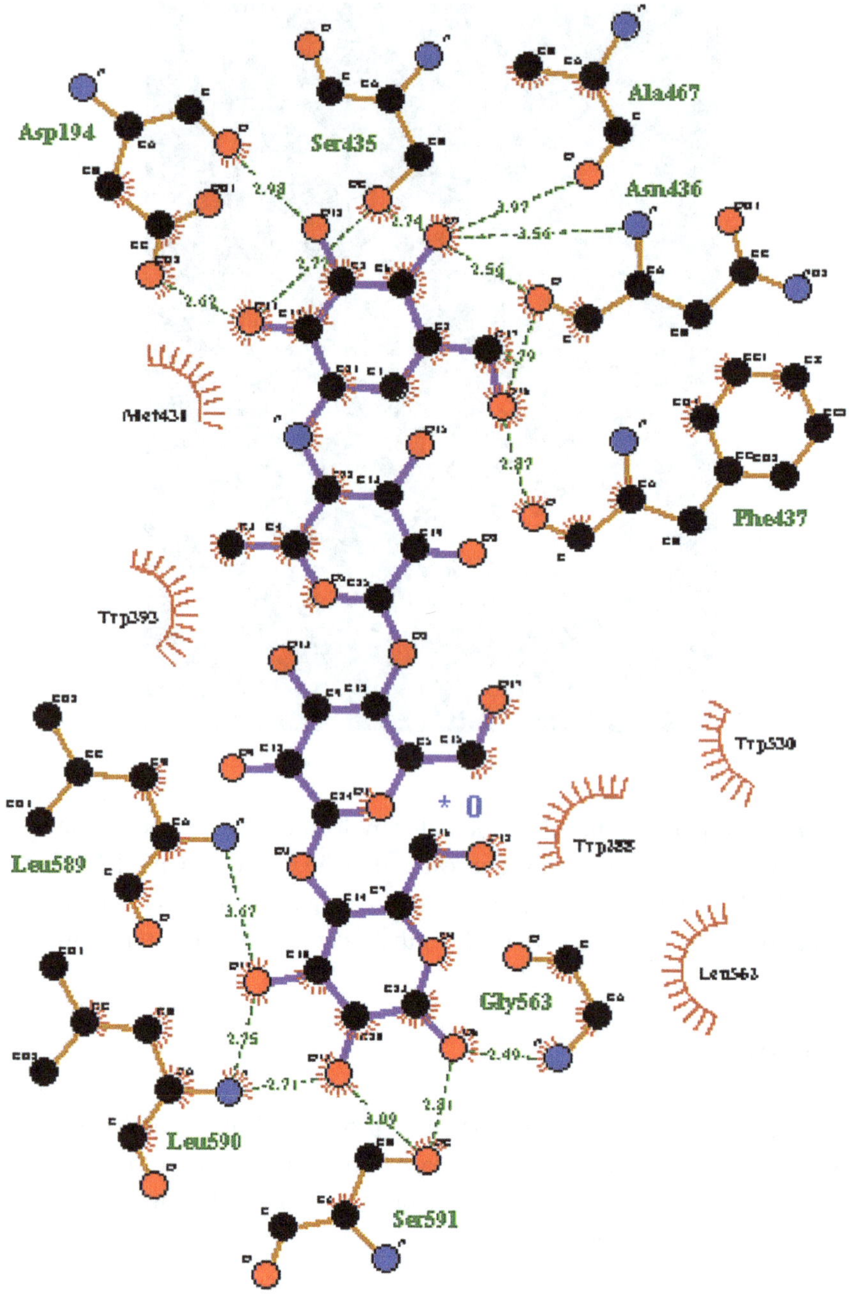

Dioscorea zingiberensis) and known to posses many medical pharmacological properties including anti-tumor, anti-hyperlipidemic, anti-fungal and antivirus activities. Yen et al. screened the available potential protein targets from humans (71), rats (7) and mice (8). The reverse docking indicated that most probable targets of dioscin were cyclin A2, calmodulin, hemoglobin subunit beta, DNA topoisomerase I, DNA

Figure 7. 3D view of modelled alpha glucosidase with the Acarbose inhibitor. This image shows the 3D binding pattern of Acarbose with alpha glucosidase enzyme.
The image is created by Pymol (The PyMOL Molecular Graphics System) and Ligplot (Wallace et al., 1995).

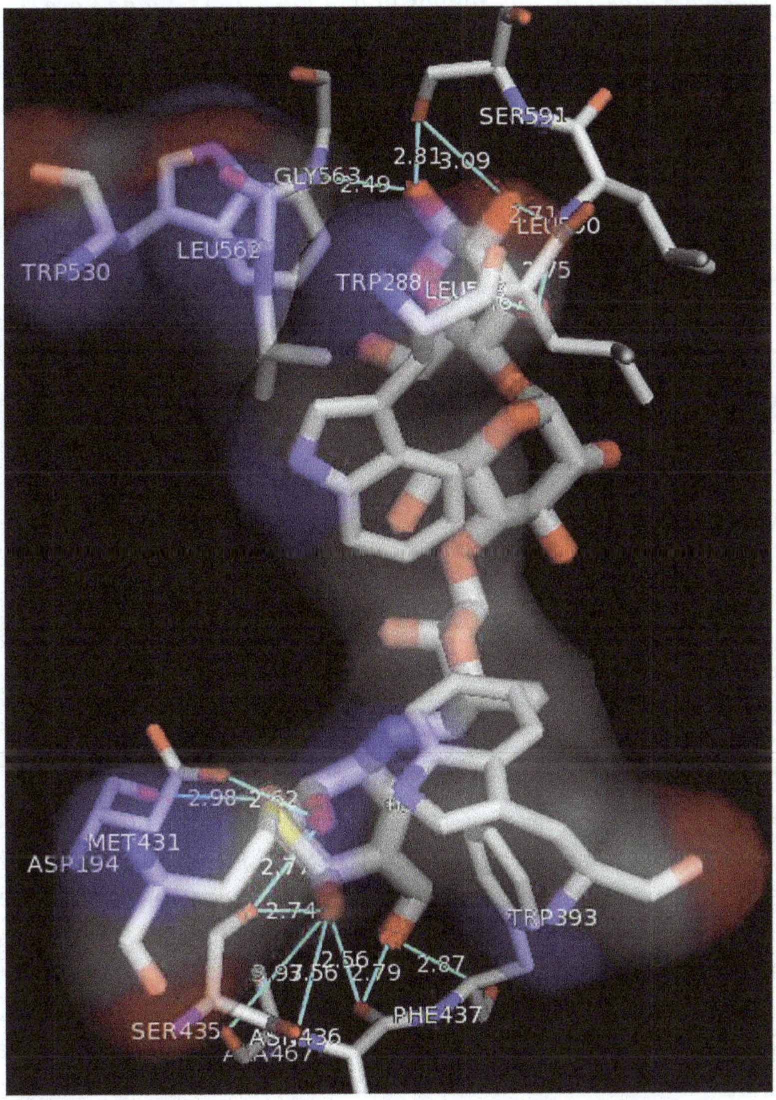

polymerase lambda, nitric oxide synthase and UDP-N-acetylhexosamine pyrophosphorylase, etc. The workflow used by (Yin et al., 2015) is depictedin Figure 9.

7.5 Adverse Drug Reaction Prediction

Prediction of Adverse drug reactions (ADRs) is essential for a pharmacologically active small molecules because adverse drug reaction produces effect which could be extremely harmful,unusual and could alter the biochemical pathways (LaBute et al., 2014). Nearly hundred thousand fatalities are reported in US

Figure 8. a) X-ray binding pose of Topotecan (II) within the Top1 DNA cleavage site (PDB code: 1K4T). The ligand is shown as magenta sticks. The enzyme and the substrate DNA are represented as green and cyan cartoons, respectively. Residues important for ligand binding are highlighted as sticks. H-bonds are depicted as dashed black lines. b) Superimposition of the X-ray binding pose (magenta sticks) of II and the docking conformation (orange sticks) as predicted by AD4. The enzyme and the substrate DNA are represented as green and cyan cartoons, respectively. Crystal and AD4 water molecules bridging between the ligand and the enzyme are shown as magenta and orange spheres, respectively.
Source: (Taliani et al., 2013).

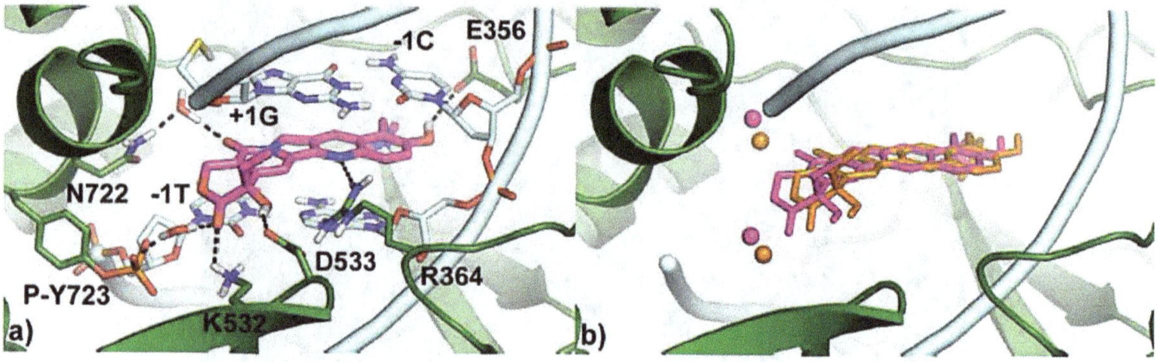

every year because of ADRs. Prediction of adverse drug reactions (ADRs) by lead compounds is very useful in drug discovery process.

Case Study: Adverse Drug Reaction Prediction Workflow

In 2014, LaBute et al. developed a computation process that enabled prediction of ADRs using molecular docking and known ADR information from DrugBank and SIDER. The workflow is shown in Figure 10.

The drug target information were obtained from Uniprot (Consortium, 2015). Nearly four thousand proteins were identified and extensive search of their crystal structure led to the discovery of 409 of them in Protein Data Bank (PDB) (Berman et al., 2000). They were used as a virtual panel. All 906 FDA-approved drugs (till date) were docked in this virtual panel using the VinaLC program (Zhang, Wong, & Lightstone, 2013). Side effect of 560 compounds were present in SIDER database which were used for building logistic regression models for ADR prediction.

This type of study immensely helped in drug development process by which the lead ADRs is accessed at earlier stage of drug discovery and could be dropped or redesigned.

This study depends on the machine learning methods which may provide various false positives and false negatives.

7.6 De-Novo Drug Design

De novo drug design is generally used to generate new drug-like compounds by linking fragments. De novo design uses library of building blocks that also contain single atoms, functional groups, and small molecular fragments. De novo drug design is widely used in the drug development process to generate new novel and potent compounds by using the drug target active site information.

Figure 9. The analysis process of the prediction of drug targets, biological activities, signal pathway and regulating networks
Source:(Yin et al., 2015).

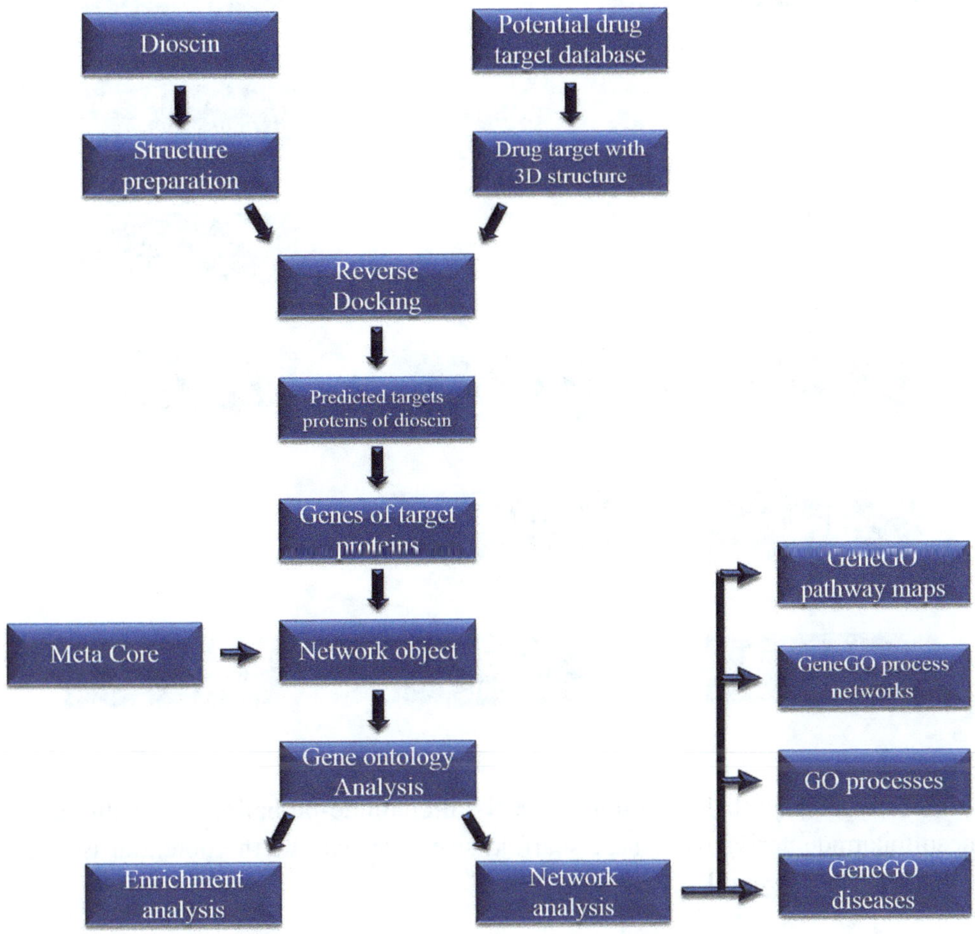

Case Study: De Nove Design

Xiang et al. used this approach to discover the isozyme-selective inhibitor scaffolds of human carbonic anhydrases (CAs) (Xiang, Xiang, Fang, Zhang, & Li, 2014). CAs are known drug targets for the treatment or prevention of disease like altitude sickness, epilepsy, glaucoma, obesity, tumor, acid–base disequilibria, and other neuromuscular diseases. Experimental procedure included the alignment of available 9 PDB structures of CAs where one of them was complexed with SBR ((R)-N-(3-indol-1-yl-2-methyl-propyl)-4-sulfamoyl-benzamide). The ligand (SBR) was docked to all 9 structures using Autodock Vina (Trott & Olson, 2010). POCKET module of LigBuilder (Yuan, Pei, & Lai, 2011) was used to identify the interaction residues and molecules were constructed for the target protein using GROW module. GROW module required a seed structure on which fragments were added to generated new hit compounds. Figure 11 depicted the interaction of SBR ligand with different CAs and seed struture with growing site. Total 200 top candidates were selected after Lipinski filter of maximal molecular weight and logP value being 500 and 5, respectively. Further top hits having multiple rings were used for further docking analysis.

Figure 10. Data integration/analysis workflow scheme

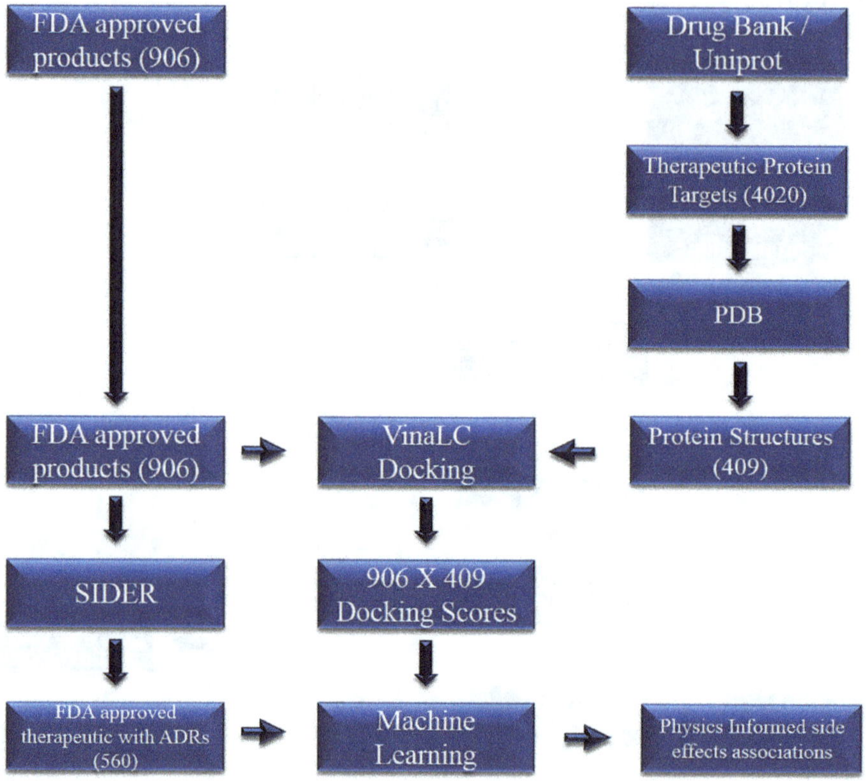

These four molecules are naphthalenesulfonamide, N-sulfonamide-phthalimide, 3-sulfonamide-acridan, and fluorenesulfonamide derivatives, these scaffold structures can be the potential isozyme-selective inhibitors against CA II, IV, and IX.

8. CHALLENGES OF MOLECULAR DOCKING

Identification of binding pattern between proteins and small molecules face substantial challenges. Majority of the docking programs are sensitive enough to predict known protein bound poses with averaged accuracies close to about 1.5–2 Å and with reported success rates in the range of 70 – 80%. However, significant improvement beyond this range seems for now unachievable. One of the fundamental challenges are scoring in which various assumptions are used in order to make calculations computationally more efficient but by using these assumptions, some very important aspects of molecular recognition like solvent effect and entropy effect got neglected. Along with this the incorporation of water molecules in the active site having interaction with ligand and receptor is also a challenge. Molecular docking process also facing limitations on the availability of the crystallographic structure of the protein. Besides all these limitations and challenges molecular docking is still a method of choice and widelly used for the prediction of ligand-receptor pose.

Figure 11. The poses of SBR in the active sites of carbonic anhydrase (CA) I, II, IV, VI, IX, XII, and XIII predicted by program AutoDock Vina, named as p1, p2,p3, p4, p5, p6, and p7, respectively. The hydrophobic half of the active site cleft is shown in yellow, while Zn(II) in blue and Mg (II) in red. (A) The superposition of SBR (in magenta) and p2 (in cyan). (B) The seed structure derived from p1-7 with the hydrogen attached to sulfur as the growing site.
Source: (Xiang et al., 2014).

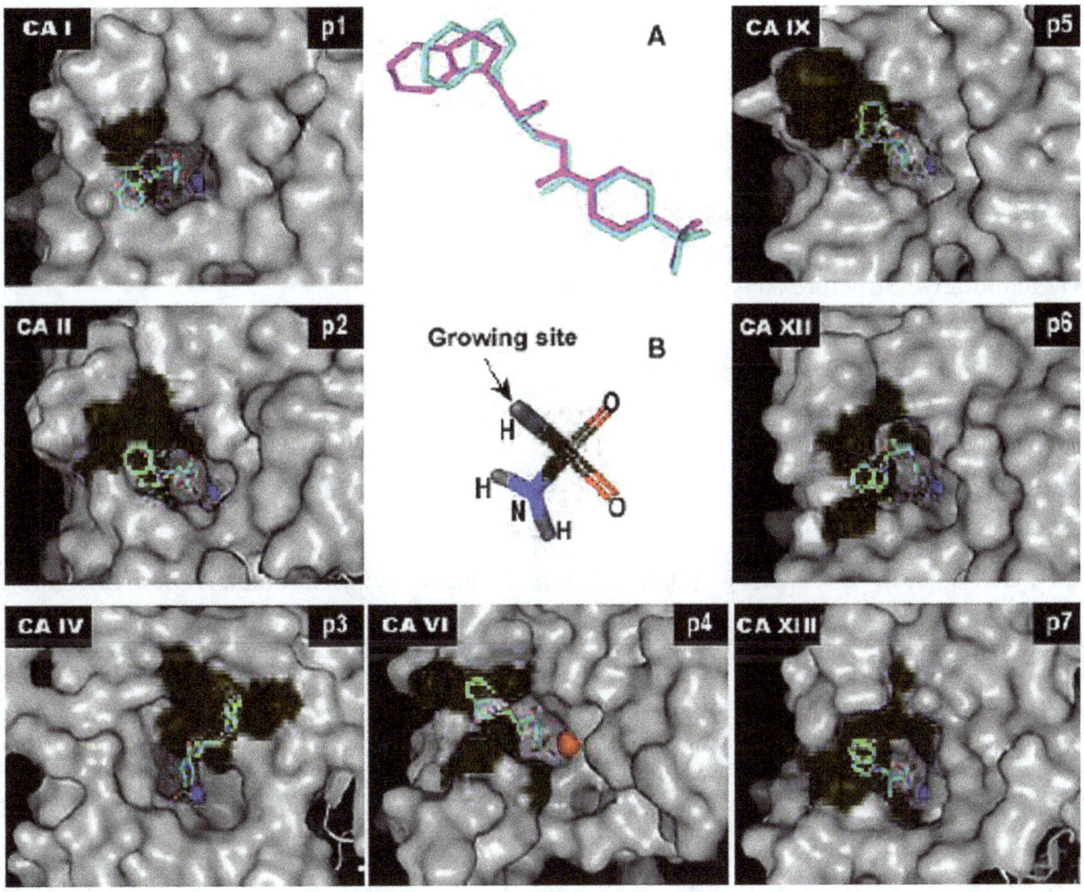

9. CONCLUSION

Herein we have provided a complete overview of diverse docking methodologies that are used in drug design and development. Our discussion typically aimed at professionals in the field of cheminformatics and bioinformatics, both in academia and in industry and aspiring scientists and students who want to take up this as a profession in the near future. We discussed several case studies, pertaining to each method. In few cases we provided step wise information to conduct the experiments. This discussion highlighted various application of molecular docking in drug discovery. We apprehend that in the years to come molecular docking in conjunction with variety of state of the art techniques *viz.* molecular biology, computer graphics and protein x-ray crystallography will be an indispensable tool in drug discovery.

List of Abbreviations

ADMET: Absorption, Distribution, Metabolism, Elimination, and Toxicity
ADRs: Adverse Drug Reactions
CADD: Computer Aided Drug Designing
CML: Chronic Myeloid Leukemia
FF: Force Field
HTS: High Throughput Screening
LBDD: Ligand Based Drug Designing
MOE: Molecular Operating Environment
NMR: Nuclear Magnetic Resonance Spectroscopy
SBDD: Structure Based Drug Designing
SWOT: Strength-Weakness-Opportunity-Threat
QSAR: Quantitative Structure Activity Relationship
VDW: Van Der Waals
VHTS: Virtual High Throughput Screening
VS: Virtual Screening

REFERENCES

Atkins, P., De Paula, J., & Walters, V. (2006). *Physical Chemistry*: Macmillan. *Higher Education.*

Berman, H. M., Westbrook, J., Feng, Z., Gilliland, G., Bhat, T. N., Weissig, H., & Bourne, P. E. et al. (2000). The Protein Data Bank. *Nucleic Acids Research, 28*(1), 235–242. doi:10.1093/nar/28.1.235 PMID:10592235

Böhm, H.-J. (1992a). The computer program LUDI: A new method for the de novo design of enzyme inhibitors. *Journal of Computer-Aided Molecular Design, 6*(1), 61–78. doi:10.1007/BF00124387 PMID:1583540

Böhm, H.-J. (1992b). LUDI: Rule-based automatic design of new substituents for enzyme inhibitor leads. *Journal of Computer-Aided Molecular Design, 6*(6), 593–606. doi:10.1007/BF00126217 PMID:1291628

Brady, G. P. Jr, & Stouten, P. F. (2000). Fast prediction and visualization of protein binding pockets with PASS. *Journal of Computer-Aided Molecular Design, 14*(4), 383–401. doi:10.1023/A:1008124202956 PMID:10815774

Chen, H. M., Liu, B. F., Huang, H. L., Hwang, S. F., & Ho, S. Y. (2007). SODOCK: Swarm optimization for highly flexible protein–ligand docking. *Journal of Computational Chemistry, 28*(2), 612–623 doi:10.1002/jcc.20542 PMID:17186483

Clark, K. P., & Ajay, . (1995). Flexible ligand docking without parameter adjustment across four ligand–receptor complexes. *Journal of Computational Chemistry, 16*(10), 1210–1226. doi:10.1002/jcc.540161004

Consortium, T. U. (2015). UniProt: A hub for protein information. *Nucleic Acids Research, 43*(D1), D204–D212. doi:10.1093/nar/gku989 PMID:25348405

Corbeil, C. R., Englebienne, P., & Moitessier, N. (2007). Docking ligands into flexible and solvated macromolecules. 1. Development and validation of FITTED 1.0. *Journal of Chemical Information and Modeling*, *47*(2), 435–449. doi:10.1021/ci6002637 PMID:17305329

Cornell, W. D., Cieplak, P., Bayly, C. I., Gould, I. R., Merz, K. M., Ferguson, D. M., & Kollman, P. A. et al. (1995). A Second Generation Force Field for the Simulation of Proteins, Nucleic Acids, and Organic Molecules. *Journal of the American Chemical Society*, *117*(19), 5179–5197. doi:10.1021/ja00124a002

Dominguez, C., Boelens, R., & Bonvin, A. M. (2003). HADDOCK: A protein-protein docking approach based on biochemical or biophysical information. *Journal of the American Chemical Society*, *125*(7), 1731–1737. doi:10.1021/ja026939x PMID:12580598

Duffy, B. C., Liu, S., Martin, G. S., Wang, R., Hsia, M. M., Zhao, H., & Kitchen, D. B. et al. (2015). Discovery of a new chemical series of BRD4(1) inhibitors using protein-ligand docking and structure-guided design. *Bioorganic & Medicinal Chemistry Letters*, *25*(14), 2818–2823. doi:10.1016/j.bmcl.2015.04.107 PMID:26022843

Fischer, E. (1894). Einfluss der Configuration auf die Wirkung der Enzyme, Ber. dtsch. Chem. Ges. 27, 2985–2993.

Forli, S., & Olson, A. J. (2012). A Force Field with Discrete Displaceable Waters and Desolvation Entropy for Hydrated Ligand Docking. *Journal of Medicinal Chemistry*, *55*(2), 623–638. doi:10.1021/jm2005145 PMID:22148468

Fradera, X., Kaur, J., & Mestres, J. (2004). Unsupervised guided docking of covalently bound ligands. *Journal of Computer-Aided Molecular Design*, *18*(10), 635–650. doi:10.1007/s10822-004-5291-4 PMID:15849994

Fradera, X., Knegtel, R., & Mestres, J. (2000). Similarity-driven flexible ligand docking. *Proteins: Structure, Function, and Bioinformatics*, *40*(4), 623–636. doi:10.1002/1097-0134(20000901)40:4<623::AID-PROT70>3.0.CO;2-I PMID:10899786

Friesner, R. A., Banks, J. L., Murphy, R. B., Halgren, T. A., Klicic, J. J., Mainz, D. T., & Perry, J. K. et al. (2004). Glide: A new approach for rapid, accurate docking and scoring. 1. Method and assessment of docking accuracy. *Journal of Medicinal Chemistry*, *47*(7), 1739–1749. doi:10.1021/jm0306430 PMID:15027865

Friesner, R. A., Banks, J. L., Murphy, R. B., Halgren, T. A., Klicic, J. J., Mainz, D. T., & Shenkin, P. S. et al. (2004). Glide: A New Approach for Rapid, Accurate Docking and Scoring. 1. Method and Assessment of Docking Accuracy. *Journal of Medicinal Chemistry*, *47*(7), 1739–1749. doi:10.1021/jm0306430 PMID:15027865

Gabb, H. A., Jackson, R. M., & Sternberg, M. J. (1997). Modelling protein docking using shape complementarity, electrostatics and biochemical information. *Journal of Molecular Biology*, *272*(1), 106–120. doi:10.1006/jmbi.1997.1203 PMID:9299341

Goodford, P. J. (1985). A computational procedure for determining energetically favorable binding sites on biologically important macromolecules. *Journal of Medicinal Chemistry*, *28*(7), 849–857. doi:10.1021/jm00145a002 PMID:3892003

Halgren, T. A. (1996). Merck molecular force field. I. Basis, form, scope, parameterization, and performance of MMFF94. *Journal of Computational Chemistry*, *17*(5-6), 490–519. doi:10.1002/(SICI)1096-987X(199604)17:5/6<490::AID-JCC1>3.0.CO;2-P

Huang, S. Y., Grinter, S. Z., & Zou, X. (2010). Scoring functions and their evaluation methods for protein-ligand docking: Recent advances and future directions. *Physical Chemistry Chemical Physics*, *12*(40), 12899–12908. doi:10.1039/c0cp00151a PMID:20730182

Jain, A. N. (2003). Surflex: Fully automatic flexible molecular docking using a molecular similarity-based search engine. *Journal of Medicinal Chemistry*, *46*(4), 499–511. doi:10.1021/jm020406h PMID:12570372

Jain, A. N. (2007). Surflex-Dock 2.1: Robust performance from ligand energetic modeling, ring flexibility, and knowledge-based search. *Journal of Computer-Aided Molecular Design*, *21*(5), 281–306. doi:10.1007/s10822-007-9114-2 PMID:17387436

Jorgensen, W. L., & Tirado-Rives, J. (1988). The OPLS [optimized potentials for liquid simulations] potential functions for proteins, energy minimizations for crystals of cyclic peptides and crambin. *Journal of the American Chemical Society*, *110*(6), 1657–1666. doi:10.1021/ja00214a001

Joseph-McCarthy, D., Thomas, B. E., Belmarsh, M., Moustakas, D., & Alvarez, J. C. (2003). Pharmacophore-based molecular docking to account for ligand flexibility. *Proteins: Structure, Function, and Bioinformatics*, *51*(2), 172–188. doi:10.1002/prot.10266 PMID:12660987

Kang, X., Shafer, R. H., & Kuntz, I. D. (2004). Calculation of ligand-nucleic acid binding free energies with the generalized-born model in DOCK. *Biopolymers*, *73*(2), 192–204. doi:10.1002/bip.10541 PMID:14755577

Kastenholz, M. A., Pastor, M., Cruciani, G., Haaksma, E. E., & Fox, T. (2000). GRID/CPCA: A new computational tool to design selective ligands. *Journal of Medicinal Chemistry*, *43*(16), 3033–3044. doi:10.1021/jm000934y PMID:10956211

Kitchen, D. B., Decornez, H., Furr, J. R., & Bajorath, J. (2004). Docking and scoring in virtual screening for drug discovery: Methods and applications. *Nature Reviews. Drug Discovery*, *3*(11), 935–949. doi:10.1038/nrd1549 PMID:15520816

Knegtel, R. M., Kuntz, I. D., & Oshiro, C. (1997). Molecular docking to ensembles of protein structures. *Journal of Molecular Biology*, *266*(2), 424–440. doi:10.1006/jmbi.1996.0776 PMID:9047373

Kozakov, D., Brenke, R., Comeau, S. R., & Vajda, S. (2006). PIPER: An FFT-based protein docking program with pairwise potentials. *Proteins: Structure, Function, and Bioinformatics*, *65*(2), 392–406. doi:10.1002/prot.21117 PMID:16933295

Kuntz, I. D., Blaney, J. M., Oatley, S. J., Langridge, R., & Ferrin, T. E. (1982). A geometric approach to macromolecule-ligand interactions. *Journal of Molecular Biology*, *161*(2), 269–288. doi:10.1016/0022-2836(82)90153-X PMID:7154081

LaBute, M. X., Zhang, X., Lenderman, J., Bennion, B. J., Wong, S. E., & Lightstone, F. C. (2014). Adverse Drug Reaction Prediction Using Scores Produced by Large-Scale Drug-Protein Target Docking on High-Performance Computing Machines. *PLoS ONE*, *9*(9), e106298. doi:10.1371/journal.pone.0106298 PMID:25191698

Laskowski, R. A. (1995). SURFNET: a program for visualizing molecular surfaces, cavities, and intermolecular interactions. *J Mol Graph, 13*(5), 307-328.

Lawrence, M. C., & Davis, P. C. (1992). CLIX: A search algorithm for finding novel ligands capable of binding proteins of known three-dimensional structure. *Proteins: Structure, Function, and Bioinformatics*, *12*(1), 31–41. doi:10.1002/prot.340120105 PMID:1313175

Lengauer, T., & Rarey, M. (1996). Computational methods for biomolecular docking. *Current Opinion in Structural Biology*, *6*(3), 402–406. doi:10.1016/S0959-440X(96)80061-3 PMID:8804827

Levitt, D. G., & Banaszak, L. J. (1992). POCKET: A computer graphics method for identifying and displaying protein cavities and their surrounding amino acids. *Journal of Molecular Graphics*, *10*(4), 229–234. doi:10.1016/0263-7855(92)80074-N PMID:1476996

Lewis, R. A., & Leach, A. R. (1994). Current methods for site-directed structure generation. *Journal of Computer-Aided Molecular Design*, *8*(4), 467–475. doi:10.1007/BF00125381 PMID:7815098

Lewis, R. A., Roe, D. C., Huang, C., Ferrin, T. E., Langridge, R., & Kuntz, I. D. (1992). Automated site-directed drug design using molecular lattices. *Journal of Molecular Graphics*, *10*(2), 66–78. doi:10.1016/0263-7855(92)80059-M PMID:1637751

Lipinski, C. A. (2004). Lead- and drug-like compounds: The rule-of-five revolution. *Drug Discovery Today. Technologies*, *1*(4), 337–341. doi:10.1016/j.ddtec.2004.11.007 PMID:24981612

Lipinski, C. A., Lombardo, F., Dominy, B. W., & Feeney, P. J. (2001). Experimental and computational approaches to estimate solubility and permeability in drug discovery and development settings1. *Advanced Drug Delivery Reviews*, *46*(1–3), 3–26. doi:10.1016/S0169-409X(00)00129-0 PMID:11259830

MacKerell, A. D. Jr, Bashford, D., Bellott, M., Dunbrack, R. L. Jr, Evanseck, J. D., Field, M. J., & Karplus, M. et al. (1998). All-Atom Empirical Potential for Molecular Modeling and Dynamics Studies of Proteins. *The Journal of Physical Chemistry B*, *102*(18), 3586–3616. doi:10.1021/jp973084f PMID:24889800

McGann, M. R., Almond, H. R., Nicholls, A., Grant, J. A., & Brown, F. K. (2003). Gaussian docking functions. *Biopolymers*, *68*(1), 76–90. doi:10.1002/bip.10207 PMID:12579581

Meiler, J., & Baker, D. (2006). ROSETTALIGAND: Protein–small molecule docking with full side-chain flexibility. *Proteins: Structure, Function, and Bioinformatics*, *65*(3), 538–548. doi:10.1002/prot.21086 PMID:16972285

Mezei, M. (2003). A new method for mapping macromolecular topography. *Journal of Molecular Graphics & Modelling*, *21*(5), 463–472. doi:10.1016/S1093-3263(02)00203-6 PMID:12543141

Miranker, A., & Karplus, M. (1995). An automated method for dynamic ligand design. *Proteins: Structure, Function, and Bioinformatics*, *23*(4), 472–490. doi:10.1002/prot.340230403 PMID:8749844

Moitessier, N., Englebienne, P., Lee, D., Lawandi, J., & Corbeil, C. R. (2008). Towards the development of universal, fast and highly accurate docking/scoring methods: A long way to go. *British Journal of Pharmacology, 153*(S1), S7–S26. doi:10.1038/sj.bjp.0707515 PMID:18037925

Molecular Operating Environment (MOE). (2013). Chemical Computing Group Inc.

Morley, S. D., & Afshar, M. (2004). Validation of an empirical RNA-ligand scoring function for fast flexible docking using RiboDock®. *Journal of Computer-Aided Molecular Design, 18*(3), 189–208. doi:10.1023/B:JCAM.0000035199.48747.1e PMID:15368919

Morris, G. M., Huey, R., Lindstrom, W., Sanner, M. F., Belew, R. K., Goodsell, D. S., & Olson, A. J. (2009). AutoDock4 and AutoDockTools4: Automated Docking with Selective Receptor Flexibility. *Journal of Computational Chemistry, 30*(16), 2785–2791. doi:10.1002/jcc.21256 PMID:19399780

Moustakas, D. T., Lang, P. T., Pegg, S., Pettersen, E., Kuntz, I. D., Brooijmans, N., & Rizzo, R. C. (2006). Development and validation of a modular, extensible docking program: DOCK 5. *Journal of Computer-Aided Molecular Design, 20*(10-11), 601–619. doi:10.1007/s10822-006-9060-4 PMID:17149653

Oshiro, C. M., Kuntz, I. D., & Dixon, J. S. (1995). Flexible ligand docking using a genetic algorithm. *Journal of Computer-Aided Molecular Design, 9*(2), 113–130. doi:10.1007/BF00124402 PMID:7608743

Österberg, F., Morris, G. M., Sanner, M. F., Olson, A. J., & Goodsell, D. S. (2002). Automated docking to multiple target structures: Incorporation of protein mobility and structural water heterogeneity in AutoDock. *Proteins: Structure, Function, and Bioinformatics, 46*(1), 34–40. doi:10.1002/prot.10028 PMID:11746701

Pei, J., Wang, Q., Liu, Z., Li, Q., Yang, K., & Lai, L. (2006). PSI-DOCK: Towards highly efficient and accurate flexible ligand docking. *Proteins: Structure, Function, and Bioinformatics, 62*(4), 934–946. doi:10.1002/prot.20790 PMID:16395666

Pettersen, E. F., Goddard, T. D., Huang, C. C., Couch, G. S., Greenblatt, D. M., Meng, E. C., & Ferrin, T. E. (2004). UCSF Chimera--a visualization system for exploratory research and analysis. *Journal of Computational Chemistry, 25*(13), 1605–1612. doi:10.1002/jcc.20084 PMID:15264254

Rarey, M., Kramer, B., Lengauer, T., & Klebe, G. (1996). A fast flexible docking method using an incremental construction algorithm. *Journal of Molecular Biology, 261*(3), 470–489. doi:10.1006/jmbi.1996.0477 PMID:8780787

Schnecke, V., & Kuhn, L. A. (2000). Virtual screening with solvation and ligand-induced complementarity. *Perspectives in Drug Discovery and Design, 20*(1), 171–190. doi:10.1023/A:1008737207775

Schneidman-Duhovny, D., Inbar, Y., Nussinov, R., & Wolfson, H. J. (2005). PatchDock and SymmDock: Servers for rigid and symmetric docking. *Nucleic Acids Research, 33*(Web Serversuppl 2), W363–W367. doi:10.1093/nar/gki481 PMID:15980490

Seifert, M. H. (2005). ProPose: Steered virtual screening by simultaneous protein-ligand docking and ligand-ligand alignment. *Journal of Chemical Information and Modeling, 45*(2), 449–460. doi:10.1021/ci0496393 PMID:15807511

Sliwoski, G., Kothiwale, S., Meiler, J., & Lowe, E. W. (2014). Computational Methods in Drug Discovery. *Pharmacological Reviews*, *66*(1), 334–395. doi:10.1124/pr.112.007336 PMID:24381236

Taliani, S., Pugliesi, I., Barresi, E., Salerno, S., Marchand, C., Agama, K., & Da Settimo, F. et al. (2013). Phenylpyrazolo[1,5-a]quinazolin-5(4H)-one: A Suitable Scaffold for the Development of Noncamptothecin Topoisomerase I (Top1) Inhibitors. *Journal of Medicinal Chemistry*, *56*(18), 7458–7462. doi:10.1021/jm400932c PMID:23987476

Taylor, J. S., & Burnett, R. M. (2000). DARWIN: A program for docking flexible molecules. *Proteins: Structure, Function, and Bioinformatics*, *41*(2), 173–191. doi:10.1002/1097-0134(20001101)41:2<173::AID-PROT30>3.0.CO;2-3 PMID:10966571

Thomsen, R., & Christensen, M. H. (2006). MolDock: A new technique for high-accuracy molecular docking. *Journal of Medicinal Chemistry*, *49*(11), 3315–3321. doi:10.1021/jm051197e PMID:16722650

Tøndel, K., Anderssen, E., & Drabløs, F. (2006). Protein Alpha Shape (PAS) Dock: A new gaussian-based score function suitable for docking in homology modelled protein structures. *Journal of Computer-Aided Molecular Design*, *20*(3), 131–144. doi:10.1007/s10822-006-9041-7 PMID:16652207

Trott, O., & Olson, A. J. (2010). AutoDock Vina: Improving the speed and accuracy of docking with a new scoring function, efficient optimization, and multithreading. *Journal of Computational Chemistry*, *31*(2), 455–461. doi:10.1002/jcc.21334 PMID:19499576

Van Drie, J. H. (2007). Computer-aided drug design: The next 20 years. *Journal of Computer-Aided Molecular Design*, *21*(10-11), 591–601. doi:10.1007/s10822-007-9142-y PMID:17989929

Verdonk, M. L., Chessari, G., Cole, J. C., Hartshorn, M. J., Murray, C. W., Nissink, J. W. M., & Taylor, R. et al. (2005). Modeling water molecules in protein-ligand docking using GOLD. *Journal of Medicinal Chemistry*, *48*(20), 6504–6515. doi:10.1021/jm050543p PMID:16190776

Vieth, M., Hirst, J. D., Kolinski, A., & Brooks, C. L. (1998). Assessing energy functions for flexible docking. *Journal of Computational Chemistry*, *19*(14), 1612–1622. doi:10.1002/(SICI)1096-987X(19981115)19:14<1612::AID-JCC7>3.0.CO;2-M

Wallace, A. C., Laskowski, R. A., & Thornton, J. M. (1995). LIGPLOT: A program to generate schematic diagrams of protein-ligand interactions. *Protein Engineering*, *8*(2), 127–134. doi:10.1093/protein/8.2.127 PMID:7630882

Wu, G., Robertson, D. H., Brooks, C. L., & Vieth, M. (2003). Detailed analysis of grid-based molecular docking: A case study of CDOCKER—A CHARMm-based MD docking algorithm. *Journal of Computational Chemistry*, *24*(13), 1549–1562. doi:10.1002/jcc.10306 PMID:12925999

Wu, G., & Vieth, M. (2004). SDOCKER: A method utilizing existing X-ray structures to improve docking accuracy. *Journal of Medicinal Chemistry*, *47*(12), 3142–3148. doi:10.1021/jm040015y PMID:15163194

Xiang, F., Xiang, J., Fang, Y., Zhang, M., & Li, M. (2014). Discovering isozyme-selective inhibitor scaffolds of human carbonic anhydrases using structural alignment and de novo drug design approaches. *Chemical Biology & Drug Design*, *83*(2), 247–258. doi:10.1111/cbdd.12234 PMID:24112770

Yang, J., Yan, R., Roy, A., Xu, D., Poisson, J., & Zhang, Y. (2015). The I-TASSER Suite: Protein structure and function prediction. *Nature Methods*, *12*(1), 7–8. doi:10.1038/nmeth.3213 PMID:25549265

Yin, L., Zheng, L., Xu, L., Dong, D., Han, X., Qi, Y., & Peng, J. et al. (2015). In-silico prediction of drug targets, biological activities, signal pathways and regulating networks of dioscin based on bioinformatics. *BMC Complementary and Alternative Medicine*, *15*(1), 41. doi:10.1186/s12906-015-0579-6 PMID:25879470

Yuan, Y., Pei, J., & Lai, L. (2011). LigBuilder 2: A Practical de Novo Drug Design Approach. *Journal of Chemical Information and Modeling*, *51*(5), 1083–1091. doi:10.1021/ci100350u PMID:21513346

Zavodszky, M. I., & Kuhn, L. A. (2005). Side-chain flexibility in protein–ligand binding: The minimal rotation hypothesis. *Protein Science*, *14*(4), 1104–1114. doi:10.1110/ps.041153605 PMID:15772311

Zhang, X., Wong, S. E., & Lightstone, F. C. (2013). Message passing interface and multithreading hybrid for parallel molecular docking of large databases on petascale high performance computing machines. *Journal of Computational Chemistry*, *34*(11), 915–927. doi:10.1002/jcc.23214 PMID:23345155

Zhao, Y., & Sanner, M. F. (2007). FLIPDock: Docking flexible ligands into flexible receptors. *Proteins: Structure, Function, and Bioinformatics*, *68*(3), 726–737. doi:10.1002/prot.21423 PMID:17523154

Zhou, T., Commodore, L., Huang, W. S., Wang, Y., Thomas, M., Keats, J., & Zhu, X. et al. (2011). Structural mechanism of the Pan-BCR-ABL inhibitor ponatinib (AP24534): Lessons for overcoming kinase inhibitor resistance. *Chemical Biology & Drug Design*, *77*(1), 1–11. doi:10.1111/j.1747-0285.2010.01054.x PMID:21118377

Zsoldos, Z., Reid, D., Simon, A., Sadjad, B. S., & Peter Johnson, A. (2006). eHiTS: An innovative approach to the docking and scoring function problems. *Current Protein & Peptide Science*, *7*(5), 421–435. doi:10.2174/138920306778559412 PMID:17073694

Chapter 2
Application of Docking Methodologies in QSAR-Based Studies

Omar Deeb
Al-Quds University, Palestine

Heidy Martínez-Pachecho
Universidad del Papaloapan, Mexico

Guillermo Ramírez-Galicia
Universidad del Papaloapan, Mexico

Ramón Garduño-Juárez
Universidad Nacional Autónoma de México, Mexico

ABSTRACT

The computational strategies permeate all aspects of drug discovery such as virtual screening techniques. Virtual screening can be classified into ligand based and structure based methods. The ligand based method such as Quantitative Structure Activity Relationship (QSAR) is used when a set of active ligand compounds is recognized and slight or no structural information is available for the receptors. In structure based drug design, the most widespread method is molecular docking. It is widely accepted that drug activity is obtained through the molecular binding of one ligand to receptor. In their binding conformations, the molecules exhibit geometric and chemical complementarity, both of which are essential for successful drug activity. The molecular docking approach can be used to model the interaction between a small drug molecule and a protein, which allow us to characterize the performance of small molecules in the binding site of target proteins as well as to clarify fundamental biochemical processes.

INTRODUCTION

The Molecular modeling expression is used to describe the use of computers to build compounds and carry out a variety of calculations on these compounds in order to predict their chemical characteristics and behavior. The computational strategies also permeate all aspects of drug discovery such as virtual screening techniques in comparison to the process of trial and error that was used in the search for novel drugs.

DOI: 10.4018/978-1-5225-0362-0.ch002

The most important challenges that medicinal chemists face today is the design of new drugs with improved potency and less side-effects for treating human diseases such as AIDS and others. Medicinal chemists begin the process by taking a lead structure and then finding analogs exhibiting the preferred biological activities. Next, they use their experience and chemical insight to eventually choose a candidate analog for further development. This process was not only difficult but also expensive and time consuming. The conventional methods of drug discovery are now being supplemented by shorter approaches made possible by the accepting of the molecular processes involved in the original disease. In this view, the preliminary point in drug design is the molecular target, which is a receptor or enzyme in the body as an option of the existence of the known lead structure.

The effective design of chemical structures with the desirable therapeutic properties is directed towards Computer Aided-Drug Design (CADD) a well-established area of Computer Aided-Molecular Design (CAMD). The main applications of CAMD are the clarification of the basic requirements for a compound to obtain a determined activity, the simulation of the binding between a ligand and the receptor, the discovery of new active compounds and the prediction of activities for non-synthesized analogues. Two major modeling strategies right now are used in the designing of new drugs. In the first strategy, the design is based on the comparative analysis of the structural features of known active and inactive molecules that are interpreted in terms of their complementarily with a supposed receptor site model. This strategy is called *ligand based design* and one of its approaches is quantitative structure activity relationships (QSAR). This discipline was promoted by Hansch and his group (Fujita, 1990). In the second strategy, the three-dimensional features of a known receptor site are directly considered and this strategy is called *structure based design* method. In structure based drug design, the most widespread method is molecular docking (Mahajan A., Gill N.S & Arora R., 2014; Xuan-Yu Meng, Hong-Xing Zhang, Mihaly Mezei & Meng Cui, 2011) The molecular docking approach which will be discussed in details in the coming section can be used to model the interaction between a small drug molecule and a protein, which allow us to characterize the performance of small molecules in the binding site of target proteins as well as to clarify fundamental biochemical processes.

This chapter deals with the two strategies that are used in designing new drugs. At first, focus will be on the ligand based design or QSAR approach. The QSAR section starts with history and the earliest efforts made in this field, types of descriptors and statistical analysis methods. Moreover, validation of QSAR models, the internal as well as the external validations. Second, the focus will be on the structure based design or molecular docking approach. The different types of docking methods will be discussed including the different software used. At the end, applications or case studies done by the authors and others will be discussed also. For example, Exploration of human serum albumin binding sites by docking and molecular dynamics, Docking study related to non-peptide HIV-1 protease inhibitors, Homology modeling, molecular dynamics and docking simulations of rat A2A receptor, Exploring the ligand recognition properties of the human vasopressin V1a receptor using QSAR and molecular modeling studies and others. These studies will introduce the reader to the field of molecular docking and its use in structure based drug design. Finally, more attention must be dedicated to the combination of the two approaches in designing new drugs.

BACKGROUND

The approach of designing new drugs can be classified into ligand based and structure based methods. The ligand based method such as quantitative structure activity relationship (QSAR) is used when a set of active ligand compounds is recognized and slight or no structural information is available for the receptors. In structure based drug design, the most widespread method is molecular docking. The molecular docking approach can be used to model the interaction between a small drug molecule and a protein, which allow us to characterize the performance of small molecules in the binding site of target proteins as well as to clarify fundamental biochemical processes.

LIGAND BASED DESIGN METHOD

Quantitative Structure Activity Relationships (QSAR)

QSAR is a way of finding and modeling a simple mathematical equation between molecular descriptors calculated from structures of a set of molecules with their reactivity in order to be able to predict the reactivity of unknown compounds.

Applications of QSAR can be extended to any molecular design purpose, including prediction of different kinds of biological activities, lead compound optimization and prediction of novel structural leads in drug discovery. QSAR consists of several steps which hopefully lead to the design of new compounds with the desired activity profile.

The QSAR approach follows these steps. The first step of building a QSAR model is to select a training set consisting of compounds with their experimental biological activities. The next step is to optimize the molecular structure by a proper technique. Afterwards, one should compute descriptors that contain sufficient relevant information about the biological phenomenon. Once descriptors have been calculated, it is necessary to pick which should be included in the QSAR model. Therefore, feature selection techniques are used to select the most relevant descriptors from a pool of descriptors. One way is simply to use the correlation coefficient method to select the descriptor with the highest correlation coefficient. Next step, a data analysis is needed to calculate the best mathematical expression linking together the descriptors and biological activities, In the final step, validation and predictions for non-tested compounds will take place. However, the predictive capability of the model first is verified using an external set of compounds which are not used in calibration. This is talented by biological testing of some additional compounds (test set) in the same way as the training set and then comparing the experimental finding with the values predicted by the QSAR model. If the QSAR predicts within acceptable restrictions, it may be used for a more extensive prediction of more compounds.

Hansch equation (Hansch, 1969) was developed to correlate physicochemical properties (descriptors) with biological activities and is given in a general form by the following equation (1):

$$\log\frac{1}{C} = a(\log P)^2 + b\,log\,P + cs + \cdots k \tag{1}$$

where *C* is the molar concentration that produces the biological effect; *P* is the octanol/water partition coefficient and *s* is the electronic Hammett constant.

Descriptors

A common question in QSAR is how to describe molecules and their physicochemical properties (descriptors). The nature of the descriptors used and the extent to which they instruct the structural properties related to the biological activity is a critical step in a QSAR study It has been estimated that over a thousand of molecular descriptors can now be calculated from a given molecular structure (Devillers & Balaban, 1999; Karelson, 2000;Todeschini, Consonni & Pavan, 2001). Most of those descriptors can be calculated by using commercial software packages such as CODESSA (Katritzky, Lobanov & Karelson, 1994), DRAGON (Todeschini, Consonni & Pavan, 2001) and others. The various descriptors in use can be largely categorized as being constitutional, topological, electrostatic, geometrical, or quantum chemical.

Statistical Methods

Many different statistical methods are available in the literature and the selection of the appropriate method is critical (Xu & Zhang, 2001). However, we will try to mention some basic techniques, one linear and one non-linear, which were extensively used in QSAR/QSPR studies.

Multiple Linear Regression (MLR) (Montgomery & Peck, 1992) can be considered as an easy interpretable linear regression method. Regression analysis correlates independent X variables or descriptors with dependent Y variables (biological data). The regression model assumes a linear relationship between *m* molecular descriptors and the biological activity variable. This relationship can be expressed with the single multiple-term linear equation as follows:

$$y = b_0 + b_1 x_1 + b_2 x_2 + \cdots + b_m x_m + e \quad (2)$$

The MLR analysis calculates the regression coefficients, b_i, by minimizing the residuals, **e**, which quantify the deviations between the data (Y) and the model (Y'), as in the case of simple linear regression.

Artificial Neural Networks (ANN) method (Duprat, Huynh & Dreyfus, 1998; Tetko, Alessandro, Villa & Livingstone, 1996;) is a non-linear technique inspired in the human brain, composed of many simple processing units called neurons. This method is also recognized as learning algorithms. The aim is to simulate the various shells of the neurons, where each neuron is connected to a number of neighboring neurons with variable coefficients of connectivity that signify the strength of these associations. Several examples of this methodology are explained in Montañez-Godinez *et al* (2014).

Validation of QSAR Models

After the model equation is obtained, or it is also significant to estimate the power and the validity of the model before using it to predict the biological activity. Validity is to establish the reliability and significance of the method for a particular use. For that reason, validation of a QSAR model must be done. There are two validation methods used for a QSAR model: internal and external validation techniques to establish the confidence and strength of the model.

Internal validation uses the data set from which the model is built and checks for internal stability. *Cross-Validation (CV) technique* is widely employed as an internal validation method of statistical models (Wold, 1991). External validation evaluates how well a built model is generalized. If a sufficiently large series of compounds with known activity are obtainable, the original data set can be divided into two subgroups, the *training set* and the *test set.* The training or calibration set is used to derive a calibration model that will be used later to predict the activities of the test or validation set compounds.

STRUCTURE BASED DESIGN METHOD

Molecular Docking

It is widely accepted that drug activity is obtained through the molecular binding of one ligand to receptor. In their binding conformations, the molecules exhibit geometric and chemical complementarity, both of which are essential for successful drug activity. The computational process of searching for a ligand that is able to fit both geometrically and energetically the binding site of a protein is called molecular docking. The docking process involves the prediction of ligand conformation and orientation within a targeted binding site. In general, there are two aims of docking studies: accurate structural modeling and correct prediction of activity.

The docking program is used to place computer-generated representations of a small molecule into the active site of an enzyme in a variety of positions, conformations and orientations. Each such docking mode is called a 'pose'. In order to identify the energetically most favorable pose each pose is evaluated or simply 'scored' based on its complementarity to the target in terms of shape and properties such as electrostatics. A good score for a given molecule indicates that it is potentially a good binder. This process is repeated for all molecules in the collection, which are subsequently rank-ordered by their scores (or their predicted affinities). Assuming that both the poses and the associated affinity scores have been predicted with reasonable accuracy, this selection will contain a relatively large proportion of active molecules

The prediction of the protein–ligand complex is usually done by searching the translational and rotational degrees of freedom of the ligand within the receptor binding site, and by searching the conformational degrees of freedom of the ligand itself. Binding conformations generated by docking programs are thus defined by both a position of the ligand on the receptor surface and a particular ligand conformer. The first docking algorithm for small molecules was developed by the Kuntz group at UCSF in 1982 (Kuntz *et al.*, 1982). Before ligands can be docked against a receptor, generally the binding site has to be identified first. This is done to limit the search space on the receptor surface and thus minimize the degrees of freedom that have to be searched. The active site is often known from crystal structures of ligand-bound receptors, but it can also be predicted. The largest cavity on a protein surface is frequently the active site, but this is not always the case and different active site prediction and analysis methods have been developed. In the absence of knowledge about the binding sites, cavity detection programs such as GRID (Goodford 1985; Kastenholz, Pastor, Cruciani, Haaksma & Fox 2000), POCKET (Levitt & Banaszak 1992), SurfNet (Laskowski 1995; Glaser, Morris, Najmanovich, Laskowski & Thornton. 2006), Docking without any assumption about the binding site is called blind docking. Various excellent reviews on docking have been published in the past (Kitchen, Decornez, Furr & Bajorath 2004; Halperin, Ma, Wolfson & Nussinov 2002; Coupez & Lewis 2006; Kontoyianni, Madhav, Suchanek & Seibel 2008;

Brooijmans & Kuntz. 2003). In the following sections, we will give more details about the docking theory including search algorithms, scoring functions, docking software's as well as docking methods.

DOCKING ALGORITHMS

Docking algorithms generally contain two search engines; one is responsible for the placement and exploration of the ligand in the binding site, while the other searches the internal degrees of freedom of the ligand. For ligands without any degrees of freedom, it is feasible to search the six translational and rotational degrees of freedom in the receptor binding site. However, as the number of rotatable bonds of the ligand increases, the number of degrees of freedom increases exponentially when finding solutions to fit into the binding site. Different algorithms are available and can be divided into three major categories. The first algorithm is a grid search that comprehensively searches the binding site by moving the rigid ligand through the available six degrees of freedom in a systematic fashion in the binding site. The original algorithm only looked for surface complementarity (Katchalski-Katzir *et al.*, 1992), but later implementations also use electrostatic complementarity, as in FTDOCK (Gabb, Jackson, & Sternberg, 1997) for example. The second algorithm and significantly more efficient algorithm place ligands in the binding site by descriptor matching (Ewing & Kuntz, 1997). Descriptors to find a solution are generally points, sometimes with properties associated to them, placed in the receptor site. Ligand atoms are matched with the receptor descriptors with some tolerance. The descriptor matching algorithm orients the ligand in the binding site, which subsequently has to be optimized and scored. Usually a large number of orientations are possible, and they all have to be explored and optimized. When binding site descriptors are associated with physical– chemical properties, such as hydrophobicity and hydrogen-bonding capabilities, the ligand atoms have to match geometrically as well as chemically. DOCK (Shoichet & Kuntz, 1993; Moustakas *et al.*, 2006) and FlexX (Rarey, Wefing & Lengauer, 1996) use descriptor matching to orient the ligand in the binding site. The third algorithm is an energy-based search method using molecular mechanics force fields to explore the energy surface of the ligand with molecular dynamics (MD) or energy minimization algorithms. The minima of the ligand on the surface have to be located and assessed in terms of their complementarity to the receptor. Energy minimization is a local search, and is only used to optimize binding conformations generated by other search engines. MD searches are global searches in principle, but also tend to get stuck in local minima. This search strategy generally explores all degrees of freedom simultaneously, and is therefore not strictly considered as orientation algorithms.

In summary ligand search algorithms are divided into three types of methods, namely systematic, stochastic, and deterministic. Systematic search methods are frequently used in rigid protein-ligand docking, where there are only six degrees of freedom, while genetic algorithms and Monte Carlo are stochastic search algorithms, and molecular dynamics and energy minimization are deterministic algorithms.

Scoring Functions

The search algorithm of a docking method generates a set of candidate dockings for a particular ligand, but some mechanism is needed to decide which ligand placements are better than others, and thereby rank conformations against each other. This mechanism is called a scoring function. Scoring functions can be classified into three categories: knowledge-based, empirical and force field-based. Knowledge-based scoring functions rely on statistical means to extract rules on preferred, and non-preferred atom pair interactions from experimentally determined protein-ligand complexes. The rules are interpreted as pair-potentials that are subsequently used to score ligand binding poses. The PMF score (Muegge & Martin 1999), for example, is a well-known knowledge-based scoring function. Empirical scoring functions sum enthalpic and entropic interactions with the relative weights of the terms based on a training set of protein-ligand complexes. The interaction terms often include Van der Waals, electrostatic interactions and hydrogen bonds. Examples of empirical scoring functions include ChemScore (Eldridge, Murray, Auton, Paolini & Mee 1997) and the FlexX (Rarey, Wefing, & Lengauer, 1996) scoring function. Force field scoring functions are similar to empirical scoring functions in that they predict the binding free energy of a protein-ligand complex by adding up individual contributions from different types of interactions. However, they differ from empirical scoring functions in that the interaction terms are derived from physical chemical phenomena as opposed to experimental affinities. Examples of force field scoring functions in docking programs include the energy score in DOCK (Shoichet and Kuntz, 1993; Moustakas *et al.*, 2006), the score function used for single ligand docking in GOLD (Jones, Willett, Glen, Leach and Taylor 1997). Several reviews on the use of various scoring functions employed in docking calculations have been published (Tame, 1999; Halperin, Ma, Wolfson & Nussinov 2002; Gohlke & Klebe 2001).

Docking Software

A large number of docking programs have been published. Each program applies a certain algorithm that explores the way the ligands are treated during docking. For example, programs that follow the algorithm in which the ligand is built up incrementally, starting from a docked 'base fragment' are DOCK (Shoichet & Kuntz, 1993; Moustakas *et al.*, 2006), and FlexX (Rarey, Wefing & Lengauer, 1996). In other programs, such as AutoDock (Morris *et al.*, 1998; Osterberg, Morris, Sanner and Olson, 2002), Genetic Optimization for Ligand Docking (GOLD) (Jones, Willett, Glen, Leach & Taylor 1997) the ligand is treated in its entirety. Few docking programs allow protein flexibility such as AutoDock (Osterberg, Morris, Sanner & Olson, 2002), FlexE (Rarey, Wefing, & Lengauer, 1996) and QXP (McMartin and Bohacek, 1997). Another measure to classify docking programs would be according to the search strategy employed. For example, DOCK (Shoichet & Kuntz, 1993; Moustakas *et al.*, 2006) and FlexX (Rarey, Wefing, & Lengauer, 1996) trying to maximize shape complementarity while AutoDock (Morris *et al.*, 1998; Osterberg, Morris, Sanner & Olson, 2002), QXP (McMartin & Bohacek, 1997) and GOLD (Jones, Willett, Glen, Leach & Taylor 1997) programs incorporating an energy-driven or stochastic algorithm.

DOCKING METHODOLOGIES

Rigid Ligand and Rigid Receptor Docking

In this type of docking, the ligand has six degrees of freedom, three translational and three rotational in which the search space is limited. Only one conformation of the ligand is considered. No internal rotations about bonds are permitted in either molecule. Although no modern docking program relies exclusively on this type, they remain an integral component of many other techniques. DOCK (Shoichet & Kuntz, 1993; Moustakas *et al.*, 2006) and FTDOCK (Gabb, Jackson & Sternberg, 1997) adopted this type docking. This method is divided into three types: Clique search, geometric hashing and pose clustering. All are based on finding complementarity between the geometry of the binding site and that of the ligand. Some early docking programs used clique detection to find docked positions of rigid ligands, and clique detection remains an element of several modern, flexible-ligand docking programs, including recent versions of DOCK. In DOCK, to place a given conformation of the ligand, the unoccupied space in the binding site of the receptor is broken down into spheres, and the docking problem is solved by matching the centers of heavy atoms in the ligand to sphere centers in the binding site using clique detection.

Flexible Ligand and Rigid Receptor Docking

Most small molecules have one or more freely rotatable bonds, allowing them to assume a variety of stable conformations, so the next logical step in the evolution of docking methods was to allow for ligand flexibility. Considering multiple ligand conformations makes it less likely that a ligand will fail to dock simply because the conformation chosen was incompatible with the receptor. Almost all the docking programs such as AutoDock (Morris *et al.*, 1998) and FlexX (Rarey, Wefing, & Lengauer, 1996) have adopted this methodology of treating the ligand as flexible while the receptor is kept rigid during docking. AutoDock incorporates Monte Carlo simulated annealing, evolutionary, genetic and Lamarckian genetic algorithm methods to model the ligand flexibility while keeping the receptor rigid. The scoring function is based on the AMBER force field, including van der Waals, hydrogen bonding, electrostatic interactions, conformational entropy and desolvation terms. Each term is weighted using an empirical scaling factor obtained from experimental data. FlexX uses an incremental construction algorithm to sample ligand conformations. The base fragment is first docked into the active site by matching hydrogen bond pairs and metal and aromatic ring interactions between the ligand and receptor. Then the remaining components are incrementally built-up in accordance with a set of predefined rotatable torsion angles to account for ligand flexibility. The current version of FlexX includes terms of electrostatic interactions, directional hydrogen bonds, rotational entropy, and aromatic and lipophilic interactions. The interactions between functional groups are also taken into account through assigning the type and geometry for groups. In summary, molecular simulations such as molecular dynamics and Monte Carlo as well as ligand fragmentation will be the approach to treat flexible ligand and rigid receptor docking methodology.

Flexible Ligand and Flexible Receptor Docking

Most proteins of pharmaceutical interest are flexible objects, constantly shifting from one stable conformation to another. When they bind a ligand, many proteins undergo a conformational change in order to better accommodate the ligand and its physical properties in their binding site. This phenomenon is called 'induced fit'. It has been demonstrated that rigid receptor methods have a high probability of failing to dock ligands to some flexible proteins when the wrong protein conformation is used (Österberg *et al.*, 2002). While some ligands readily bind to a variety of conformations, others are highly selective. Thus, in drug discovery, failure to allow induced fit and conformational change may translate into passing over viable leads. Docking methods allowing protein flexibility are therefore an active area of research (Teodoro & Kavraki, 2003; Kokh *et al.*, 2011). Incorporating the receptor flexibility is significant challenge in the field of docking. Ideally, using MD simulations could model all the degrees of freedom in the ligand-receptor complex. But MD has the problem of high computational expense, which prevents this method from being used in the screening of large chemical database. Various methods are currently available to implement the receptor flexibility. The simplest one is so-called "soft-docking" (Jiang,

Kim, 1991) decreases the van der Waals repulsion energy term in the scoring function to allow for a degree of atom-atom overlap between the receptor and ligand. AutoDock 4 (Morris *et al.*, 2009) adopts a simultaneous sample method to deal with side chain flexibility. Several side chains of the receptor can be selected by users and simultaneously sampled with a ligand using the same methods. Other portions of the receptor are treated rigidly with a grid energy map during sampling.

An extensive list of commonly used molecular docking programs, algorithms and scoring functions, can be found in Table 1 and Table 2 of the recent Review paper by Ferreira, dos Santos, Oliva & Andricopulo, (2015). Another source of the latest developments in docking programs, docking web servers, screen software, and screen webservers are listed in Table 1 of Chen's paper (2015).

APPLICATIONS OF 3D-QSAR AND DOCKING

Since (Cramer, Patterson & Bunce, 1988) introduced the 3D-QSAR concept and methodology, it has proven to be an invaluable tool in the field of drug design and activity prediction. Its impact is measured in the thousands of reports dealing with 3D-QSAR successful applications to many data sets of enzyme and receptor ligands. Excellent reviews have been written on this field (Kubinyi, Folkers & Martin, 1998, 2006; Melo-Filho, Braga and Andrade, 2014; Verma, Khedkar and Coutinho, 2010).

3D-QSAR is rooted on the hypothesis that ligands interact with their biological targets in a specific conformation, where it is assumed that identification of steric, electrostatic, and hydrophobic structural features of various drugs acting on their biological receptors can be used to explain their biological activity as well as in assisting further work on molecular design and modification.

Initially most of the research work done in this field was oriented in finding the geometric similarity among ligand and receptor. Later on, calculations of stereoelectronic properties in the binding conformation, such as comparative molecular field analysis (CoMFA) and comparative molecular similarity indices analysis (CoMSIA), along with the development of complex sophisticated docking methodologies, have allowed more accurate prediction of the biological activity of biomolecules. In the last decade there has been an explosion of reports on combining 3D-QSAR and Molecular Docking (M/D) for establishing the active conformation of a ligand in the presence of a receptor.

In 1997 a couple of articles were published on the methodology and applications of 3D-QSAR in drug design (Kubinyi, 1977a, 1977b). In 1998 (Vedani, Dobler & Zbinden) performed the first attempt to build a three dimensional receptor surface populated with atomistic properties mapped onto it. Soon after (Santos-Filho & Hopfinger, 2002) applied the 4D-QSAR idea to both receptor-independent (RI) and receptor-dependent (RD) problems. In the first method, the geometry of the receptor is not part of the data available to perform the analysis while in the second method the geometry of the receptor is available.

The first documented report of 3D-QSAR/Docking application to drug development was conducted in 2002 by (Buolamwini & Assefa) who proposed a binding mode for the cinnamoyl inhibitors at the active site of HIV-1 integrase.

In 2003 (Wu, Robertson, Books & Vieth) proposed a combined strategy involving grid docking and full force field minimization via a molecular dynamics (MD) simulated-annealing-based algorithm in order to improve docking accuracy. In their docking studies of a test set of 41 diverse protein-ligand complexes selected from the PDB, the protein was kept rigid while the ligands were treated as fully flexible and a final minimization step was used to refine the docked poses. Their results indicated that final minimization of docked structures improved the docking accuracy up to 10%.

In 2004 (Erikson, Jalaie, Robertson, Lewis & Vieth) studied the effects of ligand and protein flexibility on M/D accuracy. To examine ligand flexibility, in a test set of 41 X-ray ligand-protein complex structures they applied several docking algorithms, finding that docking accuracy decreased substantially for ligands with eight or more rotatable bonds. To examine how protein flexibility influences docking accuracy, they applied the best docking algorithm to X-ray structures of trypsin, thrombin, and HIV-1-protease in the unbound (apo) form and also bound to several ligands. For the docking experiments they positioned each ligand to one average protein structure and to the apo form as well, and compared the results to those of docking each ligand back to its original structure. The outcome indicates that the docking accuracy falls off drastically for the average or apo structure reflecting the degree to which the protein moves upon ligand binding.

Due to the impossibility of accounting for all the research papers published in the last decade that combine M/D and 3D-QSAR to propose the best pharmacophore model, below we list a number of recent reports that we have considered most relevant in accomplishing this task. Most of the research done in this area has been focused in designing potent and effective inhibitors to HIV-1 proteins, and anti-cancer drugs. However, we must give credit to other studies in drug design such as antinociceptives, anti-tuberculosis, metabolic diseases, inflammatory and allergic processes, insecticides, osteoarthritis, antivirals, anticoagulants, anti- amyloid therapies, anti-malarial and antimicrobials.

Tubulin is an αβ heterodimer protein that forms part of microtubules essential in the cell cycle. Microtubules are involved in cell division, motility and shape and when they are compromised cell division is altered or blocked. Therefore, tubulin is the target of numerous small-molecule ligands used in cancer treatment among which is colchicine extracted from the poisonous meadow saffron *Colchicum autumnale L.* In order to construct binding models for a set of structurally diverse colchicine site inhibitors (CSIs) (Nguyen *et al.* 2005) reported docking and MD studies using the αβ-tubulin:DAMA-colchicine X-ray structure as the template. The docking studies helped in determining the binding modes of the colchicine site inhibitors. They used annealing via MD to compensate for pose inaccuracies arising from target X-ray structure resolution. They also used constrained MD simulations to undock the CSIs from the colchicine site in the binding models and to reverse the resulting trajectories to produce better models of the binding event. From these results they proposed a common structure-based pharmacophore model that accounts for the multiple structural classes of CSIs.

In order to accelerate *in silico* High Throughput Screening (Cherkasov, Ban, Li, Fallahi & Hammon, 2006) reported a progressive docking method based on "inductive" QSAR descriptors which cover a broad range of molecular properties that can be used effectively to create various binary inductive QSAR classifiers. In their approach, they used previously generated docking scores of already processed compounds in order to build predictive QSAR models that could assess hypothetical target binding affinities for yet undocked items. They tested this methodology on drug-like substances from the National Cancer Institute database that have been docked into several unrelated targets. They affirm that their approach can save hundreds of hours in terms of the docking time required to process ~90 000 potential ligands, while maintaining 80-100% hit recovery rates.

In 2007 (Markus A. Lill) questioned the usefulness of 3D-QSAR alone and use of the induced fit in determining the free energy of binding and in identifying the bioactive conformation of ligand molecules, particularly in the absence of structural information of the target protein. He reviewed the predictive power of multidimensional QSAR (mQSAR), particularly 4D-QSAR, in quantifying additional contributions to the binding energy. The concepts inherent in 4D-QSAR treat the alignment issue by incorporating molecular and spatial variety by representing each molecule in different conformations, orientations, tautomers, stereoisomers or protonation states. He affirmed that in classical 3D-QSAR, even those that include CoMFA or CoMSIA, one cannot sample the entire conformational space around the binding mode due to a combinatorial explosion in the number of required QSAR simulations. Since only if the 3D structure of the target protein is known, it can be used for predicting purposes using docking for example; therefore, 4D-QSAR can be interpreted as a feasible extension of 3D-QSAR to address the uncertainties during the alignment process.

In 2008 (Santos-Filho & Cherkasov) proposed a drug design approach which included docking, molecular fingerprints-based cluster analysis, and 'induced' descriptors-based receptor-dependent 3D-QSAR. They claimed that unlike other 3D-QSAR receptor-dependent, in their approach no ambiguous alignments are required for the construction of the models and that the computational cost is relatively lower. The induced descriptors presented are R_S that defines the steric influence of a group of atoms that compose a substituent G onto a single atom j, and σ^* that defines the inductive effect of G onto the reaction center j. They suggested that induced descriptors capture intermolecular interactions between ligands and receptors. This methodology was tested in eight data sets sampled from the literature and from public databases such as: human sex hormone-binding globulin, human corticosteroid-binding globulin, anthrax lethal factor, HIV-1 reverse transcriptase, neuraminidase A, thrombin, trypsin, and *Pneumocystis carinii* dihydrofolate reductase data sets. They concluded that although the presented models are interpretable, with high statistical and predictive significance, the usage of the receptor structure is not always enough for the development of significant models.

In 2009 (Pasqualoto & Ferreira) built RD 3D-QSAR models for a set of thirty seven isoniazid (INH) derivatives bound to the enoyl-acp reductase from *M. tuberculosis* (InhA). The molecular geometries of hypothesized active conformations of each ligand, resulting from a RI 4D-QSAR analysis, were used in their study. They used the 2.7 Å X-ray structure of the complex InhA-NAD-INH as starting model for the receptor geometry, and the energy-minimized structure was used as initial structure in MD calculations. Furthermore, they used the lowest energy conformation of the InhA-NAD-INH model to dock the energy-optimized structures of all ligands. They also performed MD simulations on each complex InhA-NAD-analogue. From the MD simulations these authors found that the following energy contributions were very important to the binding process: the bound ligand solvation energy, the sum of electrostatic and hydrogen bonding energies of the unbound ligand, the bending energy of the unbound ligand, the electrostatic intermolecular ligand-receptor energy, and the change in hydrogen bonding energy upon binding.

Human serum albumin (HSA), the most abundant protein in human blood plasma, binds to different types of ligands at multiple sites. Albumin is of medical importance since it transports hormones, fatty acids, metabolites, buffers pH, and maintains osmotic pressure, among other functions. To explain the high ligand promiscuity of HSA in 2010 (Deeb, Rosales-Hernández, Gómez-Castro, Garduño-Juárez, & Correa-Basurto) performed a 5 ns MD simulation on HSA. Those HSA structures sampled every 0.5 ns were subjected to a 3D-QSAR/Docking study on 94 well known HSA ligands, including warfarin and ketoprofen, known to bind HSA sites I and II. This study provided evidence that HSA binding sites I and II interact specifically with a variety of compounds through conformational adjustments of the protein structure in conjunction with ligand conformational adaptation to these sites.

In 2011 (Taha *et al.*) proposed a docking-based comparative intermolecular contacts analysis (db-CICA) as a new 3D-QSAR concept for validating docking solutions. Their approach evaluates the docking configurations based on the observations that a set of ligands inside their corresponding binding pocket interact in such a way that potent ligands contact the binding site spots in a different way than those ligands with low affinity. Within this framework optimal dbCICA models can be transformed into valid pharmacophore models that can be used for 3-D search of new bioactive compounds. These authors applied dbCICA methodology to search for new inhibitors of candida N-myristoyl transferase as potential antifungal agents and glycogen phosphorylase (GP) inhibitors as potential antidiabetic agents.

In in an elegant study (Deeb, da Cunha, Cormanich, Ramalho & Freitas, 2012) used a multivariate image analysis applied to QSAR (MIA-QSAR) method. MIA-QSAR is a 2D image-based approach in which images are chemical structures and, where the combination and arrangement of pixels (descriptors, binaries) provide chemical information. For this work they took 60 compounds obtained from two series of non-peptides such as HIV-1 protease inhibitors reported elsewhere. Their premise was that the sum of the substructures of two compound classes can give rise to new actives can cause synergistic effects on bioactivities and enhanced pharmacokinetic parameters. The interaction modes of the ligands with the HIV-1 protease crystal structure active site were obtained with the Molegro Virtual Docker (MVD) program using the MolDock scoring function. The MIA-QSAR models were validated through LOO-CV and Y-randomization tests. The best MIA-QSAR model was the one with the lowest RMSECV value and able to predict pIC_{50} values of new compounds that have comparable to higher HIV-1 protease inhibitory activity when compared to the existing compounds used in the study. They concluded that the MIA-QSAR method is useful in showing the synergistic effect of mixing substructures to give rise of a new compound.

In 2013 (Jimenez-Botello *et al.*) performed QSAR studies on a set of experimentally tested ligands to validate a homology built 3D model of the rat adenosine receptor (rA2AR). The rA2AR model was refined by MD, in which the initial and refined 3-D structures were used for M/D simulations. The results showed that there is a hindrance effect caused by a ribose moiety attached to agonists which plays an important role in activating the receptor via formation of several hydrogen bonds; while the lack of this moiety allows blocking of the receptor. They showed that their theoretical affinity estimation had good correlation with reported experimental data.

To improve the (Nguyen *et al.*, 2005) not correctly predicted binding conformations of a few colchicine site inhibitors (CSIs), in 2013 (Da, Mooberry, Gupton & Kellogg, 2013) performed an elegant study to definitively map the αβ-Tubulin colchicine site by using 3D-QSAR, ensemble-docking and hydropathic analysis. They concluded that a combination of these techniques reveals a more detailed pharmacophore model for CSIs with a higher resolution than the ~3.6 Å available for αβ-tubulin.

In 2013 (Ul-Haq, Usmani, Shamshad, Mahmood & Halim, 2013) performed 3D-QSAR and docking studies on Danuravir derivatives, the most potent HIV-1 protease inhibitor known so far. They applied a combined study of 3D-QSAR unto the Danuravir derivatives using ligand-based and receptor-based protocols to generate CoMFA/CoMSIA models. To align a series of 102 Darunavir derivatives in the receptor-based method, they used the bound conformation of Darunavir in the crystal structure of HIV protease as a template. To validate the 3D-QSAR results, they performed docking simulations with the most active compound and with the least active compound, and to evaluate the quality of interaction as HIV-1 protease inhibitors they were docked with the wild type and mutated proteins.

In 2013 (Contreras-Romo *et al.*) used a combined approach of protein folding, MD simulations, docking, and QSAR to elucidate the detailed interaction of the vasopressin receptor V1a (V1aR) with some of its blockers. The V1aR model was constructed by I-TASSER and refined through MD simulations. For this study 134 compounds that are structurally related to benzodiazepines were used. The energy minimized 3D chemical structures of the compounds were docked at the binding pose of conivaptan on the refined V1aR structure at the 5-ns snapshot of the MD simulations. The QSAR analysis was performed upon the best binding poses of the ligands, and the accepted linear models were subjected to cross-validation (CV) analysis by the leave-one-out (LOO), leave-a-group-out (LGO) procedures and Y-scrambling test. The scrambled data set was re-examined by the MLR method. The accepted MLR models were further optimized using a feed-forward neural network (NN) with back propagation of the error algorithm. Their results suggest that ligand recognition of V1aR is mediated by π-π and π-cation interactions with aromatic residues at the binding pocket of this receptor. Other nearby residues engage in hydrogen bonds and hydrophobic interactions and that a hydrogen bond interactions with the N atom of side chain of R214 is needed for a high affinity to V1aR.

It is known that protein kinase C Related Kinase 1 (PRK1) is involved in the regulation of androgen receptor signaling and has been considered as a novel potential drug target for prostate cancer therapy. Since there is little known about the PRK1 structure and its inhibitors, in 2014 (Slynko *et al.*) used the kinase domain of the protein kinase c theta (PKC-theta), with high homology to PRK1, as template in order to generate six PRK1 homology models (hm-prk1). From a group of 84 known active kinase inhibitors obtained from *in vitro* screening, they chose six compounds among which there were potent PRK1 inhibitors as well as some moderately active inhibitors. The refinement of the hm-prk1 conformations on the binding pocket and ligand was done via receptor-based docking. They observed that an inaccurate prediction of ligand binding poses occurred when using only one protein conformation and, subsequently, in low enrichment factors. To overcome this problem they docked all six chosen compounds in all six PRK1 homology models. They used MM-PB(GB)SA, QM/MMGBSA approaches to re-score the docking solutions by calculating binding free energy scores through simple QSAR models based on descriptors such as ligand charge or number of rotatable bonds. However, their results indicate that the MM-PB(GB)SA, QM/MM-GBSA rescoring approach is not accurate enough to distinguish the compounds with similar binding affinities, but it can be used to separate strong PRK1 inhibitors from moderate or weak binders. They also demonstrate that using only one snapshot derived after the minimization is sufficient to provide a reasonable prediction of the binding free energy, but only if the correct starting conformation is used.

Due to its narrow therapeutic index the CSI colchicine has limited its applications in clinical trials as antitumor agent. As a large number of small molecules possessing significant structural diversity, have been reported to bind the colchicine site in microtubules, in 2015 a hierarchical strategy was developed for structure-based virtual screening (VS) by integrating different computational methods such as M/D, 3D-QSAR, and pharmacophore model. With this procedure, based on SAR analysis, 691 novel small molecules were designed and the binding poses were predicted by a M/D into the three-dimensional X-ray structure of human tubulin complex and generated 6920 binding poses. The binding energy of the poses was calculated by MM-GBSA. Use of 3D-QSAR modeling helped in correlating correlate compound activities with interaction fields calculated based on protein crystallography or molecule superimposition. By means of the VS strategy, sixteen novel compounds with various structural features were obtained and synthetized (Li *et al.*, 2015).

A CASE STUDY OF MOLECULAR DOCKING AND QSAR

In the last year, several studies have been performed using Molecular Docking-Molecular Dynamics and QSAR methodologies. For example, (Mei *et al*, 2015) have used the computational tools to study ROCK/PKA inhibitors. The 3D-QSAR results showed that the substitution containing positive charge attached to the phenyl ring was favored for activity and selectivity. Specially, two compounds show theses characteristics but PKA protein (PDB: 3POO) inhibitory activities were quite different. To understand this behavior, docking methodology (using Surflex-Dock module in Sybyl 2.0) were used to explain the interaction of these molecules with ROCK2 protein (PDB: 4L6Q), and finally to validate the docking results, the Molecular Dynamics (using Sybil 2.0) simulations were carried out for 5 ns.

In the same way, (Abbasi *et al*, 2015) have performed QSAR and molecular docking (AutoDock 4.2) methodology to examine 30 L-tyrosine derivatives as metalloproteinase (MMP-2, PDB: 3AYU) inhibithors. They proposed that their results could be used to design new inhibitors.

(Vats *et al*, 2015) have studied several novel flavonoid analogues as potential AChE (PDB: 4M0E) inhibitors by QSAR, Docking and MD analysis. QSAR methodology was applied by the G-QSAR module from the VLife MDS version 4.3, the docking and molecular dynamic methodologies were applied by Glide module of Schrodinger and the force field 2005 respectively, the docking was carried out only around the active site in the protein. They selected the most favorable compounds from the QSAR analysis to perform the docking and MD studied. They concluded that the stability of the complex in vivo conditions depends on the two best compounds.

(Sepehri *et al*, 2015) and (Fatemi *et al*, 2015) established a methodology, QSAR and Docking (AutoDock 4.2), to evaluate of gp41 as possible target for anti-HIV-1 (PDB: 1A1K) activity and HIV-1 protease (PDB: 4DJP) respectively. In general, docking methodology is carried out to establish the interaction between the active site and several compounds. On the other hand, (Saleh, 2015) has studied several fullerenes derivatives as HIV-1 protease (PDB: 4DJO) inhibitors. After theoretical descriptors analysis, It is concluded that the characteristics of a compound such as solubility and reactivity with the surrounding system could be sufficient to interact with the HIV-1 protease, however the docking results showed that other compounds should be more active.

(Kang *et al*, 2015) studies 36 compounds that showed inhibitory activity toward human recombinant protein of VEGFR-2 tyrosine kinase (PDB: 2IVU) by QSAR and docking (AutoDock 4.2) methodologies. Their results permitted to design seven new molecules with high predictive biological activities.

The last papers discussed before have applied the same methodology, first QSAR analysis and then molecular docking and finally, in some cases, molecular dynamic analysis. The QSAR analysis was carried out to obtain the best descriptors in the series analyzed and, in some cases, the QSAR model was used to predict new molecules with activity. The docking and molecular dynamic studies were carried out to establish the active site and the interactions between this site and the compounds.

The histone deacetylase 2 (HDAC) inhibitors are used as a new therapy for neurodegenerative diseases. Several inhibitors are in the market such as trichostatin, valproic acid, sodium butyrate, phenylbutirate, vorinostat (SAHA) (MCQouwn *et al*, 2011), however some side effects are present in its treatments. For this reason, a new compounds as HDAC2 inhibitors are required.

Our group (Martínez-Pacheco, 2012; Martínez-Pacheco, 2015; Martínez-Pacheco *et al*, 2015) has established a different methodology to find new compounds with some biological activity. First, a docking study of a series of compounds having certain biological activity took place, these compounds are exclusive of the catalytic site. In this docking analysis some new hypothetic compounds are included to identify the most probable molecules with good biological activity. After this analysis, the QSAR analysis is performed to predict the best candidate as new molecule with good biological activity.

Results

The histone deacetylase 2 (HDAC2, PDB: 4LXZ) in complex with SAHA was used in this methodology. To prepare protein input, PyMOL Molecular Graphics System was used to remove ligands and cofactors. In this case, the HDAC2 is a trimmer and only the monomer A was used. Polar hydrogens and Kollman charges were added to monomer A. In this case, a grid box of 96Åx96Åx96Å was used as search space.

From the docking results, an occupation probability was calculated by applying the following steps:

- Selecting only the catalytic sites and the Gibbs free energy
- Consider the temperature as (298 K) and the ideal gas constant (1.98 cal/mol)
- Use the occupation probability of the results obtained in order to establish a relationship between the occupation probability and Gibbs free energy.

The following equation calculates the probability of occupation considering the Gibbs free energy:

$$p_i = \frac{e^{-\frac{\Delta G_i}{RT}}}{\sum e^{-\frac{\Delta G_i}{RT}}} \quad (3)$$

where T is the temperature at 298 K, ΔG is the difference in Gibbs free energy; R is the ideal gas constant (1.98 cal/mol).

Taking ΔE *as*:

$$\Delta E = RTln\frac{p_1}{p_2} \quad (4)$$

- Establish a p> 0.7, where $p_1 = 0.7$ and $p_2 = 0.3$

Select only those catalytic sites with higher probability of success (ex ΔG>0.499 kcal/mol).

- Analyzing the catalytic sites with a p> 0.7

Using the most stable conformation of a molecule in the catalytic site with higher probability of success and the Gibbs free energy calculated previously, it is possible to perform a QSAR study and to validate the models.

Although, many authors consider a high q^2 (for instance, $q^2>0.5$) as an indicator, or even as the ultimate proof that the model is highly predictive (Golbraikh & Tropsha, (2002)), to define the relative importance of the QSAR models, we have also performed several other tests beyond the q^2. These include the Akaike's information criterion (AIC, Akaike, (1974)), the Kubinyi fitness function (FIT, Kubiny, (1994)) and calculations of the correlation coefficient for regression through the origin (RTO, R_0^2), all of which were performed for the training and predicted activity according to (Anderson and Legendre, 1999).

The tridimensional structure of HDAC2 used in this study was the PDB: 4LXZ, this structure correspond a trimmer, however, only a monomer was used in the molecular docking processes (Figure 1).

The monomer and Zn^{2+} ion (Figure 1) were used in the molecular docking by AutoDock 4.2 software, in a grid box of 96X96X96 Å, it is important to mention that the catalytic site is formed for the Zn2+ ion, Tyr306, His142, His143, Phe152 and Phe207

One hundred fourteen molecules were used in the docking and QSAR analysis; fifty four molecules (hydroxamic acid compounds, Figure 2) were taken as training set, fifty molecules (carboxylic acid compounds) as prediction set and ten commercial inhibitors of HDAC as validation set.

All compounds of the training set were docked into the catalytic site, and only fourteen compounds of the predicted set were docked into the catalytic site, they showed several conformations around the catalytic site. Figure 3 shows three differents conformations of a carboxyl acid derivative (14) around the catalytic site on the HDAC2 monomer.

Interactions between the compound 14 and the catalytic site were studied. The binding energy, the population distribution was calculated and only the interactions with probability higher than 0.7 were taken for QSAR analysis. The docking results showed that the active site is open when the lysine 36 is deacetiled from the histone, or close when this amino acid is acetiled. All our calculations were taken in the deacetiled system.

Figure 1. HDAC2 structure, only a monomer of HDAC2 and a Zn^{2+} ion (red sphere) were used in the study

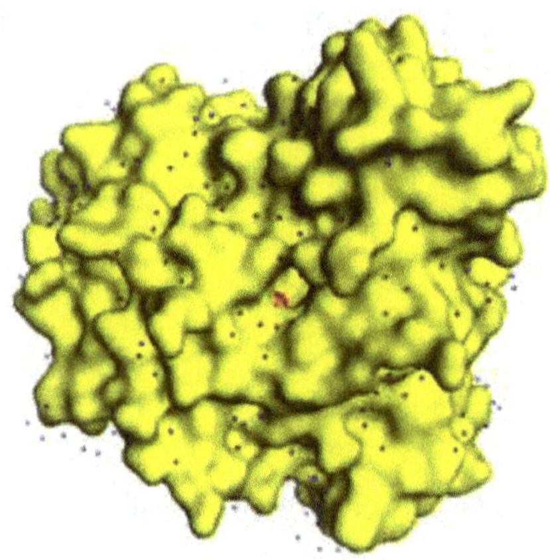

Figure 2. General structure of hydroxamic acid compounds (red color represents the lipophilic group, blue color represents the linker and the green color represents the coordination group of Zn^{2+})

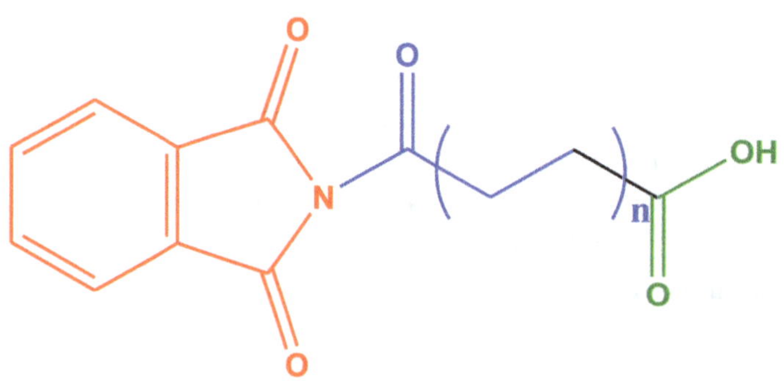

Figure 3. Three different conformations of the carboxyl acid derivative (14), of all them are probable into the catalytic site

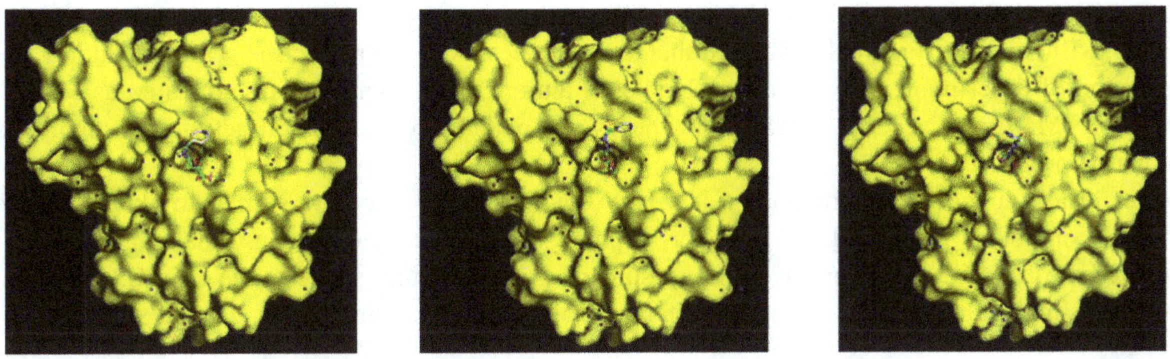

The QSAR model for the training set was built from 1239 molecular descriptors calculated by Dragon software. After statistical procedure, 151 molecular descriptors were used as a pool of descriptor. The best Multi-Lineal Regression (MLR) model (Figure 4A) is:

$$pIC_{50}^{Exp} = -0.967\ (\pm 0.583) + 0.541\ DP06\ (\pm 0.097) + 3.317\ MATS5e\ (\pm 0.963) + 0.721\ Mor18e\ (\pm 0.224) - 1.349\ Mor13p\ (\pm 0.326) + 1.433\ GATS1e\ (\pm 0.512) + 1.015\ GATS8p\ (\pm 0.421)$$

$R=0.94524$, $SD=0.23065$, $p<0.0001$, $N=52$, $SEE=2.7986$, $q^2_{LOO}=0.8498$, $q^2_{LGO}=0.7598$, $FIT=1.791$, $AIC=0.0608$, $F=60.9553$.

Figure 4. A) MLR of 52 inhibitors of HDAC2 as training set, 14 compounds proposed from molecular docking as predicted set and 10 commercial inhibitors as validation set. There are two compounds in the validation set that are a big outliers (Valproic acid and AGK2). B) Williams plot. There are three compounds of the predicted set that are including of the description region (compounds 1, 4 and 14). These compounds are very close to the leverage limit (h).*

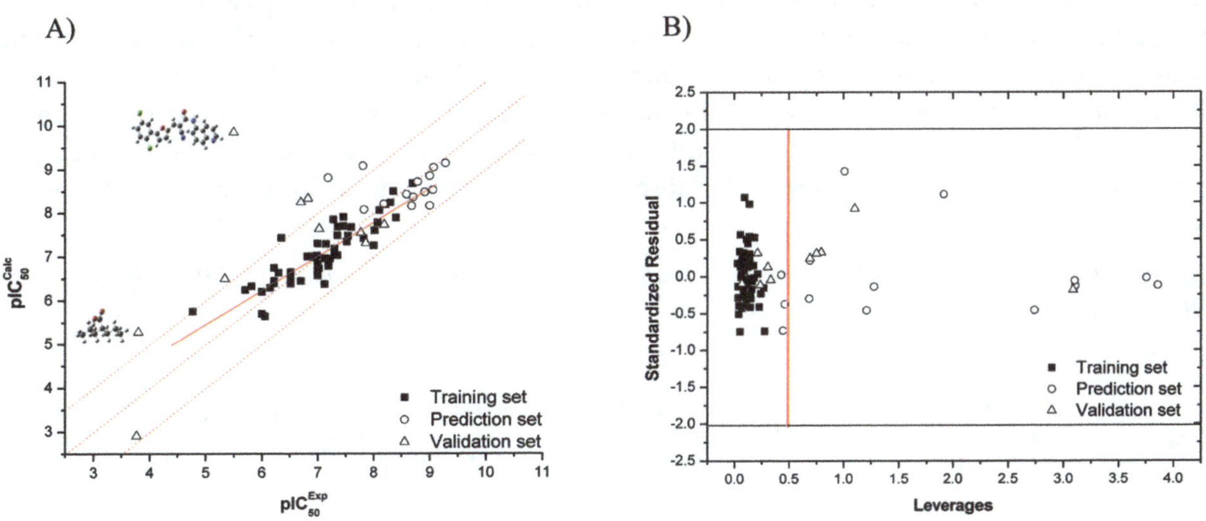

It is very interesting that the two commercial inhibitors as valproic acid (it does not present a bulky group) and AGK2 compounds are considered as outliers, this could be interpreted that these compounds are specific to HDAC8 and class III HDAC, respectively.

The Williams plot (Figure 4B) showed that three molecules of the prediction set are including into the region of description. These three molecules (Figure 5) were synthesized and tested as HDAC2 inhibitors (Martínez-Pacheco, (2015)), and they show potential experimental activity as theoretical prediction proposed.

Our results suggests that compoud 4 is a HDAC inhibitor with activity similar to phenylbutirate, whereas compound 14 is a HDAC inhibitor as phenylbutirate and trichostatin. Finally, compound 1 activated the HDAC2, however, theoretical results of this compound suggests the HDAC2 activation because its molecular docking is in a site different as the other two compounds.

CHALLENGES AND FUTURE DIRECTIONS FOR CURRENT DOCKING METHODS IN DRUG DESIGN

Despite that M/D is routinely used in virtual screening or lead optimization for drug screening and design, frequently many problems arise. These issues are related to use of the wrong binding site of the target protein, use of an inappropriate small-molecule database, the method to choose the docking pose, or even a high dock score that fails in MD simulations. Sometimes docking analysis alone can generate erroneous poses of ligand binding, and occasionally the ligand will fly away during the MD simulation, in spite of having a high dock score. Such problems have been addressed by (Chen, 2015) who has given a word of caution regarding docking results.

Sampling and scoring are two of the fundamental weaknesses in simulating the binding of proteins and small ligand molecules. Protein-ligand interactions implicate a delicate balance of competing forces that occur between flexible moieties generating far too many conformations to be sampled exhaustively. These interactions are also affected by the surrounding water molecules and ions. In order to calculate the total free energy of binding, there is the need of designing accurate force-field-based approaches able to model the many types of interactions using physics-based parameters derived either from experiments or quantum mechanical calculations. These force fields must be able to model the entropic freedom of nearby water molecules in the vicinity of a ligand-receptor, a computationally expensive process when simulating explicit water molecules, and the water screening of the electrostatic forces between charged atoms. Such scoring function, however, will add substantial computational cost in order to estimate the favorability of a protein-ligand complex. Other qualities than one may want to score for practical reasons are toxicity and properties related to absorption, distribution, metabolism, and excretion. Since

Figure 5. Compounds 1, 4 and 14. They are predicted by molecular docking and QSAR analysis as best HDAC2 inhibitors.

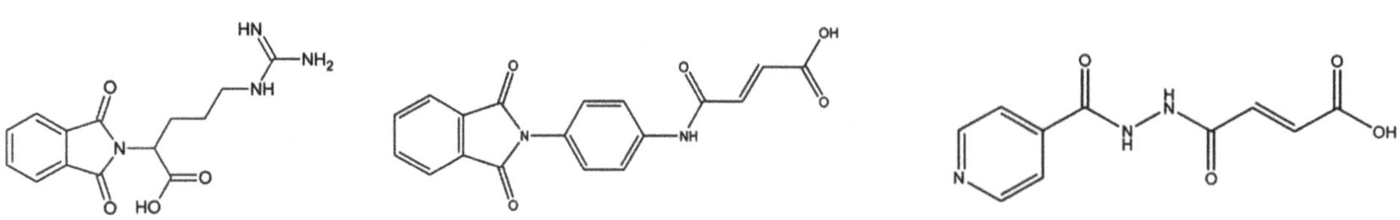

it is well-recognized that drugs may bind to different targets with significant affinities, known as drug promiscuity, another challenge is to model this tendency of ligands to bind to several sites of the protein receptor. Mainly because if a ligand binds tightly somewhere, even near the desired binding site, this does not mean that it will affect substantially the drug action, particularly important for non-traditional drug targets, such as a target found inside the interface of protein-protein interactions.

Nevertheless, some recent efforts have been oriented to help in solving the challenges mentioned above. Among these the technique known as Protein Energy Landscape Exploration (PELE) is a good example (Borrelli, Vitalis, Alcantara & Guallar, 2005) since it combines protein structure prediction algorithms and Metropolis Monte Carlo techniques in order to explore ligand diffusion on the protein energy landscape.

CONCLUSION

Computational-based rational drug design has become more common over the past decade. Most of the approaches involved in this type of design focus on ligand based drug design such as Quantitative Structure–Activity Relationship (QSAR) studies that use various molecular descriptors for chemical information training. Another branch is structure based drug design method which is molecular docking.

In this chapter, we have summarized the two methods including QSAR and Molecular Docking methodologies. Also the efforts done in 3D-QSAR and docking were presented followed by a case study.

REFERENCES

Abbasi, M., Ramezani, F., Elyasi, M., Sadeghi-Aliabadi, H., & Amanlou, M. (2015). A study on quantitative structure-activity relationship and molecular docking of metalloproteinase inhibitors based on L-tyrosine scaffold. *DARU Journal of Pharmaceutical Sciences*, *23*(1), 29. doi:10.1186/s40199-015-0111-z PMID:25925871

Akaike, H. (1974). A new look at the statistical model identification. *IEEE Transactions on Automatic Control*, *19*(6), 716–723. doi:10.1109/TAC.1974.1100705

Ambure, P. S., Gangwal, R. P., & Sangamwar, A. T. (2012). 3D-QSAR and molecular docking analysis of biphenyl amide derivatives as p38α mitogen-activated protein kinase inhibitors. *Molecular Diversity*, *16*(2), 377–388. doi:10.1007/s11030-011-9353-y PMID:22228035

Anderson, M. J., & Legendre, P. (1999). An empirical comparison of permutation methods for test of partial regression coefficients in a linear model. *Journal of Statistical Computation and Simulation*, *62*(3), 271–303. doi:10.1080/00949659908811936

Borrelli, K. W., Vitalis, A., Alcantara, R., & Guallar, V. (2005). PELE: Protein energy landscape exploration. A novel Monte Carlo based technique. *Journal of Chemical Theory and Computation*, *1*(6), 1304–1311. doi:10.1021/ct0501811 PMID:26631674

Brooijmans, N., & Kuntz, I. D. (2003). Molecular recognition and docking algorithms. *Annual Review of Biophysics and Biomolecular Structure*, *32*(1), 335–373. doi:10.1146/annurev.biophys.32.110601.142532 PMID:12574069

Buolamwini, J. K., & Assefa, H. (2002). CoMFA and CoMSIA 3D QSAR and docking studies on conformationally-restrained cinnamoyl HIV-1 integrase inhibitors: Exploration of a binding mode at the active site. *Journal of Medicinal Chemistry*, *45*(4), 841–852. doi:10.1021/jm010399h PMID:11831895

Chen, Y. C. (2015). Beware of docking! *Trends in Pharmacological Sciences*, *36*(2), 78–95. doi:10.1016/j.tips.2014.12.001 PMID:25543280

Cherkasov, A., Ban, F., Li, Y., Fallahi, M., & Hammond, G. L. (2006). Progressive docking: A hybrid QSAR/docking approach for accelerating in silico high throughput screening. *Journal of Medicinal Chemistry*, *49*(25), 7466–7478. doi:10.1021/jm060961+ PMID:17149875

Contreras-Romo, M. C., Martínez-Archundia, M., Deeb, O., Ślusarz, M. J., Ramírez-Salinas, G., Garduño-Juárez, R., & Correa-Basurto, J. et al. (2014). Exploring the ligand recognition properties of the human vasopressin v1a receptor using QSAR and molecular modeling studies. *Chemical Biology & Drug Design*, *83*(2), 207–223. doi:10.1111/cbdd.12229 PMID:24010681

Coupez, B., & Lewis, R. A. (2006). Docking and scoring--theoretically easy, practically impossible? *Current Medicinal Chemistry*, *13*(25), 2995–3003. doi:10.2174/092986706778521797 PMID:17073642

Cramer, R. D., Patterson, D. E., & Bunce, J. D. (1988). Comparative molecular field analysis (CoMFA). 1. Effect of shape on binding of steroids to carrier proteins. *Journal of the American Chemical Society*, *110*(18), 5959–5967. doi:10.1021/ja00226a005 PMID:22148765

Da, C., Mooberry, S. L., Gupton, J. T., & Kellogg, G. E. (2013). How to deal with low-resolution target structures: Using SAR, ensemble docking, hydropathic analysis, and 3D-QSAR to definitively map the αβ-tubulin colchicine site. *Journal of Medicinal Chemistry*, *56*(18), 7382–7395. doi:10.1021/jm400954h PMID:23961916

Deeb, O., da Cunha, E. F., Cormanich, R. A., Ramalho, T. C., & Freitas, M. P. (2012). Computer-assisted assessment of potentially useful non-peptide HIV-1 protease inhibitors. *Chemometrics and Intelligent Laboratory Systems*, *116*, 123–127. doi:10.1016/j.chemolab.2012.05.003

Deeb, O., Rosales-Hernández, M. C., Gómez-Castro, C., Garduño-Juárez, R., & Correa-Basurto, J. (2010). Exploration of human serum albumin binding sites by docking and molecular dynamics flexible ligand–protein interactions. *Biopolymers*, *93*(2), 161–170. doi:10.1002/bip.21314 PMID:19785033

Devillers, J., & Balaban, A. T. (1999). *Topological Indices and Related Descriptors in QSAR and QSPR* (p. 811). Amsterdam: Gordon Breach Scientific Publishers.

Duprat, A. F., Huynh, T., & Dreyfus, G. (1998). Toward a principled methodology for neural network design and performance evaluation in QSAR. Application to the prediction of log P. *Journal of Chemical Information and Computer Sciences*, *38*(4), 586–594. doi:10.1021/ci980042v PMID:9691473

Eldridge, M. D., Murray, C. W., Auton, T. R., Paolini, G. V., & Mee, R. P. (1997). Empirical scoring functions: I. The development of a fast empirical scoring function to estimate the binding affinity of ligands in receptor complexes. *Journal of Computer-Aided Molecular Design*, *11*(5), 425–445. doi:10.1023/A:1007996124545 PMID:9385547

Erickson, J. A., Jalaie, M., Robertson, D. H., Lewis, R. A., & Vieth, M. (2004). Lessons in molecular recognition: The effects of ligand and protein flexibility on molecular docking accuracy. *Journal of Medicinal Chemistry*, *47*(1), 45–55. doi:10.1021/jm030209y PMID:14695819

Ewing, T. J. A., & Kuntz, I. D. (1997). Critical evaluation of search algorithms for automated molecular docking and database screening. *Journal of Computational Chemistry*, *18*(9), 1175–1189. doi:10.1002/(SICI)1096-987X(19970715)18:9<1175::AID-JCC6>3.0.CO;2-O

Fatemi, M. H., Heidari, A., & Gharagani, S. (2015). QSAR prediction of HIV-1 protease inhibitory activities using docking derived molecular descriptors. *Journal of Theoretical Biology*, *369*, 13–22. doi:10.1016/j.jtbi.2015.01.008 PMID:25600056

Ferreira, L. G., dos Santos, R. N., Oliva, G., & Andricopulo, A. D. (2015). Molecular docking and structure-based drug design strategies. *Molecules (Basel, Switzerland)*, *20*(7), 13384–13421. doi:10.3390/molecules200713384 PMID:26205061

Fujita, T. (1990). The extrathermodynamic approach to drug design. In Comprehensive Medicinal Chemistry (vol. 4). Pergamon.

Gabb, H. A., Jackson, R. M., & Sternberg, M. J. E. (1997). Modelling protein docking using shape complementarity, electrostatics, and biochemical information. *Journal of Molecular Biology*, *272*(1), 106–120. doi:10.1006/jmbi.1997.1203 PMID:9299341

Glaser, F., Morris, R. J., Najmanovich, R. J., Laskowski, R. A., & Thornton, J. M. (2006). A method for localizing ligand binding pockets in protein structures. *Proteins*, *62*(2), 479–488. doi:10.1002/prot.20769 PMID:16304646

Gohlke, H., & Klebe, G. (2001). Statistical potentials and scoring functions applied to protein-ligand binding. *Current Opinion in Structural Biology*, *11*(2), 231–235. doi:10.1016/S0959-440X(00)00195-0 PMID:11297933

Goodford, P. J. (1985). A computational procedure for determining energetically favorable binding sites on biologically important macromolecules. *Journal of Medicinal Chemistry*, *28*(7), 849–857. doi:10.1021/jm00145a002 PMID:3892003

Halperin, I., Ma, B., Wolfson, H., & Nussinov, R. (2002). Principles of docking: An overview of search algorithms and a guide to scoring functions. *Proteins*, *47*(4), 409–443. doi:10.1002/prot.10115 PMID:12001221

Hansch, C. (1969). A quantitative approach to biochemical structure-activity relationships. *Accounts of Chemical Research*, *2*(8), 232–239. doi:10.1021/ar50020a002

Jiang, F., & Kim, S. H. (1991). "Soft docking": Matching of molecular surface cubes. *Journal of Molecular Biology*, *219*(1), 79–102. doi:10.1016/0022-2836(91)90859-5 PMID:2023263

Jimenez-Botello, L. C., Soriano-Ursúa, M. A., Deeb, O., Segura-Cabrera, A., Martínez-Archundia, M., & Correa-Basurto, J. (2013). Homology modeling, molecular dynamics and docking simulations of rat A2A receptor: A three-dimensional model validation under QSAR studies. *Current Trends in Medicinal Chemistry*, *7*, 35–52.

Jones, G., Willett, P., Glen, R. C., Leach, A. R., & Taylor, R. (1997). Development and validation of a genetic algorithm for flexible docking. *Journal of Molecular Biology*, *267*(3), 727–748. doi:10.1006/jmbi.1996.0897 PMID:9126849

Kakarla, P., Devireddy, A. R., Inupakutika, M. A., Cheeti, U. R., Floyd, J. T., Mun, M. M., & Varela, M. F. et al. (2014). Molecular modelling, 3d-qsar, and rug docking studies on the role of natural anticoagulant compounds in antithrombotic therapy. *International Journal of Pharmaceutical Sciences and Research*, *5*(10), 4141. PMID:25750914

Kang, C.-, Liu, D.-, Wang, X.-y., Dai, Y.-, Cheng, J.-, & Ly, Y.-. (2015). 3D-QSAR and docking study on 3-benzimidazol-2-ylhydroquinolin-2-one derivatives as VEGFR-2 tyrosine kinase inhibitors. *Medicinal Chemistry Research*, *24*(3), 934–943. doi:10.1007/s00044-014-1161-3

Karelson, M. (2000). *Molecular Descriptors in QSAR/QSPR*. New York: Wiley-InterScience.

Kastenholz, M. A., Pastor, M., Cruciani, G., Haaksma, E. E., & Fox, T. (2000). GRID/CPCA: A new computational tool to design selective ligands. *Journal of Medicinal Chemistry*, *43*(16), 3033–3044. doi:10.1021/jm000934y PMID:10956211

Katchalski-Katzir, E., Shariv, I., Eisenstein, M., Friesem, A. A., Aflalo, C., & Vakser, I. A. (1992). Molecular surface recognition: Determination of geometric fit between proteins and their ligands by correlation techniques. *Proceedings of the National Academy of Sciences of the United States of America*, *89*(6), 2195–2199. doi:10.1073/pnas.89.6.2195 PMID:1549581

Katritzky, A. R., Lobanov, V. S., & Karelson, M. (1994). CODESSA, Reference Manual. Gainesville, FL: University of Florida. Retrieved from http://www.semichem.com/codessa

Kitchen, D. B., Decornez, H., Furr, J. R., & Bajorath, J. (2004). Docking and scoring in virtual screening for drug discovery: Methods and applications. *Nature Reviews. Drug Discovery*, *3*(11), 935–949. doi:10.1038/nrd1549 PMID:15520816

Kokh, D. B., Wade, R. C., & Wenzel, W. (2011). Receptor flexibility in small-molecule docking calculations. *Wiley Interdisciplinary Reviews: Computational Molecular Science*, *1*(2), 298–314.

Kontoyianni, M., Madhav, P., Suchanek, E., & Seibel, W. (2008). Theoretical and practical considerations in virtual screening: A beaten field? *Current Medicinal Chemistry*, *15*(2), 107–116. doi:10.2174/092986708783330566 PMID:18220766

Kubiny, H. (1994). Variable selection in QSAR studies: 2. A highly efficient combination of systematic search and evolution. *Quantitative Structure-activity Relationships*, *13*, 393–401.

Kubinyi, H. (1997a). QSAR and 3D QSAR in drug design Part 1: Methodology. *Drug Discovery Today*, *2*(11), 457–467. doi:10.1016/S1359-6446(97)01079-9

Kubinyi, H. (1997b). QSAR and 3D QSAR in drug design Part 2: Applications and problems. *Drug Discovery Today, 2*(12), 538–546. doi:10.1016/S1359-6446(97)01084-2

Kubinyi, H., Folkers, G., & Martin, Y. C. (Eds.). (1998). *3D QSAR in drug design: Ligand-protein interactions and molecular similarity* (Vol. 2). Springer Science & Business Media.

Kubinyi, H., Folkers, G., & Martin, Y. C. (Eds.). (2006). *3D QSAR in drug design: Recent advances* (Vol. 3). Springer Science & Business Media.

Kuntz, I. D., Blaney, J. M., Oatley, S. J., Langridge, R., & Ferrin, T. E. (1982). A geometric approach to macromolecular-ligand interactions. *Journal of Molecular Biology, 161*(2), 269–288. doi:10.1016/0022-2836(82)90153-X PMID:7154081

Laskowski RA. (1995). SURFNET: a program for visualizing molecular surfaces, cavities, and intermolecular interactions. *J Mol Graph, 13*(5), 307–328.

Levitt, D. G., & Banaszak, L. J. (1992). POCKET: A computer graphics method for identifying and displaying protein cavities and their surrounding amino acids. *Journal of Molecular Graphics, 10*(4), 229–234. doi:10.1016/0263-7855(92)80074-N PMID:1476996

Li, D. D., Qin, Y. J., Zhang, X., Yin, Y., Zhu, H. L., & Zhao, L. G. (2015). Combined Molecular Docking, 3D-QSAR, and Pharmacophore Model: Design of Novel Tubulin Polymerization Inhibitors by Binding to Colchicine-binding Site. *Chemical Biology & Drug Design*. doi:10.1111/cbdd.12545 PMID:25711282

Lill, M. A. (2007). Multi-dimensional QSAR in drug discovery. *Drug Discovery Today, 12*(23), 1013–1017. doi:10.1016/j.drudis.2007.08.004 PMID:18061879

Mahajan, A., Gill, N.S., & Arora, R. (2014). *A Review on Molecular Docking*. Academic Press.

Martínez-Pacheco, H. (2012). *Evaluación in silico –Docking y QSAR- de derivados del ácido valproico sobre la HDAC8 con potencial uso antineoplástico (Evaluation in silico –Docking and QSAR- of valproic acid derivatives on HDAC8 as potential anti-neoplastic)*. Universidad del Papaloapan.

Martínez-Pacheco, H. (2015). *Diseño, síntesis y evaluación de efectos bioquímicos y conductuales de posibles inhibidores de la desacetilación de histonas (Desing, synthesis and biochemistry and behavior efects evaluation of inhibitors of histon desacetilation)*. Escuela Superior de Medicina Instituto Politécnico Nacional.

Martínez-Pacheco, H., Espinosa-Raya, J., Picazo-Picazo, O., Viñas-Bravo, O., & Ramírez-Galicia, G. (2015). *Rational design, docking and QSAR studies of a series of histone deacetylase isoforma 2 inhibitor*. In preparation.

McMartin, C., & Bohacek, R. S. (1997). QXP: Powerful, rapid computer algorithms for structure-based drug design. *Journal of Computer-Aided Molecular Design, 11*(4), 333–344. doi:10.1023/A:1007907728892 PMID:9334900

McQuown, S., Barrett, R. M., Matheos, D. P., Post, R. J., Rogge, G. A., Alenghat, T., & Wood, M. A. et al. (2011). HDAC3 is a critical negative regulator of long-term memory formation. *The Journal of Neuroscience, 31*(2), 764–774. doi:10.1523/JNEUROSCI.5052-10.2011 PMID:21228185

Mei, D., Yin, Y., Wu, F., Cui, J., Zhou, H., Sun, G., & Feng, Y. et al. (2015). Discovery of potent and selective urea-based ROCK inhibitors: Exploring the inhibitor's potency and ROCK2/PKA selectivity by 3D-QSAR, molecular docking and molecular dynamics simulations. *Bioorganic & Medicinal Chemistry*, *23*(10), 2505–2517. doi:10.1016/j.bmc.2015.03.047 PMID:25882521

Melo-Filho, C. C., Braga, R. C., & Andrade, C. H. (2014). 3D-QSAR approaches in drug design: Perspectives to generate reliable CoMFA models. *Current Computer-aided Drug Design*, *10*(2), 148–159. doi:10.2174/1573409910666140410111043 PMID:24724896

Meng, X.-Y., Zhang, H.-X., Mezei, M., & Cui, M. (2011). Molecular Docking: A powerful approach for structure-based drug discovery. *Curr Comput Aided Drug Des.*, *7*(2), 146–157. doi:10.2174/157340911795677602 PMID:21534921

Montañez-Godinez, N., Martínez-Olguin, A. C., Deeb, O., Garduño-Juárez, R., & Ramírez-Galicia, G. (2014). *QSAR/QSPR as an application of artifitial neural network. Artifitial Neuronal Networks* (pp. 319–333). Springer New York.

Montgomery, D. C., & Peck, E. A. (1992). *Introduction to linear regression analysis*. New York: Wiley.

Morris, G. M., Goodsell, D. S., Halliday, R. S., Huey, R., Hart, W. E., Belew, R. K., & Olson, A. J. (1998). Automated docking using a Lamarckian genetic algorithm and an empirical binding free energy function. *Journal of Computational Chemistry*, *19*(14), 1639–1662. doi:10.1002/(SICI)1096-987X(19981115)19:14<1639::AID-JCC10>3.0.CO;2-B

Morris, G. M., Huey, R., Lindstrom, W., Sanner, M. F., Belew, R. K., Goodsell, D. S., & Olson, A. J. (2009). AutoDock4 and AutoDockTools4: Automated docking with selective receptor flexibility. *Journal of Computational Chemistry*, *30*(16), 2785–2791. doi:10.1002/jcc.21256 PMID:19399780

Moustakas, D. T., Lang, P. T., Pegg, S., Pettersen, E., Kuntz, I. D., Brooijmans, N., & Rizzo, R. C. (2006). Development and validation of a modular, extensible docking program: DOCK5. *Journal of Computer-Aided Molecular Design*, *20*(10-11), 601–619. doi:10.1007/s10822-006-9060-4 PMID:17149653

Muegge, I., & Martin, Y. C. (1999). A general and fast scoring function for protein-ligand interactions: A simplified potential approach. *Journal of Medicinal Chemistry*, *42*(5), 791–804. doi:10.1021/jm980536j PMID:10072678

Nguyen, T. L., McGrath, C., Hermone, A. R., Burnett, J. C., Zaharevitz, D. W., Day, B. W., & Gussio, R. et al. (2005). A common pharmacophore for a diverse set of colchicine site inhibitors using a structure-based approach. *Journal of Medicinal Chemistry*, *48*(19), 6107–6116. doi:10.1021/jm050502t PMID:16162011

Osterberg, F., Morris, G. M., Sanner, M. F., Olson, A. J., & Goodsell, D. S. (2002). Automated docking to multiple target structures: Incorporation of protein mobility and structural water heterogeneity in AutoDock. *Proteins*, *46*(1), 34–40. doi:10.1002/prot.10028 PMID:11746701

Pasqualoto, K. F. M., & Ferreira, M. M. C. (2009). Molecular Modeling and Receptor-Dependent (RD) 3D-QSAR Approach to a Set of Antituberculosis Derivatives. *QSAR & Combinatorial Science*, *28*(11-12), 1455–1464. doi:10.1002/qsar.200960035

Rarey, M., Wefing, S., & Lengauer, T. (1996). Placement of medium-sized molecular fragments into active sites of proteins. *Journal of Computer-Aided Molecular Design*, *10*(1), 41–54. doi:10.1007/BF00124464 PMID:8786414

Saleh, N. A. (2015). The QSAR and docking calculations of fullerene derivatives as HIV-1 protease inhibitors. *Spectrochimica Acta: Molecular and Biomolecular Spectroscopy*, *136PC*, 1523–1529. doi:10.1016/j.saa.2014.10.045 PMID:25459714

Santos-Filho, O. A., & Cherkasov, A. (2008). Using molecular docking, 3D-QSAR, and cluster analysis for screening structurally diverse data sets of pharmacological interest. *Journal of Chemical Information and Modeling*, *48*(10), 2054–2065. doi:10.1021/ci8001952 PMID:18816024

Santos-Filho, O. A., & Hopfinger, A. J. (2002). The 4D-QSAR paradigm: Application to a novel set of non-peptidic HIV protease inhibitors. *Quantitative Structure Activity Relationships*, *21*(4), 369–381. doi:10.1002/1521-3838(200210)21:4<369::AID-QSAR369>3.0.CO;2-1

Sepehri, S., Gharagani, S., Saghaie, L., Aghasadeghi, M. R., & Fassigi, A. (2015). QSAR and docking studies of some 1,2,3,4-tetrahydropyrimidines: Evaluation of gp41 as possible target for anti-HIV-1 activity. *Medicinal Chemistry Research*, *24*(4), 1707–1724. doi:10.1007/s00044-014-1246-z

Shoichet, B. K., & Kuntz, I. D. (1993). Matching chemistry and shape in molecular docking. *Protein Engineering*, *6*(7), 723–732. doi:10.1093/protein/6.7.723 PMID:7504257

Slynko, I., Scharfe, M., Rumpf, T., Eib, J., Metzger, E., Schule, R., & Sippl, W. et al. (2014). Virtual screening of PRK1 inhibitors: Ensemble docking, rescoring using binding free energy calculation and QSAR model development. *Journal of Chemical Information and Modeling*, *54*(1), 138–150. doi:10.1021/ci400628q PMID:24377786

Taha, M. O., Habash, M., Al-Hadidi, Z., Al-Bakri, A., Younis, K., & Sisan, S. (2011). Docking-based comparative intermolecular contacts analysis as new 3-D QSAR concept for validating docking studies and in silico screening: NMT and GP inhibitors as case studies. *Journal of Chemical Information and Modeling*, *51*(3), 647–669. doi:10.1021/ci100368t PMID:21370899

Tame, J. R. (1999). Scoring functions: A view from the bench. *Journal of Computer-Aided Molecular Design*, *13*(2), 99–108. doi:10.1023/A:1008068903544 PMID:10091117

Teodoro, M., & Kavraki, L. (2003). Conformational flexibility models for the receptor in structure based drug design. *Current Pharmaceutical Design*, *9*(20), 1635–1648. doi:10.2174/1381612033454595 PMID:12871062

Tetko, I. V., Alessandro, E., Villa, P., & Livingstone, D. J. (1996). Neural network studies. 2. Variable selection. *Journal of Chemical Information and Computer Sciences*, *36*(4), 794–803. doi:10.1021/ci950204c PMID:8768768

Todeschini, R., Consonni, V., & Pavan, M. (2001). *DRAGON-Software for the Calculation of Molecular Descriptors. Release 1.12 for Windows*. Retrieved from http://www.disat.unimib.it/chm

Ul-Haq, Z., Usmani, S., Shamshad, H., Mahmood, U., & Halim, S. A. (2013). A combined 3D-QSAR and docking studies for the *In-silico* prediction of HIV-protease inhibitors. *Chemistry Central Journal, 7*(1), 88. doi:10.1186/1752-153X-7-88 PMID:23683267

Vats, C., Dhanjal, J. K., Goyal, S., Bharadvaja, N., & Grover, A. (2015). Computational design of novel flavonoid analogues as potential AChE inhibitors: Analysis using group-based QSAR, molecular docking and molecular dynamics simulations. *Structural Chemistry, 26*(2), 467–476. doi:10.1007/s11224-014-0494-3

Vedani, A., Dobler, M., & Zbinden, P. (1998). Quasi-atomistic receptor surface models: A bridge between 3-D QSAR and receptor modeling. *Journal of the American Chemical Society, 120*(18), 4471–4477. doi:10.1021/ja973976t

Verma, J., Khedkar, V. M., & Coutinho, E. C. (2010). 3D-QSAR in drug design-a review. *Current Topics in Medicinal Chemistry, 10*(1), 95–115. doi:10.2174/156802610790232260 PMID:19929826

Wold, S. (1991). Validation of QSARs. *Quantitative Structure Activity Relationship, 10*(3), 191–193. doi:10.1002/qsar.19910100302

Wu, G., Robertson, D. H., Brooks, C. L., & Vieth, M. (2003). Detailed analysis of grid-based molecular docking: A case study of CDOCKER—A CHARMm-based MD docking algorithm. *Journal of Computational Chemistry, 24*(13), 1549–1562. doi:10.1002/jcc.10306 PMID:12925999

Xu, L., & Zhang, W. J. (2001). Comparison of different methods for variable selection. *Analytica Chimica Acta, 446*(1-2), 477–483. doi:10.1016/S0003-2670(01)01271-5

Yu, H., Fang, Y., Lu, X., Liu, Y., & Zhang, H. (2014). Combined 3D-QSAR, Molecular Docking, Molecular Dynamics Simulation, and Binding Free Energy Calculation Studies on the 5-Hydroxy-2H-Pyridazin-3-One Derivatives as HCV NS5B Polymerase Inhibitors. *Chemical Biology & Drug Design, 83*(1), 89–105. doi:10.1111/cbdd.12203 PMID:23941500

KEY TERMS AND DEFINITIONS

Computer Aided-Drug Design (CADD): The effective design of chemical structures with the desirable therapeutic properties, it is a well-established area of computer aided molecular design (CAMD).

Docking Algorithm: Mechanism that two search engines; one is responsible for the placement and exploration of the ligand in the binding site, while the other searches the internal degrees of freedom of the ligand.

Molecular Docking (M/D): The computational process of searching for a ligand that is able to fit both geometrically and energetically the binding site of a protein.

Molecular Dynamics (MD): This computational method calculates the time dependent behavior of a molecular system based on Newton's second law or the equation of motion.

Quantitative Structure Activity Relationships (QSAR): A way of finding a simple equation that correlates structural molecular features (descriptors) with physicochemical properties, such as biological activities for a set of compounds by means of statistical methods.

Rigid Ligand: A drug that has no degrees of freedom to rotate around bonds.

Rigid Receptor: Protein that has only one conformation and cannot shift to another.

Scoring Function: The mechanism that is needed to decide which ligand placements (scores) are better than others, and thereby rank conformations, or candidate dockings, of a particular ligand.

Chapter 3
Molecular Docking Challenges and Limitations

Jahan B. Ghasemi
University of Tehran, Iran

Azizeh Abdolmaleki
Islamic Azad University of Touyserkan, Iran

Fereshteh Shiri
University of Zabol, Iran

ABSTRACT

Today, the development of new drugs is a challenging task of science. Researchers already applied molecular docking in the drug design field to simulate ligand- receptor interactions. Docking is a term used for computational schemes that attempt to find the "best" matching between two molecules in a complex formed from constituent molecules. It has a wide range of uses and applications in drug discovery. However, some defects still exist; the accuracy and speed of docking calculation is a challenge to explore and these methods can be enhanced as a solution to docking problem. The molecular docking problem can be defined as follows: Given the atomic coordinates of two molecules, predict their "correct" bound association. The chapter discusses common challenges critical aspects of docking method such as ligand- and receptor- conformation, flexibility and cavity detection, etc. It emphasis to the challenges and inadequacies with the theories behind as well as the examples.

INTRODUCTION

Molecular recognition plays a key role in promoting basic biomolecular experiences such as drug-protein, enzyme-substrate and drug-nucleic acid interactions. Detailed understanding of the general principles that administrate the nature of the interactions (van der Waals, hydrogen bonding, electrostatic) between the ligands and their protein or nucleic acid targets may present a conceptual support for designing the desired potency and specificity of potential drug leads for a given therapeutic target. Practical application of this knowledge requires structural information for the target of interest and a route for evaluating

DOI: 10.4018/978-1-5225-0362-0.ch003

candidate ligands. To this end, a variety of computational docking methods are available(Mohan, Gibbs, Cummings, Jaeger, & DesJarlais, 2005)

Among the vast studies, there are several reports that represent molecular docking as worthwhile strategy to predict 3D structures of complexes. Molecular docking can overcome to experimental complications surrounding the structure determination of complexes, and can give valuable structural insights on biomolecular interactions. However, the process of binding a small molecule drug to its protein target is not simple; several factors affect on the interaction between two molecules, so these methods can be improved as a solution to docking problem. Despite great developments and achievements, the widespread application of docking methods, and the accurate and rapid prediction of protein–ligand interactions is still a challenging area to explore and some downsides still exist. The flexibility or mobility of both ligand and target, the effect of the protein environment on the charge dispersal over the ligand, and their interactions with the surrounding water molecules, complicate the quantitative description of the process.

This chapter reviews the literature in the last decades to account for the various abilities, limitations and challenges of the currently widespread applied algorithms which presently characterize this methodology. It introduces some efforts of drug developers for the improvement of algorithms to overcome the shortcomings and to enhance the computations for the incorporation of both ligand and receptor flexibility in the docking process, careful exploration of the ligand conformation within the binding site and improving complementarity between them, refinement and stability evaluation of the final complexes and thus accounting for induced fit. The discussion will further be limited to protein-small ligand complexes, omitting macromolecular complexes; however, much of what is presented here is also valid for that class of complexes.

The chapter handles some critical problems of docking calculation such as the lack of speed, accuracy, protein flexibility and ligand sampling with somewhat aspects of fundamental basic concepts such as movements of side chains and the biased selection of ligands as a result of using ligand-bound protein structures during a docking process or mis-docking ligand. As mentioned before, something of the recent advances provide on the ligand sampling, protein flexibility, and scoring functions as important features in protein-ligand docking. Three types of ligand sampling algorithms will be discussed; shape matching, systematic search, and stochastic algorithms. Computational efficiency for each algorithm is represented according to their main advantages and disadvantages. Covalent docking is another issue of molecular docking which is to be considered in the last part. Authors hope that the recently published papers on covalent docking give insights on biomolecular interactions to the readers.

BACKGROUND

Mutual molecular recognition is the initial point for approximately all processes in biological systems. Today, molecular docking as a very demanding computational and algorithmic tool plays a fundamental and advanced role in structural molecular biology and drug design. These computational tools help us for understanding molecular interactions of two molecules such as protein–protein or protein–ligand that is a key for the understanding of chemical process in diseases and other life issue occurrence. Molecular docking has a wide range of potential uses and can be applied in the following fields of drug discovery:

- Structure–activity studies
- Lead optimization

- Finding potential leads by virtual screening
- Providing binding hypotheses to ease predictions for mutagenesis studies
- Helping x-ray crystallography in the fitting of substrates (Figure 1 shows a x-ray crystallography structure as PDB) and inhibitors to electron density
- Chemical mechanism studies
- Combinatorial library design

Molecular docking techniques are designed to aid the development of therapeutic agents via predicting of positioning of a small molecule (ligand) when it is bound to a protein receptor or enzyme, and to find most stable conformation of a ligand and its target macromolecule or receptor. However, in a docking process study, both molecules are flexible and may alter each other's structure as they interact:

- Hundreds to thousands of degrees of freedom
- Total possible conformations are astronomical

Docking is an important *in silico* method widely used for recognizing drug led by 3D docking ligand to the three dimensional structure of a pre-choose target site continued with the optimization of binding configurations and assessment of binding possibility based on paired molecular interactions (Gozalbes *et al.*, 2008; Kitchen, Decornez, Furr, & Bajorath, 2004). Docking process is a combination of a search algorithm and a scoring function that attempt to find the "best" matching between two molecules: a ligand and a receptor. Now a relatively big and ever growing number of search algorithms and scoring functions are accessible. Reasonably, the best solution would be to associate the ideal searching algorithm with the ideal scoring function. Figure 2. Indicates an example of the best docking corresponding to the ideal scoring function. However, numerous studies have shown that the performance of docking

Figure 1. X-ray crystallography structures deposited as PDB file: A typical PDB structure (PDB code: 3PS2)
Ghasemi, Safavi-Sohi, and Barbosa (2012)

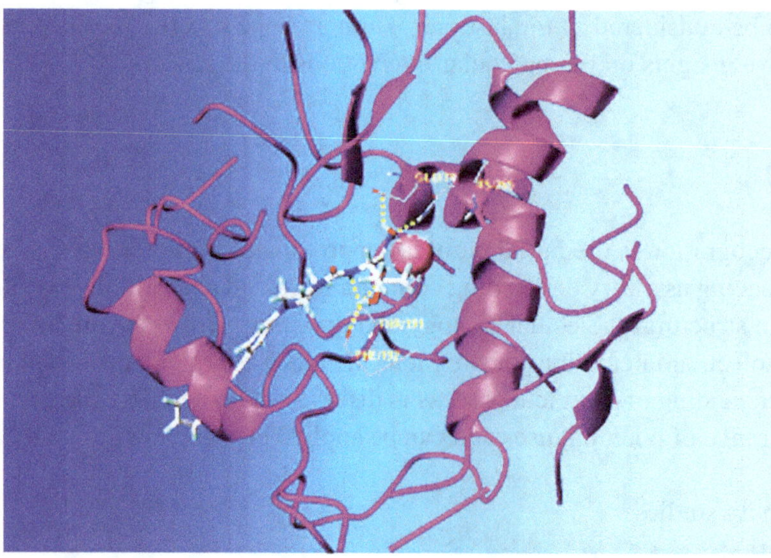

Figure 2. The best docked conformation of one of the most active compounds, (compound 141) in the active site of IN
Ardakani and Ghasemi (2013)

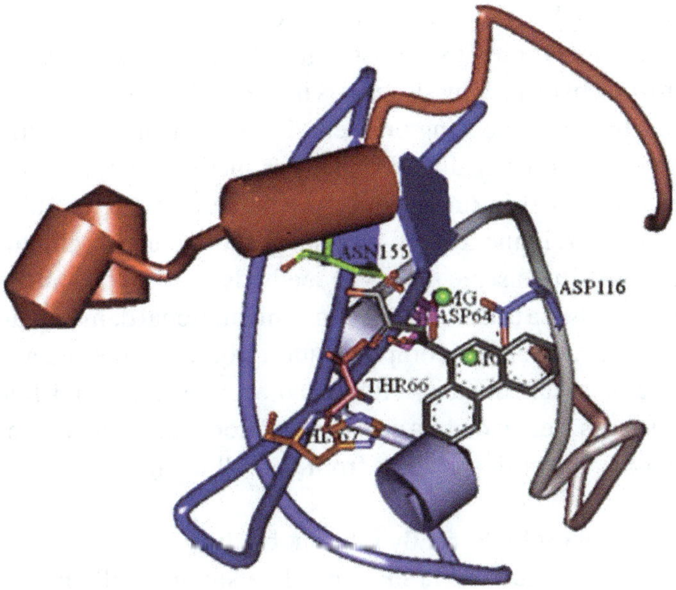

is greatly depend on the specific features of both the binding site and the ligand to be explored, and that establishing which method would be more appropriate in a precise context is nearly difficult (Sousa, Fernandes, & Ramos, 2006). The molecular docking problem can be defined as follows: Given the atomic coordinates of two molecules, predict their "correct" bound association. In its most general form, no additional data are provided.

The present field mainly adapts to dock small molecules into macromolecule target especially protein molecule and its use is increasing year by year (Pei, Yin, Ma, & Lai, 2014; Sliwoski, Kothiwale, Meiler, & Lowe, 2014; Warren, Do, Kelley, Nicholls, & Warren, 2012; Zheng *et al.*, 2013). The goal to run a docking simulation of ligand–protein is to calculate the main binding mode(s) of a ligand with a protein of known 3D structures of its own components. Ligand molecule uses its action by binding to specific molecular parts of the cell such as modulating biochemical processes in an illness adapting manner. Considering the protein-protein docking and protein-small molecule docking, the following types of problems occurs:

- **Protein-protein docking**
    - Bound docking ("rigid redocking problem"):
    - Unbound docking: side chain flexibility
- **Protein-small molecule docking**
    - Rigid receptor, rigid ligand
    - Rigid receptor, flexible ligand
    - Flexible receptor, flexible ligand

The docking process is divided into two main stages: first, the correct situation of the ligand at the protein binding-site and, second, the estimation of the ligand affinity by a scoring function. Figure 3 shows more details of the key steps in docking that are used in all protocols commonly. To determine the structure of a biomolecular complex, this technique creates a broad set of conformations of the receptor complex, and then ranks them according to their stability (Morphy & Harris, 2012).

One key area for improvement in docking systems is in the methods for scoring putative protein-ligand interactions. Scoring is central to the docking problem. Development of protein structures for potential drug leads by molecular docking is critically dependent on methods for scoring putative protein-ligand interactions. An ideal function for scoring must exhibit predictive accuracy and high computational speed, and must be tolerant of variations in the relative protein-ligand molecular alignment and conformation.

In comparison to the rapid virtual screening (VS) methods similar to pharmacophore modeling, the performance of docking methods is a trade-off between computational demands and accuracy. The extent and specificity of binding depends on the complementarity between the ligand and its target in terms of shape, polarity, and chemical functionality (Reymond & Awale, 2012). Effective docking methods search high-dimensional spaces well and use a scoring function that properly ranks candidate dockings.

The success of small molecule or ligand roots from the following facts;

1. Matches of their size for most biologically relevant binding sites;
2. Sufficient existing of the structural and functional diversity in small molecules to reach strong and specific binding to most of these binding sites (see Figure 4.); and
3. The pharmacokinetics of ligand molecule can be adjusted while target binding to support efficacy and safety *in vivo* (Bon & Waldmann, 2010; van der Horst *et al.*, 2011).

Docking method is simple and has a variety of applications but some difficulties limit its ability. Over the last decade a wide effort has been directed toward developing efficient docking methods and

Figure 3. Common key steps in typical docking protocols
A: *Choose of 3D structures of ligand and the target macromolecule.*
B: *Preparation of two structure considering the docking method being used.*
C: *Analysis of docking results and selecting the binding modes with the best scores.*

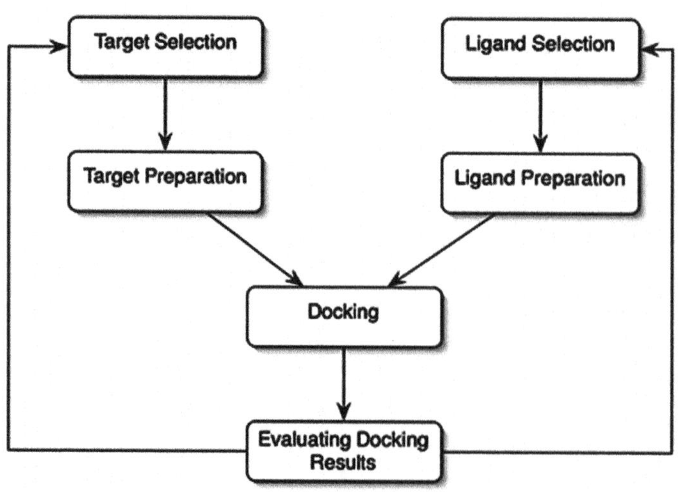

Figure 4a. Compound docked in the binding site of target factor Xa

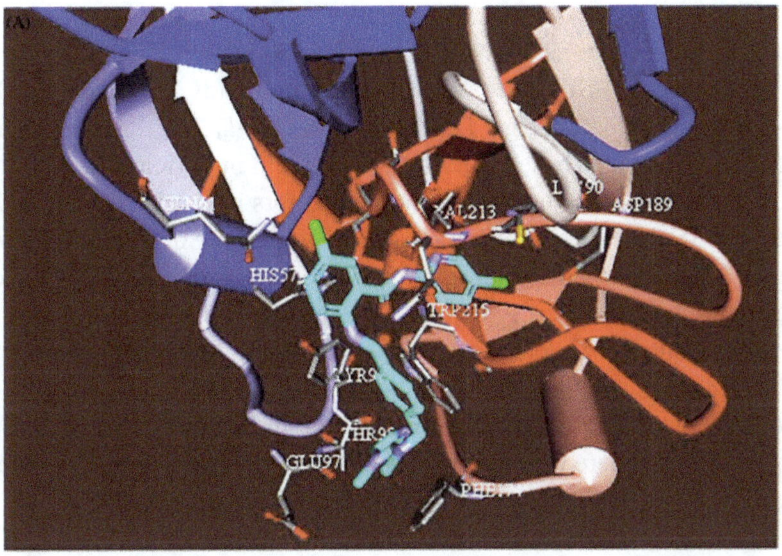

Figure 4b. The residues and pockets of the active site of factor Xa
Ghasemi and Hooshmand (2013)

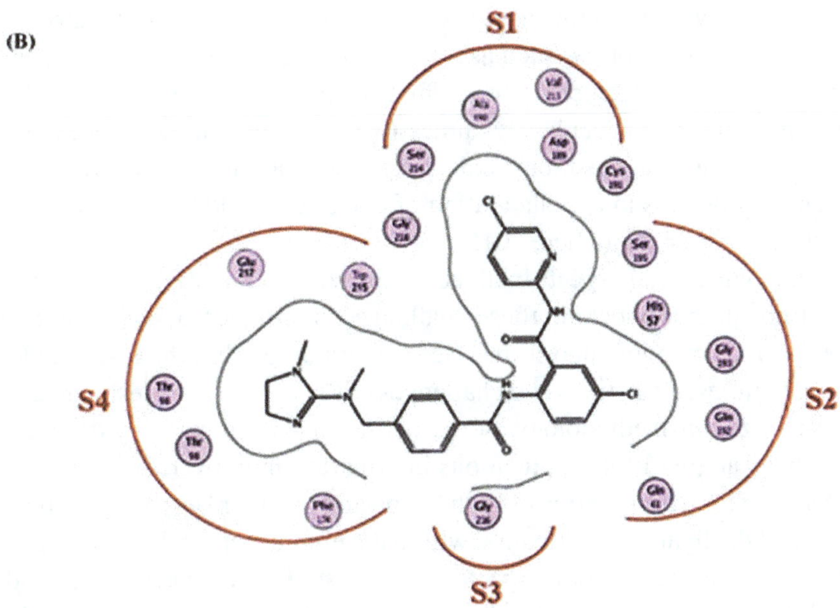

for the identification of lead compounds. Considerable progress has been made in the computational prediction of ligand-target binding modes. A number of review articles in this emerging area of research have been recently published (Mohan et al., 2005). Some of the problems of docking calculation have been considered in the present text and a number of problem solutions considering some efforts have been introduced.

COMMON DOCKING COMPLEXITY

For docking method, the equipment necessity and information of the structural aspects and molecular frameworks of active compounds is low (Alonso, Bliznyuk, & Gready, 2006; Chen, 2015), but its ability may have limited to the accounting of target flexibility (Hou *et al.,* 2013) and certain physicochemical features are vital for drug binding. These limitations arise from insufficient sampling of target conformational space, the use of gas-phase algorithms and parameters in scoring, poor modeling of solvation and entropic effects, and the use of simplified cut-off distances for non-bonded interactions. Careful calculation shows that accuracy is a major problem with docking studies, unless the docking is not approached with precision (Chen, 2015).

The complexity of computational docking increases in the following order:

1. Rigid body docking, where both the receptor and small molecule are treated as rigid.
2. Flexible ligand docking, where the receptor is held rigid, but the ligand is treated as flexible; and
3. Flexible docking, where both receptor and ligand flexibility is considered.

Thus, the most regularly used docking algorithms use the rigid receptor/flexible ligand model (Mohan et al., 2005)

Several factors influence the process of binding a small molecule to a target protein and make it quite complex; for example, the motion or flexibility of the small molecule drugs and receptor which affect the possible interactions solvent-receptor-ligand-solvent present depends on the molecular distribution of charges. There are common problems such as incorrect binding site of the target protein, the screening use of an improper small-molecule database, the selection of docking pose, and high dock score (binding affinity) but failure in molecular dynamics (MD) simulation, deficiency of clarity whether the compound is an inhibitor or agonist, or the docking results are unreliable with bioassays (Morphy & Harris, 2012). These problems in the elucidation of docking should be considered with caution and concern before performing docking (Chen, 2015).

The context covers joint critical aspects in all key steps with major emphasis to the general strengths and shortcomings. Enough theory behind these challenges is provided by avoiding many unnecessary equations, and details of calculation algorithms. Some literature within the last decades reviews to account the various capabilities, limitations and challenges of the currently widespread applied algorithms that presently characterize this methodology. Some efforts have been made by drug developers for improving algorithms by examples of its applications in drug design to overcome such shortcomings and computational enhancing for incorporation of both ligand and receptor flexibility in the docking process, extensive exploration of the ligand conformation within the binding site and improving complementarity between them (Figure 4 indicates the binding site of a case study and an active site in the same target), refinement and stability evaluation of the final complexes and thus accounting for induced fit. Discussion will further be limited to protein-small ligand complexes, omitting macromolecular complexes; however, much of what is presented here is also valid for that class of complexes. Future directions are included in the protein flexibility, ligand sampling, and scoring functions as important features in protein-ligand docking.

LIGAND CONFORMATION AND SAMPLING

Ligand - protein interaction complexes are the most important structures to get information about the mechanism of action of the chemicals compounds. A ligand molecule that can bind strongly to the target receptor should have energetically favorable interactions with the receptor with an appropriate relative geometry. In docking simulation, the problem of finding such geometry between ligand and target molecules is too difficult to be accomplished only by computational methods. To find a likely stable geometry and conformation, the ligand molecule is subjected to a series of interactive three-dimensional manipulations (rotation, translation, and bond rotation) inside the ligand binding site of the protein on the 3D-computer graphics display.

Ligand flexibility is one feature of docking simulation process. Small ligand-oriented docking schemes are either semiflexible or flexible. Small ligands are often highly flexible, with a broad range of populations, adapting their surface to optimally complement the receptor pocket. Therefore, one of the main features of drug docking schemes is enhanced ligand flexibility (Halperin, Ma, Wolfson, & Nussinov, 2002). Simulating ligand flexibility is also a computationally expensive process. Lomber and Shoichet reported that the number of possible ligand conformations increases in proportion to the power of the number of rotatable bonds (Lorber & Shoichet, 1998). They calculate that for an organic molecule with ten rotatable bonds, the number of possible conformations is 59049, if only three minima are considered per bond. However, allowing six minima per bond yields 3.48×10^9 conformations. To address the problem, one approach involves Monte Carlo simulation and simulated annealing to sample the ligand flexibility. In Goodsell et al., the binding site is rigid, whereas in Stoddard and Koshland some binding site flexibility is allowed (Halperin et al., 2002).

Rigid docking calculations are chosen when time is critical, i.e., when a large number of compounds are to be docked. Nevertheless, flexible docking methods are quite required for refinement and optimizing of poses found in an initial rigid docking process. With the development of computational resources and their efficiency, flexible docking approaches are commonly suitable. Three general algorithms are planned to give ligand flexibility: systematic methods; random or stochastic methods; and simulation methods (Sousa et al., 2006). Systematic search or systematic enumerations of conformations, and stochastic algorithms (such as genetic algorithms and Monte Carlo search method with Metropolis criterion (MCM)), shape matching and MD simulations (Figure 5 represents a typical MD calculation) are the most popular approaches for the improvement of efficiency. Systematic algorithms include ligand flexibility through a full search of a molecule's degrees of freedom. In this method, the current state of the system determines the next state. Starting from the same exact state and same set of parameters, systematic methods will yield exactly the same final state. Systematic methods can be categorized into (1) fragmentation algorithms and (2) exhaustive search algorithms (Sliwoski *et al.,* 2014).

A MD protocol was defined for docking small flexible ligands to flexible targets in water by Mangoni *et al.* (1999). They disjointed the center of mass motion of ligand from its inside and rotational motions (Mangoni, Roccatano, & Di Nola, 1999). Independent control of the different motions was conducted to permit flexible or rigid ligand and/or receptor. ROSETTALIGAND uses a knowledge-based scoring procedure with a Monte Carlo-based energy minimization scheme that reduces the number of conformations that must be sampled while providing a more rapid scoring system than offered through molecular mechanics force fields. ROSETTALIGAND through a Monte Carlo-based sampling of torsional angles includes ligand flexibility and side-chain during a high-resolution refinement stage and total torsion angles of ligand and protein are optimized (Sliwoski *et al.*, 2014).

Figure 5. Resulting from MD simulations to the most active compound, with common atom fit (a) and cluster alignment method (b)
Ghasemi et al. (2012)

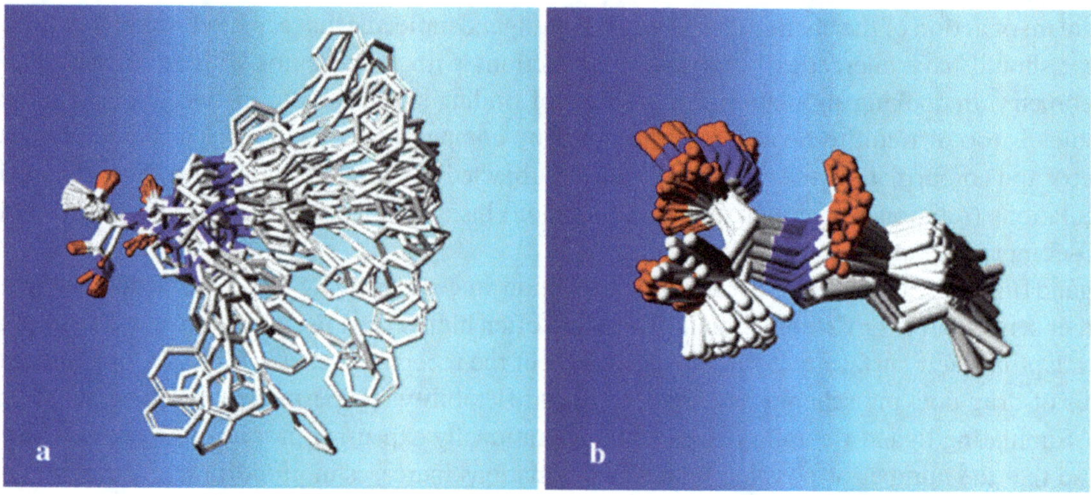

Molecular flexibility in genetic algorithms (GA) was achieved through recombination of parental conformations with child conformations and the "fittest" or best scoring conformations was retained for another round of recombination. Genetic Optimization for Ligand Docking (GOLD) discovers full ligand flexibility with limited target flexibility by a GA (Jones, Willett, Glen, Leach, & Taylor, 1997).

Docking tools samples conformations of small molecules in protein binding sites; scoring functions are used to judge which of these conformations best matches the protein binding site. Scoring functions that predict the possibility of a compound being a ligand, mainly derived from three key methods based on their characteristics:

- Knowledge based scoring methods
- Force-field or physics based scoring methods
- Regression or empirical based methods

The main interest aspect of empirical scoring functions is the typically easy computation of the numerous terms. The key disadvantage of these methods is their dependency on the experimental data set used in the parameterization process (Leach, Shoichet, & Peishoff, 2006; Sousa et al., 2006).

Venkatachalam *et al.* have described a shape-based method, LigandFit, for accurately docking ligands into protein active sites. In their study a grid-based method for assessing protein–ligand interaction energies have been used for minimizing candidate poses in the context of the active site. They reported combination of LigandFit, with LigScore, as an internally scoring function, to the thymidine kinase receptor yields very good hit rates for a ligand pool seeded with known actives (Venkatachalam, Jiang, Oldfield, & Waldman, 2003).

Another problem of a ligand during a docking process is the biased selection of ligands as an outcome of using ligand-bound protein structures that was undoubtedly revealed in a series of cross-docking analyses. Displacement of particular portions of the protein backbone, movements of side chains, and mobility of metal atoms found within the active site produce most failure (see Figure 6 shows an active

Figure 6. Active site of the CC-chemokine receptor 5 (CCR5)
Ghasemi and Nouri (2013)

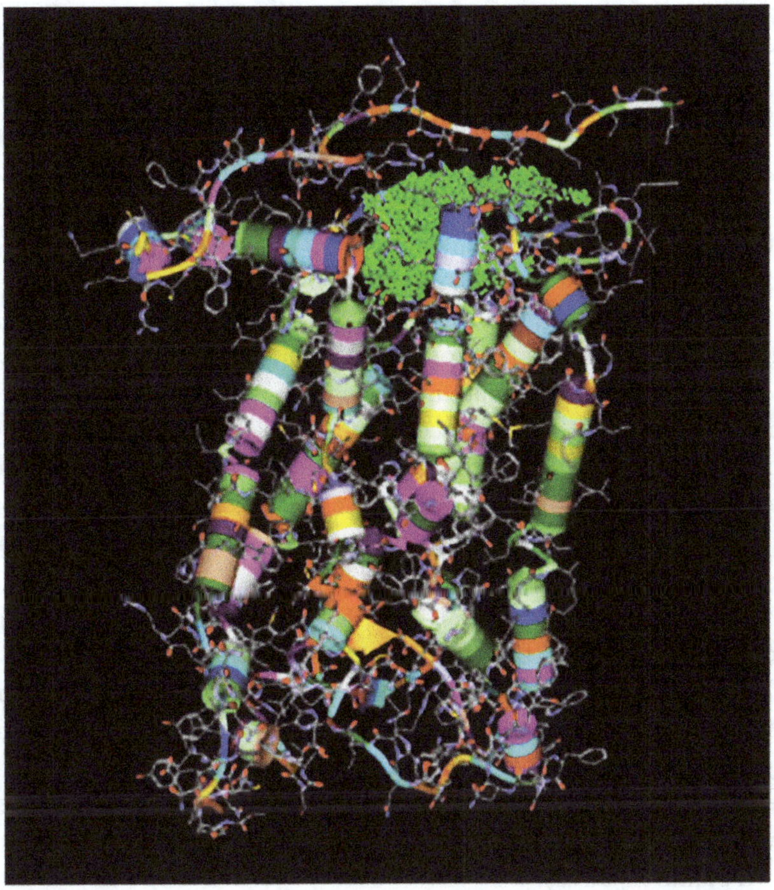

site in a case study). It was observed that the movements of side chains were usually linked to those of the backbone Ca atoms and, as a result, it was essential to consider more than sidechain flexibility to avoid mis-docking of ligands.

Therefore, careful preparation of the structure of the receptor before the docking process is very important. While structures of ligand-bound protein may offer the highest enrichments, the final consequences might be biased towards certain types of ligands.

Erickson *et al.* have evaluated the effect of some key factors on docking accuracy with CDDOCKER, DOCK, FlexX, GOLD programs. In particular, they have studied the significance of ligand flexibility. Their studies showed that the accuracy of docking considerably decreased for ligands with 8 or more rotatable bonds: DOCK (14%); FlexX (28%); GOLD (19%); CDOCKER (71%). This indicates that the capacity to differentiate between correct or incorrect conformations may decrease for the most public docking programs for ligands with large numbers of probable low-energy conformations (Erickson, Jalaie, Robertson, Lewis, & Vieth, 2004).

RECEPTOR CONFORMATION AND FLEXIBILITY

The 3D structure of both ligand and protein are essential for the application of docking process. The 3D structure of a target obtained by NMR technique or X-ray crystallography (Clore & Gronenborn, 1991; Siegal, Van Duynhoven, & Baldus, 1999) is the best starting point for docking (Sliwoski *et al.*, 2014). Docking algorithms can search the enormous conformational space of small molecule in a reasonable cost and short time. While several ligand conformational structures may be relatively easy to predict, the lowest energy conformation obtained may not match that of the bound ligand. Compared to ligand, the structures of protein is a bigger challenge. While X-ray and NMR analysis are currently routine, in some cases intrinsic difficulties exist in the planning high-throughput process. The structure of many targeted proteins in drug design is not determined by above methods experimentally and, therefore, docking studies cannot be performed straightforward. In some cases, the structure of a closely related protein homolog is predicted through computational techniques.

Other significant drawbacks for docking process is the lack, or poor flexibility of the protein, which is not allowed to adjust its conformation during ligand binding (Sgrignani *et al.*, 2009). This problem often ignored in great virtual screening (VS) calculations because of the high computational cost, but it can be managed by more focusing of docking process, and the consistency of the predicted poses can be improved (Lexa & Carlson, 2012).

Variations of existing docking program mostly treat the receptor as (more or less) rigid which is unlike of the biological facts and the ligand as flexible (Leach et al., 2006). Induced-fit docking and ensemble generated from MD simulation snapshots have been used for inducing receptor flexibility. Induced-fit algorithms agree small overlap between the ligand and receptor along with side chain movements, subsequent in elasticity. Another way to incorporate receptor flexibility in docking algorithms, known as ensemble-based screening is applied multiple fix receptor conformations (Sliwoski *et al.*, 2014). Other methods such as umbrella sampling, metadynamics, accelerated MD, etc. are used as sampling method (Sinko, Lindert, & McCammon, 2013).

ENTROPY IN BIOMOLECULAR INTERACTIONS

Binding affinity is a function of the free energy of binding which is composed of entropy and enthalpy terms dependent on the properties of an ensemble of structures. In a computational manner to scoring drug affinity, entropy of fixation can be expressed in terms of the number of ways in which a rotatable bond can be incorporated, $\Delta S= RTlnW$. If we assume that there are 3 orientations per rotatable bond then $\Delta G= RTln\ (3^N) = \sim N$ kcals/mole. Quantitation of these effects requires temperature dependent measurements of the free energy of complex formation.

The processes involving ligand binding are extremely complicated, both ligand and protein are flexible molecules, and the energy inventory between the bound and unbound states must be considered in aqueous solution. The most common handling of receptor flexibility in structure-based drug design is to ignore it and to design ligands to bind to more rigid regions of the receptor (Morton & Matthews, 1995). This has the advantage that the entropic costs of immobilising a receptor on binding a ligand are minimized.

There is an entropic cost when two molecules associate in any biomolecular interaction process. The entropic cost is a result of degrees of freedom of motion that lost when two molecules are strictly constrained within a complex. If the binding process involves capturing a flexible molecule, whose internal rotations about single bonds must also be restricted, the consequence is a further adverse entropic penalty that results in a reduction in binding constant ***K***, according to the classical relationships

$\Delta G = \Delta H - T\Delta S$ and $\Delta G = -RT\ln K$.

If we are to begin to predict binding constants for bimolecular associations in a manner that might develop useful rules for understanding the forces that drive molecular recognition,' then we need to be able to semiquantitate the adverse cost in free energy of generating conformational order from associations of flexible molecules(Searle & Williams, 1992).

Approximate methods can be used to estimate the loss of configurational entropy of the ligand by counting the number of torsional angles that have become restricted upon binding to the protein, and if desired, this entropy loss can be subtracted from the total energy change(van den Berg et al., 1995) (Burgen, Roberts, & Feeney, 1975).

The estimation of the entropic cost is difficult for several reasons.

- The entropies of molecules in solution are subject to uncertainties.
- The amount of residual motion present in a complex formed by association in solution is also uncertain.
- This amount of residual motion in a complex may vary as a function of the nature and extent of the intermolecular forces (enthalpy term) involved in binding.

Moreover, the issue is complicated by the fact that the accounting of the balance of entropy changes may be done in different ways. For example, it may be considered that the degrees of vibrational freedom associated with residual motion in the complex are translational and rotational entropy of the separate components which are not lost upon association. Otherwise, this vibrational freedom can be regarded as a benefit of the association, appearing after the interacting components have lost all of their translational and rotational entropy (Searle & Williams, 1992).

STRUCTURAL WATER MOLECULES

The handling of structural water molecules is a particular problem in molecular docking. Recently, structural water molecules in ligand protein recognition have intensely been studied and its importance is a well-known topic. However, handling the difficulty of structural water molecule in docking designs is still problematic and different guidelines have been suggested for managing water molecules during docking (Sgrignani, Novati, Colombo, & Grazioso, 2015). Now, some methodologies are applied to tackle this problem including: penalizing the existence of water molecules, awarding rational situation of structural water molecules in an active location or cavity and no amendment for water. Single water molecules in the interface may be particularly important in small ligand docking, mediating hydrogen bonds. The water solvent is described by a soft sphere Langevin dipole model, with discrete dipoles that interact with the electric field of the protein but subject to random thermal fluctuations that reduce the effective electric field at the dipole itself. van der Waals and field-dependent hydrophobic terms are also included(Halperin et al., 2002).

A number of currently available programs suffer the lack of satisfying to account for structural water molecules which can be either displaced in ligand binding or mediate protein-ligand interactions. Some of research aims is data generation to enable the worldwide work of evolving scoring functions in docking programs to account for contributing of structural water molecules. This is done by validating the performance of docking using a simple model system (such as cytochrome C peroxidase (CCP) W191G) containing four well-arranged structural water molecules which are known to either interact with a ligand or to be moved upon ligand binding. The goal of working on molecular dynamics simulations of Sotriffer's group is estimation the enthalpic and entropic influences of the structural water molecules in CCP W191G to ligand binding affinity (Z. Li & Lazaridis, 2003; Olano & Rick, 2004). In a similar study on molecule docking in the framework of structural water molecules, the performance was certified by the following methods:

- Comparison of the binding modes with the docked ligands and the crystal structures,
- Comparing docking scores with the ligands in the different situations,
- Improving of the ligands from a database of decoys
- predicting new ligands from the decoy database (Wahlström, 2009).

The GOLD docking program lets water molecules to be "changed on and off" with a penalty for enabled waters thus preferring water movement by working with the first approach (Verdonk *et al.*, 2005). Sgrignani *et al* in study of boronic acid have reported, GOLD 5.2 provides a correct computational process for management water molecules in binding sites, permitting to alteration water placing during docking calculations (Sgrignani *et al.*, 2015). The FlexX program puts the second tactic by awarding interactions of the ligand with the water molecule and penalizing for unoccupied interaction sites at the water molecule (Kramer, Rarey, & Lengauer, 1999). In an extension to DOCK 3.5, interactions of structural water molecules with a ligand were scored in similar mode as interactions with protein atoms (Huang & Shoichet, 2008). For the entropically disapproving fixation of water molecules in the active site, no penalties were added.

Recent surveys pay attention more carefully to account selected *water molecules* in docking calculations (Roberts & Mancera, 2008; Weill, Therrien, Campagna-Slater, & Moitessier, 2014). These reports show the undiscriminating inclusion of all the water molecules placed in or around the binding site can lead to a noticeable decrease of the predictive performances of the docking simulation process (Kumar & Zhang, 2013). Sgrignani *et al.* have evaluated the significance of the protein flexibility and catalytic site hydration during docking simulations in order to tune up a robust covalent docking practice for boronic acids. They set out to identify the solvent molecules more often present in the Amp/ boronic acid crystal structures, for identifying the minimum number of water molecules to provide accurate results. Their results show a reasonable cooperation between computational speed and structural accuracy (Sgrignani *et al.*, 2015).

This difficulty of predicting and scoring the above parameters, binding modes unambiguously generates false negative and positive scoring compounds as it has been made obvious previously. One example was Warren *et al.* who docked 1303 known ligands into 7 receptors using 10 different docking programs and compared the docking poses to the structurally determined binding modes. The result for the different algorithms performance ranged from 0% to 90% satisfactorily predicted docking modes (Wong *et al.*, 2006).

CAVITY DETECTION

A variety of algorithms have been proposed with the aim to identify cavity detection. The vast majority of cavity detection algorithms have been developed to treat static structures like crystal structures of proteins available in the protein data bank (PDB). However, this represents a serious limitation to account for the intrinsic plasticity of the binding pocket. Protein dynamics act on a multitude of aspects in protein function. For instance, side chain flipping or domain motions can obstruct or free internal cavities or channels that allow migration of ligands and reshape the binding sites (Bidon-Chanal et al., 2006; Carrillo & Orozco, 2008; Spyrakis et al., 2011). In turn, these findings raise challenging questions about the impact of protein flexibility on the topological features of cavities and their binding properties.

Few works have tried to account for the dynamical behavior of proteins in the identification of binding cavities and tunnels. The interplay between protein dynamics and ligand migration pathways can be characterized by tools that rely on prior molecular dynamics (MDs) simulations and further post-processing of the trajectory. VOIDOO (Kleywegt & Jones, 1994) allows internal cavity and volume calculations, but it is rather time consuming and its use is not straightforward. CAVER (Beneš et al., 2010) is a PyMOL plugin that allows internal channel detection on MD trajectories. CAVER was improved to a software called MOLE using computational geometry principles instead of grid-based calculations (Petřek, Košinová, Koča, & Otyepka, 2007). Wolfson and coworkers have proposed a method called MolAxis for detection of channels from the interior of the protein to the bulk solvent (Yaffe, Fishelovitch, Wolfson, Halperin, & Nussinov, 2008). These methods are designed to detect channels on static structures or conformational ensembles of the protein. MDpocket is another generic pocket detection program. The aim of MDpocket is to identify and characterize binding sites and channels that might be transiently formed in the protein from the analysis of conformational ensembles generated by MDs or other sources. The core of this program is the recently published open-source platform Fpocket (Le Guilloux, Schmidtke, & Tuffery, 2009), which is a very fast geometry-based cavity detection algorithm.

Although many softwares and algorithms to discern different types of cavity are introduced and guidelines for assessing the accuracy and improving the comparability of cavity calculations are given, we need an accurate protocol to find the best cavity.

COVALENT DOCKING

Many methods of target-based drug design have been established for prediction of the protein/ligand complex structure and, they are used to this aim, (Warren *et al.*, 2006; Warren *et al.*, 2012; Weill *et al.*, 2014; Wong *et al.,* 2006). In the docking field, a number of methods mainly focus on the studying of docking between the two molecules through non-covalent interactions (the electrostatics interaction, van der Waals interaction and hydrogen bonding) or non-covalent complexes, while covalent docking is a crucial aspect in the docking field. Only a few methods are accomplished to predict the 3D structure of complexes in which the ligand is covalently bound. However, except non-covalently drugs, there exist other classes of drugs, i.e. the covalent drugs that bind non-covalently to the active site. Covalent ligands are not given significant attention in the traditional drug discovery process due its off-target reactivity besides of toxicity profile. However two FDA-approved covalent inhibitors targeting HCV, namely telaprevir and boceprevir highlighted the focus on covalent inhibitors. *GOLD* and *Autodock* software support feature of covalent docking, but they have major limitations. However, this problem has been addressed in *CovDock* and *CovalentDock* by automatic preparation of ligand files.

In the frame of these considerations, Sgrignani *et al.* have assessed the importance of catalytic site hydration and the protein flexibility during docking calculations for boronic acids with GOLD program. GOLD permits water molecules to be rotated round their three principal axes (to optimize hydrogen bonding) during docking. Moreover, GOLD can switch automatically whether a water molecule should be bound or displaced by the ligand during docking. To certify the accurate geometry of the bound ligand, GOLD 5.2 applies the angle-bending potential from the Tripos Force Field, merged in the fitness function during covalent docking process (Sgrignani et al., 2015).

SOLUTIONS AND RECOMMENDATIONS

Some of drug developers focused their attention on the enhancement of computational protocols. These improvements include several methods. A number of target flexibility approximation methods have been used to tackle the risks connected with some searching methods to take a proper ranking of the final complexes in order to improve docking performance. For performance increasing, the molecular dynamics-enhanced docking approach has been used to try to account for the target flexibility (Morphy & Harris, 2012). The most common of approximate methods include:

- Combining docking and MD simulation
- Soft docking
- Side chains flexibility
- Rotamer-based flexible docking (conformation of binding-site residues are tested by rotamer libraries as backbone-dependent or backbone-independent)
- Molecular relaxation
- Joined protein grid docking (several structures and conformations are combined into an energy-weighted or geometry-weighted average depiction)
- Protein ensemble docking
- United description of receptor docking (multiple structures and conformations are overlaid and an average structure is made from conserved features)
- Conformational sampling of receptor side-chain
- Conformational sampling of combined ligand, and
- Receptor backbone energy minimization

More details on the some popular methods of the above list have been presented in the next sections.

Currently, *in silico* structure-based methods have lower throughput and poor consideration of solvation and conformational flexibility. Then, ligand-based virtual screening is considered as more robust hit generation for multi-target drugs in future development of improving of the speed and fidelity of docking-based methods. Conformational flexibility in both ligand and protein can adopt different conformations in binding the same ligand (Wong *et al.*, 2006) and the same ligand can bind to different targets in very different conformations (Smid *et al.,* 2005).

The development of search algorithms for better and efficiently sampling from conformational space of the protein–ligand and inclusion of solvation and rotational entropy contributions in the scoring function are attempts to improve the level of success of docking (Sousa *et al.,* 2006).

There are three key properties in an ideal scoring function:

1. Accuracy in predicting binding affinities
2. Speed
3. Tolerance to inaccurate ligand pose.

The first two are obvious, but the end requires some discussion. A geometric search engine attempting to generate suitable poses for a ligand given a protein binding site will create the 'correct' pose within some tolerance with some probability within some amount of time. As the tolerance increases, so does the probability of generating a good enough pose quickly. So, the degree to which a scoring function is able to compute an accurate binding affinity given an inaccurate pose will have a significant impact on its utility in such a system.

The deficiencies in the scoring function are the main limiting factor. Scoring functions in docking program routinely make some assumptions and simplifications to let a more computationally efficient assessment of ligand affinity, but logically at the cost of accuracy (Sousa *et al.*, 2006). Unfortunately, current scoring functions achieve predicting binding free energies poorly. Improved scoring functions would increase the capability to distinguish binders from non-binders. However, the enhancement of scoring functions has been stuck by the lack of a hard formalism for obtaining binding free energies from molecular docking.

COMBINED DOCKING AND MD SIMULATION

In recent years in order to improve docking performance, and have a reliable model of the complex ligand-receptor docking methods have been joined with MD simulations. Docking is applied to search for conformations of the ligands, and in the long run contains the screening of giant libraries of drug-like compounds in a short period of time. The problem is that the conformation of the receptor is not permitted to change in the binding process. Here, MD is a tool that can augment some flexibility to the receptor, ligand allow some modification upon the binding experience. Different procedures have been established for the MD insertion before or after the docking process. Multiple conformations of the receptor can be produced during the receptor structure preparation and then subjected to docking as an ensemble (Alonso *et al.*, 2006). MD simulations can help us to refine the resulting docking complexes. The dynamics can augment some flexibility to the ligand and receptor, refine their assembly and improve the intermolecular interactions between two molecules. Furthermore, the stability of the complex structure during a simulation path shows that the complex is true and it was appropriately docked. The interaction of the solvent with the complex is another significant feature of MD running after docking that the solvent molecules influence on the stability of the ligand's docked pose. MD simulations can be used to improve a docking procedure for the most of docking algorithms do not consider the contribution of the mobility of the receptor or the interactions with the molecules in the explicit solvent (Carlson, Masukawa, & McCammon, 1999).

SOFT DOCKING

Soft docking is a simpler procedure for dealing with protein flexibility in an indirect way. Despite handling the receptor as a rigid object, the repulsive expressions of the Lennard-Jones potential can be diminished by making a *soft* interaction. Thus, this method involve allowing the small molecular drug to partially penetrate the binding site surface of the protein in order to account for localized changes that could occur in a more flexible environment (Morphy & Harris, 2012).

Soft docking describes the molecular surface and volume as a "cube representation". This cube representation mentioned implicit conformational variations by way of size/shape complementarity, close packing and, most importantly, liberal steric overlap. Soft docking method containing of two steps, geometric docking and interaction scoring, is able to accommodate conformational charges associated with the molecular complexing process, and to identify the correct complex model or a small number of complex models, which can be further screened visually or based on other experimental results to recognize the correct complex model(Jiang & Kim, 1991). The soft docking concept has evolved primarily toward use in protein-protein docking (Fernández-Recio, Totrov, & Abagyan, 2002; C. H. Li, Ma, Chen, & Wang, 2003; C. H. Li, Ma, Zu Chen, & Wang, 2003; Palma, Krippahl, Wampler, & Moura, 2000; Schneidman-Duhovny et al., 2003) and protein-receptor modeling combined with experimental NMR data (X. Morelli, Czjzek, et al., 2000; X. Morelli, Dolla, et al., 2000; X. J. Morelli, Palma, Guerlesquin, & Rigby, 2001).

SIDE CHAIN FLEXIBILITY

The mobility of some residues mainly those within an enzyme active site can be managed in a different and more comprehensive tactic either during the docking process or after the ligand has been approximately placed. The receptor is generally held rigid in studies in which docking is required to be fast. However, some methods permit explicit treatment of side-chain (and a few selected backbone) torsions, e.g. by Monte Carlo (MC) (Abagyan, Totrov, & Kuznetsov, 1994) or Conformational search (CS)(Leach, 1994) procedures, and more computationally demanding molecular dynamics approaches allow binding site back-bone and side chain mobility to be simulated.

In study of receptor-ligand interactions, commonly two methods to treat side-chain flexibility during docking have been examined:

1. Systematic CS of discrete side-chain rotamers. The number of possible states is defined and therefore partition functions and free energies can be computed.
2. Stochastic MC exploration of a continuous probability distribution of side-chain rotamers.

A set of rotamer libraries can be used to discover the conformational space of selected side chains. Side-chain flexibility increases the computational cost but it lets localized protein movement and improved fit of the ligand in outcomes (Källblad & Dean, 2003).

COMBINED PROTEIN GRID

Some alternative structures of the protein receptor can be combined into a single representation of the ensemble to account for bigger conformational changes that may be critical for the binding process. The averaging can be done over atom coordinates, to generate a final average structure, or over the grid representation of all receptor conformations to produce an average docking grid.

In a similar approach, Wei *et al.* (Wei, Weaver, Ferrari, Matthews, & Shoichet, 2004) used a modified version of the program DOCK to incorporate receptor flexibility during the docking process. In this case, the interactions between a given configuration of the ligand and different flexible parts of the receptor were calculated independently. Then, those flexible regions that present the best interaction energies with the ligand were recombined into a final representation of the protein receptor.

The inclusion of protein flexibility during the docking procedure using a united description of the protein with rigid conserved regions and alternative flexible parts can improve the screening process and produce new ''hits'' in a shorter time than simple sequential docking against each protein structure. For obtaining an adequate ranking of the final complexes, the internal energy of the receptor must be taken into account, as high-energy conformations of the protein may lead to unrealistic low-energy positioning of a ligand.

Briefly, docking methods that have combined multiple protein structures, whether from NMR, X-ray complexes, or MD simulations, into a single grid representation of the target molecule have provided better results than grids from single structures. The choice of combination procedure, however, has an even greater impact on the final outcome, with weighted-average protocols providing better results than simple average combination of structures.

FREE ENERGY CALCULATIONS

For a docking process to be successful, it is necessary that both the right conformation of the ligand–receptor complex is predicted, and that the ranking of final structures will be correct. The procedure needs to be able to differentiate among similar conformations of the same system, as well as to predict the relative stability of different complexes.

As most contain empirically fitted parameters, their performance on any particular problem will depend on the set of structures used for the calibration. So far, no scoring function has proven to be reliable for every docking case tested. The main constraint on their improvement rests with the need for speed; when ranking hundreds, if not thousands, of complexes a compromise in accuracy must be made. Knowledge-based functions used in the ranking of molecular interactions may not be general and accurate enough, because of the limited number of interactions that can be inferred from crystal structures and the inadequate description of repulsive forces.

Molecular dynamics simulations present an attractive alternative for structural refinement of the final docked complexes. They incorporate flexibility of both ligand and receptor, improving interactions and enhancing complementarity between them, and thus accounting for induced fit.

Recently developed approaches called MM/PBSA and MM/GBSA improves ligand ranking and, marginally, binding affinity prediction (Yuriev, Agostino, & Ramsland, 2011; Yuriev & Ramsland,

2013). The correlation between MM/GBSA energies and experimentally determined binding affinities was found to degrade with greater variability in the affinity data and/or increased structural faults in the complex structures, either experimental or computational. These findings demonstrate that improving pose prediction methods could improve the computation of binding affinities using implicit solvation methods.

Some of drug developers have focused their attention on the enhancement of computational protocols. These improvements include: incorporation of protein flexibility in the docking process, extensive exploration of the ligand conformation within the binding site, refinement and stability evaluation of the final complexes, and estimation of the binding free energies. The refinement of docked complexes deals with the applications of MD for the optimization and validation of the final complexes.

SUMMARIZES

This chapter handles some of crucial aspects such as the lack of accuracy, speed, ligand sampling, protein flexibility and basic concepts such as movements of side chains, displacement of particular portions of the protein backbone, and mobility of metal atoms within the active site and the biased selection of ligands as a result of using ligand-bound protein structures during a docking process or mis-docking ligand. Meanwhile, it describes approximate methods such as soft docking and side chains flexibility and protein ensemble docking to tackle the risks associated with some searching methods to obtain an adequate ranking of the final complexes. The solvent effect and the direct contribution of water molecules in protein–ligand interactions, the restricted resolutions of crystallographic receptors, and obviously, protein flexibility, both in terms of intrinsic structural flexibility and in terms of conformational alterations upon ligand binding, are some of the most crucial of docking calculations (Sousa *et al.,* 2006).

Here, briefly three types of ligand sampling algorithms are discussed; shape matching, systematic search, and stochastic algorithms. Computational efficiency for each algorithm represents according to their main advantages and disadvantages.

CONCLUDING REMARKS

Molecular docking is a computational method for predicting the location of ligands in the binding sites of their receptor(s). Ruling principle of molecular docking is essential in drug discovery process. Despite considerable advances and success the widespread application of docking methods, accurate and rapid prediction of protein–ligand interactions is still a challenge area to explore and still holds several hidden weaknesses. The most docking algorithms do not explanation for receptor flexibility or the effect of explicit water molecules, MD simulations can be used to complement and improve a docking protocol. In order to improving docking performance most common of approximate methods including combining docking and MD simulation, soft docking, side chains flexibility, rotamer-based flexible docking, molecular relaxation, joined protein grid docking, protein ensemble docking, conformational sampling of receptor side-chain and conformational sampling of combined ligand are applied.

REFERENCES

Abagyan, R., Totrov, M., & Kuznetsov, D. (1994). Docking and Structure Prediction from the Distorted Native Conformation. *Journal of Computational Chemistry*, *15*(5), 488–506. doi:10.1002/jcc.540150503

Alonso, H., Bliznyuk, A. A., & Gready, J. E. (2006). Combining docking and molecular dynamic simulations in drug design. *Medicinal Research Reviews*, *26*(5), 531–568. doi:10.1002/med.20067 PMID:16758486

Ardakani, A., & Ghasemi, J. B. (2013). Identification of novel inhibitors of HIV-1 integrase using pharmacophore-based virtual screening combined with molecular docking strategies. *Medicinal Chemistry Research*, *22*(11), 5545–5556. doi:10.1007/s00044-013-0545-0

Bidon-Chanal, A., Martí, M. A., Crespo, A., Milani, M., Orozco, M., Bolognesi, M., & Estrin, D. A. et al. (2006). Ligand-induced dynamical regulation of NO conversion in Mycobacterium tuberculosis truncated hemoglobin-N. *Proteins: Structure, Function, and Bioinformatics*, *64*(2), 457–464. doi:10.1002/prot.21004 PMID:16688782

Bon, R. S., & Waldmann, H. (2010). Bioactivity-guided navigation of chemical space. *Accounts of Chemical Research*, *43*(8), 1103–1114. doi:10.1021/ar100014h PMID:20481515

Burgen, A., Roberts, G., & Feeney, J. (1975). *Binding of flexible ligands to macromolecules*. Academic Press.

Carlson, H. A., Masukawa, K. M., & McCammon, J. A. (1999). Method for including the dynamic fluctuations of a protein in computer-aided drug design. *The Journal of Physical Chemistry A*, *103*(49), 10213–10219. doi:10.1021/jp991997z

Carrillo, O., & Orozco, M. (2008). GRID-MD—A tool for massive simulation of protein channels. *Proteins: Structure, Function, and Bioinformatics*, *70*(3), 892–899. doi:10.1002/prot.21592 PMID:17803238

Chen, Y.-C. (2015). Beware of docking! *Trends in Pharmacological Sciences*, *36*(2), 78–95. doi:10.1016/j.tips.2014.12.001 PMID:25543280

Clore, G. M., & Gronenborn, A. M. (1991). Structures of larger proteins in solution: Three-and four-dimensional heteronuclear NMR spectroscopy. *Science*, *252*(5011), 1390–1399. doi:10.1126/science.2047852 PMID:2047852

Erickson, J. A., Jalaie, M., Robertson, D. H., Lewis, R. A., & Vieth, M. (2004). Lessons in molecular recognition: The effects of ligand and protein flexibility on molecular docking accuracy. *Journal of Medicinal Chemistry*, *47*(1), 45–55. doi:10.1021/jm030209y PMID:14695819

Fernández-Recio, J., Totrov, M., & Abagyan, R. (2002). Soft protein–protein docking in internal coordinates. *Protein Science*, *11*(2), 280–291. doi:10.1110/ps.19202 PMID:11790838

Ghasemi, J. B., & Hooshmand, S. (2013). 3D-QSAR, docking and molecular dynamics for factor Xa inhibitors as anticoagulant agents. *Molecular Simulation*, *39*(6), 453–471. doi:10.1080/08927022.2012.741235

Ghasemi, J. B., & Nouri, M. (2013). Docking and CoMFA study on novel human CCR5 receptor antagonists. *Medicinal Chemistry Research*, *22*(3), 1356–1364. doi:10.1007/s00044-012-0118-7

Ghasemi, J. B., Safavi-Sohi, R., & Barbosa, E. G. (2012). 4D-LQTA-QSAR and docking study on potent gram-negative specific LpxC inhibitors: A comparison to CoMFA modeling. *Molecular Diversity*, *16*(1), 203–213. doi:10.1007/s11030-011-9340-3 PMID:22127637

Gozalbes, R., Simon, L., Froloff, N., Sartori, E., Monteils, C., & Baudelle, R. (2008). Development and experimental validation of a docking strategy for the generation of kinase-targeted libraries. *Journal of Medicinal Chemistry*, *51*(11), 3124–3132. doi:10.1021/jm701367r PMID:18479119

Halperin, I., Ma, B., Wolfson, H., & Nussinov, R. (2002). Principles of docking: An overview of search algorithms and a guide to scoring functions. *Proteins: Structure, Function, and Bioinformatics*, *47*(4), 409–443. doi:10.1002/prot.10115 PMID:12001221

Hou, X., Du, J., Zhang, J., Du, L., Fang, H., & Li, M. (2013). How to improve docking accuracy of AutoDock4. 2: A case study using different electrostatic potentials. *Journal of Chemical Information and Modeling*, *53*(1), 188–200. doi:10.1021/ci300417y PMID:23244516

Huang, N., & Shoichet, B. K. (2008). Exploiting ordered waters in molecular docking. *Journal of Medicinal Chemistry*, *51*(16), 4862–4865. doi:10.1021/jm8006239 PMID:18680357

Jiang, F., & Kim, S.-H. (1991). "Soft docking": Matching of molecular surface cubes. *Journal of Molecular Biology*, *219*(1), 79–102. doi:10.1016/0022-2836(91)90859-5 PMID:2023263

Jones, G., Willett, P., Glen, R. C., Leach, A. R., & Taylor, R. (1997). Development and validation of a genetic algorithm for flexible docking. *Journal of Molecular Biology*, *267*(3), 727–748. doi:10.1006/jmbi.1996.0897 PMID:9126849

Källblad, P., & Dean, P. M. (2003). Efficient conformational sampling of local side-chain flexibility. *Journal of Molecular Biology*, *326*(5), 1651–1665. doi:10.1016/S0022-2836(03)00083-4 PMID:12595271

Kitchen, D. B., Decornez, H., Furr, J. R., & Bajorath, J. (2004). Docking and scoring in virtual screening for drug discovery: Methods and applications. *Nature Reviews. Drug Discovery*, *3*(11), 935–949. doi:10.1038/nrd1549 PMID:15520816

Kleywegt, G. J., & Jones, T. A. (1994). Detection, delineation, measurement and display of cavities in macromolecular structures. *Acta Crystallographica. Section D, Biological Crystallography*, *50*(2), 178–185. doi:10.1107/S0907444993011333 PMID:15299456

Kramer, B., Rarey, M., & Lengauer, T. (1999). Evaluation of the FLEXX incremental construction algorithm for protein–ligand docking. *Proteins: Structure, Function, and Bioinformatics*, *37*(2), 228–241. doi:10.1002/(SICI)1097-0134(19991101)37:2<228::AID-PROT8>3.0.CO;2-8 PMID:10584068

Kumar, A., & Zhang, K. Y. (2013). Investigation on the effect of key water molecules on docking performance in CSARdock exercise. *Journal of Chemical Information and Modeling*, *53*(8), 1880–1892. doi:10.1021/ci400052w PMID:23617355

Le Guilloux, V., Schmidtke, P., & Tuffery, P. (2009). Fpocket: An open source platform for ligand pocket detection. *BMC Bioinformatics*, *10*(1), 168. doi:10.1186/1471-2105-10-168 PMID:19486540

Leach, A. R. (1994). Ligand docking to proteins with discrete side-chain flexibility. *Journal of Molecular Biology*, *235*(1), 345–356. doi:10.1016/S0022-2836(05)80038-5 PMID:8289255

Leach, A. R., Shoichet, B. K., & Peishoff, C. E. (2006). Prediction of protein-ligand interactions. Docking and scoring: Successes and gaps. *Journal of Medicinal Chemistry*, *49*(20), 5851–5855. doi:10.1021/jm060999m PMID:17004700

Lexa, K. W., & Carlson, H. A. (2012). Protein flexibility in docking and surface mapping. *Quarterly Reviews of Biophysics*, *45*(03), 301–343. doi:10.1017/S0033583512000066 PMID:22569329

Li, C. H., Ma, X. H., Chen, W. Z., & Wang, C. X. (2003). A soft docking algorithm for predicting the structure of antibody-antigen complexes. *Proteins: Structure, Function, and Bioinformatics*, *52*(1), 47–50. doi:10.1002/prot.10382 PMID:12784367

Li, C. H., Ma, X. H., Zu Chen, W., & Wang, C. X. (2003). A protein–protein docking algorithm dependent on the type of complexes. *Protein Engineering*, *16*(4), 265–269. doi:10.1093/proeng/gzg035 PMID:12736369

Li, Z., & Lazaridis, T. (2003). Thermodynamic contributions of the ordered water molecule in HIV-1 protease. *Journal of the American Chemical Society*, *125*(22), 6636–6637. doi:10.1021/ja0299203 PMID:12769565

Lorber, D. M., & Shoichet, B. K. (1998). Flexible ligand docking using conformational ensembles. *Protein Science*, *7*(4), 938–950. doi:10.1002/pro.5560070411 PMID:9568900

Mangoni, M., Roccatano, D., & Di Nola, A. (1999). Docking of flexible ligands to flexible receptors in solution by molecular dynamics simulation. *Proteins: Structure, Function, and Bioinformatics*, *35*(2), 153–162. doi:10.1002/(SICI)1097-0134(19990501)35:2<153::AID-PROT2>3.0.CO;2-E PMID:10223288

Mohan, V., Gibbs, A. C., Cummings, M. D., Jaeger, E. P., & DesJarlais, R. L. (2005). Docking: Successes and challenges. *Current Pharmaceutical Design*, *11*(3), 323–333. doi:10.2174/1381612053382106 PMID:15723628

Morelli, X., Czjzek, M., Hatchikian, C. E., Bornet, O., Fontecilla-Camps, J. C., Palma, N. P., & Guerlesquin, F. et al. (2000). Structural Model of the Fe-Hydrogenase/Cytochromec 553 Complex Combining Transverse Relaxation-optimized Spectroscopy Experiments and Soft Docking Calculations. *The Journal of Biological Chemistry*, *275*(30), 23204–23210. doi:10.1074/jbc.M909835199 PMID:10748163

Morelli, X., Dolla, A., Czjzek, M., Palma, P. N., Blasco, F., Krippahl, L., Moura, J.J.G., Guerlesquin, F. (2000). Heteronuclear NMR and soft docking: an experimental approach for a structural model of the cytochrome c 553-ferredoxin complex. *Biochemistry, 39*(10), 2530-2537.

Morelli, X. J., Palma, P. N., Guerlesquin, F., & Rigby, A. C. (2001). A novel approach for assesing macromolecular complexes combining soft-docking calculations with NMR data. *Protein Science*, *10*(10), 2131–2137. doi:10.1110/ps.07501 PMID:11567104

Morphy, J. R., & Harris, C. J. (2012). *Designing multi-target drugs* (Vol. 21). Royal Society of Chemistry. doi:10.1039/9781849734912

Morton, A., & Matthews, B. W. (1995). Specificity of ligand binding in a buried nonpolar cavity of T4 lysozyme: linkage of dynamics and structural plasticity. *Biochemistry, 34*(27), 8576-8588.

Olano, L. R., & Rick, S. W. (2004). Hydration free energies and entropies for water in protein interiors. *Journal of the American Chemical Society*, *126*(25), 7991–8000. doi:10.1021/ja049701c PMID:15212549

Palma, P. N., Krippahl, L., Wampler, J. E., & Moura, J. J. (2000). BiGGER: A new (soft) docking algorithm for predicting protein interactions. *Proteins: Structure, Function, and Bioinformatics*, *39*(4), 372–384. doi:10.1002/(SICI)1097-0134(20000601)39:4<372::AID-PROT100>3.0.CO;2-Q PMID:10813819

Pei, J., Yin, N., Ma, X., & Lai, L. (2014). Systems biology brings new dimensions for structure-based drug design. *Journal of the American Chemical Society*, *136*(33), 11556–11565. doi:10.1021/ja504810z PMID:25061983

Petřek, M., Košinová, P., Koča, J., & Otyepka, M. (2007). MOLE: A Voronoi diagram-based explorer of molecular channels, pores, and tunnels. *Structure (London, England)*, *15*(11), 1357–1363. doi:10.1016/j.str.2007.10.007 PMID:17997961

Reymond, J.-L., & Awale, M. (2012). Exploring chemical space for drug discovery using the chemical universe database. *ACS Chemical Neuroscience*, *3*(9), 649–657. doi:10.1021/cn3000422 PMID:23019491

Roberts, B. C., & Mancera, R. L. (2008). Ligand-protein docking with water molecules. *Journal of Chemical Information and Modeling*, *48*(2), 397–408. doi:10.1021/ci700285e PMID:18211049

Schneidman-Duhovny, D., Inbar, Y., Polak, V., Shatsky, M., Halperin, I., Benyamini, H., & Nussinov, R. et al. (2003). Taking geometry to its edge: Fast unbound rigid (and hinge-bent) docking. *Proteins: Structure, Function, and Bioinformatics*, *52*(1), 107–112. doi:10.1002/prot.10397 PMID:12784375

Searle, M. S., & Williams, D. H. (1992). The cost of conformational order: Entropy changes in molecular associations. *Journal of the American Chemical Society*, *114*(27), 10690–10697. doi:10.1021/ja00053a002

Sgrignani, J., Bonaccini, C., Grazioso, G., Chioccioli, M., Cavalli, A., & Gratteri, P. (2009). Insights into docking and scoring neuronal α4β2 nicotinic receptor agonists using Molecular Dynamics simulations and QM/MM calculations. *Journal of Computational Chemistry*, *30*(15), 2443–2454. doi:10.1002/jcc.21251 PMID:19360794

Sgrignani, J., Novati, B., Colombo, G., & Grazioso, G. (2015). Covalent docking of selected boron-based serine beta-lactamase inhibitors. *Journal of Computer-Aided Molecular Design*, *29*(5), 441–450. doi:10.1007/s10822-015-9834-7 PMID:25676821

Siegal, G., Van Duynhoven, J., & Baldus, M. (1999). Biomolecular NMR: Recent advances in liquids, solids and screening. *Current Opinion in Chemical Biology*, *3*(5), 530–536. doi:10.1016/S1367-5931(99)00004-6 PMID:10508666

Sinko, W., Lindert, S., & McCammon, J. A. (2013). Accounting for Receptor Flexibility and Enhanced Sampling Methods in Computer-Aided Drug Design. *Chemical Biology & Drug Design*, *81*(1), 41–49. doi:10.1111/cbdd.12051 PMID:23253130

Sliwoski, G., Kothiwale, S., Meiler, J., & Lowe, E. W. (2014). Computational methods in drug discovery. *Pharmacological Reviews*, *66*(1), 334–395. doi:10.1124/pr.112.007336 PMID:24381236

Smid, P., Coolen, H. K., Keizer, H. G., van Hes, R., de Moes, J. P., den Hartog, A. P., & Kruse, C. G. et al. (2005). Synthesis, Structure-Activity Relationships, and Biological Properties of 1-Heteroaryl-4-[ω-(1 H-indol-3-yl) alkyl] piperazines, Novel Potential Antipsychotics Combining Potent Dopamine D2 Receptor Antagonism with Potent Serotonin Reuptake Inhibition. *Journal of Medicinal Chemistry*, *48*(22), 6855–6869. doi:10.1021/jm050148z PMID:16250644

Sousa, S. F., Fernandes, P. A., & Ramos, M. J. (2006). Protein–ligand docking: Current status and future challenges. *Proteins: Structure, Function, and Bioinformatics*, *65*(1), 15–26. doi:10.1002/prot.21082 PMID:16862531

Spyrakis, F., Faggiano, S., Abbruzzetti, S., Dominici, P., Cacciatori, E., Astegno, A., & Luque, J. et al. (2011). Histidine E7 dynamics modulates ligand exchange between distal pocket and solvent in AHb1 from Arabidopsis thaliana. *The Journal of Physical Chemistry B*, *115*(14), 4138–4146. doi:10.1021/jp110816h PMID:21428382

van den Berg, B., Tessari, M., Boelens, R., Dijkman, R., de Hass, G. H., Kaptein, R., & Verheij, H. M. (1995). NMR structures of phospholipase A2 reveal conformational changes during interfacial activation. *Nature Structural & Molecular Biology*, *2*(5), 402–406. doi:10.1038/nsb0595-402 PMID:7664098

van der Horst, E., E. Peironcely, J., J. P. van Westen, G., O. van den Hoven, O., R. J. D. Galloway, W., R. Spring, D., & Bender, A. et al. (2011). Chemogenomics approaches for receptor deorphanization and extensions of the chemogenomics concept to phenotypic space. *Current Topics in Medicinal Chemistry*, *11*(15), 1964–1977. doi:10.2174/156802611796391230 PMID:21470175

Venkatachalam, C. M., Jiang, X., Oldfield, T., & Waldman, M. (2003). LigandFit: A novel method for the shape-directed rapid docking of ligands to protein active sites. *Journal of Molecular Graphics & Modelling*, *21*(4), 289–307. doi:10.1016/S1093-3263(02)00164-X PMID:12479928

Verdonk, M. L., Chessari, G., Cole, J. C., Hartshorn, M. J., Murray, C. W., Nissink, J. W. M., & Taylor, R. et al. (2005). Modeling water molecules in protein-ligand docking using GOLD. *Journal of Medicinal Chemistry*, *48*(20), 6504–6515. doi:10.1021/jm050543p PMID:16190776

Wahlström, R. (2009). *Validation of docking performance in the context of a structural water molecule using model system*. Academic Press.

Warren, G. L., Andrews, C. W., Capelli, A.-M., Clarke, B., LaLonde, J., Lambert, M. H., & Head, M. S. et al. (2006). A critical assessment of docking programs and scoring functions. *Journal of Medicinal Chemistry*, *49*(20), 5912–5931. doi:10.1021/jm050362n PMID:17004707

Warren, G. L., Do, T. D., Kelley, B. P., Nicholls, A., & Warren, S. D. (2012). Essential considerations for using protein–ligand structures in drug discovery. *Drug Discovery Today*, *17*(23), 1270–1281. doi:10.1016/j.drudis.2012.06.011 PMID:22728777

Wei, B. Q., Weaver, L. H., Ferrari, A. M., Matthews, B. W., & Shoichet, B. K. (2004). Testing a flexible-receptor docking algorithm in a model binding site. *Journal of Molecular Biology*, *337*(5), 1161–1182. doi:10.1016/j.jmb.2004.02.015 PMID:15046985

Weill, N., Therrien, E., Campagna-Slater, V., & Moitessier, N. (2014). Methods for docking small molecules to macromolecules: A user's perspective. 1. The theory. *Current Pharmaceutical Design*, *20*(20), 3338–3359. doi:10.2174/13816128113199990605 PMID:23947645

Wong, T. W., Lee, F. Y., Yu, C., Luo, F. R., Oppenheimer, S., Zhang, H., & Vite, G. D. et al. (2006). Preclinical antitumor activity of BMS-599626, a pan-HER kinase inhibitor that inhibits HER1/HER2 homodimer and heterodimer signaling. *Clinical Cancer Research*, *12*(20), 6186–6193. doi:10.1158/1078-0432.CCR-06-0642 PMID:17062696

Yaffe, E., Fishelovitch, D., Wolfson, H. J., Halperin, D., & Nussinov, R. (2008). MolAxis: Efficient and accurate identification of channels in macromolecules. *Proteins: Structure, Function, and Bioinformatics*, *73*(1), 72–86. doi:10.1002/prot.22052 PMID:18393395

Yuriev, E., Agostino, M., & Ramsland, P. A. (2011). Challenges and advances in computational docking: 2009 in review. *Journal of Molecular Recognition*, *24*(2), 149–164. doi:10.1002/jmr.1077 PMID:21360606

Yuriev, E., & Ramsland, P. A. (2013). Latest developments in molecular docking: 2010–2011 in review. *Journal of Molecular Recognition*, *26*(5), 215–239. doi:10.1002/jmr.2266 PMID:23526775

Zheng, M., Liu, X., Xu, Y., Li, H., Luo, C., & Jiang, H. (2013). Computational methods for drug design and discovery: Focus on China. *Trends in Pharmacological Sciences*, *34*(10), 549–559. doi:10.1016/j.tips.2013.08.004 PMID:24035675

Chapter 4
Application of Molecular Docking in Studies on the Binding Mechanism of Three Enzymes with Natural Products

Hua-jin Zeng
Zhengzhou University, China

Ran Yang
Zhengzhou University, China

Ling-bo Qu
Zhengzhou University, China & Henan University of Technology, China

ABSTRACT

Enzymes play an important role in many biologically relevant processes and are some attractive targets in the therapy and pharmaceutical research. The interaction between drugs and enzymes in vitro might account for a variety of biological processes and has attracted scientists' great interest for several decades. Investigation of the interaction can explore their mechanism of biological activities and provide useful knowledge for optimizing molecular structure of drug, prescriptions and route of administration and it can also provide the information for their bioavailability and bioactivity. In this chapter, the bindings of natural products (including flavionoids and coumarins) with three enzymes, including pepsin, hyaluronidase and acetylcholinesterase, were investigated by fluorescence spectroscopy and molecular docking. The present studies provide direct evidence at a molecular level to understand the mechanism of inhibitory effect of natural products against enzymes.

1. INTRODUCTION

Enzymes play an important role in many biologically relevant processes and are some attractive targets in the therapy and pharmaceutical research (Li, Zhang, Xu, & Ji, 2011). The nature and magnitude of drug-enzyme interaction significantly influences the biological behaviors of the drug, such as the metabolism, distribution, toxicity and the effectiveness of drug. The study on the drug-enzyme interaction is very

DOI: 10.4018/978-1-5225-0362-0.ch004

useful to explore the action mechanism and metabolic process of drug, which can benefit for providing useful knowledge for optimizing molecular structure of drug, prescriptions and route of administration, and the information for their bioavailability and bioactivity (Ma, Yin, Liu, & Xie, 2011). In recent years, several public and scientific interests have been focused on the interactions of drugs with some enzymes (Yadava, Singh, & Roychoudhury, 2013; Cui, Yang, & Li, 2015; Masood et al., 2015; Yadava, Gupta, & Roychoudhury, 2015; Shanmugaraj, Anandakumar, & Ilanchelian, 2015; Yadava, Shukla, Roychoudhury, & Kumar, 2015), including pepsin (Shen et al., 2015; Ying et al., 2015), hyaluronidase (Liu et al., 2013) and acetylcholinesterase (Sinko, Brglez, & Kovarik, 2010; Dounin, Constantinof, Schulze, Bachmann, & Kerman, 2011; Puiatti et al., 2013).

Flavonoids belong to a major group of polyphenols that possess a common diphenylpropane ($C_6C_3C_6$) skeleton. According to the variations in their heterocyclic C_3-ring, flavonoids can be categorized mainly into flavones, flavonols, flavanones, flavanol, isoflavones, chalcones, aurones and anthocyanidins (Teillet, Boumendjel, Boutonnat, & Ronot, 2008). Until now, more than 8000 varieties of flavonoids have been identified in fruits, vegetables and other plant sources, many of which are responsible for the attractive colors of flowers, fruits and leaves (de Groot, & Rauen, 1998). Flavonoids exhibit extensive biological effects and broad-spectrum pharmacological activities, and intake of plants and their products which are rich in flavonoids has been associated with a reduced risk of various diseases (Hertog, Feskens, Hollman, Katan, & Kromhout, 1993; Blot, McLaughlin, & Chow, 1997; Duthie, Duthie, & Kyle, 2000). Recently, several scientists have found that some flavonids can inhibit the activity of enzymes in some biological processes and can be used as inhibitors of these enzymes, such as xanthine oxidase (Lin, Zhang, Liao, Pan, & Gong, 2015), beta-ketoacyl acyl carrier protein synthase I(Ghalia, & Noura, 2015), human lactate dehydrogenase (Bader et al., 2015), deoxyxylulose phosphate reductoisomerase (Tritsch, Zingle, Rohmer, & Grosdemange-Billiard, 2015), carbonyl reductase 1 (Arai et al., 2015) and aldose reductase (Utpal, Tanusree, Debanjan, & Sudhan, 2015). Flavonoids can bind to enzymes according to different mechanisms owing to their multiple binding possibilities: hydrogen bonding, hydrophobic interactions, metal chelation and π-π stacking, with a tendency to occupy hydrophobic pocket (Dangles, & Dufore, 2005). However, to the best of our knowledge, little concerns were placed on the binding of flavonoids to pepsin and hyaluronidase and their effect on the activity of these enzymes.

The general methods for the investigation of interaction between drug and biomacromolecules *in vitro* mainly include fluorescence (Hui, Quan, Jian, Jian, & Ming, 2008; Yang, Hu, Fan, & Shen, 2008; Stan et al., 2009), fourier transform infrared (FT-IR) (Qin, Xie, & Liu, 2007; Li, Yao, Jin, Chen, & Hu, 2007), UV absorption (Gentili, Ortica, & Favaro, 2008) and circular dichroism (CD) spectra (Mahesha, Singh, Srinivasan, & Rao, 2006), nuclear magnetic resonance (NMR) (Richard, Lefeuvre, Descendit, Quideau, & Monti, 2006), mass spectrometry (MS) (Liu, Wang, Cai, & Lee, 2008), capillary electrophoresis (CE) (Lu, Ba, & Chen, 2008), electrochemistry (EC) (Lin et al., 2008), etc. Each method can characterize the mechanism of drug-protein interactions on one or several aspects. FT-IR and CD spectra are usually used to analyze protein conformation. NMR is the primary analytical tool to obtain structural information relating to protein-ligand binding sites, dissociation and binding constants at atomic level and protein conformation changes induced by complex formation. However, the NMR instrument is expensive and complicated and is not suitable for most scientists. MS is emerging as powerful tools for studying non-covalent interactions including protein interactions with drugs, metal ions, or other peptides in recent years. But MS does not provide direct structural data. Although high sensitivity and small amounts of samples have been obtained in CE and EC methods, a drawback of these methods appears to be that they need some special experimental conditions with poor repeatability. In contrary, due to its simplification, reliability and sensitivity, fluorescence spectroscopy seems to be the most common technique used to study the interaction of drug with biomacromolecules. As the tryptophan and tyrosine in protein/enzyme have intrinsic fluorescence, the fluorescence quenching of these amino acid residues by drugs can provide rich information for understanding the interaction mechanism (Ma et al., 2011). Some binding parameters including binding constant, binding number, binding distance and conformation change can be easily obtained. However, the obtained knowledge through fluorescence method is just the chemical characters, which is limited for us to know the probable situation of the interaction. With the development of computer technology, computational approaches are adopted to take part in the study of the ligand-receptor interactions. The molecular docking approach can be used to model the interaction between a small molecule and a protein at the atomic level, which allow us to characterize the behavior of small molecules in the binding site of target proteins as well as to elucidate fundamental biochemical processes. A successful docking methodology must be able to correctly predict the native ligand pose within the receptor binding site and the associated physical-chemical molecular interactions. Generally, the basic tools of a docking methodology include a search algorithm and an energy scoring function for generating and evaluating ligand poses. Search algorithms are used to explore the free energy landscape to find the best ligand poses. If the thermodynamics of the system, i.e. the enthalpic and entropic effects, are correctly modeled by the energy function, the global minimum of the energy landscape will correspond to the experimental receptor-ligand conformation, i.e. the native binding mode, and local minima will correspond to alternative binding modes. Unfortunately, considering the entropic effects is not straightforward, and current docking methods employ rough approximations. Therefore, it is not guaranteed that the global minimum associated with the energy landscape investigated by a docking methodology will correspond to the native binding mode. The scoring functions used evaluate the quality of these docking poses, guiding the search methods towards relevant ligand conformations. A scoring function must be able to distinguish the experimentally observed binding modes—associating them with the lowest energy values of the energy landscape—from all the other poses explored by the search algorithm (pose prediction). The second goal of the scoring functions is to properly classify active and inactive ligands (virtual screening). The third and most critical issue is that they predict the affinity constants and correctly rank several compounds according to their potency (binding affinity estimation) (Yuriev, Holien, & Ramsland, 2014). Through these two steps, we could predict the interaction mode and obtain many binding information, such as interaction site, possible binding amino acids and corresponding binding force, which are very useful for us to interpret the chemical characteristic obtained by these experimental methods. In this case, molecular docking and fluorescence method often are combined to investigate the molecular interactions.

In recent years, discovery the lead compound from the natural products has become a hot research. Study on the interaction of the active components with the action objects is a very effective method to screen out the star molecule with high activities. As the mention above, a number of investigations have revealed that many enzymes are the targets for the therapeutically active of these natural products. So, in this chapter, the fluorescence method and molecular docking were combined to investigate the interactions between some enzymes (pepsin, hyaluronidase and acetylcholinesterase) and some natural active compounds. The binding characteristic and molecular mechanism were investigated in details. The obtained knowledge is very helpful for us to explore the biological behaviors of these natural compounds and it may open up new avenues for the design of the most suitable active molecules with structure variants.

2. MOLECULAR DOCKING INVESTIGATION ON THE INTERACTION BETWEEN DRUGS AND PEPSIN

Pepsin is an aspartic protease that acts in food digestion in mammal's stomach. It is stored as pepsinogen so it will only be released when needed, and does not digest the body's own proteins in the stomach' lining (Campos, & Sancho, 2003; Huang et al., 2010). Its three-dimensional structure has been determined at high resolution in 1997 (Shintani, Nomura, & Ichishima, 1997). Pepsin is composed of two structurally homologous domains, an N-terminal domain (residues 1-172) and a C-terminal domain (residues 173-326). Both the two domains consist almost entirely of β-sheets (Jin et al., 2008). The active binding site of pepsin is located in the cleft between the domains, and is formed by two aspartate residues, Asp32 and Asp215. One of the two aspartate residues has to be protonated, while the other is deprotonated, for the protein to be active (Pande, Kumari, Dubey, Tripathi, & Jagannadham, 2009). The optimum pH for the enzymatic activity of pepsin is in a range from 1.7 to 2.0, but even in a range from 4.5 to 5.0, pepsin still reveals a proteolytic activity (Ciolkowski, Rozanek, Bryszewska, & Klajnert, 2013). When drug enters the human stomach, the digestive proteases may be the indirect binding targets. However, in this process, the drugs are very probable to interact with pepsin and may affect the activity of pepsin and result in some adverse effects. Therefore, in order to improve the safety of drug usage in clinical, it is very necessary for us to learn about the knowledge that whether the drug could interact with the pepsin, what the mechanism of this action was in this process and would such an interaction affect the activity of pepsin. In this section, the interactions between pepsin and some natural products have been investigated.

2.1 Molecular Docking Investigation on the Interaction Behavior between Silybin and Pepsin

Silybin (structure shown in Figure 1) is the major biologically active component of an extract from the seed of the milk thistle (*Silybum marianum* (L.) Gaertn) known as silymarin (Gazak et al., 2010). Silybin possesses hepatoprotective ability and has been widely used as a natural remedy in therapy of various acute and chronic liver diseases as well as in protection of the liver from numerous hepatotoxins and mycotoxins (Flora, Hahn, Rosen, & Benner, 1998; Schümann et al., J., 2003; Gazák, Walterová, & Kren, 2007). However, based on the data on file with the manufacture of silybin, some adverse effects, including vomiting, hiccup singultation and so on, would occur in some patients even if they took the normal dosage (about 3.5-7.0 mg/kg). The reason for this might be the digestive damage caused by silybin. Pepsin is the most importantly responsible enzyme for the digestion. In this case, the interaction of silybin and

Figure 1. Molecular structure of silybin

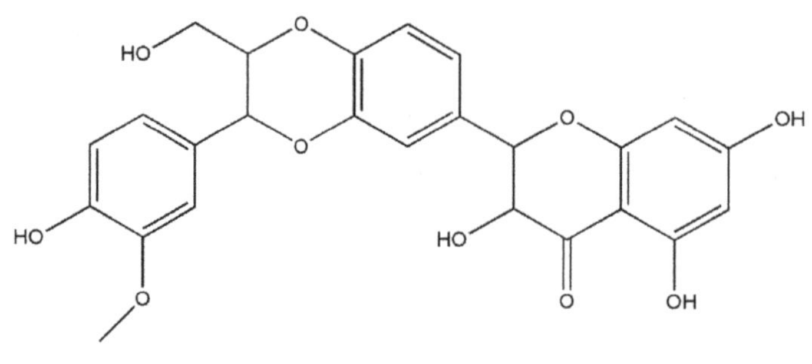

pepsin in this section was studied by fluorescence quenching and molecular docking. According to the results, the molecular mechanism for its side effects was also explored.

In order to investigate the interaction between silybin and pepsin, the fluorescence emission spectra were recorded in the range of 300-500 nm upon excitation at 280 nm. The effect of silybin on pepsin fluorescence intensity is shown in Figure 2. It is obvious that pepsin has a strong fluorescence emission peak at 344 nm after being excited with a wavelength of 280 nm, while silybin emitted little fluorescence under the same experimental conditions. When a fixed concentration of pepsin was titrated with different amount of silybin, a remarkable intrinsic fluorescence decrease of pepsin was observed. This result revealed that the silybin quenched the intrinsic fluorescence of pepsin.

Since the intrinsic fluorescence of pepsin can be quenched by silybin, the quenching mechanism was further explored. As usual, quenching mechanisms are divided into dynamic quenching and static quenching. Since higher temperature results in larger diffusion coefficients, the dynamic quenching constants will increase with increasing temperature. In contrast, the increase of temperature is likely to result in decreased stability of complexes, thus the values of the static quenching constants are expected to be smaller (Zhang et al., 2009; Zeng, Hu, You, Yang, & Qu, 2015). In order to distinguish static quenching from dynamic quenching by the temperatures, fluorescence tests were operated at different temperature. The quenching constant at each temperature was obtained using Stern-volmer equation (Yang et al., 2012):

$$F_0/F=1+K_{sv}[Q]=1+K_q\tau_0[Q] \quad (1)$$

where F_0 and F are the fluorescence intensity in the absence and presence of the quencher, [Q] is the concentration of the quencher, τ_0 is the fluorescence lifetime in the absence of quencher and its value is 10^{-8} s (Lakowicz, & Weber, 1973), K_q is the quenching rate constant of the biological macromolecule and K_{sv} is the Stern-Volmer quenching constant. Hence, Equation (1) was applied to determine K_{sv} and K_q by

Figure 2. Effect of silybin on pepsin fluorescence

Conditions: *$C_{Pepsin}=2.5\times10^{-5}$ mol·L^{-1}; silybin ($\times10^{-6}$ mol·L^{-1}):(a) 0, (b) 4.0, (c) 8.0, (d) 12.0, (e) 16.0, (f) 20.0, (g) 25.0, (h) 30.0, (i) 35.0, (j) 40.0; (k): silybin only, 40.0$\times10^{-6}$ mol·L^{-1}; T=293K.*

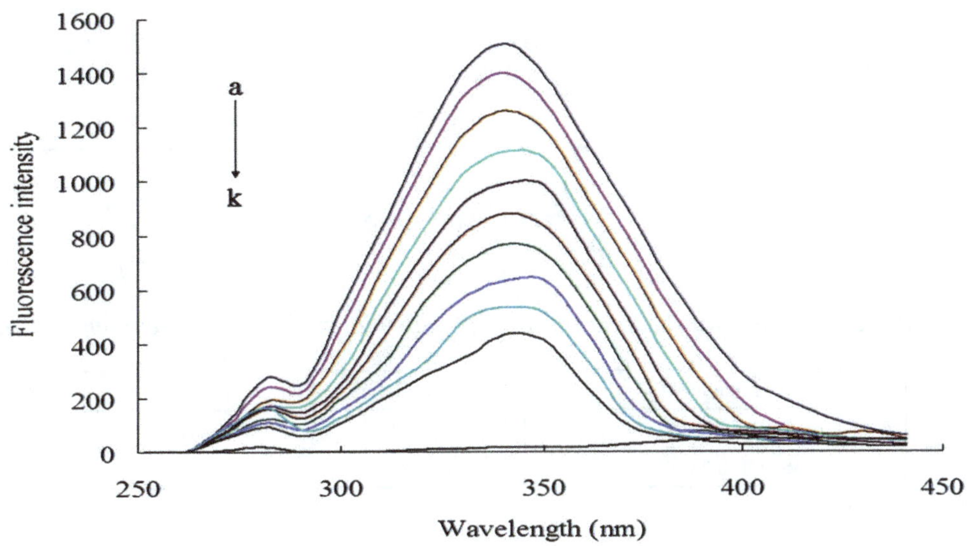

linear regression of F_0/F-1 versus [Q]. Figure 3 shows the Stern-Volmer plots for the pepsin fluorescence quenching by the silybin and the calculated K_{sv} and K_q values are summarized in Table 1. The K_{sv} values decrease with increasing temperature, indicating that the probable quenching process is static quenching mechanism rather than dynamic quenching mechanism. Moreover, the maximum dynamic quenching constant K_q of the various quenchers is 2.0×10^{10} L mol^{-1} s^{-1} (Chi, Liu, & Zhang, 2010). However, the values of K_q at 293 K and 310 K are greater than 2.0×10^{10} L mol^{-1} s^{-1}. Thus, the results indicated that the overall quenching was dominated by a static quenching mechanism, and in this process a silybin-pepsin complex was formed.

For the static quenching interaction, the binding constant (K_a) and the number of binding sites (n) can be determined when small molecules bind independently to a set of equivalent sites on a protein (Liu, Xi, Chen, Xu, & Zeng, 2008). The relationship between the fluorescence intensity and the quenching medium can be deduced as the following equation (Bi et al., 2004):

$$\log[(F_0\text{-}F)/F]=n\cdot\log K_a-n\cdot\log\{1/[[Q_d]\text{-}(F_0\text{-}F)[Q_p]/F_0]\} \quad (2)$$

Figure 3. Sterm-Volmer plots for the quenching of pepsin by silybin at different temperatures (n=3)

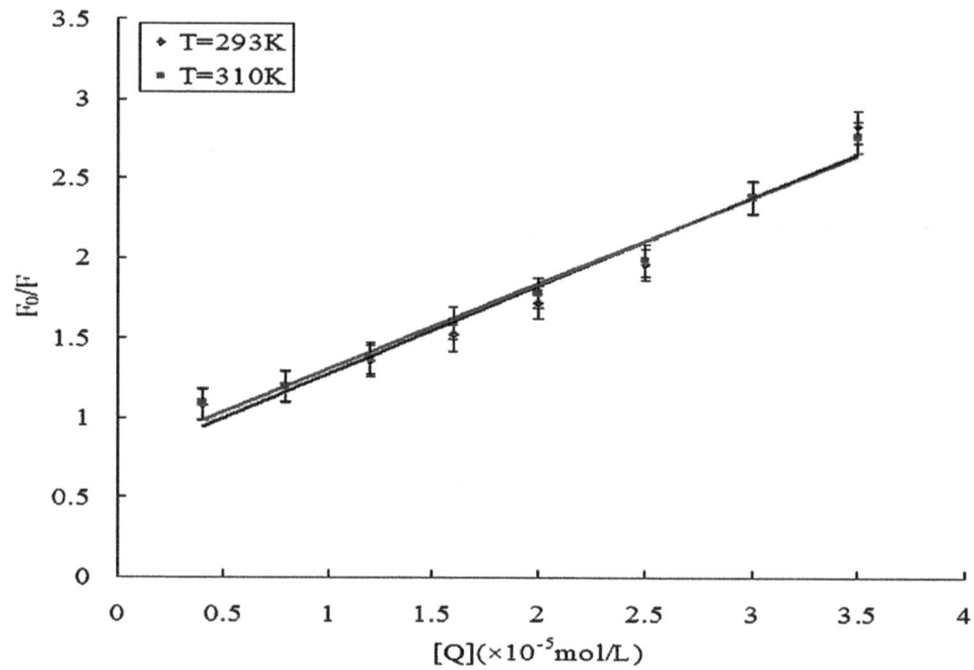

Table 1. Stern-Volmer constants for the interaction of pepsin with silybin at different temperatures (n=3)

T (K)	Equations	K_{sv} (L mol^{-1})	K_q (L mol^{-1})	R[a]	SD[b]
293	F_0/F =0.5510[Q]+0.7218	5.51×10^4	5.51×10^{13}	0.9830	0.12
310	F_0/F =0.5364[Q]+0.7628	5.36×10^4	5.36×10^{13}	0.9906	0.09

[a]The correlation coefficient.
[b]The standard deviation.

where $[Q_d]$ and $[Q_p]$ are the concentrations of drug molecule and protein, respectively. K_a is the binding constant. By plot of $\log(F_0\text{-}F)/F$ versus $\log\{1/[[Q_d]\text{-}(F_0\text{-}F)[Q_p]/F_0]\}$ (shown in Figure 4), the number binding sites n and binding constant K_a of the interaction between silybin and pepsin can be calculated and the results are summarized in Table 2. From Table 2, it can be found that the values of n at the experimental temperatures were approximately equal to one, indicating that the existence of just a single binding site in pepsin for silybin. The value of K_a is 8.2155×10^4 L mol^{-1} in room temperature, indicating that a strong interaction exists between silybin and pepsin.

As is well known, there are four types of interactions between small molecule ligands and biological macromolecules: hydrophobic forces, hydrogen bonds, van der Waals' interactions and electrostatic forces (Shi, Zhu, Wang, Chen, & Shen, 2013). The noncovalent interaction forces between proteins and small molecules can be determined by thermodynamic parameters. Ross and Subramanian have summed up the thermodynamic laws to determine the types of binding with various interactions (Ross, & Subramanian,

Figure 4. Double-log plots of silybin quenching effect on pepsin fluorescence at different temperatures (n=3)

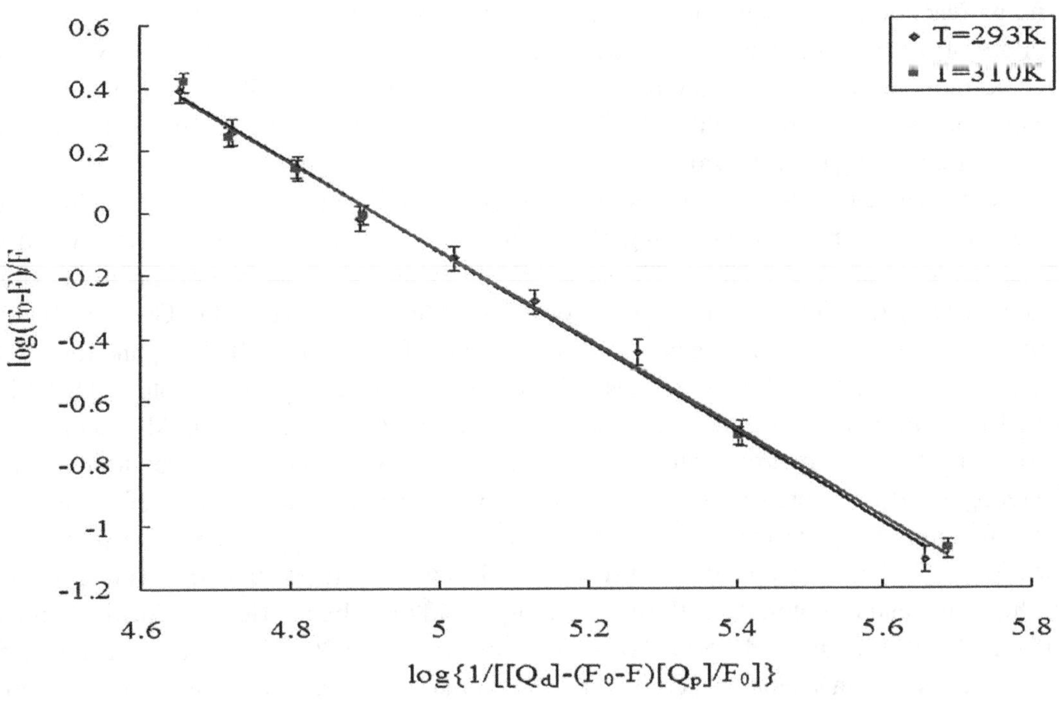

Table 2. The binding constant K_a and relative thermodynamic parameters of the silybin-pepsin system (n=3)

T (K)	K_a (L mol^{-1})	n	R [a]	SD [b]	ΔH (kJ mol^{-1})	ΔG (kJ mol^{-1})	ΔS (J mol^{-1} K^{-1})
293	8.2155×10^4	1.4454	0.9977	0.03	0.663	-27.57	93.71
310	8.1953×10^4	1.4219	0.9956	0.05	0.663	-29.16	93.71

[a]The correlation coefficient.
[b]The standard deviation.

1981). If $\Delta H°<0$ and $\Delta S°<0$, van der Waals and hydrogen bond interactions play the main roles in the binding reaction. If $\Delta H°>0$ and $\Delta S°>0$, hydrophobic interactions are dominant. If $\Delta H°<0$ and $\Delta S°>0$, the main force is an electrostatic force. The enthalpy change ($\Delta H°$) and entropy change ($\Delta S°$) can be calculated by the Van Hoff equations:

$$\ln(K_2/K_1) = (1/T_1 - 1/T_2)(\Delta H / R) \quad (3)$$

$$\Delta G° = -RT\ln K \quad (4)$$

$$\Delta G° = \Delta H° - T\Delta S° \quad (5)$$

where K_1 and K_2 are the binding constants (analogous to K_a in Equation (2)) at T_1 and T_2, and R is the universal gas constant. When the change of temperature is small, ΔH can be considered a constant, and can be approximated from Equation (3). The free-energy change ($\Delta G°$) and the entropy change ($\Delta S°$) of the binding reaction follow Equations (4) and (5), respectively (Khan et al., 2008).

The values of the thermodynamic parameters were $\Delta H°=0.663$ kJ mol^{-1}, $\Delta G°=-27.57$ kJ mol^{-1} and $\Delta S°=93.71$ kJ mol^{-1} at ambient temperature (shown in Table 2). Negative $\Delta G°$ means that the interaction process was spontaneous and the positive $\Delta H°$ and positive $\Delta S°$ indicated that hydrophobic interactions play the major role during the interaction.

In order to further determine the preferred binding sites of silybin on pepsin and the binding structure of silybin and pepsin, the molecular docking of silybin with pepsin was carried out using Autodock 4.0 (Zeng, You, Liang, Yang, & Qu, 2014). The structure of silybin was generated by Chem 3D Ultra and optimized by density functional theory at B3lyp/6-31+g (d) level implemented in Gaussian 03 until all egienvalue of the Hesssian matrix were positive. With the aid of AutoDock 4.0, the ligand root of silybin was detected and rotatable bonds were defined. The crystal structure of pepsin (protein ID: 5PEP) was downloaded from Protein Data Bank (http://www.rcsb.org/pdb/home/home.do). All water molecules were removed and the polar hydrogen and the Gasteiger charges were added at the beginning of docking study. To recognize the binding sites in pepsin, docking was carried out with setting of grid box size 90 Å×100 Å×90 Å along x, y, z axes covering whole protein with Kollman charges. The grid center was set at –17.802 Å, 40.376 Å and 86.848 Å. At first, AutoGrid was run to generate the grid map of various atoms of the ligand and receptor. After the completion of grid map, ligand flexible docking simulations were performed with 100 runs and 2,500,000 energy evaluations, 27000 numbers of generations, 50 GA populations and root mean square cluster tolerance 2.0 Å per run. Finally, the lowest energy conformation was used for docking analysis.

From the docking simulation the observed free energy change of binding (ΔG) for the silybin-pepsin complex was found to be -24.55 kJ mol^{-1}, which was not extremely close to the experimental data (-27.57 and -29.16 kJ mol^{-1}). A possible explanation may be that the X-ray structure of the protein from crystals differs from that of the aqueous system in this study. As usual, the total binding free energy can be divided into electrostatic energy and nonelectrostatic energy, such as hydrophobic, polar and hydrogen bond. As shown in Figure 5, the silybin was located in the hydrophobic cavity of pepsin and was surrounded by the hydrophilic side chains, such as Ser-35, Thr-74, Thr-75, Thr-77, Tyr-189, Thr-218, Ser-219, and Thr-222. Therefore, it can be concluded that the interaction between silybin and pepsin is mainly hydro-

Figure 5. Binding mode between silybin and pepsin
A: *The cartoon ribbons of pepsin with silybin.*
B: *Molecular modeling of the interaction between silybin and pepsin. Green molecule displays silybin; broken lines display hydrogen bonds.*

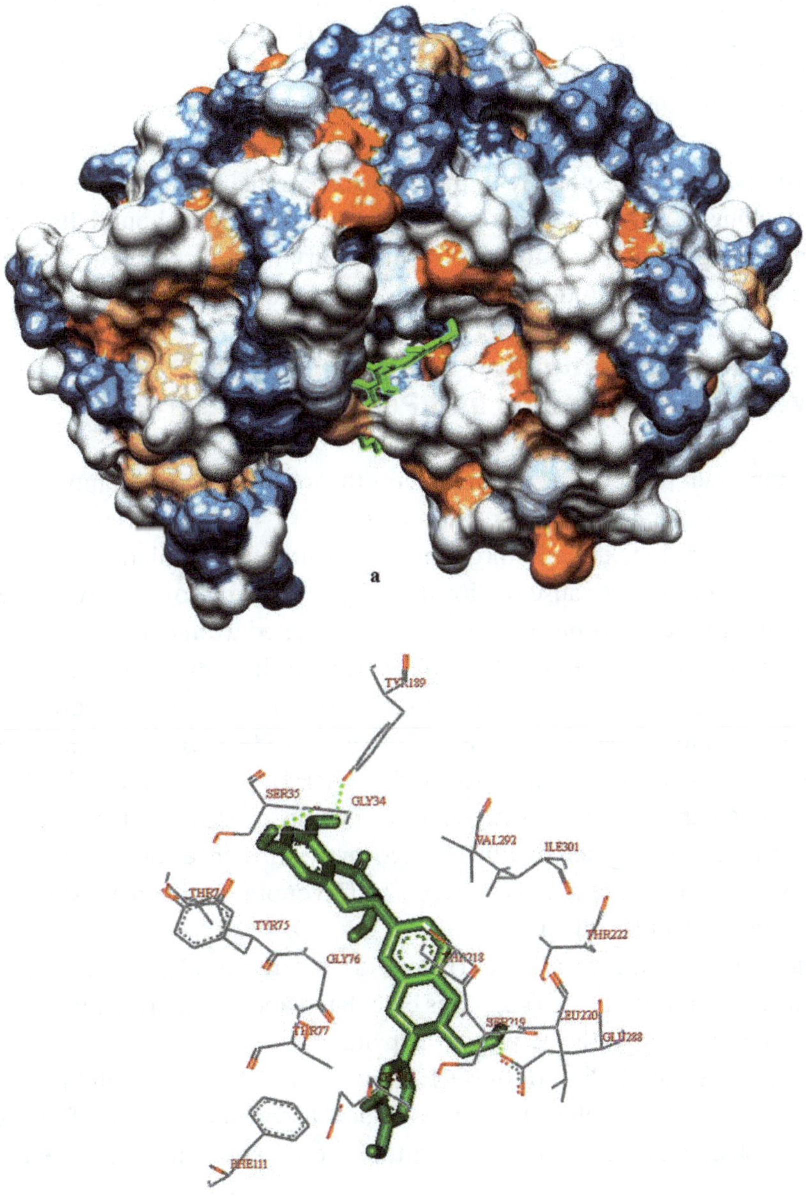

phobic interactions in nature. The result is consistent with the results obtained from the thermodynamic parameter analysis. Moreover, due to the presence of several ionic and polar groups like Gly-34, Gly-76, Phe-111, Leu-220, Glu-288, Val-292 and Ile-301 near the probe molecule, there are also considerable number of hydrogen bonding and electrostatic forces between silybin and pepsin. Molecular docking indicated that the distance of hydrogen bonding between the -OH of silybin and Gly-34, Tyr-189 and Glu288 is 2.384 Å, 2.384 Å and 2.121 Å, respectively.

In order to reveal whether silybin can affect the activity of pepsin after it enters the organism through food and drug, the effect of silybin on the pepsin activity in rats was investigated. As shown in Table 3, at a low dosage of silybin (5 mg kg^{-1}), the amount of gastric juice and the activity of pepsin were not changed obviously. However, when the dosages exceeded 10 mg kg^{-1} of silybin, the amount of gastric juice and the activity of pepsin were decreased significantly. The results indicated that the silybin can affect the secretion of gastric juice and inhibit the activity of pepsin. However, as shown in Figure 5, silybin did not cause the protonation of any of these two aspartate residues (Asp32 and Asp215), therefore, based on the docking and synchronous fluorescence results, silybin bound directly into the enzyme cavity site and the binding of silybin into the enzyme cavity influenced the microenvironment of the pepsin activity site which resulted in the reduced pepsin activity.

2.2 Molecular Docking Investigation on the Interaction Behavior between Ten Flavonoids and Pepsin

Flavonoids are the important phytonutrient components that occur in edible plants, vegetables, fruits and plant-originated foodstuffs (Graf, Mibury, & Blumberg, 2005). Therefore, flavonoids may be a class of natural compounds that people take daily through the consumption of plant food. Generally, the trace mount of flavonoids in food do not cause obviously indigestive symptoms. However, due to exhibiting broad pharmaceutical activities, several flavonoids were extracted from plants as the main component of drugs in clinic. According to published results and the data on file with the manufacture of these drugs, some adverse effects on digestion would occur in some patients even if they took the normal dosage, such as Silybin Capsules and Lpriflavone Tablets. Therefore, to further reveal the reason of indigestion caused by flavonoids, in this section the inhibitory effect of 10 flavonoids (including baicalein, apigenin, luteolin, keampferol, quercetin, morin, liquritigenin, naringenin, daidzein and genistein, structures shown in Figure 6) on pepsin was investigated *in vitro*. Moreover, to further reveal the mechanism of digestion caused by these flavonoids, the interactions between 10 flavonoids and pepsin were studied by fluorescence spectroscopy and molecular docking.

As shown in Figure 7, with the increasing of flavonoid concentration, the inhibitive rate was increased significantly and the 50% inhibitive rate (IC_{50}) was calculated and summarized in Table 4. These results indicate that flavonoids can inhibit the activity of pepsin.

Similarly, to investigate the binding of these 10 flavonoids with pepsin, the fluorescence emission spectra were recorded in the range of 300-500 nm. Under these conditions, no fluorescence spectra of flavonoid itself were observed. Figure 8 illustrated the fluorescence emission spectra of pepsin in the

Table 3. The effect of silybin on pepsin activity in rat (n=5)

Dosage (mg kg^{-1})	Amount of Gastric Juice (mL)	Pepsin Activity (μg mL^{-1} min^{-1})
Water	4.3±1.2	3.23±0.55
5.0	3.7±1.7	3.29±0.46
10.0	2.8±0.9	2.92±0.39
20.0	2.4±1.1	2.52±0.47

Figure 6. The molecular structures of the tested flavonoids

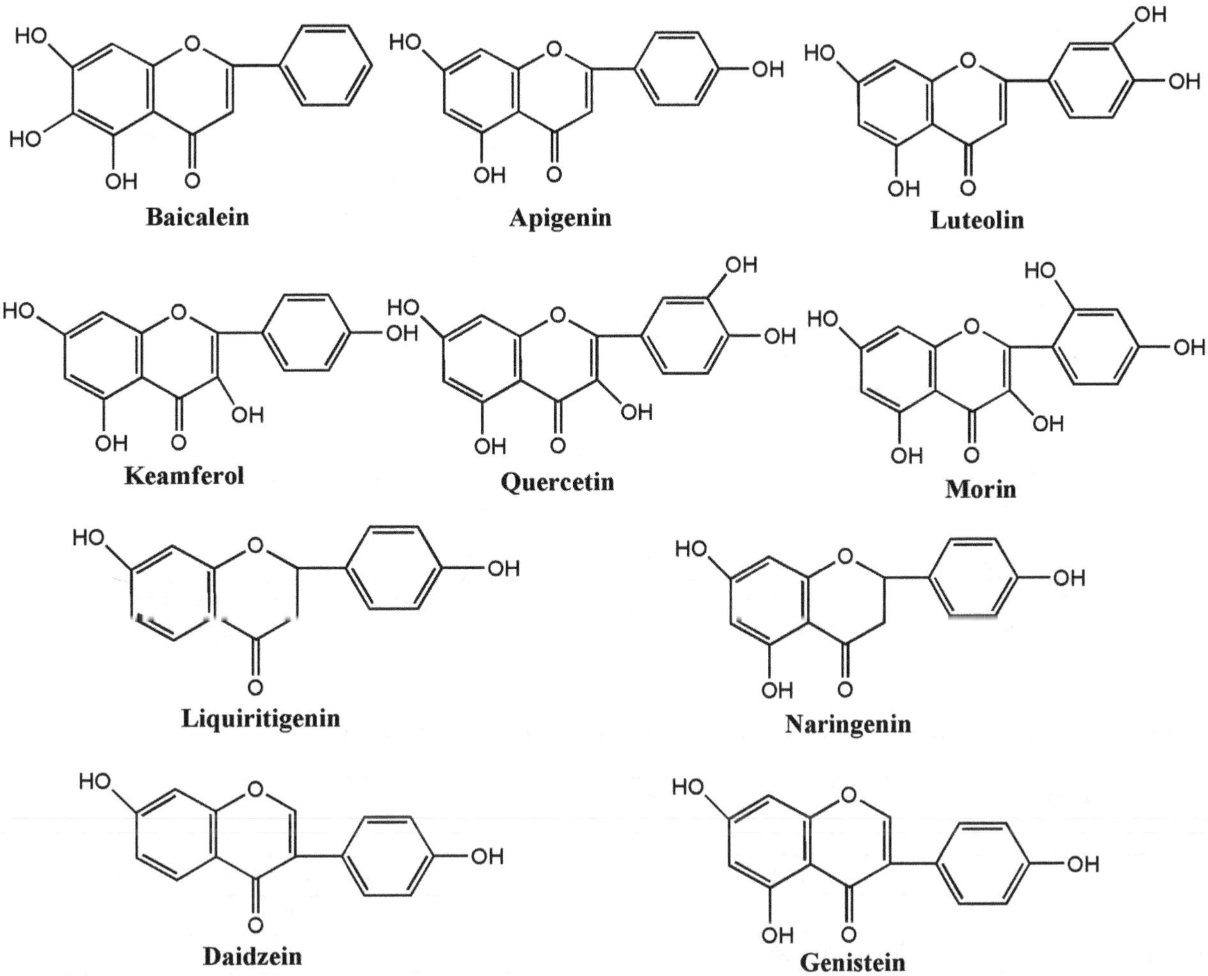

presence of various concentrations of flavonoids at 293 K. As shown in Figure 8, it was found that the addition of flavonoids not only led to a significant reduction in the fluorescence signal, but a light wavelength shift of the maximum (λ_{max}) occurred in some flavonoid-pepsin systems. Blue wavelength shifts of λ_{max} were observed in keampferol-pepsin (from 340 to 335 nm), quercetin-pepsin (from 340 to 333 nm) and morin-pepsin (from 340 to 334 nm) systems, and red wavelength shifts were discovered in baicalein-pepsin (from 340 to 355 nm), liquiritigenin-pepsin (from 340 to 345 nm) and naringenin-pepsin (from 340 to 347 nm) systems. The shift in the position of emission maximum corresponds to the changes of the polarity around the chromophore molecule, suggesting that flavonoids have formed non-covalent compounds with pepsin, which could quench the intrinsic fluorescence of pepsin.

According to Equation (1), the plots of Stern-Volmer at different temperatures were shown in Figure 9 and the K_{sv} and K_q values were presented in Table 5. From Figure 9 and Table 5, it can be found that there are good linearity relationship between F_0/F and $[Q]_{flavonoid}$ and the most of K_{sv} values decrease with increasing temperature, indicating that the probable quenching process is static quenching mechanism rather than dynamic quenching mechanism. Moreover, all K_q values are much greater than the maximum

Figure 7. The effect of flavonoids on pepsin activity in vitro

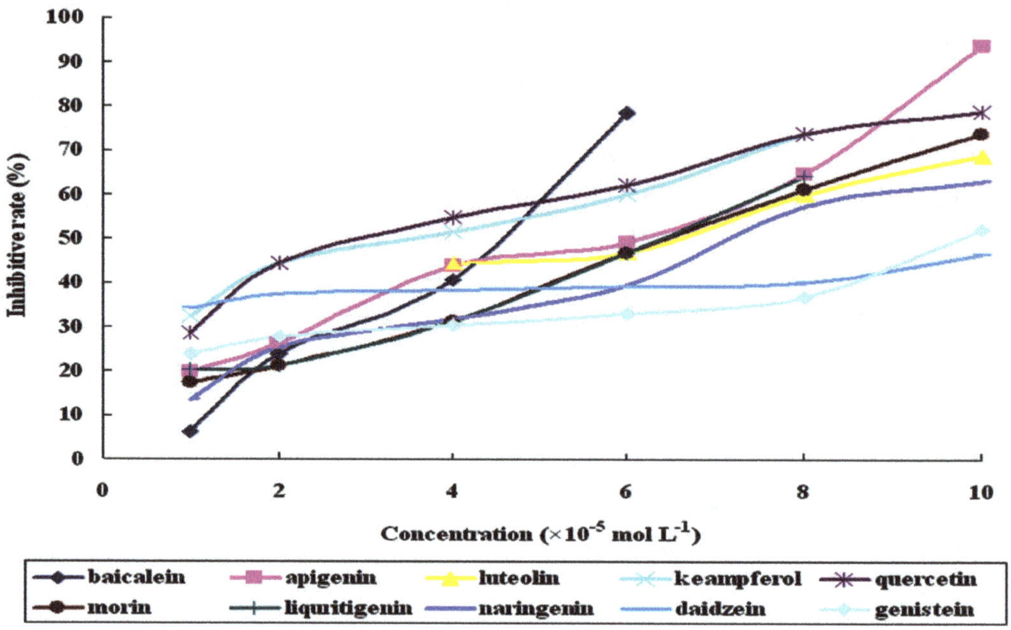

Table 4. The effect of the tested flavonoids on pepsin activity in vitro (n=3)

Flavonoid	Equation	r	IC_{50} (mol L^{-1})
Baicalein	$y=11.653x-20.971$	0.9770	6.09×10^{-5}
Apigenin	$y=7.2454x+11.541$	0.9835	5.31×10^{-5}
Luteolin	$y=4.3146x+24.756$	0.9701	5.85×10^{-5}
Kaempferol	$y=5.3807x+29.884$	0.9856	3.74×10^{-5}
Quercetin	$y=4.3931x+36.407$	0.9946	3.09×10^{-5}
Morin	$y=6.4776x+8.4415$	0.9968	6.42×10^{-5}
Liquiritigenin	$y=6.4051x+9.8018$	0.9830	6.28×10^{-5}
Naringenin	$y=5.3542x+10.792$	0.9861	7.32×10^{-5}
Daidzein	$y=0.0731x^3-1.1127x^2+5.6188x+29.599$	0.9974	10.59×10^{-5}
Genistein	$y=0.1171x^3-1.6166x^2+7.9139x+17.488$	0.9986	9.81×10^{-5}

diffusion collision quenching rate constant of pepsin with a variety of quenchers, further indicating that the quenching of pepsin in the presence of flavonoids is not caused by dymanic collision but by the formation of flavonoid-pepsin complexes.

By plot of $\log(F_0-F)/F$ versus $\log\{1/[[C_d]-(F_0-F)[C_p]/F_0]\}$ (shown in Figure 10), the number binding sites n and binding constant K_a of the interaction between pepsin and flavonoids can be calculated and the results are summarized in Table 6. From Table 6, it can be seen that the values of n at the experimental temperatures were approximately equal to one, indicating that the existence of just a single binding site in pepsin for each flavonoid during their interaction. The value of K_a is of the order of 10^4 L mol^{-1}, indicating that a strong interaction exists between each flavonoid and pepsin.

Figure 8. Effect of flavonoids on pepsin fluorescence

Conditions: *Peak from up to down* $C_{baicalein}=(0, 3.0, 6.0, 9.0, 12.0, 15.0,20.0, 25.0, 30.0, 35.0)\times10^{-6}\ mol\cdot L^{-1}$, $C_{apigenin}=(0, 3.0, 6.0, 9.0, 12.0, 15.0,18.0, 21.0, 25.0, 30.0)\times10^{-6}\ mol\cdot L^{-1}$, $C_{Luteoline}=(0, 2.0, 4.0, 6.0, 9.0, 12.0,15.0, 18.0, 20.0, 25.0)\times10^{-6}\ mol\cdot L^{-1}$, $C_{kaempferol}=(0, 5.55, 11.1, 16.65, 19.425, 22.2,27.75, 33.3, 38.85, 44.4)\times10^{-6}\ mol\cdot L^{-1}$, $C_{quercetin}=(0, 5.0, 10.0, 15.0, 20.0, 25.0,30.0, 35.0, 40.0, 45.0)\times10^{-6}\ mol\cdot L^{-1}$, $C_{morin}=(0, 5.0, 10.0, 15.0, 20.0, 25.0,30.0, 35.0, 40.0, 45.0)\times10^{-6}\ mol\cdot L^{-1}$, $C_{liquritigenin}=(0, 5.0, 10.0, 15.0, 20.0, 25.0,30.0, 35.0, 40.0, 45.0)\times10^{-6}\ mol\cdot L^{-1}$, $C_{naringenin}=(0, 3.0, 5.0, 8.0, 11.0, 14.0,17.0, 20.0, 25.0, 30.0)\times10^{-6}\ mol\cdot L^{-1}$, $C_{daidzein}=(0, 10.0, 15.0, 20.0, 25.0, 30.0,35.0, 40.0, 45.0, 50.0)\times10^{-6}\ mol\cdot L^{-1}$, $C_{genistein}=(0, 3.7, 7.4, 11.1, 12.95, 14.8,18.5, 22.2, 25.9, 29.6)\times10^{-6}\ mol\cdot L^{-1}$; $C_{pepsin}=2.510^{-5}\ mol\cdot L^{-1}$, $T=293\ K$.

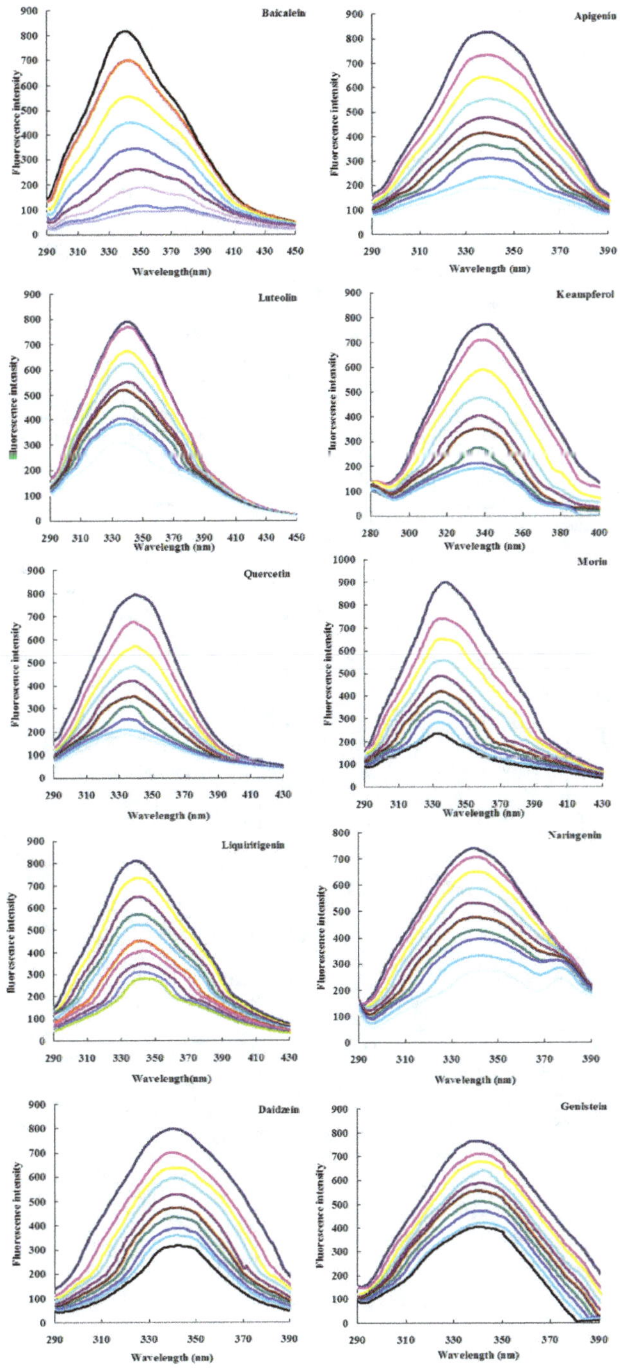

Figure 9. Sterm-Volmer plots for the quenching of pepsin by flavonoids at different temperature

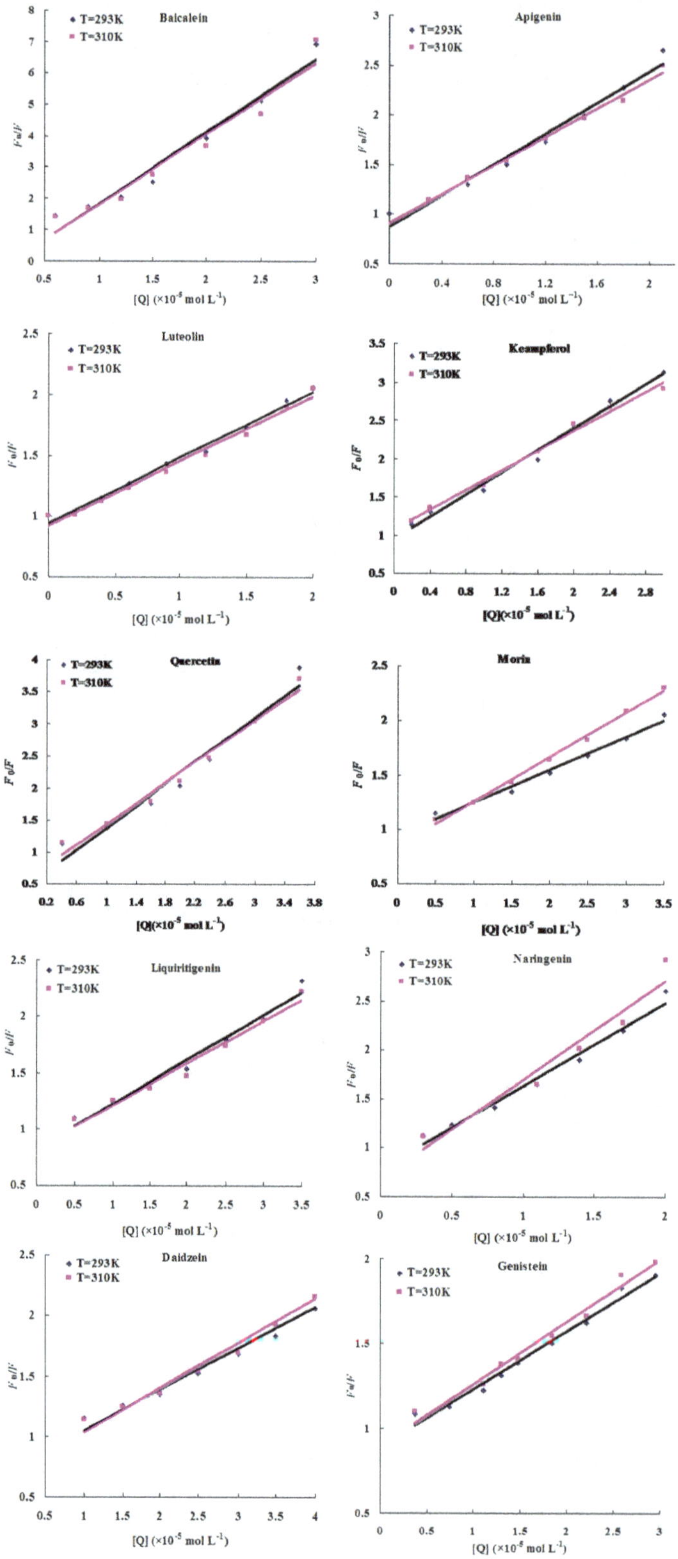

Table 5. Stern-Volmer constants for the interaction of pepsin with flavonoids at different temperatures (n=3)

Flavonoid	T (K)	Equations	K_{sv} (L mol^{-1})	K_q (L mol^{-1})	R[a]	SD[b]
Baicalein	293	F_0/F =0.2307[Q]+0.9206	2.31x10^4	2.31x10^{12}	0.9668	0.14
	310	F_0/F =0.2258[Q]+0.9186	2.26x10^4	2.26x10^{12}	0.9696	0.15
Apigenin	293	F_0/F =0.5825[Q]+0.8758	5.83x10^4	5.83x10^{13}	0.9886	0.09
	310	F_0/F =0.5203[Q]+0.9086	5.20x10^4	5.20x10^{13}	0.9956	0.05
Luteolin	293	F_0/F =0.5408[Q]+0.9421	5.41x10^4	5.41x10^{12}	0.9912	0.07
	310	F_0/F =0.5308[Q]+0.919	5.31x10^4	5.31x10^{12}	0.9840	0.10
Keampferol	293	F_0/F =0.7264[Q]+0.9415	7.26x10^4	7.26x10^{12}	0.9653	0.14
	310	F_0/F =0.6451[Q]+0.9807	6.45x10^4	6.45x10^{12}	0.9386	0.23
Quercetin	293	F_0/F =0.7187[Q]+0.6927	7.19x10^4	7.19x10^{12}	0.9774	0.03
	310	F_0/F =0.5759[Q]+0.7892	5.76x10^4	5.76x10^{12}	0.9664	0.02
Morin	293	F_0/F =0.4096[Q]+0.98371	4.10x10^4	4.10x10^{12}	0.9904	0.07
	310	F_0/F =0.3023[Q]+0.9426	3.02x10^4	3.02x10^{12}	0.9941	0.06
Liquiritigenin	293	F_0/F =0.396[Q]+0.8901	3.96x10^4	3.96x10^{12}	0.9895	0.09
	310	F_0/F =0.3721[Q]+0.8967	3.72x10^4	3.72x10^{12}	0.9835	0.12
Naringenin	293	F_0/F =0.3441[Q]+0.7936	3.44x10^4	3.44x10^{12}	0.9693	0.21
	310	F_0/F =0.3240[Q]+0.7767	3.24x10^4	3.24x10^{12}	0.9797	0.26
Daidzein	293	F_0/F =0.3061[Q]+0.8322	3.06x10^4	3.06x10^{12}	0.9816	0.10
	310	F_0/F =0.3282[Q]+0.8055	3.28x10^4	3.28x10^{12}	0.9778	0.12
Genistein	293	F_0/F =0.3212[Q]+0.9289	3.21x10^4	3.21x10^{12}	0.9890	0.05
	310	F_0/F =0.3453[Q]+0.9314	3.45x10^4	3.45x10^{12}	0.9894	0.05

[a]The correlation coefficient.
[b]The standard deviation.

The thermodynamic parameters for the interaction of pepsin with flavonoids were obtained from the linear relationship between lnK_a and the reciprocal absolute temperature, and the results were listed in Table 6. The negative sign for $\Delta G°$ means that the interactions between flavonoids and pepsin are spontaneous processes. The electrostatic interactions are the main driving force in the binding of pepsin with baicalein, apigenin, luteolin, keampferol, quercetin, morin, liquritigenin and naringenin and hydrophobic interactions are playing major roles in daidzein and genistein binding to pepsin. However, as shown in Table 6, the value of $\Delta H°$ in quercetin-pepsin and naringenin-pepsin systems are close to zero, indicating that hydrophobic interaction also play a very important role in the formation of these two complexes. Meanwhile, the different amounts of hydroxyls exist in molecular structure of flavonoids, thus the hydrogen bonding formation might also participate in the binding processes.

To identify the precise binding sites on pepsin, a molecular modeling investigation was performed to simulate the binding mode between pepsin and flavonoids. From the docking calculation, the lowest energy-ranked results of flavonoid-pepsin conformations were summarized in Table 7. It can be seen from Table 7 that the observed free energy change of binding for flavonoid-pepsin systems was not extremely close to the experimental data from Table 6. The reason for this result may also be the differences between the X-ray structure of the pepsin in crystal and that of the aqueous system. The exact binding sites of flavonoids on pepsin with the lowest binding free energy were presented in Figure 11.

Figure 10. Double-log plots of flavonoid quenching effect on pepsin fluorescence at different temperature

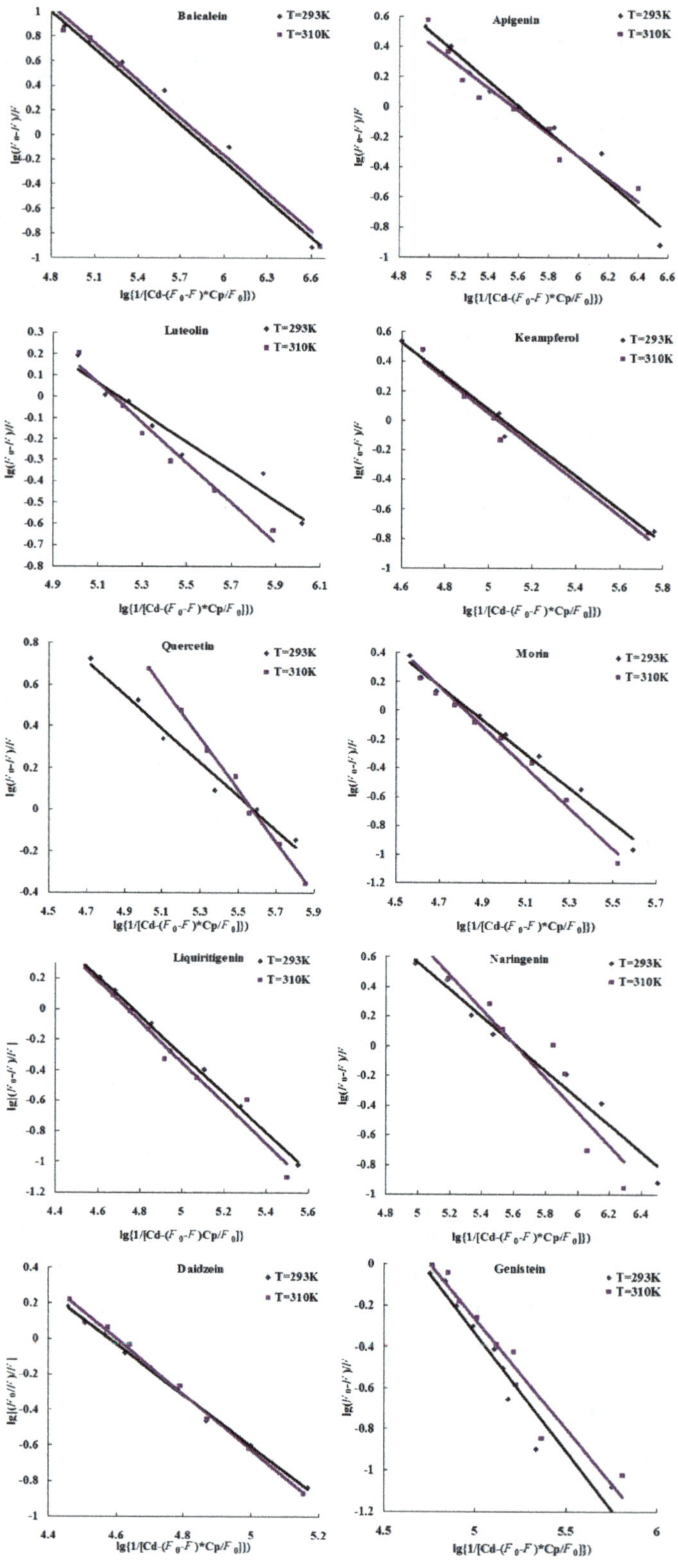

Table 6. The binding constant K_a and relative thermodynamic parameters of the flavonoid-pepsin systems (n=3)

Flavonoid	T (K)	K_a (L mol^{-1})	*n*	R [a]	SD[b]	$\Delta H°$ (kJ mol^{-1})	$\Delta G°$ (kJ mol^{-1})	$\Delta S°$ (J mol^{-1} K^{-1})
Baicalein	293	5.81×10^4	0.99	0.9736	0.15	-1.19	-33.58	108.69
	310	5.48×10^4	0.97	0.9713	0.19		-34.86	
Apigenin	293	6.91×10^4	1.02	0.9822	0.09	-1.90	-31.41	73.40
	310	6.54×10^4	1.21	0.9311	0.18		-32.66	
Luteolin	293	10.45×10^4	0.82	0.9561	0.10	-2.69	-29.64	90.45
	310	9.48×10^4	1.09	0.9680	0.10		-30.68	
Keampferol	293	13.60×10^4	1.22	0.9927	0.06	-2.41	-27.14	85.31
	310	13.19×10^4	1.21	0.9959	0.11		-28.58	
Quercetin	293	8.45×10^4	0.91	0.9945	0.05	-0.31	-28.58	105.73
	310	7.78×10^4	0.90	0.9811	0.08		-29.54	
Morin	293	4. 18×10^4	1.18	0.9923	0.05	-2.41	-27.14	85.31
	310	4.16×10^4	1.41	0.9894	0.07		-28.58	
Liquiritigenin	293	3.97×10^4	1.27	0.9982	0.03	-2.61	-26.72	82.27
	310	3.77×10^4	1.33	0.9863	0.07		-28.12	
Naringenin	293	5.58×10^4	0.93	0.9869	0.08	-0.55	-32.03	105.63
	310	5.09×10^4	1.02	0.9499	0.08		-33.29	
Daidzein	293	4.18×10^4	1.56	0.9983	0.02	2.28	-25.67	95.00
	310	4.16×10^4	1.43	0.9969	0.03		-27.29	
Genistein	293	5.71×10^4	1.06	0.9685	0.09	4.09	-26.40	104.06
	310	5.06×10^4	1.14	0.9669	0.10		-28.70	

[a]The correlation coefficient.
[b]The standard deviation.

Table 7. The lowest energy-ranked results of flavonoid-pepsin binding conformations

Flavonoid	Binding Energy (kcal mol^{-1})	Inhib Constant (μM)	Ligand Efficiency	Internal Energy
Baicalein	-6.53 (-27.33 kJ mol^{-1})	16.36	-0.33	-6.7
Apigenin	-6.57 (-27.50 kJ mol^{-1})	15.40	-0.33	-6.93
Luteolin	-6.39 (-26.75 kJ mol^{-1})	20.68	-0.3	-7.5
Keampferol	-6.57 (-27.50 kJ mol^{-1})	15.37	-0.31	-7.72
Quercetin	-6.66 (-27.88 kJ mol^{-1})	13.18	-0.3	-8.1
Morin	-6.13 (-25.66 kJ mol^{-1})	32.34	-0.28	-7.09
Liquiritigenin	-5.86 (-24.53 kJ mol^{-1})	50.26	-0.31	-6.41
Naringenin	-5.83 (-24.40 kJ mol^{-1})	53.31	-0.29	-6.24
Daidzein	-5.9 (-24.70 kJ mol^{-1})	47.17	-0.31	-6.45
Genistein	-5.98 (-25.03 kJ mol^{-1})	41.51	-0.3	-6.81

Figure 11. Docked pose corresponding to the minimum energy conformation for flavonoid binding to pepsin cavity. Detailed illustration of the amino acid residues lining the binding site in the pepsin cavity. Green molecule displays flavonoids; broken lines display hydrogen bonds.

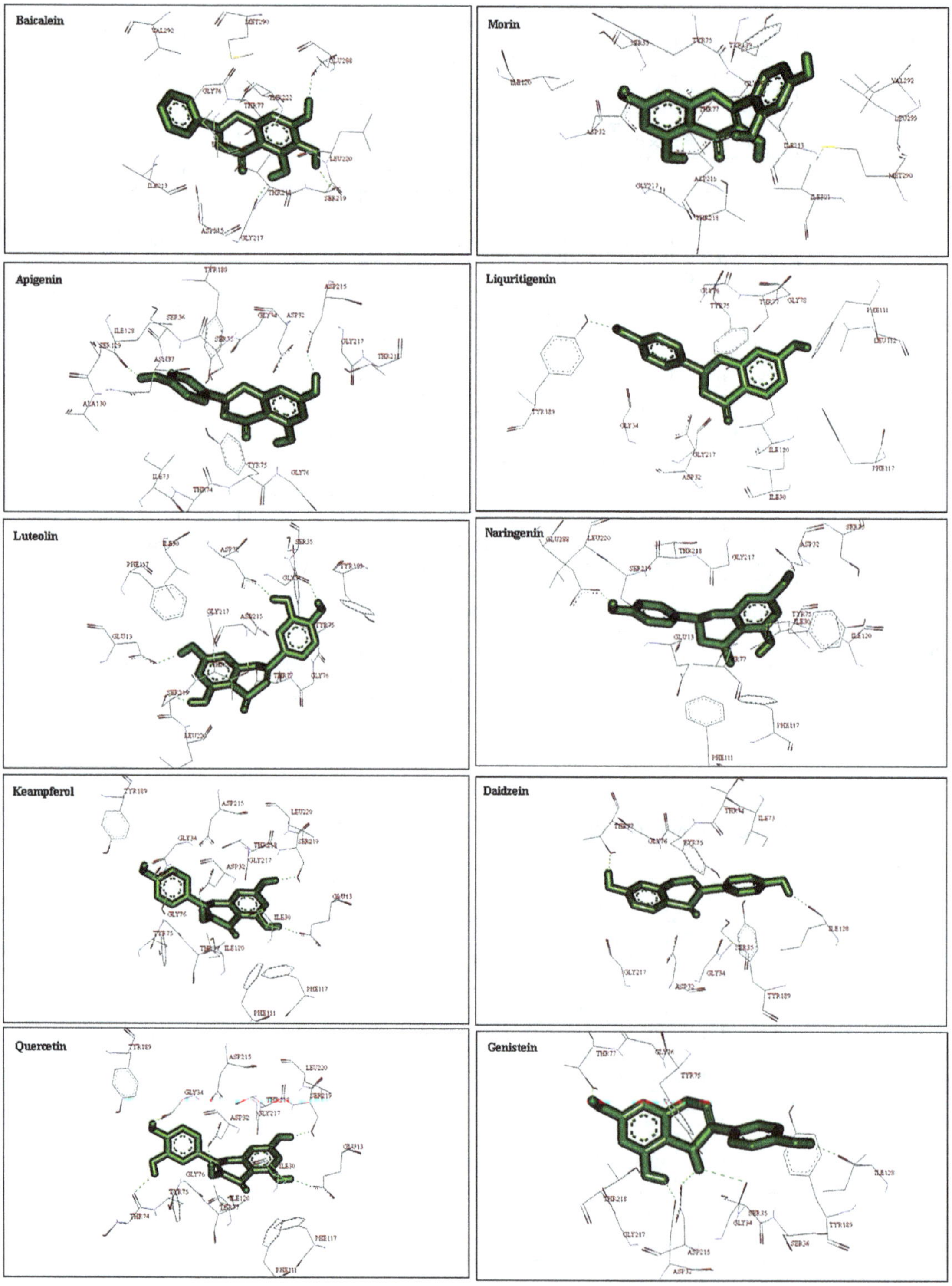

As shown in Figure 11, all flavonoids were located in the hydrophobic cavity of pepsin and were surrounded by the hydrophobic and hydrophilic amino acid residues (shown in Table 8). Therefore, it can be concluded that the interactions between flavonoids and pepsin were mainly electrostatic forces and hydrophobic interactions in nature, which were consistent with the results obtained from the thermodynamic parameter analysis.

As mentioned above, the catalytic site of pepsin was formed by Asp32 and Asp215. As shown in Table 8, both two residues appeared in baicalein-, apigenin-, luteolin-, keampferol-, quercetin- and genistein-pepsin systems, however, only Asp32 residue appeared in morin-, liquritigenin-, naringenin- and daidzein-pepsin systems. These results can be explained why the binding constant of the former six systems was larger than that of the latter four systems in the binding processes.

In addition, due to the presence of several ionic and polar groups near the flavonoid, there was also considerable number of hydrogen bonds in flavonoid-pepsin systems. Therefore, hydrogen bonding was also an important force in the binding process. As shown in Table 8, in some flavonoid-pepsin systems one of the hydrogen bonds was formed with the Asp32 and/or Asp215 residues, which might be associated with the activity of pepsin.

Table 8. The amino acid residues lining the binding site in pepsin cavity and hydrogen bonds between flavonoid and pepsin

Flavonoids	Amino Acid Residues Lining the Binding Site		Hydrogen Bond
	Hydrophobic Amino Acid	Hydrophilic Amino Acid	
Baicalein	Gly76, Ile213, Gly217, Ser219, Leu220, Met290, Val292, Ile301,	Thr77, Asp215, Thr218, Thr222, Glu288,	Gly217(1.9977 Å), Ser219(1.9447 Å), Glu288(1.6709 Å)
Apigenin	Gly34, Asn37, Ile73, Tyr75, Gly76, Ile128, Ser129,Ala130, Tyr189, Gly217,	Asp32, Ser35, Ser36, Thr74, Asp215, Thr218,	Ile128(1.9095Å), Asp215(2.0511 Å)
Luteolin	Ile30, Gly34, Tyr75, Gly76, Phe117, Tyr189, Gly217, Leu220	Glu13, Asp32, Ser35, Thr77, Asp215, Thr218, Ser219,	Glu13 (2.1360Å), Asp32 (1.6854Å), Gly34 (1.9878Å), Ser219 (2.1041Å),
Keampferol	Ile30, Gly34, Gly76, Tyr75, Phe111, Phe117, Ile120, Tyr189, Gly217,Leu220,	Glu13, Asp32, Ser35, Thr77, Asp215, Thr218, Ser219,	Glu13 (1.9260Å), Asp32 (2.1851Å), Gly34 (1.8082Å), Asp215 (1.9501Å),
Quercetin	Ile30, Gly34, Tyr75, Gly76, Phe111, Phe117, Ile120, Tyr189, Gly217, Leu220,	Glu13, Asp32, Thr74, Thr77, Asp215, Thr218, Ser219,	Glu13 (2.1371Å), Asp32 (2.3384Å), Gly34 (1.9804Å), Ser219 (2.0050Å),
Morin	Tyr75, Gly76, Ile120, Tyr189, Ile213, Gly217, Met290, Val292, Leu299, Ile301	Asp32, Ser35, Thr77, Asp215, Thr218,	Asp32 (1.7936Å), Thr77 (2.1695Å), Asp215 (1.7527Å)
Liquritigenin	Ile30, Gly34, Tyr75, Gly76, Gly78, Phe111, Leu112, Phe117, Ile120, Tyr189, Gly217	Asp32, Thr77,	Tyr189 (2.2666Å)
Naringenin	Ile30, Tyr75, Phe111,Phe117, Ile120, Gly217, Ser219, Leu220	Glu13, Asp32, Ser35, Thr77, Thr218, Glu288,	Asp32 (1.8544Å), Glu288 (1.7079Å)
Daidzein	Gly34, Ile73, Tyr75, Gly76, Ile128, Tyr189, Gly217	Asp32, Ser35, Thr74, Thr77,	Gly76 (2.1137Å), Thr77 (1.8501Å), Ile128 (1.8483Å)
Genistein	Gly34,Tyr75, Gly76, Ile128, Tyr189, Gly217,	Ser35, Ser36, Asp32, Thr77, Asp215, Thr218,	Asp32 (2.1439Å), Gly34 (2.8586Å), Thr77 (1.8065Å), Asp215 (3.0830Å), Ile218 (1.9646Å)

3. MOLECULAR DOCKING INVESTIGATION ON THE INTERACTION BETWEEN FLAVONOIDS AND HYALURONIDASE

Initially, the inflammation is a normal response to infection and tissue injury, but without an appropriate and timely treatment it can lead to the development of chronic diseases in human bodies (Gautam, & Jachak, 2009; Yahaya, Don, & Yahaya, 2014). It was reported that there were several enzymes known to be involved in a promoting inflammatory pathway (Duthie, & Chain, 1939; Cameron, Pauling, & Leibovitz, 1979; Kakegawa, Matsumoto, Satoh, 1999). One of the most important enzymes in this process are hyaluronidase (HAase), which are a group of homologous enzymes that depolymerise hyaluronan (HA) and play an important role in the control of the size and concentration of HA chains, which modulating activity of many pathological processes strongly depends on their length (Stern, Asari, & Sugahara, 2006). The enzyme HAase and its substrate HA are known to be involved in fundamental, physiological and pathological processes such as embryological development, migration, adhesion, proliferation and differentiation of cells, immune surveillance, wound healing, angiogenesis, tumorigenesis, virulency, venom and inflammation (Escalante, Franceschi, Rucavado, & Gutierrez, 2000). Therefore, it can be deduced that the anti-inflammatory agents had a strong inhibitory effect on the activity of HAase. This result seemed to indicate that the potent HAase inhibitory substances might have anti-inflammatory effects, and could become leading compounds in the development of new anti-inflammatory drugs. On the basis of this information, recently many researchers have devoted to searching and detecting the inhibitory effects of some natural products against HAase (Lee, & Choi, 1999; Shibata et al., 2002; Furusawa, Narita, Iwai, Fukunaga, & Nakagiri, 2011; Liu et al., 2013).

Flavonoids, a group of important polyphenols that are widely present in many anti-inflammatory traditional Chinese medicines, such as *Lysimachia christinae* Hance, *Tussilage farfara* L., *Ginkgo biloba* L., *Lonicera japonica* Thunb., *Sophora japonica* L., *Bupleurum chinense* DC., and *Pueraria lobata* (willd) Ohwi., have been proved to exhibit broad pharmaceutical activities (Heim, Tagliaferro, & Bobilya, 2002; Graf et al., 2005). In recent years, many flavonoids, including apigenin (Wang et al., 2014), luteolin (Megumi, Kei, Kenji, Tadahiko, & Tadashi, 2011; Jeon et al., 2014), keampferol (Peramaiyan et al., 2014), quercetin (Yilmaz et al., 2014), morin (Galvez et al., 2001; Jung, Kim, Song, Park, & Lim, 2010), naringenin (Jayaraman, Jesudoss, Menon, & Namasivayam, 2012; Tsai et al., 2012), daidzein (Liu, Lin, Sheu, & Sun, 2009) and genistein (Verdrengh, Jonsson, Holmdahl, & Tarkowski, 2003; Ji et al., 2011), have also been shown to possess anti-inflammatory and anti-allergic activities. These results suggested that flavonoids might have potential ability to inhibit the activity of hyaluronidase. And this view has been supported by previous studies (Kuppusamy, & Das, 1991; Moon, Kim, & Lee, 2009; Lee, & Kim, 2010; Liu et al., 2013), which have proved that some flavonoids, including apigenin, luteolin, keampferol, quercetin and naringenin, had the inhibitory effect on HAase. However, these researches were limited to the enzymatic activity assay and the inhibitory mechanism of flavonoids on HAase has not been investigated. In addition, many herbal drugs, such as *Lysimachia christinae* Hance, *Tussilage farfara* L., *Ginkgo biloba* L., *Lonicera japonica* Thunb., *Sophora japonica* L., *Bupleurum chinense* DC., and *Pueraria lobata* (willd) Ohwi., have been widely used in clinic to treat various inflammations and allergies in China and have been officially listed in the China Pharmacopoeia as traditional Chinese medicines (TCMs) [China Pharmacopoeia Committee, 2010]. According to the China Pharmacopoeia, flavonoids are one of the major anti-inflammatory and anti-allergic ingredients of these herbal drugs. Therefore, it is very significant for clarifying the anti-inflammatory mechanism of traditional Chinese medicines to investigate on the interaction of flavonoids with HAase. In this part, some flavonoids with anti-inflammatory activity were selected and the interactions between them and HAase were studied by fluorescence method and molecular docking. The obtained knowledge is very helpful for us to explore their molecular mechanism of anti-inflammatory activity.

Figure 12 illustrated the fluorescence emission spectra of HAase in the absence and presence of flavonoids. As shown in Figure 12, it could be seen that the fluorescence intensities of HAase decreased regularly with an increasing concentration of flavonoids, which indicated that flavonoids can bind to and alter the structure of HAase. Moreover, conspicuous changes of peak shape in the emission spectra of HAase were also found in morin-HAase system (red shift from 338 nm to 345 nm) and naringenin-HAase system (red shift from 338 nm to 355 nm), which suggested that the microenvironment of amino acid residues was changed after addition of these two flavonoids.

Similarly, according to Equation (1) and (2), the Stern-Volmer plots and double-log plots of the fluorescence of HAase quenched by flavonoids at different temperatures were shown in Figure 13 and Figure 14, respectively. The Stern-Volmer constants and binding parameters at two different temperatures were calculated and listed in Table 9 and Table 10, respectively. As shown in Table 9, the Stern-Volmer dynamic quenching constant K_{sv} decreased with the rinsing temperature, indicating that the fluorescence quenching of HAase resulted from complex formation is predominant; while from dynamic collision can be negligible. The rate constants (K_q) for the quenching of HAase caused by flavonoids are all greater than the K_q for the scatter mechanism. This confirms that the fluorescence quenching is not the result of dynamic collision quenching, rather a consequence of static quenching. From Table 10, it can be seen that the numbers of binding sites *n* for flavonoids were all approximately to one at different temperatures, which suggested that one molecule of the protein combined with one molecule of flavonoids. The values of K_a for all flavonoids were of the order of 10^4 L mol^{-1}, indicating that a strong interaction exists between HAase and flavonoids. Moreover, by comparison of the binding constants of different flavonoids with HAase, it could be concluded that the amount and position of hydrogen-bond, double bond between C_2 and C_3 were the main structure characters that influenced their binding potencies which were consistent with the structure features that determine the inhibitory effects on HAase (Kuppusamy et al., 1991; Pessini et al., 2001). This gives a clue that the determination of protein binding to structurally related compounds is a valuable tool for identifying the groups of a drug molecule.

According to thermodynamic equations (Equations 3, 4, and 5), the values of $\Delta S°$, $\Delta H°$ and $\Delta G°$ for flavonoids binding to HAase are listed in Table 10. The negative value of $\Delta G°$ revealed that binding interactions between the flavonoids and HAase were spontaneous processes. The electrostatic interactions are the main driving force in the binding of HAase with seven flavonoids, and hydrophobic interaction was playing a major role in quercetin binding to HAase. Meanwhile, the different amounts of hydroxyls exist in molecular structure of these flavonoids, therefore, the hydrogen bonding formation might also participate in the binding processes.

In order to study the binding between flavonoid and HAase systematically, to identify the precise binding sites and the exact conformation of flavonoid at the binding sites, a docking program was run to simulate the binding mode between flavonoid and HAase (Zeng, Ma, Yang, You, & Qu, 2015). Docking calculations were carried out on a HAase model (PDB code 2PE4). To recognize the binding sites in HAase, blind docking was carried out and grid maps of 126 Å×126 Å×126 Å grid points and 0.375 Å spacing were generated. The AutoDocking parameters used were, GA population size: 100; maximum number of energy evaluations: 250000. The conformation with the lowest binding free energy was used for further analysis. From the docking calculation, the lowest energy-ranked results of flavonoid-HAase conformations and the exact binding sites of flavonoids on pepsin with the lowest binding free energy

Figure 12. Effect of flavonoids on HAase fluorescence

Conditions: *Peak from up to down* $C_{Apigenin}=(0, 4, 8, 12, 16, 20, 24)\times10^{-6}\ mol{\cdot}L^{-1}$, $C_{Luteolin}=(0, 4, 10, 16, 24, 30, 36)\times10^{-6}\ mol{\cdot}L^{-1}$, $C_{Keampferol}=(0, 2, 4, 10, 16, 20, 24, 30)\times10^{-6}\ mol{\cdot}L^{-1}$; $C_{Quercetin}=(0, 4, 10, 16, 20, 24, 30, 36)\times10^{-6}\ mol{\cdot}L^{-1}$; $C_{Morin}=(0, 4, 10, 16, 24, 30, 36)\times10^{-6}\ mol{\cdot}L^{-1}$; $C_{Naringenin}=(0, 6, 12, 18, 24, 30, 36, 42)\times10^{-6}\ mol{\cdot}L^{-1}$; $C_{Daidzein}=(0, 10, 16, 20, 30, 40, 50)\times10^{-6}\ mol{\cdot}L^{-1}$; $C_{Genistein}=(0, 4, 8, 12, 16, 22, 24)\times10^{-6}\ mol{\cdot}L^{-1}$; $C_{HAase}=0.1\times10^{-5}\ mol{\cdot}L^{-1}$, $T=293K$.

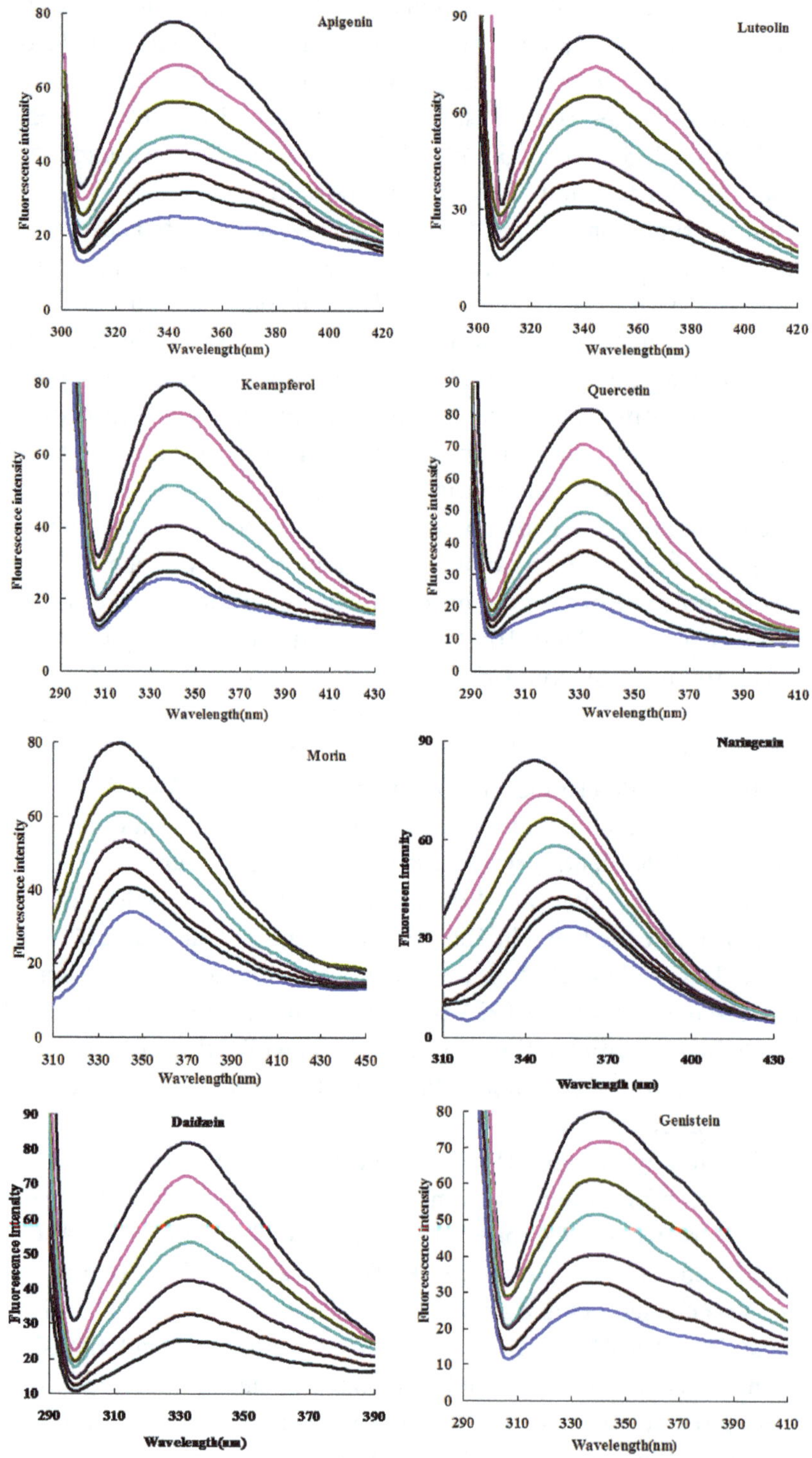

Figure 13. Sterm-Volmer plots for the quenching of HAase by flavonoids at different temperature (n=3)

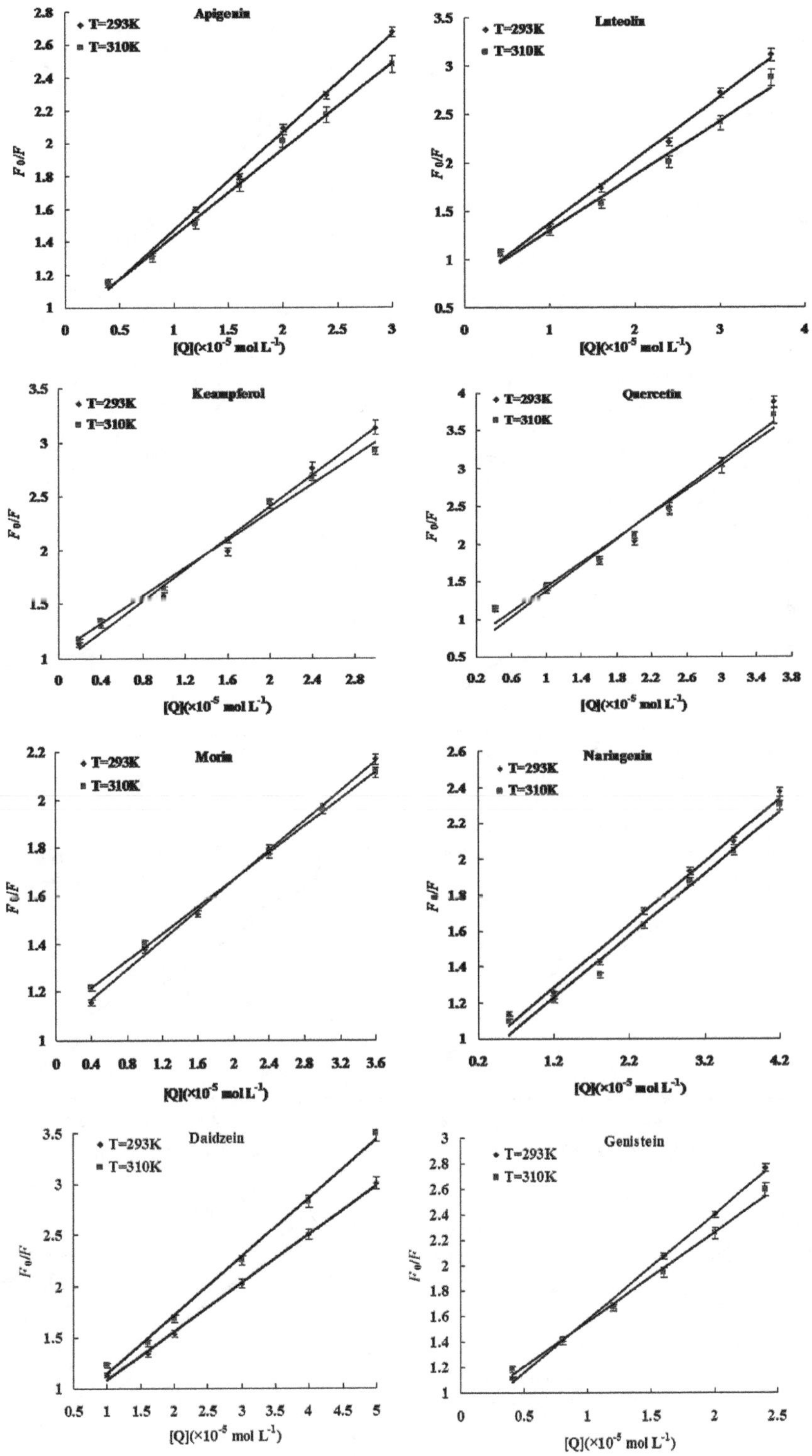

Figure 14. Double-log plots of flavonoid quenching effect on HAase fluorescence at different temperature (n=3)

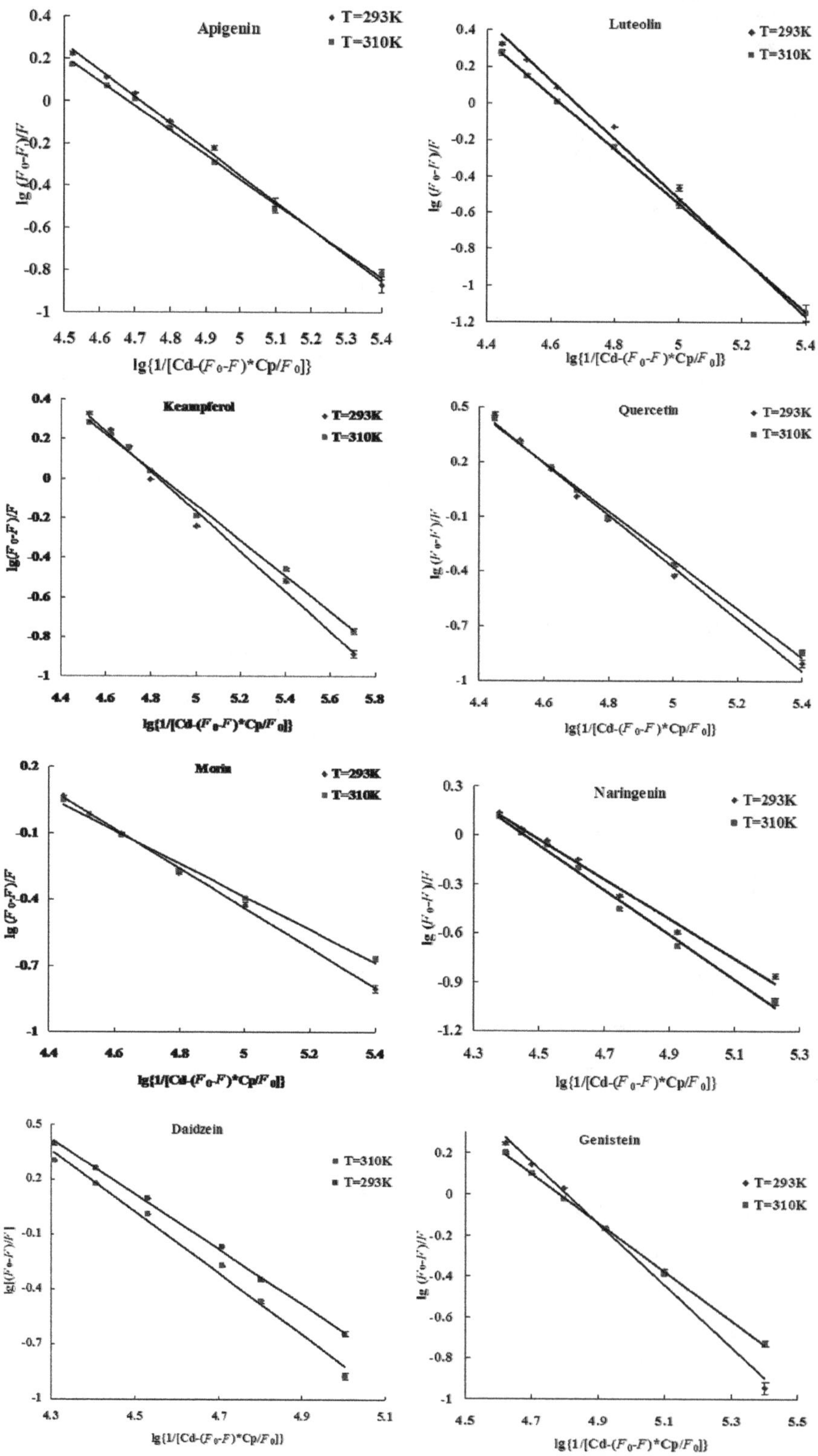

Table 9. Stern-Volmer constants for the interaction of hyaluronidase with flavonoids at different temperatures (n=3)

Flavonoid	T (K)	Equations	K_{sv} (L mol^{-1})	K_q (L mol^{-1})	R[a]	SD[b]
Apigenin	293	F_0/F =5.9640[Q]+0.8740	5.96×10^4	5.96×10^{12}	0.9991	0.09
	310	F_0/F =5.2605[Q]+0.9106	5.26×10^4	5.26×10^{12}	0.9979	0.11
Luteolin	293	F_0/F =6.5263[Q]+0.7259	6.53×10^4	6.53×10^{12}	0.9973	0.07
	310	F_0/F =5.6440[Q]+0.7416	5.64×10^4	5.64×10^{12}	0.9923	0.08
Keampferol	293	F_0/F =7.2644[Q]+0.9458	7.26×10^4	7.26×10^{12}	0.9947	0.13
	310	F_0/F =6.4506[Q]+1.0628	6.45×10^4	6.45×10^{12}	0.9955	0.15
Quercetin	293	F_0/F =8.5658[Q]+0.5285	8.57×10^4	8.57×10^{12}	0.9613	0.16
	310	F_0/F =8.0290[Q]+0.6234	8.03×10^4	8.03×10^{12}	0.9769	0.11
Morin	293	F_0/F =3.1002[Q]+1.0437	3.10×10^4	3.10×10^{12}	0.9984	0.12
	310	F_0/F =2.8055[Q]+1.106	2.81×10^4	2.81×10^{12}	0.9987	0.11
Naringenin	293	F_0/F =3.5028[Q]+0.8625	3.50×10^4	3.50×10^{12}	0.9952	0.10
	310	F_0/F =3.4442[Q]+0.8156	3.44×10^4	3.44×10^{12}	0.9937	0.11
Daidzein	293	F_0/F =5.7221[Q]+0.5714	5.72×10^4	5.72×10^{12}	0.9960	0.10
	310	F_0/F =4.7483[Q]+0.6104	4.75×10^4	4.75×10^{12}	0.9986	0.13
Genistein	293	F_0/F =8.2006[Q]+0.7607	8.20×10^4	8.20×10^{12}	0.9966	0.13
	310	F_0/F =7.0044[Q]+0.8612	7.00×10^4	7.00×10^{12}	0.9948	0.14

[a]The correlation coefficient.
[b]The standard deviation.

Table 10. The binding constant K_a and relative thermodynamic parameters of the flavonoid-hyaluronidase systems (n=3)

Flavonoid	T (K)	K_a (L mol^{-1})	n	R[a]	SD[b]	$\Delta H°$ (kJ mol^{-1})	$\Delta G°$ (kJ mol^{-1})	$\Delta S°$ (J mol^{-1} K^{-1})
Apigenin	293	5.22×10^4	1.25	0.9989	0.04	-3.89	-26.46	77.02
	310	4.78×10^4	1.16	0.9985	0.06	-3.89	-27.77	77.02
Luteolin	293	5.90×10^4	1.61	0.9961	0.09	-2.17	-26.76	83.94
	310	5.62×10^4	1.45	0.9999	0.07	-2.17	-28.19	83.94
Keampferol	293	6.82×10^4	1.01	0.9951	0.07	-1.77	-27.11	86.49
	310	6.55×10^4	0.90	0.9970	0.05	-1.77	-28.58	86.49
Quercetin	293	5.42×10^4	1.42	0.9934	0.08	1.10	-26.55	94.36
	310	5.55×10^4	1.33	0.9961	0.05	1.10	-28.16	94.36
Morin	293	3.23×10^4	0.75	0.9939	0.05	-2.82	-25.28	76.70
	310	3.03×10^4	0.90	0.9982	0.04	-2.82	-26.34	76.70
Naringenin	293	3.01×10^4	1.22	0.9947	0.09	-2.24	-25.12	78.09
	310	2.86×10^4	1.37	0.9962	0.04	-2.24	-26.45	78.09
Daidzein	293	3.82×10^4	1.49	0.998	0.06	-6.88	-25.70	87.68
	310	3.27×10^4	1.68	0.9905	0.05	-6.88	-26.79	87.68
Genistein	293	6.37×10^4	1.50	0.9916	0.07	-2.44	-26.95	91.96
	310	6.03×10^4	1.19	0.9994	0.06	-2.44	-28.37	91.96

[a]The correlation coefficient.
[b]The standard deviation.

were summarized in Table 11 and presented in Figure 15, respectively. As shown in Figure 15, all flavonoids were located in the hydrophobic cavity of HAase and were surrounded by the hydrophobic and hydrophilic amino acids (shown in Table 12). Therefore, it can be concluded that the interaction between flavonoid and HAase is mainly electrostatic forces and hydrophobic interactions in nature. However, due to more hydrophobic amino acids lining quercetin in the binding site, hydrophibic interaction was more prominent than electrostatic force in quercetin-HAase system. These results were consistent with the results obtained from the thermodynamic parameter analysis. As shown in Table 12, several amino acid residues, such as Tyr75, Asp129, Glu131, Tyr202, Tyr247, Tyr286, Gln288, Asp292 and Trp321, appeared in the binding of each flavonoid with HAase. These results indicated that these residues would play a very important role in the interaction between flavonoid and HAase and might constitute the catalytic site of HAase. Moreover, due to the presence of several ionic and polar groups, there are also considerable numbers of hydrogen bonds in flavonoid-HAase systems. As shown in Table 12, one of the hydrogen bonds is formed with the Asp129 and/or Asp292 residues in all flavonoid-HAase systems. It can be assumed from these results that the catalytic site of HAase, like pepsin (Antonov et al., 1978) and rennin (Danser, & Deinum, 2005), might be also formed by two aspartate residues (Asp129 and Asp292) and had the same activate mode, namely, one of which has to be protonated, and the other deprotonated, for the enzyme to be active.

4. MOLECULAR DOCKING INVESTIGATION ON THE INTERACTION BETWEEN COUMARINS AND ACETYLCHOLINESTERASE

Alzheimer's disease (AD) is a progressive, degenerative disorder of the brain and is the most common form of dementia among the elderly especially in industrialized countries. AD is associated with a loss of cholinergic system with decreased levels of acetylcholine in the brain areas dealing with learning, memory, behavior and emotional responses. Neuropathologically, AD is characterized by the presence of beta-amyloid plaques, neurofibrillary tangles and degeneration/atrophy of the basal forebrain cholinergic neurons (Riberson, & Harrell, 1997; Anand, Singh, & Singh, 2012). The loss of basal forebrain cholinergic cells results in reduction of synaptic availability of acetylcholine (ACh) leading to the cog-

Table 11. The lowest energy-ranked results of flavonoid-HAase binding conformations

Flavonoid	Binding Energy (kcal mol^{-1})	Ligand Efficiency	Internal Energy
Apigenin	-5.9 (-24.70 kJ mol^{-1})	-0.3	-6.73
Luteolin	-5.98 (-25.03 kJ mol^{-1})	-0.28	-7.02
Keampferol	-6.15 (-25.74 kJ mol^{-1})	-0.29	-7.14
Quercetin	-6.46 (-27.04 kJ mol^{-1})	-0.29	-7.41
Morin	-5.74 (-24.03 kJ mol^{-1})	-0.26	-6.61
Naringenin	-5.8 (-24.28 kJ mol^{-1})	-0.29	-7.06
Daidzein	-6.47 (-27.08 kJ mol^{-1}	-0.34	-7.02
Genistein	-6.66 (-27.88 kJ mol^{-1})	-0.33	-7.18

Figure 15. Docked pose corresponding to the minimum energy conformation for flavonoid binding to HAase cavity. Detailed illustration of the amino acid residues lining the binding site in the HAase cavity. Green molecule displays flavonoids; broken lines display hydrogen bonds.

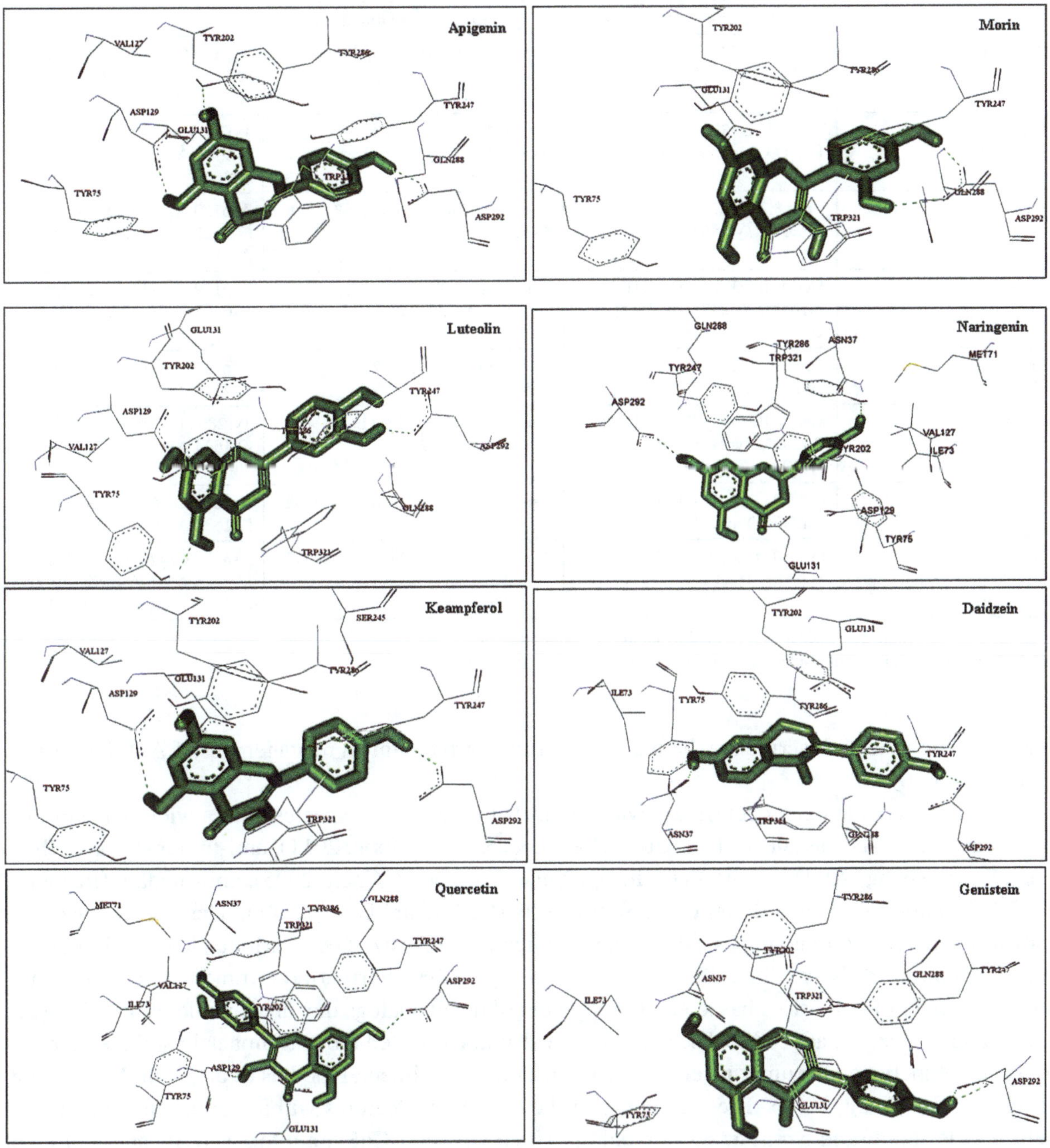

nitive impairment in AD (Bartus, Dean, Beer, & Lippa, 1982; Dunnett, & Fibiger, 1993). Accordingly, the most promising approach for the symptomatic treatment of AD is to increase the synaptic levels of ACh in the brain by inhibiting the acetylcholinesterase (AChE) enzyme, which is primarily responsible for its hydrolysis and termination of action. Therefore, AChE inhibitors, such as galanthamine, done-

Table 12. The amino acid residues lining the binding site in HAase cavity and hydrogen bonds between flavonoid and HAase

Flavonoids	Amino Acid Residues Lining the Binding Site		Hydrogen Bond
	Hydrophobic Amino Acid	Hydrophilic Amino Acid	
Apigenin	Tyr75*, Val127, Tyr202*, Tyr247*, Tyr286*, Trp321*	Asp129*, Glu131*, Gln288*, Asp292*	Asp129 (1.8624 Å), Tyr286 (2.3539 Å), Asp292 (1.9174 Å)**
Luteolin	Tyr75*, Val127, Tyr202*, Tyr247*, Tyr286*, Trp321*	Asp129*, Glu131*, Gln288*, Asp292*	Tyr75 (1.9603 Å), Tyr286 (1.9700 Å), Asp292 (1.8364 Å)**
Keampferol	Tyr75*, Val127, Tyr202*, Tyr247*, Tyr286*, Trp321*	Asp129*, Glu131*, Gln288*, Asp292*	Asp129 (1.9654 Å), Tyr286 (2.2922 Å), Asp292 (1.8954 Å)**
Quercetin	Met71, Ile73, Tyr75*, Val127, Tyr202*, Tyr247*, Tyr286*, Trp321*	Asn37, Asp129*, Glu131*, Gln288*, Asp292*	Asp129 (1.6714 Å), Tyr286 (1.6805 Å), Asp292 (2.2390 Å)**
Morin	Tyr75*, Tyr202*, Tyr247*, Tyr286*, Trp321*	Glu131*,Gln288*,Asp292*	Asp292 (1.8984 Å)**, Asp292 (2.4276 Å)**
Naringenin	Met71, Ile73, Tyr75*, Val127, Tyr202*, Tyr247*, Tyr286*, Trp321*	Asn37, Asp129*, Glu131*, Gln288*, Asp292*	Tyr286 (1.7105 Å), Asp292 (2.1644 Å)**
Daidzein	Ile73, Tyr75*, Tyr202*, Tyr247*, Tyr286*, Trp321*	Asn37, Glu131*, Gln288*, Asp292*	Asn37 (1.8764 Å), Asp292 (2.1536 Å)**
Genistein	Ile73, Tyr75*, Tyr202*, Tyr247*, Tyr286*, Trp321*	Asn37, Glu131*, Gln288*, Asp292*	Asn37 (1.7493 Å), Asp292 (2.3171 Å)**, Asp292 (2.1005 Å)**

* The common amino acid residues lining the flavonoids in the binding site.

** The common amino acid residue formating the hydrogen bond with flavonoids.

pezil, rivastigmine and tacrine are the main stay drugs for the clinical management of AD (Osborn, & Saunders, 2010).

Coumarins are naturally occurring phytochemicals in many plant species with a wide range of biological activities, such as anti-inflammatory (Lee, Lee, Kim, & Moon, 2011), anti-tumor (Huang, Shan, Zhai, Su, & Zhang, 2011), anti-HIV-1 (Huang, Yuan, Yu, Lee, & Chen, 2005), anti-oxidant (Razavi et al., 2013), anti-microbial and anti-depressant effects (Sashidhara et al., 2011). The studies have also shown that naturally occurring as well as the chemically synthesized coumarin analogs exhibit potent AChE inhibitory activity (Zhou, Wang, Wang, & Kong, 2008). Furthermore, functionalization of the aromatic center of coumarins has led to development of novel analogs that are capable of inhibiting Aβ aggregation (Piazzi et al., 2008). The studies have also documented the anti-amnestic and the memory restorative functions of coumarin derivatives in different experimental models of amnesia (Wu, Chang, Hsieh, Lin, & Ching, 2007). The recognition of key structural features within coumarin template has helped in designing and synthesizing new alalogs with improved AChE inhibitory activity and additional pharmacological activities including beta secretase inhibition associated with decreased Aβ aggregation (Kontogiorgis, Xu, Hadjipavlou-Litina, & Luo, 2007; Soto-Ortega et al., 2011). Therefore, as a promising natural drug for the cure of AD, it is very essential to learn about the inhibition mechanism of courmarins on AChE. In this section, the interaction of AChE and 13 coumarins (structures shown in Figure 16) were investigated by fluorescence approach and molecular docking. The binding capacity

Figure 16. The molecular structures of the 13 coumarins

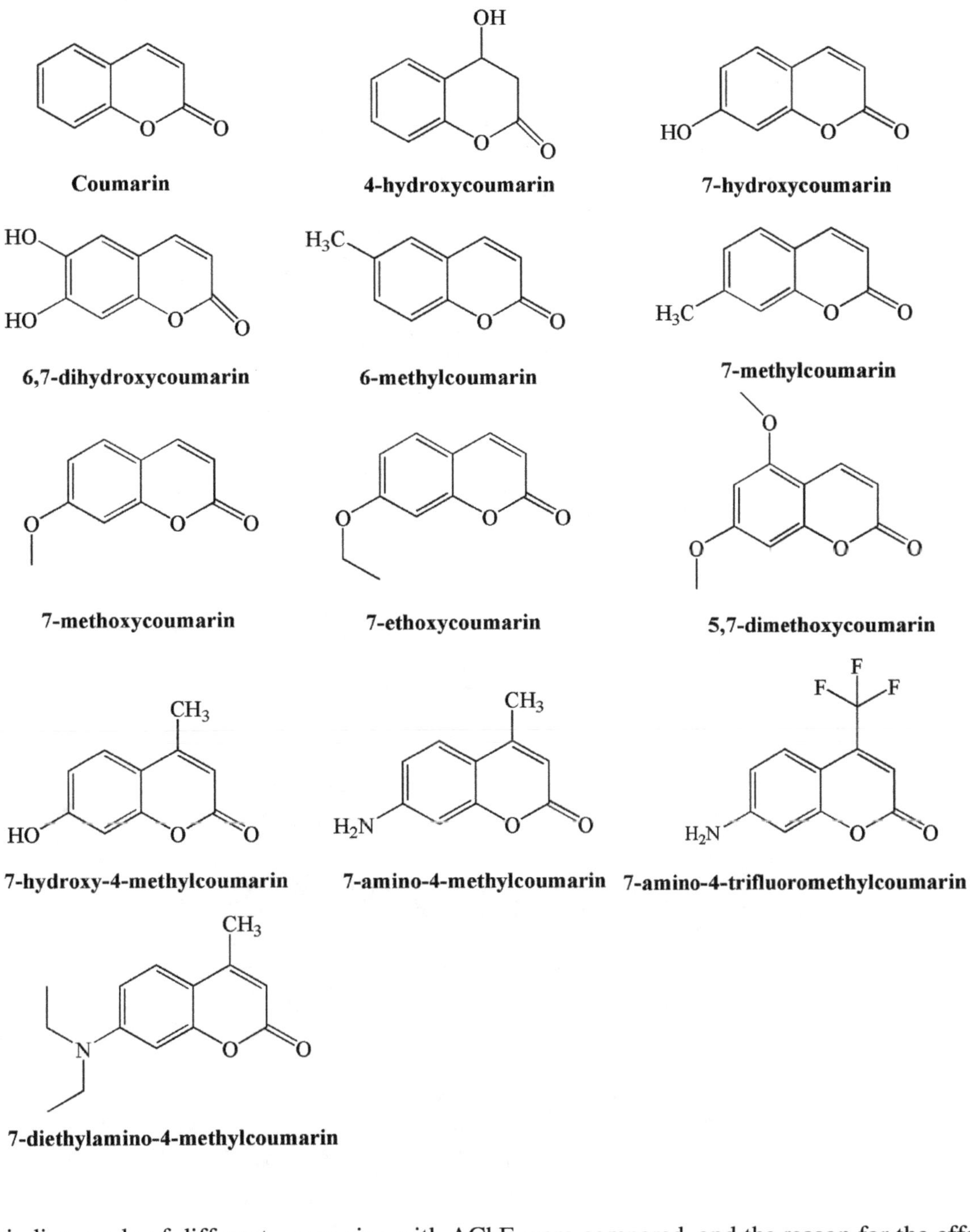

and binding mode of different coumarins with AChE were compared, and the reason for the effect of structure of coumarins on their interactions was explored. The obtained knowledge is very useful for us to understand the molecular mechanism of AChE inhibitory activity of coumarins and design the most suitable coumarins derivatives with structure variants.

The fluorescence spectra of AChE at different concentrations of coumarins were shown in Figure 17. As shown in Figure 17, it can be seen that the fluorescence intensities of AChE decreased regularly

Figure 17. Effect of coumarins on AChE fluorescence
Conditions: *Peak from up to down* $C_{coumarins}$*=0, 0.5, 1.0, 1.5, 2.0, 2.5, 3.0, 3.5, 4.0, 4.5×*10^{-5} *mol/L,* C_{AChE}*=1.0×*10^{-5} *mol·*L^{-1}*, T=292K.*

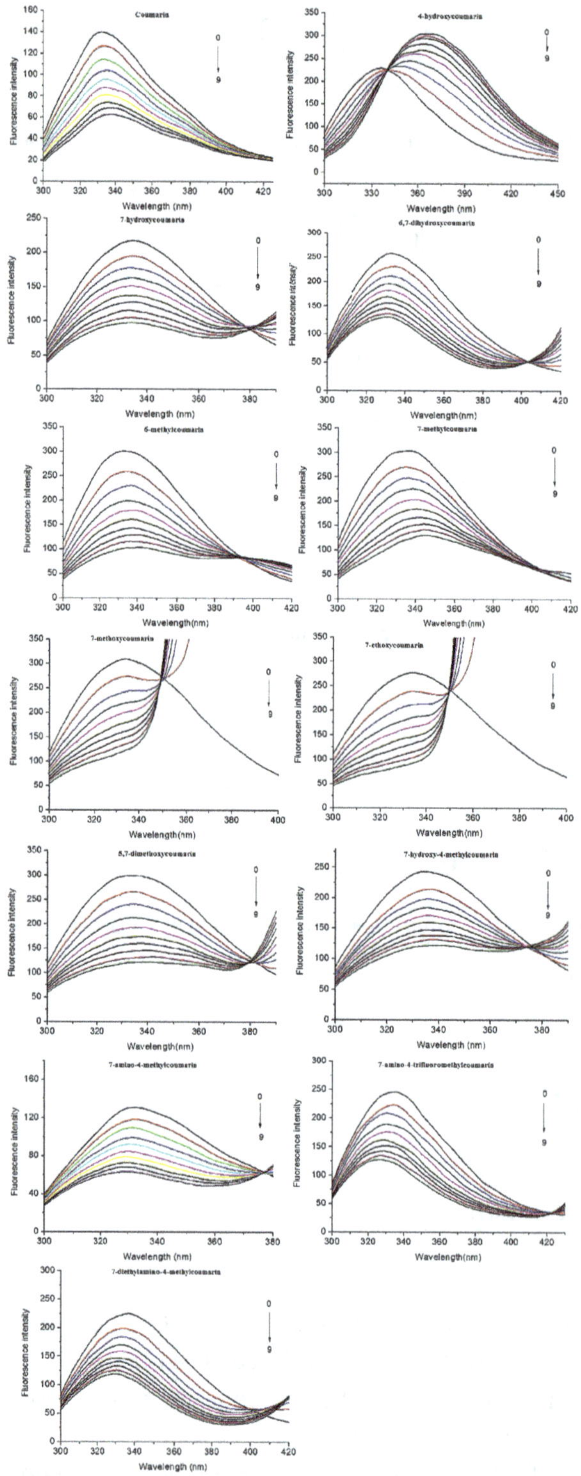

with an increasing concentration of coumarin. Furthermore, distinct changes in the emission spectra of AChE were also observed, indicating that these observations may be attributed to a strong binding of coumarins to AChE and a radiationless energy transfer between coumarins and AChE.

According to the Stern-Volmer plot and double-log plot, the binding parameters and thermodynamic parameters at different temperatures were calculated and listed in Table 13 and Table 14, respectively. It can be found from Table 13 that all of the K_{sv} values decreased with the increasing temperature and all of the K_q values were much higher than the limiting diffusion constant of the bio-molecule, indicating that the fluorescence quenching processes were mainly static quenching and the complexes between coumarins and AChE have been formed. As seen from Table 14, the numbers of binding sites *n* for all coumarins were approximately to one at different temperatures, which suggested that one molecule of the protein combined with one molecule of coumarin. The values of K_a for all coumarins were of the order of 10^4 L mol^{-1}, indicating that a strong interaction exists between coumarins and AChE.

Table 13. Stern-Volmer constants for the interaction of AChE with coumarins at different temperatures

Compound	T (K)	K_{sv} (L mol^{-1})	K_q (L mol^{-1})	r
Coumarin	292	3.22x10^4	3.22x10^{12}	0.9915
	310	3.05x10^4	3.05x10^{12}	0.9924
4-hydroxycoumarin	292	4.65x10^4	4.65x10^{12}	0.9958
	310	3.89x10^4	3.89x10^{12}	0.9956
7-hydroxycoumarin	292	2.89x10^4	2.89x10^{12}	0.9946
	310	2.60x10^4	2.60x10^{12}	0.9961
6,7-dihydroxycoumarin	292	1.95x10^4	1.95x10^{12}	0.9995
	310	1.61x10^4	1.61x10^{12}	0.9981
6-methylcoumarin	292	4.19x10^4	4.19x10^{12}	0.9939
	310	3.43x10^4	3.43x10^{12}	0.9961
7-methylcoumarin	292	2.98x10^4	2.98x10^{12}	0.9963
	310	2.44x10^4	2.44x10^{12}	0.9989
7-methoxycoumarin	292	3.76x10^4	3.76x10^{12}	0.9964
	310	3.40x10^4	3.40x10^{12}	0.9963
7-ethoxycoumarin	292	4.80x10^4	4.80x10^{12}	0.9998
	310	4.48x10^4	4.48x10^{12}	0.9949
5,7-dimethoxycoumarin	292	3.67x10^4	3.67x10^{12}	0.9976
	310	3.23x10^4	3.23x10^{12}	0.9978
7-hydroxy-4-methylcoumarin	292	2.14x10^4	2.14x10^{12}	0.9926
	310	1.72x10^4	1.72x10^{12}	0.9993
7-amino-4-methylcoumarin	292	4.51x10^4	4.51x10^{12}	0.9996
	310	4.02x10^4	4.02x10^{12}	0.9968
7-amino-4-trifluoromethylcoumarin	292	2.78x10^4	2.78x10^{12}	0.9996
	310	2.34x10^4	2.34x10^{12}	0.9990
7-diethylamino-4-methylcoumarin	292	2.51x10^4	2.51x10^{12}	0.9993
	310	1.45x10^4	1.45x10^{12}	0.9995

The thermodynamic parameters for the interaction between coumarins and AChE were calculated from the linear relationship between lnK_a and the reciprocal absolute temperature, and the results were listed in Table 14. The $\Delta G°$ at 292 K and 310 K were all negative, indicating that the binding process was spontaneous. Based on the thermodynamic view, Van der Waals or hydrogen bonding interactions were the main driving forces in the binding of AChE with coumarin, 7-diethylamino-4-methylcoumarin, 7-amino-4-trifluoromethylcoumarin, 4-hydroxycoumarin, 7-methylcoumarin, 7-methoxycoumarin, 7-ethoxycoumarin, 5,7-dimethoxycoumarin, and hydrophobic interactions were playing a major role in other coumarins binding to AChE. Like flavonoids, there are several hydroxyls exist in molecular structure of these coumarins, therefore, the hydrogen bonding formation might also participate in the binding processes.

Table 14. The binding constant K_a and relative thermodynamic parameters of the coumarin-AChE systems (n=3)

Compound	T (K)	K_a (L mol^{-1})	n	r	$\Delta H°$/(kJ·mol^{-1})	$\Delta S°$/ (J·mol^{-1}·K^{-1})	$\Delta G°$/(kJ·mol^{-1})
Coumarin	292	3.88×10^4	1.04	0.9969	-45.15	14.83	-24.62
	310	3.08×10^4	1.06	0.9953		14.83	-25.86
4-hydroxycoumarin	292	5.46×10^4	0.90	0.9937	-7.46	63.80	-26.09
	310	4.50×10^4	0.84	0.9991		63.80	-27.24
7-hydroxycoumarin	292	2.88×10^4	1.01	0.9968	-34.41	-34.34	-24.38
	310	2.69×10^4	1.09	0.9998		-34.34	-23.76
6,7-dihydroxycoumarin	292	2.11×10^4	0.93	0.9993	-64.95	-138.18	-24.60
	310	1.80×10^4	1.09	0.9997		-138.18	-22.12
6-methylcoumarin	292	4.55×10^4	1.04	0.9976	-26.82	-6.63	-24.89
	310	3.72×10^4	1.11	0.9990		-6.63	-24.77
7-methylcoumarin	292	3.20×10^4	1.05	0.9963	-9.79	50.04	-24.45
	310	2.73×10^4	1.00	0.9996		50.04	-25.35
7-methoxycoumarin	292	4.25×10^4	0.95	0.9970	-4.21	73.17	-25.58
	310	4.03×10^4	0.79	0.9900		73.17	-26.89
7-ethoxycoumarin	292	6.74×10^4	0.78	0.9948	-7.82	65.65	-26.99
	310	5.59×10^4	0.81	0.9907		65.65	-28.17
5,7-dimethoxycoumarin	292	3.55×10^4	1.06	0.9997	-13.93	34.06	-23.88
	310	3.53×10^4	1.10	0.9991		34.06	-24.49
7-hydroxy-4-methylcoumarin	292	2.29×10^4	0.85	0.9955	-52.33	-90.64	-25.86
	310	2.37×10^4	1.09	0.9985		-90.64	-24.23
7-amino-4-methylcoumarin	292	5.07×10^4	1.07	0.9975	-45.15	-69.79	-24.77
	310	4.83×10^4	1.14	0.9986		-69.79	-23.52
7-amino-4-trifluoromethylcoumarin	292	2.40×10^4	0.99	0.9995	-21.88	6.95	-23.91
	310	2.00×10^4	1.15	0.9986		6.95	-24.03
7-diethylamino-4-methylcoumarin	292	2.12×10^4	0.88	0.9998	-20.29	16.22	-25.03
	310	1.73×10^4	0.94	0.9988		16.22	-25.32

In the same way, in order to investigate the binding mechanism of coumarins with AChE and to identify the precise binding sites, molecular docking studies were carried out to simulate the binding mode between coumarins and AChE. The 3D structures of coumarins were generated in Chem 3D Ultra 8.0, and the crystal structure of AChE (PDB ID: 1QTI) was retrieved from the RCSB Protein Data Bank. To carry out docking simulations, a grid box was defined to enclose the active site with dimensions of 82 Å×90 Å×80 Å and a grid spacing of 0.375 Å. The grid maps for energy scoring were calculated using AutoGrid. Docking calculations were performed using the Lamarckian genetic algorithm and the search parameters were set to 100 times. From the docking results, the best scoring docked model (the lowest energy conformation) of a compound was chosen to represent its most favorable binding mode predicted by AutoDock.

Until now, a lot of 3D structures of AChE bound with different inhibitors have been determined by X-ray crystallography from native *Torpedo Californica Acetylcholinesterase* (TcAChE) and humans (hAChE) (Harel et al., 1993; Greenblatt, Kryger, Lewis, Silman, & Sussman, 1999; Kryger, Silman, & Sussman, 1999; Rydberg et al., 2006; Carletti et al., 2010). All these structures significantly enhance our understanding of the structural elements of AChE. The crystal structure of TcAChE revealed that its active site is buried at the bottom of a narrow gorge, about 20 Å deep (shown in Figure 18), lined with conserved aromatic residues and the active site gorge is only ~5 Å wide at a bottleneck (BN) formed by the van der Waals surfaces of Tyr121 and Phe330 (Sussman et al., 1991). Several kinetic models for AChE proposed that the binding pocket of AChE consists of two substrate-binding sites, the catalytic anionic site (CAS), near the bottom of the active site gorge, and the peripheral anionic site (PAS), which contains three principal amino acids, Trp279, Tyr70 and Asp72, near its entrance. The binding of ligands

Figure 18. Schematic description of the active site of the AChE and the corresponding subsites

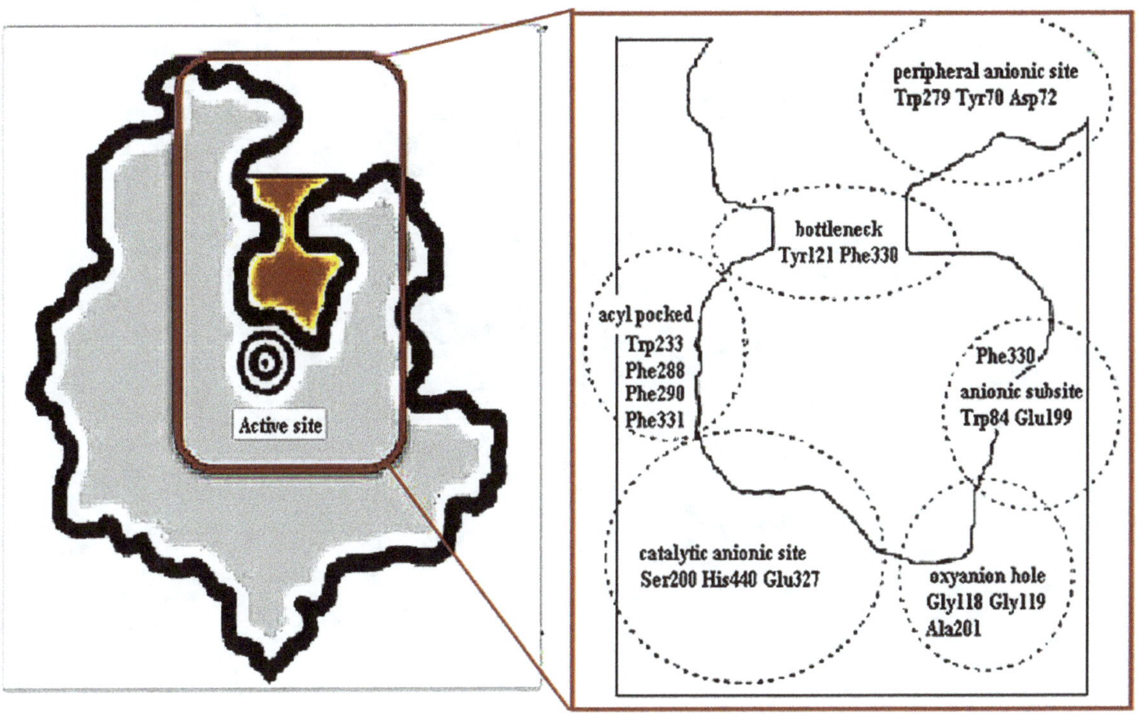

Figure 19. Docked pose corresponding to the minimum energy conformation for coumarins binding to AChE cavity. Detailed illustration of the amino acid residues lining the binding site in the AChE cavity. Green molecule displays c; broken lines display hydrogen bonds.

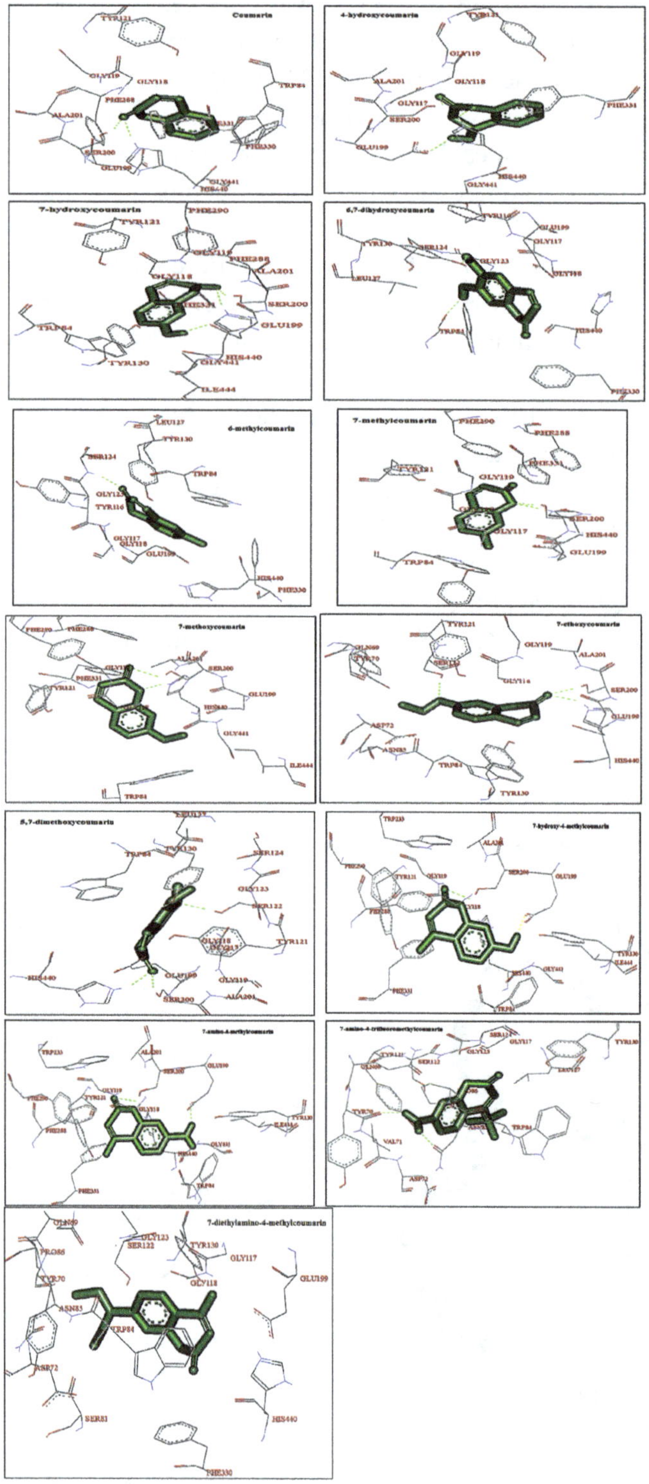

at the PAS affects catalytic activity (Silman, & Sussman, 2008). In the catalytic anionic subsite (AS), it has been proposed that the choline moiety of Ach is stabilized principally via a cation-π interaction with Trp84, and it also interacts with Glu199 and Phe330. The catalytic site is the binding site of classical AChE inhibitors, such as tacrine and huperzine, which have been studied thoroughly (Sussman et al., 1991). The esteratic subsite in TcAChE contains a typical serine-hydrolase catalytic triad, Ser200-His400-Glu327. A substantial contribution to Ach binding within the active site also arises from stabilization of the carbonyl oxygen within the oxyanion hole (OH), Gly118, Gly119, and Ala201, and of the acetyl group in the "acyl-pocked (AP)", Trp233, Phe288, Phe290 and Phe331 (Harel et al., 1993; Deb, Sharma, Piplani, & Akkinepally, 2012).

Figure 19 illustrated the exact binding sites of coumarins on AChE with the lowest binding free energy (shown in Table 15). As shown in Figure 19, all coumarins were located in the AChE cavity and surrounded by some amino acid residues from different sites and subsites in the gorge (shown in Table 16). As a consequence, the AChE catalytic activity was inhibited by the binding of coumarins. It can be seen from Table 15 and 16 that hydrophobic amino acid residues are much more than that of hydrophilic residues, and the van der Waals interactions between coumarins and AChE are obviously stronger than electrostatic forces, therefore, the essential driving forces of coumarins binding to the site were mainly hydrophobic and van der Waals interactions in nature, which were consistent with the results obtained from the thermodynamic parameter analysis. As the mention above, different amounts of hydroxyls exist in the structure of coumarins, when coumarins enter into the cavity of AChE, these groups would form hydrogen bonds with amino acid residues. Consequently, hydrogen bonding is also an important force in the binding process.

Table 15. The lowest energy-ranked results of coumarin-AChE binding conformations

Compound	Inhib Constant (μM)	Binding Energy	Internal Energy	Vdw-hb-Desolve Energy	Electrostatic Energy
Coumarin	98.38	-5.47	-5.47	-5.35	-0.12
4-hydroxycoumarin	107.7	-5.41	-5.85	-5.39	-0.45
7-hydroxycoumarin	65.24	-5.71	-5.98	-5.68	-0.31
6,7-dihydroxycoumarin	71.12	-5.66	-6.21	-5.99	-0.22
6-methylcoumarin	28.53	-6.2	-6.2	-6.15	-0.05
7-methylcoumarin	52.75	-5.84	-5.84	-5.71	-0.13
7-methoxycoumarin	64.71	-5.71	-5.98	-5.8	-0.18
7-ethoxycoumarin	56.29	-5.8	-6.34	-6.11	-0.22
5,7-dimethoxycoumarin	61.45	-5.75	-6.29	-6.06	-0.23
7-hydroxy-4-methylcoumarin	31.78	-6.14	-6.41	-6.04	-0.37
7-amino-4-methylcoumarin	59.48	-5.76	-6.04	-5.85	-0.18
7-amino-4-trifluoromethylcoumarin	65.77	-5.71	-6.25	-6.18	-0.07
7-diethylamino-4-methylcoumarin	12.9	-6.67	-7.41	-7.34	-0.07

Table 16. The amino acid residues lining the binding site in AChE cavity and hydrogen bonds between coumarins and AChE

Compounds	Amino Acid Residues Lining the Binding Site							Hydrogen Bond (Å)
	CAS	AS	AP	OH	BN	PAS	Other Residues	
Coumarin	SER200,HIS440	TRP84, GLU199	PHE288, PHE331	GLY118,GLY119,ALA201	TYR121, PHE330	---	GLY441	SER200(1.99), HIS440(2.21)
4-hydroxycoumarin	SER200,HIS440	GLU199,	PHE331	GLY118,GLY119, ALA201	TYR121	---	GLY117, GLY441	GLU199(2.12), SER200(2.13), HIS440(1.89)
7-hydroxycoumarin	SER200,HIS440	TRP84, GLU199	PHE288, PHE290, PHE331	GLY118,GLY119, ALA201	TYR121	---	TYR130, GLY441	SER200(2.20), GLU199(2.07), HIS440(2.40)
6,7-dihydroxycoumarin	HIS440	TRP84, GLU199		GLY118	PHE330	---	TYR116, GLY117, GLY123, LEU127, SER124, TYR130	TRP84(1.85), TYR130(2.07)
6-methylcoumarin	HIS440	TRP84, GLU199		GLY118,	PHE330	---	TYR116, GLY117, GLY123, LEU127, SER124, TYR130	SER124(2.04)
7-methylcoumarin	SER200, HIS440	TRP84, GLU199	PHE288, PHE290, PHE331,	GLY118, GLY119	TYR121	---	GLY117, TYR130	SER200(1.84), HIS440(1.88)
7-methoxycoumarin	SER200,HIS440	TRP84, GLU199	PHE288, PHE290, PHE331	GLY118,GLY119, ALA201	TYR121	---	GLY441, ILE444	SER200(2.40), HIS440(2.14)
7-ethoxycoumarin	SER200, IS440	TRP84, GLU199		GLY118,GLY119, ALA201	TYR121	TYR70, ASP72,	GLN69, ASN85, SER122, TYR130	SER122(1.94), SER200(1.94), HIS440(2.12)
5,7-dimethoxycoumarin	SER200,HIS440	TRP84, GLU199	PHE288, PHE290, PHE331	GLY118,GLY119, ALA201	TYR121, PHE330	---	TRP233	GLY119(2.18), TYR121(1.88), SER200(1.90)
7-hydroxy-4-methylcoumarin	SER200,HIS440	TRP84, GLU199	PHE288, PHE290, PHE331	GLY118,GLY119, ALA201	TYR121	---	TYR130, TRP233, GLY441, ILE444	GLU199(1.73), SER200(2.29)
7-amino-4-methylcoumarin	SER200,HIS440	TRP84, GLU199	HE288, PHE290, PHE331	GLY118,GLY119, ALA201	TYR121	---	TYR130, TRP233, PGLY441, ILE444	GLU199(1.88), SER200(2.29)
7-amino-4-trifluoromethylcoumarin	---	TRP84	---	---	TYR121	TYR70, ASP72	GLN69, VAL71, ASN85, PRO86, GLY117, SER122, GLY123, SER124, LEU127, TYR130	TYR70(1.77), ASN85(1.76)
7-diethylamino-4-methylcoumarin	HIS440	TRP84, GLU199		GLY118	PHE330	TYR70, ASP72	GLN69, SER81, ASN85, PRO86, GLY117, SER122, GLY123, TYR130,	------

5. CONCLUSION

In pharmacology, knowledge of the binding of inhibitors with protease can not only help to understand the metabolic process of inhibitors in body, but also help to reveal the inhibitory mechanism of protease inhibitors and accelerate the development of new inhibitor. Thus, in this chapter, the interactions between some natural active drugs and enzymes, including pepsin, hyaluronidase and acetylcholinesterase, were investigated by using a combination of experimental and computational methods. Currently, molecular docking is a standard computational tool that has been successfully employed in drug design and discovery studies. Nonetheless, some theoretical and computational challenges remain to be overcome, doing so would increase the predictive power and widen the applications of this important computational tool. All these experimental results and theoretical data in this chapter would be helpful in understanding the mechanism of inhibitory effects of drugs against enzymes. It hopes that this chapter will stimulate more pioneer works on this area.

REFERENCES

Anand, P., Singh, B., & Singh, N. (2012). A review on coumarins as acetylcholinesterase inhibitors for alzheimer's disease. *Bioorganic & Medicinal Chemistry*, *20*(3), 1175–1180. doi:10.1016/j.bmc.2011.12.042 PMID:22257528

Antonov, V. K., Ginodman, L. M., Kapitannikov, Y. V., Barshevskaya, T. N., Gurova, A. G., & Rumsh, L. D. (1978). Mechanism of pepsin catalysis: General base catalysis by the active-site carboxylate ion. *FEBS Letters*, *88*(1), 87–90. doi:10.1016/0014-5793(78)80613-9 PMID:346376

Arai, Y., Endo, S., Miyagi, N., Abe, N., Miura, T., Nishinaka, T., & El-Kabbani, O. et al. (2015). Structure-activity relationship of flavonoids as potent inhibitors of carbonyl reductase 1 (CBR1). *Fitoterapia*, *101*, 51–56. doi:10.1016/j.fitote.2014.12.010 PMID:25549925

Bader, A., Tuccinardi, T., Granchi, C., Martinelli, A., Macchia, M., Minutolo, F., & Braca, A. et al. (2015). Phenylpropanoids and flavonoids from Phlomis kurdica as inhibitors of human lactate dehydrogenase. *Phytochemistry*, *116*, 262–268. doi:10.1016/j.phytochem.2015.03.007 PMID:25890391

Bartus, R. T., Dean, R. L., Beer, B., & Lippa, A. S. (1982). The cholinergic hypothesis of geriatric memory dysfunction. *Science*, *217*(4558), 408–414. doi:10.1126/science.7046051 PMID:7046051

Bi, S. Y., Ding, L., Tian, Y., Song, D. Q., Zhou, X., Liu, X., & Zhang, H. Q. (2004). Investigation of the interaction between flavonoids and human serum albumin. *Journal of Molecular Structure*, *703*(1-3), 37–45. doi:10.1016/j.molstruc.2004.05.026

Blot, W. J., McLaughlin, J. K., & Chow, W. H. (1997). Cancer rates among drinkers of black tea. *Critical Reviews in Food Science and Nutrition*, *37*(8), 739–760. doi:10.1080/10408399709527800 PMID:9447273

Cameron, E., Pauling, L., & Leibovitz, B. (1979). Ascordic acid and cancer: A review. *Cancer Research*, *39*, 663–681. PMID:371790

Campos, L. A., & Sancho, J. (2003). The active site of pepsin is formed in the intermediate conformation dominant at mildly acidic pH. *FEBS Letters*, *538*(1-3), 89–95. doi:10.1016/S0014-5793(03)00152-2 PMID:12633859

Carletti, E., Colletier, J. P., Dupeux, F., Trovaslet, M., Masson, P., & Nachon, F. (2010). Structural evidence that human acetylcholinesterase inhibited by tabun ages through O-dealkylation. *Journal of Medicinal Chemistry*, *53*(10), 4002–4008. doi:10.1021/jm901853b PMID:20408548

Chen, G. F., Ma, X., Meng, F. B., & Li, G. X. (2008). The electron transfer reactivity of kaempferol and its interaction with amino acid residues. *Bioelectrochemistry (Amsterdam, Netherlands)*, *72*(2), 169–173. doi:10.1016/j.bioelechem.2008.02.001 PMID:18321794

Chi, Z. X., Liu, R. T., & Zhang, H. (2010). Noncovalent Interaction of Oxytetracycline with the Enzyme Trypsin. *Biomacromolecules*, *11*(9), 2454–2459. doi:10.1021/bm100633h PMID:20681619

China Pharmacopoeia Committee. (2010). *Pharmacopoeia of the People's Republic of China*. Beijing: Chemical Industry Press.

Ciolkowski, M., Rozanek, M., Bryszewska, M., & Klajnert, B. (2013). The influence of PAMAM dendrimers surface groups on their interaction with porcine pepsin. *Biochimica et Biophysica Acta*, *1834*(10), 1982–1987. doi:10.1016/j.bbapap.2013.06.020 PMID:23851144

Cui, F. C., Yang, K. C., & Li, Y. Q. (2015). Investigate the binding of catechins to trypsin using docking and molecular dynamics simulation. *PLoS ONE*, *10*(5), e0125848. doi:10.1371/journal.pone.0125848 PMID:25938485

Dangles, O., & Dufore, C. (2005). Flavonoid-protein interactions. In Ø. M. Andersen & K. R. Markham (Eds.), *Flavonoids: chemistry, biochemistry and applications* (pp. 443–469). CRC Press. doi:10.1201/9781420039443.ch9

Danser, A. H., & Deinum, J. (2005). Renin, prorenin and the putative (pro) renin receptor. *Hypertension*, *46*(5), 1069–1076. doi:10.1161/01.HYP.0000186329.92187.2e PMID:16186442

de Groot, H., & Rauen, U. (1998). Tissue injury by reactive oxygen species and the protective effects of flavonoids. *Fundamental & Clinical Pharmacology*, *12*(3), 249–255. doi:10.1111/j.1472-8206.1998.tb00951.x PMID:9646056

Deb, P. K., Sharma, A., Piplani, P., & Akkinepally, R. R. (2012). Molecular docking and receptor-specific 3D-QSAR studies of acetylcholinesterase inhibitors. *Molecular Diversity*, *16*(4), 803–823. doi:10.1007/s11030-012-9394-x PMID:22996404

Dounin, V., Constantinof, A., Schulze, H., Bachmann, T. T., & Kerman, K. (2011). Electrochemical detection of interaction between thioflavin T and acetylcholinesterase. *Analyst (London)*, *136*(6), 1234–1238. doi:10.1039/c0an00743a PMID:21264414

Dunnett, S. B., & Fibiger, H. C. (1993). Role of forebrain cholinergic systems in learning and memory: Relevance to the cognitive deficits of aging and Alzheimer's dementia. *Progress in Brain Research*, *98*, 413–420. doi:10.1016/S0079-6123(08)62425-5 PMID:8248529

Duthie, E. S., & Chain, E. A. (1939). A mucolytic enzyme in tests extract. *Nature*, *144*(3658), 977–978. doi:10.1038/144977b0

Duthie, G. G., Duthie, S. J., & Kyle, J. A. M. (2000). Plant polyphenols in cancer and heart disease: Implications as nutritional antioxidants. *Nutrition Research Reviews*, *13*(01), 79–106. doi:10.1079/095442200108729016 PMID:19087434

Escalante, T., Franceschi, A., Rucavado, A., & Gutierrez, J.M. (2000). *Effectiveness of batimastat, a synthetic inhibitor of matrix metalloproteinases, in neutralizing local tissue damage induced by BaP1, a hemorrhagic metalloproteinase from the venom of the snake* Bothrops asper. Academic Press.

Flora, K., Hahn, M., Rosen, H., & Benner, K. (1998). Milk thistle (Silybum marianum) for the therapy of liver disease. *The American Journal of Gastroenterology*, *93*(2), 139–143. doi:10.1111/j.1572-0241.1998.00139.x PMID:9468229

Furusawa, M., Narita, Y., Iwai, K., Fukunaga, T., & Nakagiri, O. (2011). Inhibitory effect of a hot water extract of coffee 'silverskin' on hyaluronidase. *Bioscience, Biotechnology, and Biochemistry*, *75*(6), 1205–1207. doi:10.1271/bbb.110106 PMID:21670509

Galvez, J., Coelho, G., Crespo, M. E., Cruz, T., Rodriguez-Cabezas, M. E., Concha, A., & Zarzuelo, A. et al. (2001). Intestinal anti-inflammatory activity of morin on chronic experimental colitis in the rat. *Alimentary Pharmacology & Therapeutics*, *15*(12), 2027–2039. doi:10.1046/j.1365-2036.2001.01133.x PMID:11736735

Gautam, R., & Jachak, S. M. (2009). Recent developments in anti-inflammatory natural products. *Medicinal Research Reviews*, *29*(5), 767–820. doi:10.1002/med.20156 PMID:19378317

Gazak, R., Purchartova, K., Marhol, P., Zivna, L., Sedmera, P., Valentova, K., & Kren, V. et al. (2010). Antioxidant and antiviral activities of silybin fatty acid conjugates. *European Iournal of Medicinal Chemistry*, *45*(3), 1059–1067. doi:10.1016/j.ejmech.2009.11.056 PMID:20036447

Gazák, R., Walterová, D., & Kren, V. (2007). Silybin and silymarin—new and emerging applications in medicine. *Current Medicinal Chemistry*, *14*(3), 315–338. doi:10.2174/092986707779941159 PMID:17305535

Gentili, P. L., Ortica, F., & Favaro, G. (2008). Static and dynamic interaction of a naturally occurring photochromic molecule with bovine serum albumin studied by UV-visible absorption and fluorescence spectroscopy. *The Journal of Physical Chemistry B*, *112*(51), 16793–16801. doi:10.1021/jp805922g PMID:19367911

Ghalia, S., & Noura, B. (2015). Docking studies of flavonoid compounds as inhibitors of beta-ketoacyl acyl carrier protein synthase I (Kas I) of Escherichia coli. *Journal of Molecular Graphics & Modelling*, *61*, 214–223. doi:10.1016/j.jmgm.2015.07.005 PMID:26292066

Graf, B. A., Mibury, P. E., & Blumberg, J. B. (2005). Flavonols, flavones, flavanones, and human health: Epidemiological evidence. *Journal of Medicinal Food*, *8*(3), 281–290. doi:10.1089/jmf.2005.8.281 PMID:16176136

Greenblatt, H. M., Kryger, G., Lewis, T., Silman, I., & Sussman, J. L. (1999). Structure of acetylcholinesterase complexedwith (-)-galanthamine at 2.3 Å resolution. *FEBS Letters*, *463*(3), 321–326. doi:10.1016/S0014-5793(99)01637-3 PMID:10606746

Harel, M., Schalk, I., Ehret-sabatier, L., Bouet, F., Goeldner, M., Hirth, C., & Sussman, J. L. et al. (1993). Quaternary ligand binding to aromatic residues in the active-site gorge of acetylcholinesterase. *Proceedings of the National Academy of Sciences of the United States of America*, *90*(19), 9031–9035. doi:10.1073/pnas.90.19.9031 PMID:8415649

Heim, K. E., Tagliaferro, A. R., & Bobilya, D. J. (2002). Flavonoid antioxidants: Chemistry, metabolism and structure-activity relationships. *The Journal of Nutritional Biochemistry*, *13*(10), 572–584. doi:10.1016/S0955-2863(02)00208-5 PMID:12550068

Hertog, M. G. L., Feskens, E. J. M., Hollman, P. C. H., Katan, M. B., & Kromhout, D. (1993). Dietary antioxidant flavonoids and risk of coronary heart disease: The Zutphen elderly study. *Lancet*, *342*(8878), 1007–1011. doi:10.1016/0140-6736(93)92876-U PMID:8105262

Huang, L., Yuan, X., Yu, D., Lee, K. H., & Chen, C. H. (2005). Mechanism of action and resistant profile of anti-HIV-1 coumarin derivatives. *Virology*, *332*(2), 623–628. doi:10.1016/j.virol.2004.11.033 PMID:15680427

Huang, X. Y., Shan, Z. J., Zhai, H. L., Su, L., & Zhang, X. Y. (2011). Study on the anticancer activity of coumarin derivatives by molecular modeling. *Chemical Biology & Drug Design*, *78*(4), 651–658. doi:10.1111/j.1747-0285.2011.01195.x PMID:21791009

Huang, Y. B., Yan, J., Liu, B. Z., Yu, Z., Gao, X. Y., Tang, Y. C., & Zi, Y. Q. (2010). Investigation on interaction of prulifloxacin with pepsin: A spectroscopic analysis. *Spectrochimica Acta. Part A: Molecular and Biomolecular Spectroscopy*, *75*(3), 1024–1029. doi:10.1016/j.saa.2009.12.044 PMID:20045662

Hui, C., Quan, L., Jian, S., Jian, B. X., & Ming, X. (2008). Comparing the affinities of flavonoid isomers with protein by fluorescence spectroscopy. *Analytical Letters*, *41*(4), 521–532. doi:10.1080/00032710801910486

Jayaraman, J., Jesudoss, V. A. S., Menon, V. P., & Namasivayam, N. (2012). Anti-inflammatory role of naringenin in rats with ethanol induced liver injury. *Toxicology Mechanisms and Methods*, *22*(7), 568–576. doi:10.3109/15376516.2012.707255 PMID:22900548

Jeon, I. H., Kim, H. S., Kang, H. J., Lee, H. S., Jeong, S. I., Kim, S. J., & Jang, S. I. (2014). Anti-inflammatory and antipruritic effects of luteolin from Perilla (*P. frutescens* L.) leaves. *Molecules (Basel, Switzerland)*, *19*(6), 6941–6951. doi:10.3390/molecules19066941 PMID:24871572

Ji, G. Y., Yang, Q. H., Hao, J., Guo, L. N., Chen, X., Hu, J. P., & Jiang, Z. Q. et al. (2011). Anti-inflammatory effect of genistein on non-alcoholic steatohepatitis rats induced by high fat diet and its potential mechanisms. *International Immunopharmacology*, *11*(6), 762–768. doi:10.1016/j.intimp.2011.01.036 PMID:21320636

Jin, K. S., Rho, Y., Kim, J., Kim, H., Kim, I. J., & Ree, M. (2008). Synchrotron small-angle X-ray scattering studies of the structure of porcine pepsin under various pH conditions. *Journal of Physical Chemistry*, *112*(49), 15821–15827. doi:10.1021/jp805940d PMID:19367902

Jung, H. J., Kim, S. J., Song, Y. S., Park, E. H., & Lim, C. J. (2010). Evaluation of the antiangiogenic, anti-inflammatory, and antinociceptive activities of morin. *Planta Medica*, *76*(03), 273–275. doi:10.1055/s-0029-1186079 PMID:19774509

Kakegawa, H., Matsumoto, H., & Satoh, T. (1999). Inhibitory effects of some natural products on the activation of hyaluronidase and their anti-allergic action. *Chemical & Pharmaceutical Bulletin*, *40*(6), 1439–1442. doi:10.1248/cpb.40.1439 PMID:1382876

Khan, S. N., Islam, B., Yennamalli, R., Sultan, A., Subbarao, N., & Khan, A. U. (2008). Interaction of mitoxantrone with human serum albumin: Spectroscopic and molecular modeling studies. *European Journal of Pharmaceutical Sciences*, *35*(5), 371–382. doi:10.1016/j.ejps.2008.07.010 PMID:18762252

Kontogiorgis, C. A., Xu, Y., Hadjipavlou-Litina, D., & Luo, Y. (2007). Coumarin derivatives protection against ROS production in cellular models of A beta toxicities. *Free Radical Research*, *41*(10), 1168–1180. doi:10.1080/10715760701447884 PMID:17886039

Kryger, G., Silman, I., & Sussman, J. L. (1999). Structure of acetylcholinesterase complexed with E2020 (Aricept): Implications for the design of new anti-Alzheimer drugs. *Structure (London, England)*, *7*(3), 297–307. doi:10.1016/S0969-2126(99)80040-9 PMID:10368299

Kuppusamy, U. R., & Das, N. P. (1991). Inhibitory effects of flavonoids on several venom hyaluronidases. *Experientia*, *47*(11-12), 1196–1200. doi:10.1007/BF01918384 PMID:1765130

Lakowicz, J. R., & Weber, G. (1973). Quenching of protein fluorescence by oxygen. Detection of structural fluctuations in proteins on the nanosecond time scale. *Biochemistry*, *12*(21), 4171–4179. doi:10.1021/bi00745a021 PMID:4200894

Lee, J. H., & Kim, G. H. (2010). Evaluation of antioxidant and inhibitory activities for different subclasses flavonoids on enzymes for *Rheumatoid arthritis*. *Journal of Food Science*, *75*(7), H212–H217. doi:10.1111/j.1750-3841.2010.01755.x PMID:21535545

Lee, K. K., & Choi, J. D. (1999). The effects of *Areca catechu* L. extract on anti-inflammation and anti-melanogensis. *International Journal of Cosmetic Science*, *21*(4), 275–284. doi:10.1046/j.1467-2494.1999.196590.x PMID:18503445

Lee, S. J., Lee, U. S., Kim, W. J., & Moon, S. K. (2011). Inhibitory effect of esculetin on migration, invasion and matrix metalloproteinase-9 expression in TNF-alpha-induced vascular smooth muscle cells. *Molecular Medicine Reports*, *4*, 337–341. PMID:21468574

Li, D. J., Zhang, T., Xu, C., & Ji, B. M. (2011). Effect of pH on the interaction of baicalein with lysozyme by spectroscopic approaches. *Journal of Photochemistry and Photobiology. B, Biology*, *104*(3), 414–424. doi:10.1016/j.jphotobiol.2011.04.009 PMID:21570863

Li, Y., Yao, X., Jin, J., Chen, X., & Hu, Z. (2007). Interaction of rhein with human serum albumin investigation by optic spectroscopic technique and modeling studies. *Biochimca et Biophysica Acta-general Subjects*, *1774*(1), 51–58. doi:10.1016/j.bbapap.2006.09.020 PMID:17174615

Lin, L., Wei, D. Q., Wang, J. F., & Chou, K. C. (2007). Computational studies of the binding mechanism of calmodulin with chrysin. *Biochemical and Biophysical Research Communications*, *358*(4), 1102–1107. doi:10.1016/j.bbrc.2007.05.053 PMID:17521610

Lin, S. Y., Zhang, G. W., Liao, Y. J., Pan, J. H., & Gong, D. M. (2015). Dietary flavonoids as xanthine oxidase inhibitors: Structure-affinity and structure-activity relationships. *Journal of Agricultural and Food Chemistry*, *63*(35), 7784–7794. doi:10.1021/acs.jafc.5b03386 PMID:26285120

Liu, J., Wang, X. R., Cai, Z. W., & Lee, F. S. C. (2008). Effect of tanshinone HA on the noncovalent interaction between warfarin and human serum albumin studied by electrospray ionization mass spectrometry. *Journal of the American Chemical Society*, *19*, 1568–1575. PMID:18657993

Liu, M., Yin, H., Dong, J. J., Xiao, L., Liu, G., Qian, Z. H., & Miao, J. L. (2013). Inhibition and interaction with hyaluronidase by compounds from Hop (*Humulus iupulus* L.) flowers. *Asian Journal of Chemistry*, *25*, 10262–10266.

Liu, M. H., Lin, Y. S., Sheu, S. Y., & Sun, J. S. (2009). Anti-inflammatory effects of daidzein on primary astroglial cell culture. *Nutritional Neuroscience*, *12*(3), 123–134. doi:10.1179/147683009X423274 PMID:19356315

Liu, X. H., Xi, P. X., Chen, F. J., Xu, Z. H., & Zeng, Z. Z. (2008). Spectroscopic studies on binding of 1-phenyl-3-(coumarin-6-yl) sulfonylurea to bovine serum albumin. *Journal of Photochemistry and Photobiology. B, Biology*, *92*(2), 98–102. doi:10.1016/j.jphotobiol.2008.04.008 PMID:18571426

Lu, Q. H., Ba, C. D., & Chen, D. Y. (2008). Investigating noncovalent interactions of rutin-serum albumin by capillary electrophoresis-frontal analysis. *Journal of Pharmaceutical and Biomedical Analysis*, *47*(4-5), 888–891. doi:10.1016/j.jpba.2008.03.007 PMID:18436413

Ma, J., Yin, Y. M., Liu, H. L., & Xie, M. X. (2011). Interactions of flavonoids with biomacromolecules. *Current Organic Chemistry*, *15*(15), 2627–2640. doi:10.2174/138527211796367345

Mahesha, H. G., Singh, S. A., Srinivasan, N., & Rao, A. G. A. (2006). A spectroscopic study of the interaction of isoflavones with human serum albumin. *The FEBS Journal*, *273*(3), 451–467. doi:10.1111/j.1742-4658.2005.05071.x PMID:16420470

Masood, M. M., Pillalamarri, V. K., Irfan, M., Aneja, B., Jairajpuri, M. A., Zafaryab, M., & Abid, M. et al. (2015). Diketo acids and their amino acid/dipeptidic analogues as promising scaffolds for the development of bacterial methionine aminopeptidase inhibitors. *RSC Advances*, *5*(43), 34173–34183. doi:10.1039/C5RA03354C

Megumi, F. T., Kei, N., Kenji, T., Tadahiko, M., & Tadashi, K. (2011). Anti-inflammatory activity of structurally related flavonoids, apigenin, luteolin and fisetin. *International Immunopharmacology*, *11*(9), 1150–1159. doi:10.1016/j.intimp.2011.03.012 PMID:21443976

Moon, S. H., Kim, K. T., & Lee, N. K. (2009). Inhibitory effects of naringenin and its novel derivatives on hyaluronidase. *Food Science and Biotechnology*, *18*, 267–270.

Osborn, G. G., & Saunders, A. V. (2010). Current treatments for patients with Alzheimer disease. *The Journal of the American Osteopathic Association*, *110*, s16–s26. PMID:20926739

Pande, M., Kumari, N. K. P., Dubey, V. K., Tripathi, P., & Jagannadham, M. V. (2009). Stability and unfolding studies on alkaline denatured state (I_p) of pepsin. *Process Biochemistry*, *44*(8), 906–911. doi:10.1016/j.procbio.2009.04.016

Peramaiyan, R., Thamaraiselvan, R., Natarajan, N., Rajendran, P., Yutaka, N., & Ikuo, N. (2014). Kaempferol, a potential cytostatic and cure for inflammatory disorders. *European Journal of Medicinal Chemistry*, *86*, 103–112. doi:10.1016/j.ejmech.2014.08.011 PMID:25147152

Pessini, A. C., Takao, T. T., Cavalheiro, E. C., Vichnewski, W., Sampaio, S. V., Giglio, J. R., & Arantes, E. C. (2001). A hyaluronidase from *Tityus serrulatus* scorpion venom: Isolation, characterization and inhibition by flavonoids. *Toxicon*, *39*(10), 1495–1504. doi:10.1016/S0041-0101(01)00122-2 PMID:11478957

Piazzi, L., Cavalli, A., Colizzi, F., Belluti, F., Bartolini, M., Mancini, F., & Rampa, A. et al. (2008). Multi-target-direct coumarin derivatives: hAChE and BACE 1 inhibitors as potential anti-Alzheimer compounds. *Bioorganic & Medicinal Chemistry Letters*, *18*(1), 423–426. doi:10.1016/j.bmcl.2007.09.100 PMID:17998161

Puiatti, M., Borioni, J. L., Vallejo, M. G., Cabrera, J. L., Agnese, A. M., Ortega, M. G., & Pierini, A. B. (2013). Study of the interaction of *Huperzia saururus* Lycopodium alkaloids with the acetylcholinesterase enzyme. *Journal of Molecular Graphics & Modelling*, *44*, 136–144. doi:10.1016/j.jmgm.2013.05.009 PMID:23827878

Qin, C., Xie, M. X., & Liu, Y. (2007). Characterization of the myricetin-human serum albumin complex by spectroscopic and molecular modeling approaches. *Biomacromolecules*, *8*(7), 2182–2189. doi:10.1021/bm070319c PMID:17559264

Razavi, S. F., Khoobi, M., Nadri, H., Sakhteman, A., Moradi, A., Emami, S., & Shafiee, A. et al. (2013). Synthesis and evaluation of 4-substituted coumarins as novel acetylcholinesterase inhibitors. *European Journal of Medicinal Chemistry*, *64*, 252–259. doi:10.1016/j.ejmech.2013.03.021 PMID:23644208

Riberson, M. R., & Harrell, L. E. (1997). Cholinergic activity and amyloid precursor protein metabolism. *Brain Research. Brain Research Reviews*, *25*(1), 50–69. doi:10.1016/S0165-0173(97)00016-7 PMID:9370050

Richard, T., Lefeuvre, D., Descendit, A., Quideau, S., & Monti, J. P. (2006). Recognition characters in peptide-polyphenol complex formation. *Biochimica et Biophysica Acta-general Subjects*, *1760*(6), 951–958. doi:10.1016/j.bbagen.2006.01.005 PMID:16527409

Ross, P. D., & Subramanian, S. (1981). Thermodynamics of protein association reactions: Forces contributing to stability. *Biochemistry*, *20*(11), 3096–3102. doi:10.1021/bi00514a017 PMID:7248271

Rydberg, E. H., Brumshtein, B., Greenblatt, H. M., Wong, D. M., Shaya, D., Williams, L. D., & Sussman, J. L. et al. (2006). Complexes of alkylene-linked tacrine dimers with *Torpedo californica* acetylcholinesterase: Binding of bis(5)-tacrine produces a dramatic rearrangement in the active-site gorge. *Journal of Medicinal Chemistry*, *49*(18), 5491–5500. doi:10.1021/jm060164b PMID:16942022

Sashidhara, K. V., Kumar, A., Chatterjee, M., Rao, K. B., Singh, S., Verma, A. K., & Palit, G. (2011). Discovery and synthesis of novel 3-phenylcoumarin derivatives as antidepressant agents. *Bioorganic & Medicinal Chemistry Letters*, *21*(7), 1937–1941. doi:10.1016/j.bmcl.2011.02.040 PMID:21377878

Schümann, J., Prockl, J., Kiemer, A. K., Vollmar, A. M., Bang, R., & Tiegs, G. (2003). Silibinin protects mice from T cell-dependent liver injury. *Journal of Hepatology*, *39*(3), 333–340. doi:10.1016/S0168-8278(03)00239-3 PMID:12927918

Shanmugaraj, K., Anandakumar, S., & Ilanchelian, M. (2015). Probing the binding interaction of thionine with lysozyme: A spectroscopic and molecular docking investigation. *Dyes and Pigments*, *112*, 210–219. doi:10.1016/j.dyepig.2014.07.003

Shen, L. L., Xu, H., Huang, F. W., Li, Y., Xiao, H. F., Yang, Z., & Li, Y. N. et al. (2015). Investigation on interaction between ligupurpuroside A and pepsin by spectroscopic and docking methods. *Spectrochimica Acta. Part A: Molecular and Biomolecular Spectroscopy*, *135*, 256–263. doi:10.1016/j.saa.2014.06.087 PMID:25078459

Shi, J. H., Zhu, Y. Y., Wang, J., Chen, J., & Shen, Y. J. (2013). Intermolecular interaction of prednisolone with bovine serum albumin: Spectroscopic and molecular docking methods. *Spectrochimica Acta. Part A: Molecular and Biomolecular Spectroscopy*, *103*, 287–294. doi:10.1016/j.saa.2012.11.034 PMID:23261625

Shibata, T., Fujimoto, K., Nagayama, K., Yamaguchi, K., Yamaguchi, K., & Nakamura, T. (2002). Inhibitory activity of brown algal phlorotannins against hyaluronidase. *International Journal of Food Science & Technology*, *37*(6), 703–709. doi:10.1046/j.1365-2621.2002.00603.x

Shintani, T., Nomura, K., & Ichishima, E. (1997). Engineering of porcine pepsin alteration of S1 substrate specificity of pepsin to those of fungal aspartic proteinases by site-directed mutagenesis. *The Journal of Biological Chemistry*, *272*(30), 18855–18861. doi:10.1074/jbc.272.30.18855 PMID:9228062

Silman, I., & Sussman, J. L. (2008). Acetylcholinesterase: How is structure related to function. *Chemico-Biological Interactions*, *175*(1-3), 3–10. doi:10.1016/j.cbi.2008.05.035 PMID:18586019

Sinko, G., Brglez, J., & Kovarik, Z. (2010). Interactions of pyridinium oximes with acetylcholinesterase. *Chemico-Biological Interactions*, *187*(1-3), 172–176. doi:10.1016/j.cbi.2010.04.017 PMID:20417629

Soto-Ortega, D. D., Murphy, B. P., Gonzalez-Velasquez, F. J., Wilson, K. A., Xie, F., Wang, Q., & Moss, M. A. (2011). Inhibition of amyloid-β aggregation by coumarin analogs can be manipulated by functionalization of the aromatic center. *Bioorganic & Medicinal Chemistry*, *19*(8), 2596–2602. doi:10.1016/j.bmc.2011.03.010 PMID:21458277

Stan, D., Matei, I., Mihailescu, C., Savin, M., Matache, M., Hillebrand, M., & Baciu, I. (2009). Spectroscopic investigations of the binding interaction of a new indanedione derivative with human and bovine serum albumins. *Molecules (Basel, Switzerland)*, *14*(4), 1614–1626. doi:10.3390/molecules14041614 PMID:19396020

Stern, R., Asari, A. A., & Sugahara, K. N. (2006). Hyaluronan fragments: An information-rich system. *European Journal of Cell Biology*, *85*(8), 699–715. doi:10.1016/j.ejcb.2006.05.009 PMID:16822580

Sussman, J. L., Harel, M., Frolow, F., Oefner, C., Goldman, A., Toker, L., & Silman, I. (1991). Atomic structure of acetylcholinesterase from *Torpedo californica*: A prototypic acetylcholine-binding protein. *Science*, *253*(5022), 872–879. doi:10.1126/science.1678899 PMID:1678899

Teillet, F., Boumendjel, A., Boutonnat, J., & Ronot, X. (2008). Flavonoids as RTK inhibitors and potential anticancer agents. *Medicinal Research Reviews*, *28*(5), 715–745. doi:10.1002/med.20122 PMID:18080331

Tritsch, D., Zingle, C., Rohmer, M., & Grosdemange-Billiard, C. (2015). Flavonoids: True or promiscuous inhibitors of enzyme? The case of deoxyxylulose phosphate reductoisomerase. *Bioorganic Chemistry*, *59*, 140–144. doi:10.1016/j.bioorg.2015.02.008 PMID:25800132

Tsai, S. J., Huang, C. S., Mong, M. C., Kam, W. Y., Huang, H. Y., & Yin, M. C. (2012). Anti-inflammatory and antifibrotic effects of naringenin in diabetic mice. *Journal of Agricultural and Food Chemistry*, *60*(1), 514–521. doi:10.1021/jf203259h PMID:22117528

Utpal, C.D., Tanusree, D., Debanjan, S., & Sudhan, D. (2015). Three-dimensional quantitative structure-activity relationships and docking studies of some structurally diverse flavonoids and design of new aldose reductase inhibitors. *Journal of Advanced Pharmaceutical technology and Research*, *6*, 13-18.

Verdrengh, M., Jonsson, I. M., Holmdahl, R., & Tarkowski, A. (2003). Genistein as an anti-inflammatory agent. *Inflammation Research*, *52*(8), 341–346. doi:10.1007/s00011-003-1182-8 PMID:14504672

Wang, Y. C., & Huang, K. M. (2013). In vitro anti-inflammatory effect of apigenin in the helicobacter pylori-infected gastric adenocarcinoma cells. *Food and Chemical Toxicology*, *53*, 376–383. doi:10.1016/j.fct.2012.12.018 PMID:23266501

Wu, C. R., Chang, C. L., Hsieh, P. Y., Lin, L. W., & Ching, H. (2007). Psoralen and isopsoralen, two coumarins of Psoraleae fructus, can alleviate scopolamine-induced amnesia in rats. *Planta Medica*, *73*(3), 275–278. doi:10.1055/s-2007-967127 PMID:17318785

Yadava, U., Gupta, H., & Roychoudhury, M. (2015). Stabilization of microtubules by taxane diterpenoids: Insight from docking and MD simulations. *Journal of Biological Physics*, *41*(2), 117–133. doi:10.1007/s10867-014-9369-5 PMID:25542396

Yadava, U., Shukla, B. K., Roychoudhury, M., & Kumar, D. (2015). Pyrazolo[3,4-*d*]pyrimidines as novel inhibitors of O-acetyl-L-serine sulfhydrylase of *Entamoeba histolytica*: An in silico study. *Journal of Molecular Modeling*, *21*(4), 7–13. doi:10.1007/s00894-015-2631-3 PMID:25799964

Yadava, U., Singh, M., & Roychoudhury, M. (2013). Pyrazolo [3,4-*d*] pyrimidines as inhibitor of anti-coagulation and inflammation activities of phospholipase A2: Insight from molecular docking studies. *Journal of Biological Physics*, *39*(3), 419–438. doi:10.1007/s10867-013-9299-7 PMID:23860918

Yahaya, Y. A., Don, M. M., & Yahaya, A. S. (2014). The effect of culture conditions on the growth of *T. lactinea* and anti-inflammatory activities via in vitro inhibition of hyaluronidase and lipoxygenase enzyme activities. *Journal of the TAIWAN Institute of Chemical Engineers*, *45*(5), 2054–2059. doi:10.1016/j.jtice.2014.04.003

Yang, R., Yu, L. L., Zeng, H. J., Liang, R. L., Chen, X. L., & Qu, L. B. (2012). The interaction of flavonoid-lysozyme and the relationship between molecular structure of flavonoids and their binding activity to lysozyme. *Journal of Fluorescence*, *22*(6), 1449–1459. doi:10.1007/s10895-012-1082-9 PMID:22772804

Yang, Y. M., Hu, Q. L., Fan, Y. L., & Shen, H. S. (2008). Study of the binding of luteolin to bovine serum albumin. *Spectrochimica Acta. Part A: Molecular and Biomolecular Spectroscopy*, *69*(2), 432–436. doi:10.1016/j.saa.2007.04.019 PMID:17719269

Yilmaz, M. Z., Guzel, A., Torun, A. C., Okuyucu, A., Salis, O., Karli, R., & Alacam, H. et al. (2014). The therapeutic effects of anti-oxidant and anti-inflammatory drug quercetin on aspiration-induced lung injury in rats. *Journal of Molecular Histology*, *45*(2), 195–203. doi:10.1007/s10735-013-9542-3 PMID:24122261

Ying, M., Huang, F. W., Ye, H. D., Xu, H., Shen, L. L., Huan, T. W., & Zhou, K. et al. (2015). Study on interaction between curcumin and pepsin by spectroscopic and docking methods. *International Journal of Biological Macromolecules*, *79*, 201–208. doi:10.1016/j.ijbiomac.2015.04.057 PMID:25940524

Yuriev, E., Holien, J., & Ramsland, P. A. (2014). Improvements, trends, and new ideas in molecular docking: 2012-2013 in review. *Journal of Molecular Recognition*, *28*(10), 581–604. doi:10.1002/jmr.2471 PMID:25808539

Zeng, H. J., Hu, G. Z., You, J., Yang, R., & Qu, L. B. (2015). Spectroscopic and molecular modeling investigation on the interactions between hyaluronidase and baicalein and chrysin. *Process Biochemistry*, *50*(5), 738–744. doi:10.1016/j.procbio.2015.02.007

Zeng, H. J., Ma, J., Yang, R., You, J., & Qu, L. B. (2015). Molecular interactios of flavonoids to hyaluronidase: Insights from spectroscopic and molecular modeling studies. *Journal of Fluorescence*, *25*(4), 941–959. doi:10.1007/s10895-015-1576-3 PMID:26006100

Zeng, H. J., You, J., Liang, H. L., Yang, R., & Qu, L. B. (2014). Investigation on the binding interaction between silybin and pepsin by spectral and molecular docking. *International Journal of Biological Macromolecules*, *67*, 105–111. doi:10.1016/j.ijbiomac.2014.02.051 PMID:24608028

Zhang, X. X., Wang, G. J., Gurley, E. C., & Zhou, H. P. (2014). Flavonoid apigenin inhibits lipopolysaccharide-induced inflammatory response through multiple mechanisms in macrophages. *PLoS ONE*, *9*, e107072. PMID:25192391

Zhang, Y. Z., Zhou, B., Zhang, X. P., Huang, P., Li, C. H., & Liu, Y. (2009). Interaction of malachite green with bovine serum albumin: Determination of the binding mechanism and binding site by xpectroxcopic methods. *Journal of Hazardous Materials*, *163*(2-3), 1345–1352. doi:10.1016/j.jhazmat.2008.07.132 PMID:18786760

Zhou, X., Wang, X. B., Wang, T., & Kong, L. Y. (2008). Design, synthesis, and acetylcholinesterase inhibitory activity of novel coumarin analogues. *Bioorganic & Medicinal Chemistry*, *16*(17), 8011–8021. doi:10.1016/j.bmc.2008.07.068 PMID:18701305

Chapter 5
Molecular Docking of Biologically Active Substances to Double Helical Nucleic Acids:
Problems and Solutions

Kateryna V. Miroshnychenko
O. Ya. Usikov Institute for Radiophysics and Electronics of NAS of Ukraine, Ukraine

Anna V. Shestopalova
O. Ya. Usikov Institute for Radiophysics and Electronics of NAS of Ukraine, Ukraine

ABSTRACT

Molecular docking of ligands to DNA-targets is of great importance for the design of new anticancer drugs. Unfortunately, most docking programs were developed for protein-ligand docking which raises a question about their applicability for the DNA-ligand docking. In this study, the popular docking programs AutoDock Vina, AutoDock4 and AutoDock3 were compared for a test set of 50 DNA-ligand complexes taken from the Nucleic Acid Database. It was shown that the version 3.05 of the AutoDock program was the most successful in reproducing the structures of intercalation and minor-groove complexes. The program AutoDock4 was able to re-dock to within 2 Å RMSD most of the intercalation complexes of the test set, but showed poor performance for minor groove binders. While Vina, on the contrary, failed to construct six intercalation complexes of the test set, but showed satisfactory results for DNA-ligand minor-groove complexes when small search space was used.

INTRODUCTION

Molecular docking method allows one to construct the optimal complex - the complex with the minimum energy – of macromolecule-target and small biologically active substance (BAS)-ligand. Proteins are used usually as macromolecules-targets nucleic acids-targets are less common. At the same time, there is no doubt that nucleic acids (NA) are attractive targets for small molecules (Dailey et al., 2009; Hermann & Tor, 2005). One can hardly overestimate the role they play in biological processes, with the

DOI: 10.4018/978-1-5225-0362-0.ch005

most important being probably the storage and propagation of genetic information. Many anticancer drugs exert their biological activity by binding to DNA (Brana, Cacho, Gradillas, de Pascual-Teresa, & Ramos, 2001). Drugs targeting NA can be even more effective then those targeting proteins in view of the fact that number of NA-targets is much less compared to protein-targets (Holt, Buscaglia, Trent, & Chaires, 2011). Each DNA gene-target is transcribed into multiple mRNAs-targets, and each mRNA-target is translated into multiple protein-targets. So blocking of a single gene will cause blocking of multiple protein-targets.

The docking of nucleic acids and BAS has a number of peculiarities compared to protein docking. The reason of them is primarily the differences in the structure of proteins and nucleic acids. Unlike the proteins, there are no clearly confined binding pockets for ligands in DNA duplexes. The polyanionic NA-targets have higher charge density compared to proteins and the treatment of solvent plays an important role in the evaluation of NA-ligand interactions. Another problem of NA docking is that the empirical energy scoring functions in many docking programs were calibrated using a set of protein-ligand complexes. Taking into account that the contribution of individual interactions (van der Waals, hydrogen bonds, electrostatic) to the total energy of docking complex may be different for BAS-protein complexes and BAS-NA complexes, the use of such programs for docking of NA and BAS does not always allow one to obtain the optimal complex for the given target and ligand. Nevertheless, in the literature, there are a number of successful examples of BAS-NA docking using programs designed for proteins (Chen, Shafer, & Kuntz, 1997; Detering & Varani, 2004, Evans & Neidle, 2006; Holt, Chaires, & Trent, 2008; Ricci & Netz, 2009; Li et al., 2010; Srivastava, Chourasia, Kumar, & Sastry, 2011). Protein docking programs that were successfully used for docking of ligands to NA-targets are listed in Table 1. At the same time, the application of protein-designed scoring functions for NA-ligand docking revealed some scoring problems: for the majority of studied complexes the crystal ligand pose had lower score than the best scored pose and the hydrogen bonding pattern was different in the crystal complex and in the docking top-ranked complex (Gilad & Senderowitz, 2014; Deligkaris, Ascone, Sweeney, & Greene, 2014). Recently several docking programs and scoring functions appeared that were specially designed for RNA-targets: MORDOR (Guilbert & James, 2008), RiboDock (Morley & Afshar, 2004), DrugScore^{RNA} (Pfeffer & Gohlke, 2007), DOCK6 (Lang et al., 2009). Unfortunately, there is a lack of scoring functions designed for DNA-duplexes.

This study was focused on docking of small ligands to DNA-duplexes. Since the most of the reported DNA-ligand docking studies were done using AutoDock software (Morris et al., 1998), this program

Table 1. Protein docking programs that were successfully applied to NA-ligand docking

Program	Description	Reference
AutoDock	Free program based on Lamarckian genetic algorithm and empirical force-field based scoring function	Morris et al., 1998; Morris et al., 2009
DOCK	Free anchor-and-grow based docking program	Lang et al., 2009
Surflex	Commercial docking program using molecular similarity-based search algorithm	Jain, 2003
GLIDE	Commercial exhaustive search-based docking program	Friesner et al., 2004
GOLD	Commercial genetic algorithm-based docking program	Verdonk, Cole, Hartshorn, Murray, & Taylor, 2003
CDOCKER	Commercial CHARMm-based molecular dynamics docking program	Wu, Robertson, Brooks III, & Vieth, 2003

was chosen for docking. AutoDock is one of the most popular non-commercial docking programs and, during many years, it remains the most-cited docking software (Mihasan, 2012; Chen, 2015). It uses the Lamarckian genetic algorithm and the empirical force-field based free energy function. The current version of the program is AutoDock4.2 (AD4) (Morris et al., 2009). The AD4 scoring function includes five energy terms: the van der Waals term, hydrogen bonding term, electrostatic term, desolvation term and the term for the loss of torsional entropy of ligand upon binding (Huey, Morris, Olson, & Goodsell, 2007). Each term is weighted by the coefficient that was obtained upon calibration of the AD4 scoring function for the test set of protein-ligand complexes with experimentally determined binding constants. AD4 was shown to be successful in re-docking experiments of protein-ligand complexes for ligands with up to 10 rotatable bonds (Huey et al., 2007) and in several DNA-ligand docking studies (Holt et al., 2008; Ricci & Netz, 2009; Gilad & Senderowitz, 2014). There is also the MPI/OpenMP-parallelized version of AD4.2 that makes this program an attractive choice for virtual screening (Norgan, Coffman, Kocher, Katzmann, & Sosa, 2011).

The previous version of AutoDock – AutoDock3.05 (AD3) was also used with success for NA-ligand docking (Detering & Varani, 2004, Evans & Neidle, 2006). The AD3 scoring function differs from the AD4 in the form of desolvation and hydrogen bonding terms and in the values of empirical coefficients standing before energy terms (Morris et al., 1998). Only serial version of AD3 is available.

Another program that was developed in the AutoDock lab is the AutoDock Vina (Vina) (Trott & Olson, 2010). It is not a new version of AutoDock but rather a new generation program with a new scoring function and a new search algorithm. It uses the Iterated Local Search global optimizer and the hybrid scoring function combining the advantages of empirical and knowledge-based scoring functions. The Vina scoring function consists of the three steric terms identical for all atom pairs, the hydrophobic term for hydrophobic atoms and the hydrogen-bonding term for atoms participating in hydrogen bonding. The terms are weighted by the empirical coefficients. Vina was shown to be superior to AD4 both in the accuracy of the binding mode predictions for protein-ligand complexes and in the speed of calculations. Further speed-up of Vina can be achieved by using multithreading on multi-core machines. Due to its speed and accuracy, Vina is widely used for protein-ligand docking (Chang, Ayeni, Breuer, & Torbett, 2010). On the other hand, there is no information on application of Vina for DNA-ligand docking.

In this study, the comparative analysis of DNA-ligand docking using programs Vina, AD4 and AD3 will be carried out. To the best of authors' knowledge, there was no comparison of these programs for DNA-ligand docking before. At the same time, AD4 and Vina are one of the most-used docking programs. All these programs were trained for protein-ligand complexes. Therefore it would be interesting to compare their performance with respect to the DNA-ligand docking. Vina is the fastest of the three programs considered, whereas AD3 – is the slowest because only serial version of it is available. The OpenMP-parallelized version of AD4 was used in this study.

DOCKING OF BAS TO DNA USING AUTODOCK3, AUTODOCK4 AND VINA

Computational Methods

In order to obtain a comprehensive DNA-ligand docking test set, two main types of DNA-ligand complexes should be considered: intercalation and minor-groove complexes. This study was aimed at DNA-duplexes without chemical modifications. The ligands of different size and flexibility were chosen in order to compare the effectiveness of sampling of docking programs for flexible ligands. Where possible, multiple structures of DNA complexes with one ligand were considered so that conclusions could be made for each ligand class. 50 structures of DNA-BAS complexes satisfying the above conditions were taken from the Nucleic Acid Database (NDB) (Berman et al., 1992) for the docking test set. 26 of them were DNA-ligand intercalation complexes and the rest 24 – minor-groove complexes (Table 1). The number of active torsions in ligands ranged from 0 to 13. All ligands except one (actinomycin D) were positively charged.

The methodology of the docking procedure is summarized in the flowchart in Figure 1. First water molecules and ions were removed from the pdb file of DNA-ligand complex and DNA-target and ligand were saved as separate files. For each ligand type, only the structure with the best resolution was considered. Hydrogen atoms were added to ligand structures according to their appropriate protonation state and the obtained structures were optimized in ORCA (Neese, 2012) at the B3LYP/def2-SVP level of theory. The structure of the ligand actinomycin D was optimized using semi-empirical AM1 method due to the large size of ligand. The partial atomic charges for ligands were calculated using AM1-BCC method in the module Antechamber of AmberTools 14 program package (Case et al., 2014). The further preparation of ligands for docking was done using AutoDockTools (Morris et al., 2009).

Hydrogen atoms were added to the DNA-targets using Leap module of AmberTools package and their positions were optimized in Sander module of AmberTools. The Kollman atomic charges were used for AD3 DNA-targets. Kollman charges are also known as RESP charges used in the AMBER family of force fields, they were obtained as a result of fitting to the quantum mechanical electrostatic potential (Cornell et al., 1995). To AD4 DNA-targets, Gasteiger atomic charges were added by Open Babel (O'Boyle et al., 2011). Gasteiger charges are topology-based atomic charges that are derived through an

Figure 1. The flowchart of the docking procedure

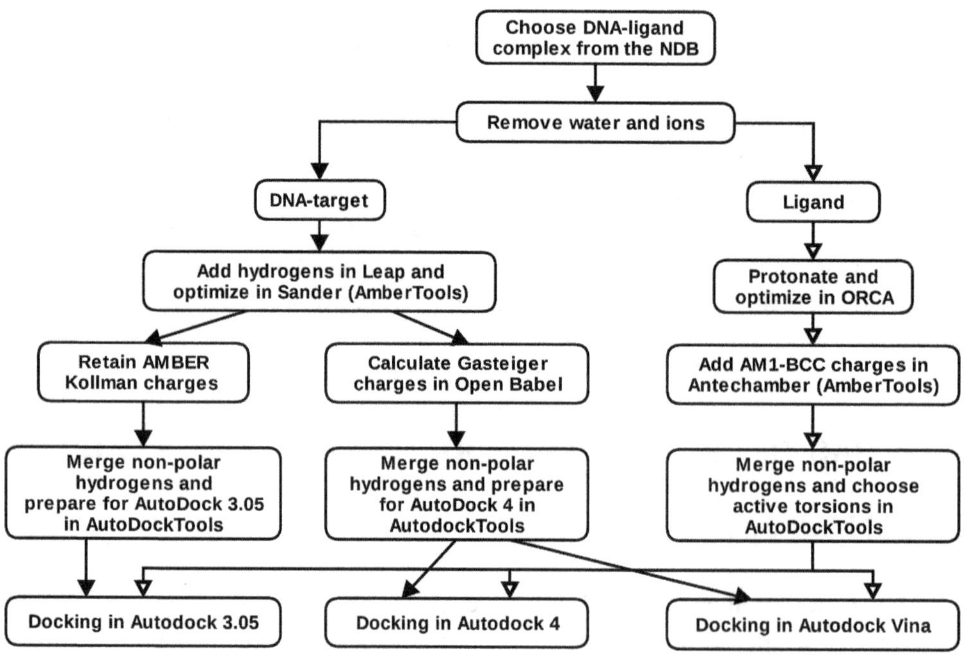

iterative procedure of partial equalization of orbital electronegativity (Gasteiger & Marsili, 1980). The assignment of partial atomic charges is not needed for Vina, but its files formats are compatible with AD4, therefore the target and ligand files prepared for AD4 were used also for Vina.

The Lamarckian genetic algorithm was used for docking in AD3 and AD4 with following parameters: maximum number of energy evaluations = 10^7, maximum number of generations = 27000, population size = 300 individuals. For each system, 100 docking runs were performed and the results were clustered using a 2 Å tolerance. The default values were used for the other docking parameters. In Vina, one cannot control the number of energy evaluations, it is determined heuristically in the course of calculation. Each Vina docking consists of a number of independent runs that is defined by the exhaustiveness parameter. Unlike AD3 and AD4, each run in Vina can produce more than one docking pose. In this study, the parameter exhaustiveness was set to 100.

The docking poses of the top-ranked clusters of AD3 and AD4 and the best-scored model of Vina were analyzed and their root-mean-square deviation (RMSD) with respect to the crystal pose was calculated. To account for ligand symmetry, the best-correspondence RMSD instead of the ordinary RMSD was calculated according to the Hungarian algorithm as implemented in DOCK (Allen & Rizzo, 2014). All RMSD values given in this study refer to the best-correspondence RMSD. The program VMD (Humphrey, Dalke, & Schulten, 1996) was used for the visualization of docking complexes.

Re-Docking of DNA-Ligand Intercalation Complexes

One of the most important types of non-covalent ligand binding to DNA is an intercalation. It is an insertion of planar part of ligand molecule between the base pairs into the DNA double helix. Upon intercalation, the structure of DNA double helix undergoes significant changes: it is lengthened and unwound. Many anticancer antibiotics bind to DNA in this way (Brana et al., 2001). In this study, seven different intercalators were considered: proflavine, cryptolepine, daunomycin, doxorubicin, idarubicin, nogalamycin and actinomycin D. Their structural formulas are shown in Figure 2. The NDB indices of DNA-drug intercalation complexes taken for the docking test set are given in Table 2. There were two intercalation sites in some structures. For each intercalation site, the docking of corresponding ligand was performed. The size of the search space was defined by the AutoDock grid of 70x70x70 points, with grid spacing of 0.375 Å, which corresponded roughly to 27Åx27Åx27Å search space in Vina. The grid center was chosen in the geometric center of intercalation site base pairs. The comparison of AD3, AD4 and Vina docking results will be discussed for each intercalator.

Proflavine Complexes

Proflavine (PF) is a small planar symmetric molecule that is known to have a mutagenic effect on DNA. It intercalates in a parallel mode between the DNA base pairs with its amino-groups looking into the major groove. Most structures of PF complexes deposited in NDB are its complexes with dinucleoside monophosphates. There is only one deposited PF-DNA hexamer complex in NDB (dd0103) in which two PF molecules are inserted in the hexamer terminal steps. Three PF-dinucleoside monophosphate structures and the PF-hexamer complex were considered in this study. Two amino-groups of PF were allowed to rotate during docking.

The calculated best correspondence RMSD values with respect to the NDB structures for top ranking docking poses are shown in Table 3. One can see that RMSD values for AD3 and AD4 PF complexes are

Figure 2. Structural formulas of intercalators

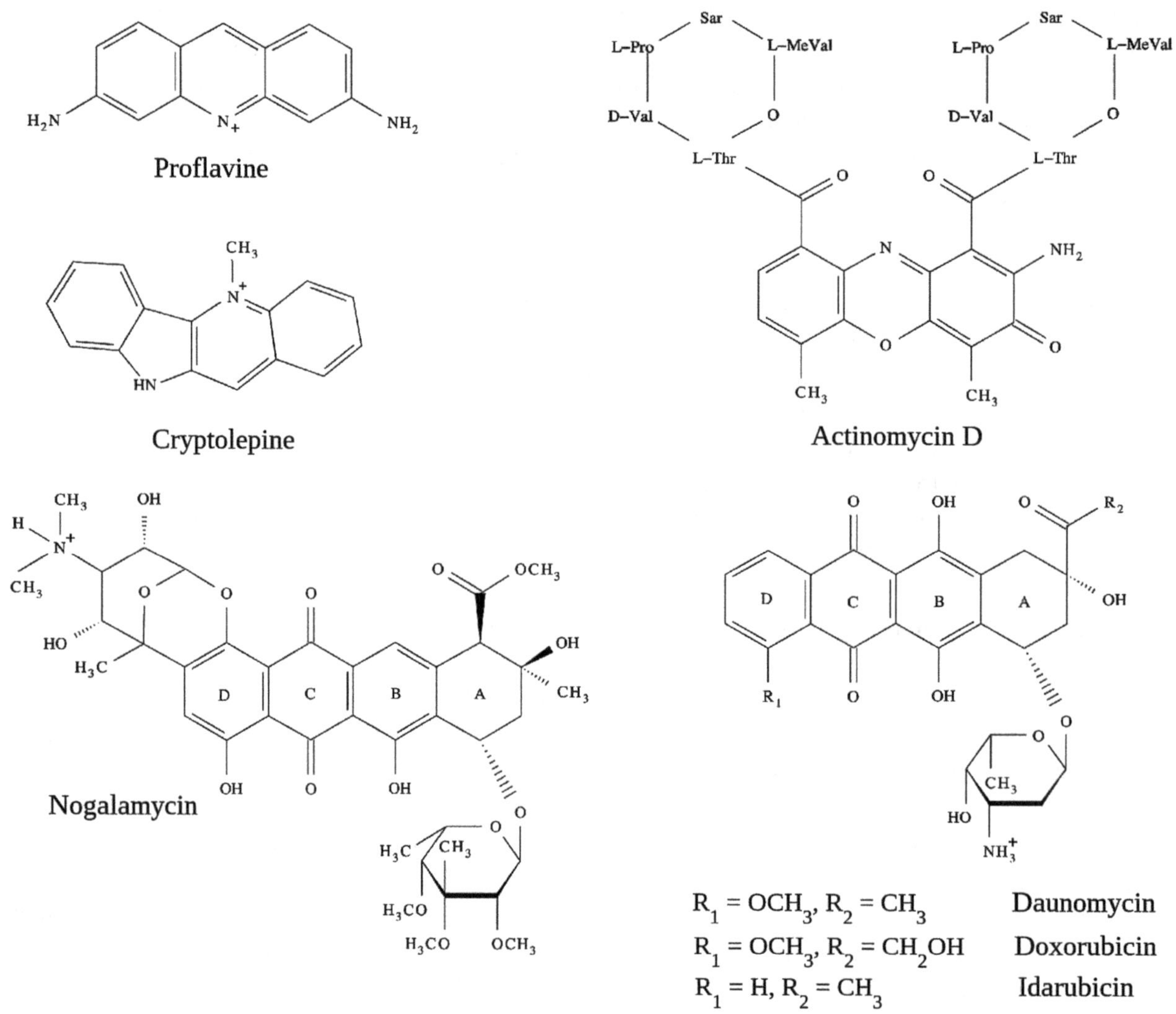

in range of 1.5-2.4 Å. These values are relatively high in view of the fact that PF is quite a small molecule. Visual inspection of complexes reveals that while the PF chromophore is docked correctly parallel to the long base pair axis, its orientation is opposite to that of the NDB structures: the amino-groups of PF are directed towards the minor groove (Figure 3a, 3b). On the other hand, in the modeling study of Sasikala and Mukherjee (2013), it was shown that such alternate orientation of PF in the intercalated state is possible and it is even slightly more stable than the crystal orientation.

The RMSD values for Vina top docked structures are higher than those of AD3 and AD4 and fall in range of 2.1-10.0 Å. The PF chromophore in most Vina complexes adopts intermediate position between parallel and perpendicular insertion (Figure 3c), which causes higher RMSD. Besides, Vina failed to insert the PF into the first intercalation site of dd0103 DNA hexamer structure; the minor groove complex of PF was obtained instead (Figure 3d).

Table 2. Docking test set DNA-ligand complexes

Ligand	NDB Code	PDB Code	Resolution, Å	Ligand	NDB Code	PDB Code	Resolution, Å
Intercalation Complexes				**Minor-Groove Complexes**			
Proflavine	ddb009	-	0.83	**Berenil**	gdl016	1d63	2.00
	ddb033	-	0.89		gdl009	2dbe	2.50
	ddb034	-	0.93		dd0081	2gvr	1.65
	dd0103	3ft6	1.12	**DAPI**	gdl008	1d30	2.40
Cryptolepine	dd0047	1k9g	1.40		dd0002	432d	1.89
Daunomycin	ddf039	110d	1.90	**Hoechst 33258**	gdl022	127d	2.00
	ddf045	152d	1.40		gdl010	1d43	2.00
	ddf020	1d10	1.50		gdl012	1d45	1.90
	ddf001	1d11	1.18		gdl002	1dnh	2.25
	ddf018	1da0	1.50		gdl026	264d	2.44
	ddf031	1vth	1.60		gdl028	296d	2.25
	ddf032	1vti	1.70		gdl006	8bna	2.20
Doxorubicin	ddf044	151d	1.40	**Netropsin**	gdl014	121d	2.20
	ddf019	1d12	1.70		gdl030	195d	2.30
	ddf040	1da9	1.70		gdl004	1dne	2.40
Idarubicin	ddf050	198d	1.97		gdl001	1vtj	2.40
	ddf029	1d38	1.70		bd0078	1z8v	1.75
	ddf038	1d67	1.60		pd0678	1ztt	1.85
Nogalamycin	ddf049	182d	1.80		gdj046	261d	2.40
	ddf052	224d	1.40		2lwh	2lwh	NMR
Actinomycin D	1fja	1fja	NMR	**Distamycin A**	dd0042	1jtl	1.85
	dd0053	1mnv	2.60		dd0046	1k2z	2.38
	na2074	4hiv	2.60		gdl003	2dnd	2.20
	ddh048	173d	3.00		gdh060	378d	2.40
	1dsc	1dsc	NMR				
	ddh037	2d55	3.00				

Summarizing, AD4 and AD3 are able to reproduce the structure of most PF intercalation complexes within the resolution of ~2 Å, and they perform definitely better than Vina in PF-DNA docking.

Cryptolepine-DNA Complex

Cryptolepine is an antimalarial drug that, unlike most intercalators, interacts with the non-alternating CC sites (Lisgarten, Coll, Portugal, Wright, & Aymami, 2002). The drug intercalates parallel to the long DNA base pair axis in an asymmetric way: its left six-membered aromatic ring stacks between the two

Table 3. The RMSDs (in Å) of the ligand-intercalators top-ranked docking poses with respect to the NDB structures. The RMSD with respect to the flipped crystal pose is given in parentheses.

Ligand/Number of Active Torsions	Complex NDB Code		AD3	AD4	Vina
Proflavine 2 active torsions	ddb009		2.07	2.01	2.17
	ddb033		2.04	1.94	2.78
	ddb034		2.36	1.64	3.56
	dd0103	site1	1.55	1.62	10.02
		site2	1.54	1.60	3.11
Cryptolepine 0 active torsions	dd0047	site1	2.63	2.86	12.28
		site2	2.64	2.87	12.27
Daunomycin 9 active torsions	ddf039	site1	1.45	1.73	1.28
		site2	1.43	4.51 (1.99)	1.23
	ddf045	site1	0.84	1.03	0.83
		site2	0.84	1.04	0.82
	ddf020	site1	0.96	1.12	0.65
		site2	0.97	1.09	0.66
	ddf001	site1	1.03	1.14	0.48
		site2	1.00	1.26	0.66
	ddf018	site1	0.70	4.60 (1.89)	0.42
		site2	0.71	4.73 (2.09)	0.74
	ddf031	site1	0.99	1.61	0.80
		site2	0.99	1.63	0.73
	ddf032	site1	0.96	1.08	1.54
		site2	0.97	4.50 (1.78)	1.55
Doxorubicin 11 active torsions	ddf044	site1	0.73	1.28	0.50
		site2	0.69	1.27	0.56
	ddf019	site1	0.76	1.10	0.51
		site2	0.81	4.80 (1.81)	0.51
	ddf040	site1	1.26	4.38 (1.68)	1.19
		site2	1.28	4.42 (1.69)	1.22
Idarubicin 8 active torsions	ddf050	site1	0.61	0.82	0.85
		site2	0.61	1.08	0.93
	ddf029	site1	0.91	1.06	0.70
		site2	0.91	1.02	0.69
	ddf038	site1	0.56	1.01	0.33
		site2	0.62	1.02	0.64
Nogalamycin 13 active torsions	ddf049	site1	0.87	4.40 (1.64)	10.54
		site2	0.80	4.40 (1.65)	8.43
	ddf052	site1	0.93	4.32 (1.94)	10.64
		site2	0.85	1.54	8.46

continued on following page

Table 3. Continued

Ligand/Number of Active Torsions	Complex NDB Code		AD3	AD4	Vina
Actinomycin D 9 active torsions	1fja	site1	0.93	0.69	0.79
		site2	0.83	0.74	0.86
	dd0053	site1	0.81	0.72	2.32 (1.33)
		site2	2.29 (0.79)	2.32 (0.72)	2.33 (0.98)
	na2074	site1	2.57 (1.92)	1.67	2.50 (1.69)
		site2	2.32 (1.76)	2.36 (1.74)	2.33 (1.49)
	ddh048		0.94	2.27 (0.83)	7.11
	1dsc		0.89	1.08	5.59
	ddh037		1.01	1.23	2.81 (1.82)

Figure 3. The top-ranked PF docking poses for dd0103 structure: a) AD3 at intercalation site2; b) AD4 at intercalation site 2; c) VINA at intercalation site 2; and d) VINA at intercalation site 1. Drug docking pose is shown in cyan (gray) with filled chromophore, crystal pose is shown in black.

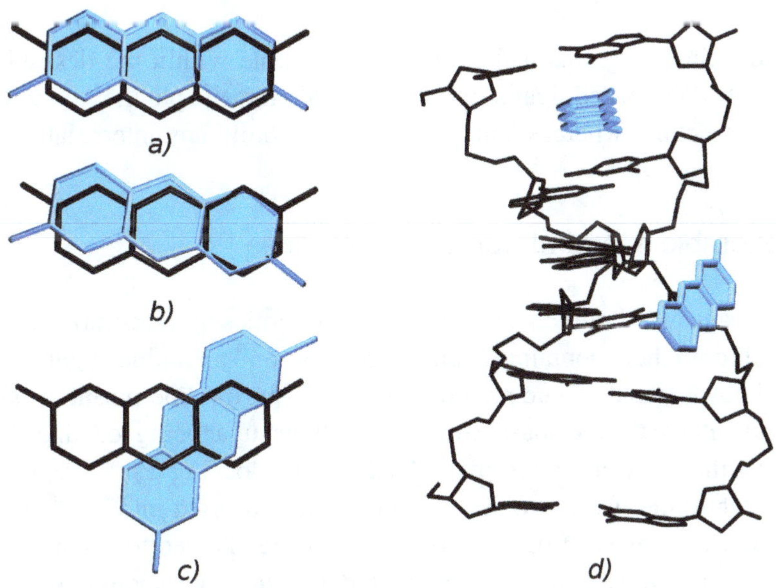

cytosines and the right part, consisting of two fused six-membered rings, stacks between the guanines. The structure of the deposited in the NDB cryptolepine-DNA hexamer complex contains two intercalation sites. As one can see from the Table 3, the RMSD values of the top-ranked AD3 and AD4 cryptolepine complexes equal to 2.6 and 2.9 Å, correspondingly. Such high values are caused by the incorrect drug orientation in the intercalation site (Figure 4a, 4b). The docked cryptolepine orientation is a mirror image of the crystal one, but due to the asymmetry of the intercalation site, these two orientations cannot be considered equivalent to each other. The RMSD values of the top-ranked Vina complexes are extremely high and equal to 12.3 Å. Despite the presence of the intercalation cavity, Vina docked ligand to the minor groove of the DNA hexamer (Figure 4c).

Figure 4. The top-ranked cryptolepine docking poses for intercalation site1 of dd0047: a) AD3; b) AD4; and c) VINA. Color scheme is the same as in Figure 3.

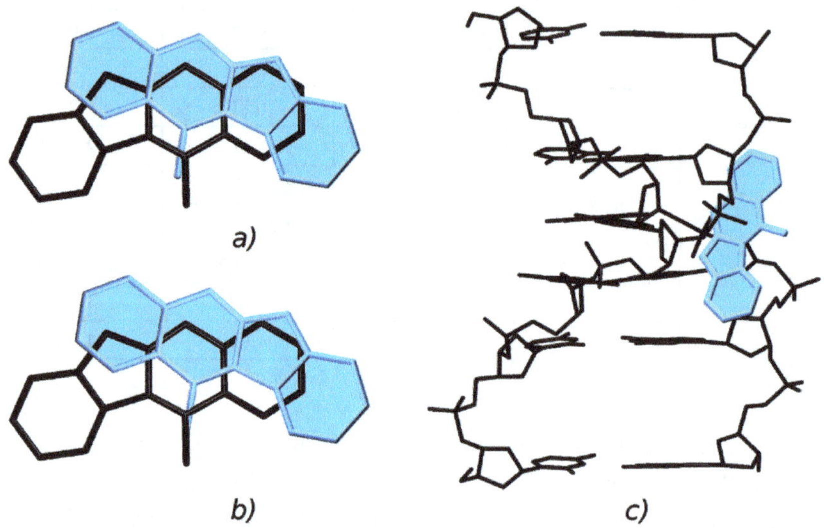

So, none of the docking programs had docked cryptolepine within the desired accuracy of 2 Å. Nevertheless, AD3 and AD4 docked ligand into the intercalation site parallel to the long base pair axis within 3 Å of crystal structure, whereas Vina was unable to build any intercalation complex of DNA and cryptolepine.

Daunomycin, Doxorubicin, and Idarubicin Complexes

Daunomycin, doxorubicin and idarubicin are anticancer antibiotics of the anthracycline family. They all possess an aglicone chromophore containing four fused rings and an amino sugar attached to it (Figure 2). These drugs are known to intercalate preferentially to 5'-pyrimidine-purine-3' sites inserting their chromophore perpendicular to the long base pair axis with their sugar being localized in the minor groove of DNA. All structures that were chosen from the NDB for the docking test set represent the complexes of a drug with a DNA hexamer were two drug molecules are inserted into the terminal steps of DNA helix from both sides. The number of ligand torsions that were allowed to change during docking was equal to 9, 11 and 8 for daunomycin, doxorubicin and idarubicin, correspondingly.

The RMSD values obtained for the top-ranked docking poses of AD3 and Vina are less than 1.6 Å for all NDB structures (Table 3). As one can see (Figure 5a, 5c), there is an excellent agreement between the position of drug in the docking complex and in the X-ray structure. For the AD4 top-ranked docking complexes, there are two ranges of RMSD: the RMSD is less than 1.8 Å for the majority of NDB targets indicating good fit to the crystal structures, but for some targets, RMSD falls in range 4.3 - 4.8 Å. In these complexes, the orientation of drug is flipped with respect to the crystal structure and the drug amino sugar is located from the other side of the chromophore plane (Figure 5b). Taking into account the symmetry (or pseudo symmetry) of the intercalation site, the existence of such complexes is quite possible. Therefore, the RMSD for these docking complexes was recalculated with respect to the

Figure 5. The top-ranked daunomycin docking poses for intercalation site2 of ddf039: a) AD3; b) AD4; and c) VINA. Color scheme is the same as in Figure 3.

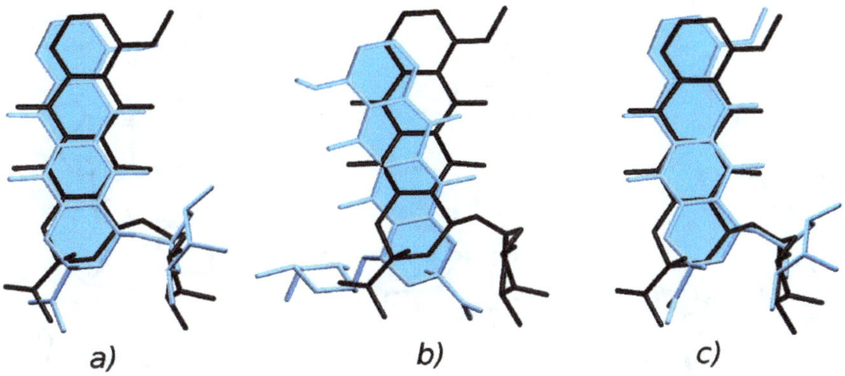

flipped crystal ligand structures. The recalculated RMSD values are less than 2.1 Å for all of the AD4 top-ranked docked structures. Summarizing, all the docking programs reproduce the X-ray structures of daunomycin, doxorubicin and idarubicin-DNA intercalation complexes reasonably well (RMSD <2.1 Å) with Vina and AD3 being more accurate than AD4.

Nogalamycin Complexes

The antibiotic nogalamycin also belongs to the anthracycline family but in contrast to daunomycin it has two sugar substituents on both ends of the aglicone chromophore. An uncharged nogalose sugar is attached to the A-ring of the drug chromophore and a positively charged bicyclo amino sugar is attached to the D-ring of chromophore (Figure 2). Upon intercalation the nogalamycin chromophore inserts perpendicular to the long axis of the DNA base pairs with its nogalose sugar located in the minor groove and bicyclo amino sugar being in the major groove of DNA. In the crystal complexes from databank, the drug is intercalated into the two terminal steps of the DNA hexamer. The number of nogalamycin active torsions that were allowed to rotate during docking was equal to 13.

The top-ranked docking poses of AD3 reproduce very well the orientation of nogalamycin in the X-ray complex with RMSD being less than 1 Å (Table 3, Figure 6a). In all except one AD4 top-ranked complexes, the drug is in flipped orientation with respect to the crystal structure resulting in RMSD ~ 4.4 Å (Figure 6b). Assuming that such drug orientation is equally probable to the crystal one, the RMSD for AD4 flipped complexes was recalculated with respect to the flipped crystal orientation. In this case all AD4 RMSD values were less than 2 Å. The RMSD values for Vina top-ranked complexes were above 8 Å. Such high values were obtained because Vina failed to construct the intercalation complexes of nogalamycin and DNA and built the minor-groove complexes instead (Figure 6c). So, despite the presence of the same aglycon chromophore in structures of nogalamycin and daunomycin, Vina that performed extremely well in docking of daunomycin was unable to reproduce the nogalamycin intercalation complexes. Both AD3 and AD4 docked nogalamycin into the DNA intercalation site, but the AD3 complexes were closer to the X-Ray structure.

Figure 6. The top-ranked nogalamycin docking poses for intercalation site1 of ddf049: a) AD3; b) AD4; and c) VINA. Color scheme is the same as in Figure 3.

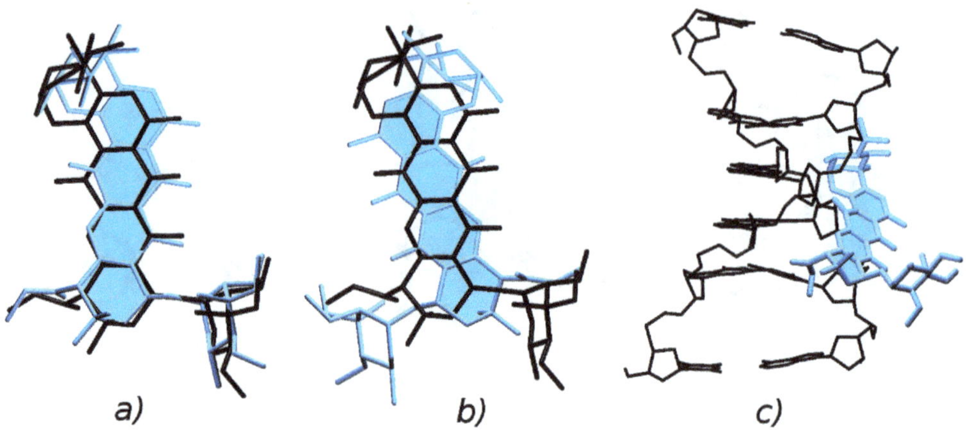

Actinomycin D Complexes

Actinomycin D (AMD) is an antitumor antibiotic that has been used for cancer treatment in medical practice for many years. It consists of the phenoxazone chromophore and two identical cyclic pentapeptide lactones (Figure 2). The drug is known to intercalate preferentially into the GC sites with its side chains being in the DNA minor groove. In structures deposited in NDB, either one AMD molecule is intercalated into central GC step or two drug molecules insert symmetrically into non-terminal GC sites from both sides of the helix. 9 torsions in AMD were allowed to change during docking.

There were three ranges of RMSD values calculated for the top-ranked docking complexes: RMSD < 2Å, 2 Å < RMSD < 3 Å and RMSD > 5 Å (Table 3). Complexes with RMSD < 2 Å reproduced the structure of the crystal AMD complexes very well (Figure 7a). Visual inspection of complexes with RMSD in range 2-3 Å revealed that, in these structures, the AMD was in flipped state with respect to the crystal orientation (Figure 7b). When symmetry of the intercalation site was taken into consideration, the recalculated RMSD values for these complexes dropped below 2 Å. RMSD values higher than 5 Å were obtained for two structures by Vina. For NDB structures ddh048 and 1dsc Vina was unable to insert the AMD into the intercalation cavity of DNA and the minor-groove complexes of AMD were obtained instead (Figure 7c). So, AD3 and AD4 were able to reproduce all the NDB AMD intercalation complexes within the accuracy of 2 Å RMSD while Vina could not construct intercalation complexes for two structures from the AMD test set.

Summarizing the results for all intercalators, the AD3 and AD4 top-ranked docking poses were within 2 Å of crystal structures for all intercalators except cryptolepine. These two programs are a good choice for docking of ligands to intercalation targets and they can be applied for the de novo design of intercalators. At the same time, program Vina reproduced properly the structures of intercalation complexes only for a number of antibiotics of the anthracycline family (daunomycin, doxorubicin and idarubicin) and failed to construct the intercalation complexes for nogalamycin and cryptolepine, and for PF and AMD with some targets from the test set. Thus, in general, Vina should not be recommended for the construction of drug-DNA intercalation complexes.

Figure 7. The top-ranked AMD docking poses for ddh048: a) AD3; b) AD4; and c) VINA. Color scheme is the same as in Figure 3.

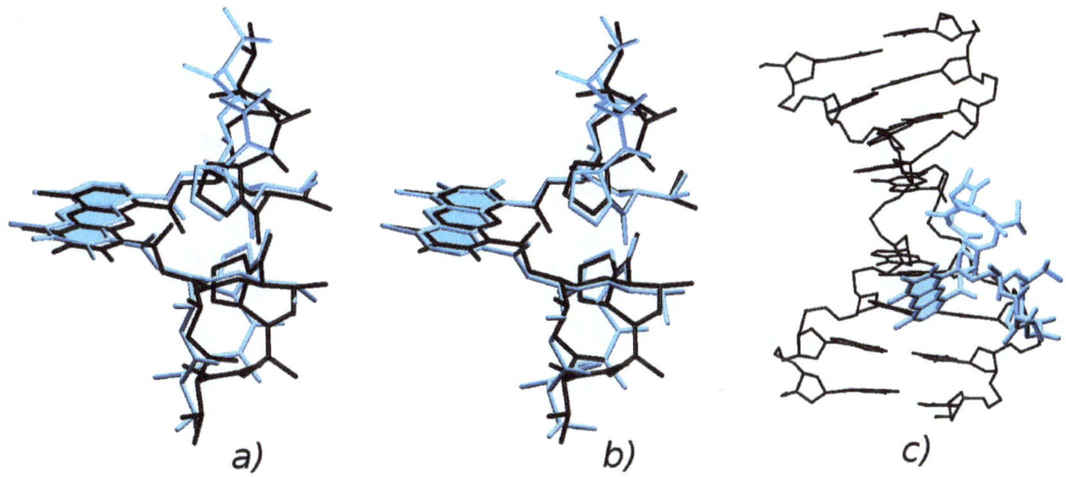

Re-Docking of DNA-Ligand Minor-Groove Complexes

Another important type of non-covalent ligand-DNA binding is the binding of ligands in the DNA minor groove (Baraldi et al., 2004). Under physiological conditions, DNA exists as a double helix composed of two anti-parallel strands that twist around each other. Strands form grooves: major and minor. In the B-form of DNA duplex the minor groove is much narrower than the major one being an attractive target for binding of small ligands. Unlike intercalators, minor groove binders usually don't disturb much the structure of DNA helix upon binding. The common features of minor groove binders are a positive charge and a crescent shape resembling the geometry of the DNA minor groove. Most of them bind preferentially to AT-rich DNA sequences (Neidle, 2001). In this study, five minor groove binders were considered: berenil, DAPI, Hoechst 33258, netropsin and distamycin A (Figure 8). The NDB indices of the drug-DNA minor-groove complexes that were taken for the docking test set are given in Table 2. All DNA-targets contained central AT-tract consisting of 4-6 base pairs. The center of the grid, defining the search space, was chosen in the geometric center of bound ligand in crystal structure. First, the docking in a large grid (27Åx27Åx27Å) enclosing the most part of the DNA-target was performed. All programs obtained minor-groove complexes for all DNA-targets of the test set (except for the two distamycin complexes built by Vina). But the calculated best correspondence RMSD values of these minor groove complexes were unexpectedly high due to the fact that ligands were displaced along the DNA minor groove with respect to the crystal pose. The length of the AT-tract favorable for ligand binding usually exceeds the size of ligand. Therefore there can be multiple binding sites for a ligand in the minor groove of DNA-target. Unlike the proteins, one can hardly define the exact boundaries of the binding site for ligand in the DNA minor groove. In order to compare directly the docked ligand pose with the crystal one, the appropriate smaller search space was defined for each ligand; large enough to allow ligand rotate freely in it. The results of docking in small search space are given below. The symmetry of the DNA-target was taken into consideration upon RMSD calculation: if the docking ligand pose was flipped with respect to the crystal pose, the RMSD of the docking pose was calculated with respect to the flipped crystal drug orientation.

Figure 8. Structural formulas of minor groove binders

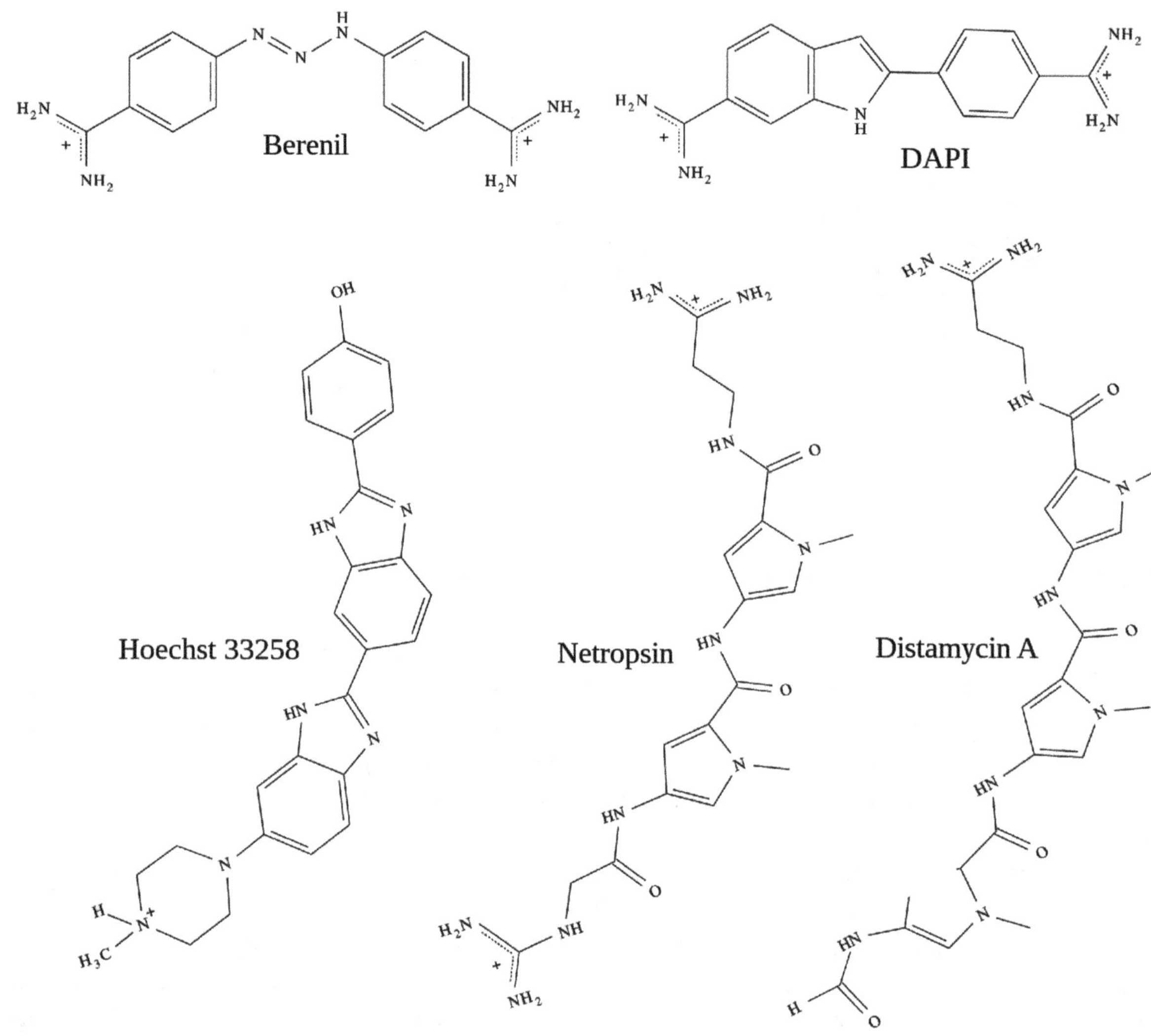

Berenil Complexes

Berenil is an aromatic diamidine that is being used as an antitrypanosomal drug for animals. Three structures of berenil complexes were taken from the NDB to the docking test set. The size of the AutoDock grid used for docking was 42x42x42 points, with grid spacing of 0.375 Å, which corresponded to the 16Åx16Åx16Å search space of Vina. There were 5 active torsions in ligand that were allowed to rotate during docking. The RMSD of the top-ranked docking poses was less than 2 Å for all DNA-targets of AD3 (Figure 9a) and for two targets of Vina (Table 4). The Vina top-ranked pose for target dd0081 had RMSD of 3.6 Å. In this case, the drug was in similar conformation to the crystal structure but despite the small search space, it was displaced along the groove for one base pair with respect to crystal pose (Figure 9c). The top-ranked berenil complexes obtained by AD4 had RMSD > 3 Å due to the incorrect

Figure 9. The top-ranked berenil docking poses for dd0081: a) AD3; b) AD4; and c) VINA. Color scheme is the same as in Figure 3.

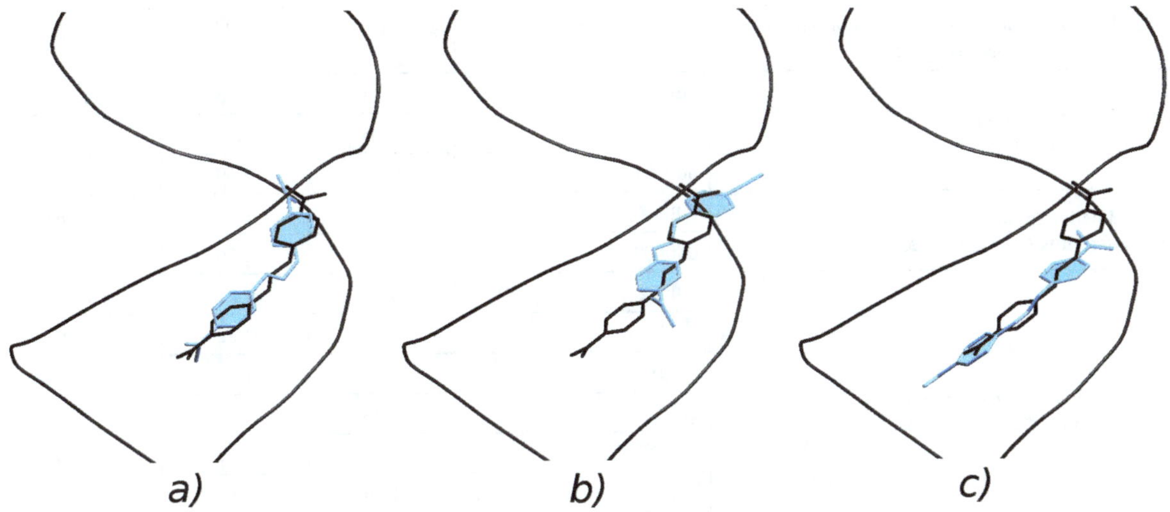

drug orientation: unlike the crystal pose, the amidine groups of berenil were pointing outside the minor groove (Figure 9b). Summarizing, the poses predicted by AD3 and Vina were within 2 Å from crystal poses, whereas AD4 didn't find the correct orientation of ligand in the binding site.

DAPI Complexes

DAPI (4',6-diamidino-2-phenylindole) is a synthetic antibiotic that is mainly used as a DNA stain in fluorescent microscopy. Two NDB structures of DAPI were considered. The AutoDock grid of 40x40x40 points was used that corresponded to 15Åx15Åx15Å box of Vina. The number of active torsions in DAPI that could be changed was equal to 3. The top-ranked complexes of AD3 and Vina reproduced the crystal structures to within 2 Å RMSD (Figure 10a, 10c), whereas the RMSD of the best scored AD4 poses was higher than 2 Å (Table 4). As in the case of berenil, the incorrect orientation of ligand was obtained by AD4: the amidine groups of the docking pose were directed outside the groove unlike the crystal pose (Figure 10b).

Hoechst 33258 complexes

Hoechst 33258 is a bis-benzimidazole derivative that is used as a DNA fluorescent cytological stain. Seven X-ray structures were taken from the NDB for the Hoechst docking test set. The grid used for docking had 58x58x58 points that corresponded to 22Åx22Åx22Å search space in Vina. Four torsion angles of Hoechst were allowed to change during docking. The RMSD less than 2 Å was obtained for all the AD3 top-ranked docking poses and for all except one AD4 top-ranked poses (Table 4, Figure11a, 11b). The best-scored Vina poses for 4 of 7 DNA-targets had RMSD in range 2.8-6.2 Å compared to the crystal poses. In these complexes ligand was displaced along the groove and interacted with GC base pairs outside the AT-tract (Figure 11c). So, AD3 and AD4 performed better than Vina in the docking of Hoechst.

Table 4. The RMSDs (in Å) of the ligands-minor groove binders top-ranked docking poses with respect to the NDB structures

Ligand/ Number of Active Torsions/Size of Small Search Space	Complex NDB Code		AD3	AD4	Vina
Berenil / 5 active torsions/ 42x42x42 points / 16Åx16Åx16Å	gdl016		0.83	3.45	0.75
	gdl009		1.26	3.40	0.54
	dd0081		0.97	3.69	3.61
DAPI / 3 active torsions/ 40x40x40 points / 15Åx15Åx15Å	gdl008		1.77	2.21	1.68
	dd0002		0.79	2.50	0.53
Hoechst 33258 / 4 active torsions/ 58x58x58 points / 22Åx22Åx22Å	gdl022		0.95	1.26	2.79
	gdl010		1.21	1.22	4.56
	gdl012		1.32	1.20	4.24
	gdl002		0.96	0.89	6.17
	gdl026		1.17	1.07	1.16
	gdl028		1.69	1.85	0.97
	gdl006		1.99	2.26	0.67
Netropsin / 9 active torsions/ 58x58x58 points / 22Åx22Åx22Å	gdl014		1.48	3.75	5.87
	gdl030		1.39	3.72	0.85
	gdl004		3.52	5.06	2.00
	gdl001		1.34	4.02	1.69
	bd0078		1.60	3.64	0.76
	pd0678		1.59	3.51	1.79
	gdj046		1.15	4.06	1.27
	2lwh		1.47	3.29	0.73
Distamycin A / 10 active torsions/ 58x58x58 points / 22Åx22Åx22Å	dd0042		1.65	4.21	5.32
	dd0046		3.28	4.66	3.14
	gdl003		2.48	6.02	6.69
	gdh060	site1	4.09	5.28	3.70
		site2	3.57	5.94	3.55

Figure 10. The top-ranked DAPI docking poses for dd0002: a) AD3; b) AD4; and c) VINA. Color scheme is the same as in Figure 3.

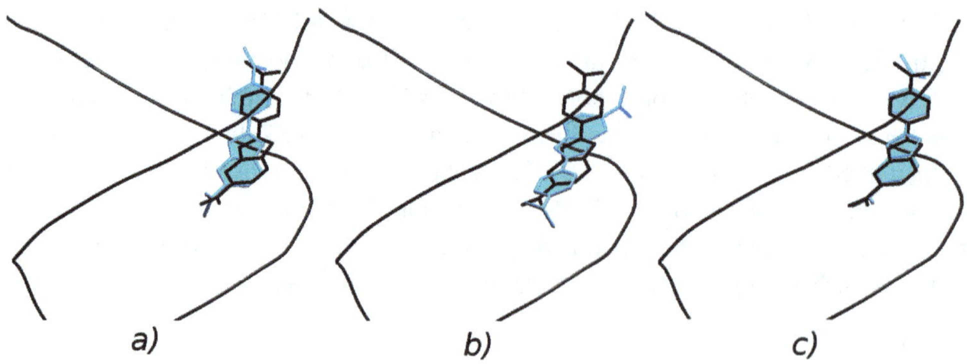

Figure 11. The top-ranked Hoechst docking poses for gdl002: a) AD3; b) AD4; and c) VINA. Color scheme is the same as in Figure 3.

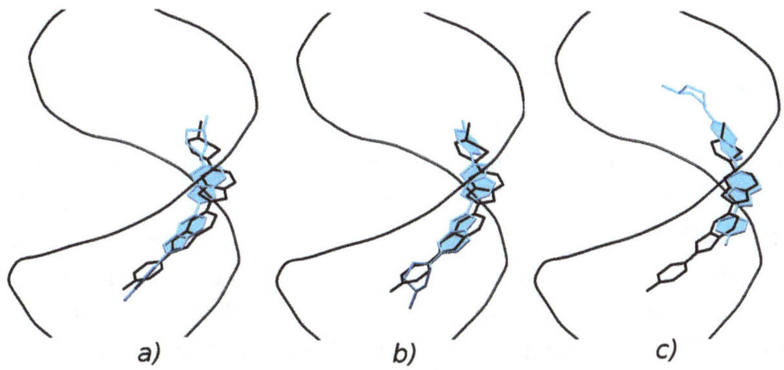

Netropsin Complexes

Netropsin is a naturally occurring dicationic oligopeptide that has antibiotic and antiviral activity. Eight netropsin complexes were chosen from the NDB for the docking test set. There were 9 active torsions in netropsin that could be changed during docking. Preliminary dockings of netropsin with a large grid (70x70x70 points) showed that the obtained clusters were poorly populated due to the insufficient sampling. Therefore further dockings of netropsin in smaller search space were performed with an increased number of energy evaluations (up to 10^8) in AD3 and AD4 and an increased exhaustiveness in Vina (up to 1000). The maximum number of generations in AD3 and AD4 was also increased up to 100000 so that the maximum number of energy evaluations would be the program stop criterion. Due to the increased computational cost, the number of docking runs was lowered to 20. The size of the small grid used for netropsin docking was the same as for Hoechst (58x58x58 points).

As one can see from the Table 4, AD3 and Vina were able to predict the crystal pose of netropsin to within 2 Å RMSD for 7 of 8 NDB complexes (Figure 12a, 12c). At the same time, the RMSDs of the top-ranked AD4 poses were above 3 Å for all complexes of the test set. In the AD4 complexes, netrop-

Figure 12. The top-ranked netropsin docking poses for 2lwh: a) AD3; b) AD4; and c) VINA. Color scheme is the same as in Figure 3.

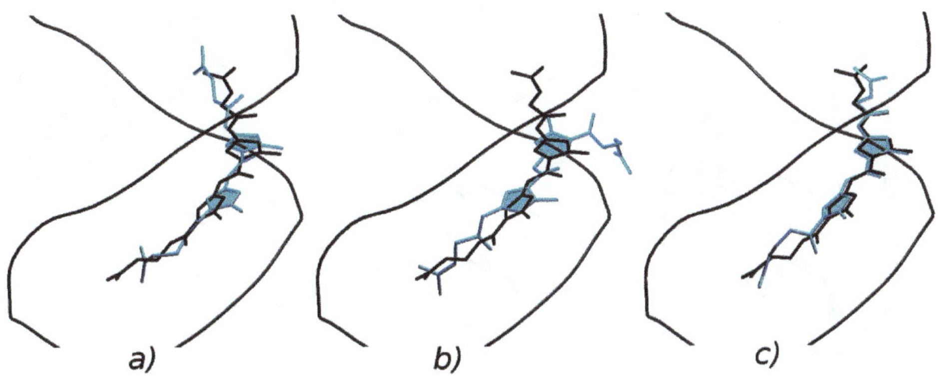

sin was in a conformation different from the crystal pose: one of the ligand tails was bent and directed outside the minor groove (Figure 12b). So, AD3 and Vina showed better results in netropsin docking than AD4 did.

Distamycin A Complexes

Distamycin A is a monocationic oligopeptide antibiotic consisting of three methylpyrole rings. There were 4 structures of distamycin complexes in the docking test set. One of them (gdh060) had two bound distamycin molecules. Two binding sites centered on each bound ligand were considered for this structure. 10 torsions of ligand were allowed to rotate during docking. Two distamycin major-groove complexes were obtained as top-ranked structures by Vina for DNA-targets dd0042 and dd0046 during preliminary dockings in a large grid (70x70x70 points). Due to the large number of active torsions in ligand, the parameters of docking sampling for further dockings were increased. The same docking protocol as for netropsin was used for distamycin docking in small grid (58x58x58 points). The top-ranked docking complexes obtained in the small search space by all programs were minor-groove complexes of distamycin. Nevertheless, the RMSD of only one AD3 top-ranked complex for 1jtl target was less than 2 Å (Table 4, Figure 13a). All the other best-scored complexes of AD3, AD4 and Vina had RMSD higher than 2 Å. Visual inspection of complexes revealed that in complexes of AD3 and Vina ligand was in conformation similar to the crystal one but it was displaced along the groove with respect to the crystal pose and extended beyond the AT-tract (Figure 13c). The distamycin conformation of AD4 top-ranked poses was different from the crystal one: as in the case of netropsin, one of the ligand tails was bent and directed outside the groove (Figure 13b). Summarizing the results, none of the docking programs was able to reproduce the distamycin X-ray complexes within the desired accuracy of 2 Å. The reason for this is most likely the large number of active torsions that were allowed to change during docking in distamycin molecule.

Taking into account the results of docking for all minor groove binders, one can conclude that AD3 program was the best in reproducing X-ray structures of DNA-ligand minor-groove complexes, Vina showed good results for 3 of 5 ligands and AD4 was able to reproduce only the complexes of Hoechst to within 2 Å RMSD.

Figure 13. The top-ranked distamycin docking poses for dd0042: a) AD3; b) AD4; and c) VINA. Color scheme is the same as in Figure 3.

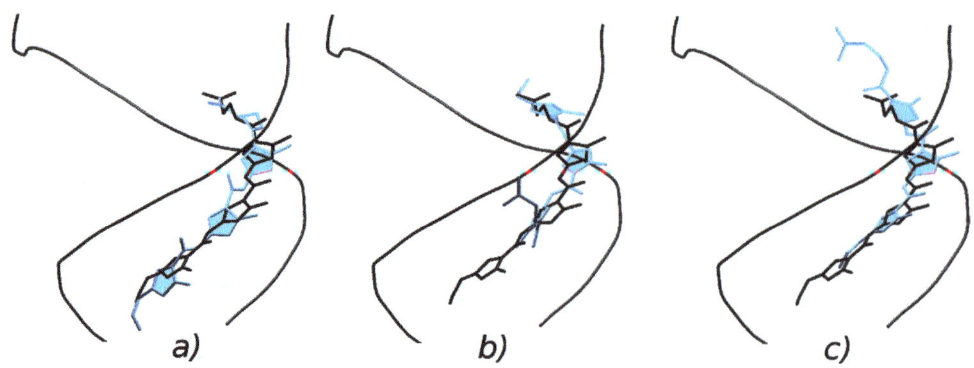

The performance of AD3, AD4 and Vina on docking of intercalators and minor groove binders to DNA-duplexes is summarized in Table 5. As one can see, the AD3 program is the best choice for docking of ligands to DNA duplexes. If the ligand binding mode is unknown, this program (AD3) should be recommended for NA-ligand docking as it gives accurate results both for intercalators and minor groove binders. The disadvantage of AD3 is being too slow due to the absence of parallel version. Therefore for ligands-intercalators, AD4 would be better choice for the construction of DNA-ligand complexes than AD3 because of its speed (parallel version) and accuracy. At the same time, for the docking of DNA minor groove binders, AD4 should be not recommended due to its poor performance for this type of DNA-ligand complexes. The main advantage of the program Vina is its speed and ease of use. Unfortunately, docking of ligands-intercalators to DNA in Vina should be avoided due to the large number of false complexes obtained by Vina for this class of ligands in the docking test set. On the other hand, Vina can be used for docking of minor groove binders if docking search space is confined to the proper part of DNA minor groove. Ligands with large number of rotatable bonds (more than 8) should be taken with caution. In this study, the successful docking of ligands doxorubicin and nogalamycin containing 11 and 13 rotatable bonds, respectively, was performed by AD3 and AD4. At the same time, even a tenfold increase of sampling parameters didn't help to reproduce the structures of X-ray complexes of ligand distamycin containing 10 rotatable bonds. Thus, the complexity of the docking task depends not only on the number of ligand rotatable bonds but also on the structures of particular target and ligand. In the case of doxorubicin and nogalamycin, most of the ligand active torsions rotated small radicals attached to the chromophore, and there was a clearly defined binding site for ligand in target - an intercalation gap. At the same time, the linear shape of the distamycin molecule resulted in a greater conformational

Table 5. Success rate of prediction of DNA-ligand crystallographic poses within 2 Å by AD3, AD4 and Vina (in %)

Ligand (Number of Active Torsions)		AD3	AD4	Vina
Intercalators	**Proflavine (2)**	40	80	0
	Cryptolepine (0)	0	0	0
	Daunomycin (9)	100	93	100
	Doxorubicin (11)	100	100	100
	Idarubicin (8)	100	100	100
	Nogalamycin (13)	100	100	0
	Actinomycin D (9)	100	100	78
	Average for all intercalators	77	82	54
Minor Groove Binders	**Berenil (5)**	100	0	67
	DAPI (3)	100	0	100
	Hoechst 33258 (4)	100	86	43
	Netropsin (9)	88	0	88
	Distamycin A (10)	20	0	0
	Average for all minor groove binders	82	17	60
Average for all ligands		80	50	57

freedom of ligand, and in the DNA minor groove, there were probably more ways to bind for ligand than in the intercalation cavity. One of the possible solutions to the large ligands docking problem is the reduction of ligand torsional degrees of freedom (by freezing some of rotatable bonds) or the use of specially designed docking protocols such as DINC – an AutoDock-based incremental protocol for docking large ligands (Dhanik, McMurray, & Kavraki, 2013).

THE INFLUENCE OF THE ELECTROSTATIC INTERACTIONS ON THE RESULTS OF LIGAND-DNA DOCKING

An important aspect of nucleic acids-BAS docking is the account of electrostatic interactions. Nucleic acids carry large negative charge; ligands that bind to them are usually positively charged. The choice of method for the calculation of partial atomic charges of target and ligand may have a critical influence on the results of docking. At the same time, in calculation of target and ligand partial charges, it is important to follow the methodology used by the authors of the docking program during its calibration. The program AD4 was calibrated using set of protein-ligand complexes with partial atomic charges of target and ligand calculated by Gasteiger method (Gasteiger & Marsili, 1980). This method advantage is being very fast. However, it was shown that the accuracy of AD4 docking with respect to the complex geometry could be significantly increased when semi-empirical PM6 partial charges were used both for protein and ligand (Bikadi & Hazai, 2009). Unfortunately, there is no information in literature on application of this method for docking in AD4 when nucleic acids are used as targets. Besides, the drawback of PM6 method is being computationally expensive for large target molecules.

The earlier version of AutoDock program (AD3) used Kollman method for the calculation of partial charges on atoms of macromolecule-target. The Kollman charges are used also in AMBER force field that is widely applied for molecular dynamics simulations of nucleic acids (Cornell et al., 1995). It would be interesting thus to compare the results of AD4 docking with Kollman charges on atoms of DNA-targets with the results of docking to DNA-targets with Gasteiger charges.

The Kollman partial charges are being more polar compared to the Gasteiger ones. In the AD4 scoring function (Huey et al., 2007), the charge values are needed for the calculation of two terms: the electrostatic and the desolvation term. The last one includes an atomic solvation parameter that is calculated as:

$$S_i = ASP_i + QASP \times |q_i| \tag{1}$$

where q_i is the atomic charge and ASP_i and $QASP$ are the solvation parameters. The parameter ASP_i was calibrated for six atom types (C, A, N, O, S, H) and the single parameter $QASP$ was calibrated over the set of atom charges on all atom types (it was equal to 0.01097 in the original AD4 function). In this study, in order to account for bigger absolute values of Kollman charges compared to Gasteiger ones, the modified value of QASP parameter equal to 0.006383 was used for dockings of DNA-targets with Kollman partial charges. This value was obtained by Wang et al. (2011) for AD4 when Kollman charges were applied for protein-targets and AM1-BCC charges were used for ligands. Besides, in that work, the recalibration of AD4 scoring function using this charge model for protein-ligand complexes was performed, that was shown to improve the docking results both in structure and energy predictions.

In this study, the comparison of AD4 DNA-ligand docking results was done for three models: the DNA-target with Gasteiger charges and the original AD4 function abbreviated as GAD4, the DNA-target with Kollman partial charges and the original AD4 function (with modified QASP parameter) abbreviated as KAD4, and the DNA-target with Kollman charges and the recalibrated by Wang AD4 function (with modified QASP parameter) abbreviated as KWAD4. The ligand partial atomic charges were obtained by AM1-BCC method in all three models. The structures of complexes of intercalators PF (dd0103) and idarubicin (ddf038), and minor groove binder berenil (dd0081) were chosen for this docking test set. The size of the AutoDock grid was equal to 70x70x70 points in the case of intercalation complexes with grid center being localized at the geometric center of intercalation site base pairs. For berenil complex, the small grid of 42x42x42 points was used with grid center being in the geometric center of crystal bound ligand. The grid spacing was equal to 0.375 Å. The same docking protocol as previously described for these ligands was used.

The RMSDs of the top-ranked structures with respect to the crystal pose for all three models are given in Table 6. As one can see, docking with model KAD4 failed to construct the intercalation complexes of PF and idarubicin, complexes of external binding type were obtained instead (Figure 14a, 14b). Models GAD4 and KWAD4 were able to reproduce the crystal structures of intercalation complexes of PF and idarubicin to within 2 Å RMSD. At the same time, no improvement was observed for KWAD4 model with respect to GAD4 model. The orientation of PF in both complexes KWAD4 and GAD4 was incorrect compared to the crystal structure: the NH_2-groups of ligand were directed to the DNA minor groove. The top-ranked berenil poses obtained by all three models had RMSD higher than 3 Å with respect to the X-ray structure. The orientation of drug in the DNA minor groove was incorrect: its amidino-groups were pointing outside the groove unlike the crystal structure. The top-scored berenil poses of GAD4 and KWAD4 models were the same, therefore only KWAD4 pose is shown in Figure 14c. So, none of the considered models was able to reproduce the X-ray structure of the minor groove berenil-DNA complex within the desired accuracy.

The results of docking of PF and idarubicin with KAD4 model indicate that the change of the charge model without the recalibration of scoring function may have a critical impact on the docking results. Summarizing, both models GAD4 and KWAD4 could be used for docking of intercalators to DNA-targets, whereas none of the models should be recommended for docking of DNA and minor groove binders. The recalibration of the AD4 scoring function with respect to the DNA-ligand complexes is needed in order to improve results for minor groove binders.

Table 6. The RMSDs (in Å) of the PF, idarubicin, and berenil top-ranked AD4 docking poses for different charge models

Ligand/NDB Code		GAD4	KAD4	KWAD4
Proflavine/ dd0103	site1	1.62	8.98	1.60
	site2	1.60	9.17	1.60
Idarubicin/ ddf038	site1	1.01	10.98	1.04
	site2	1.02	11.01	1.05
Berenil/dd0081		3.69	3.87	3.70

Figure 14. The top-ranked AD4 docking poses for different charge models: a) PF; b) idarubicin; and c) berenil. The top-ranked docking pose for KWAD4 model is shown in cyan (light gray), KAD4 – in orange (dark gray), GAD4 – in black. For berenil, the top-ranked poses of GAD4 and KWAD4 were almost the same, therefore only the KWAD4 pose is shown and, instead of GAD4 pose, crystal pose is shown in black for comparison.

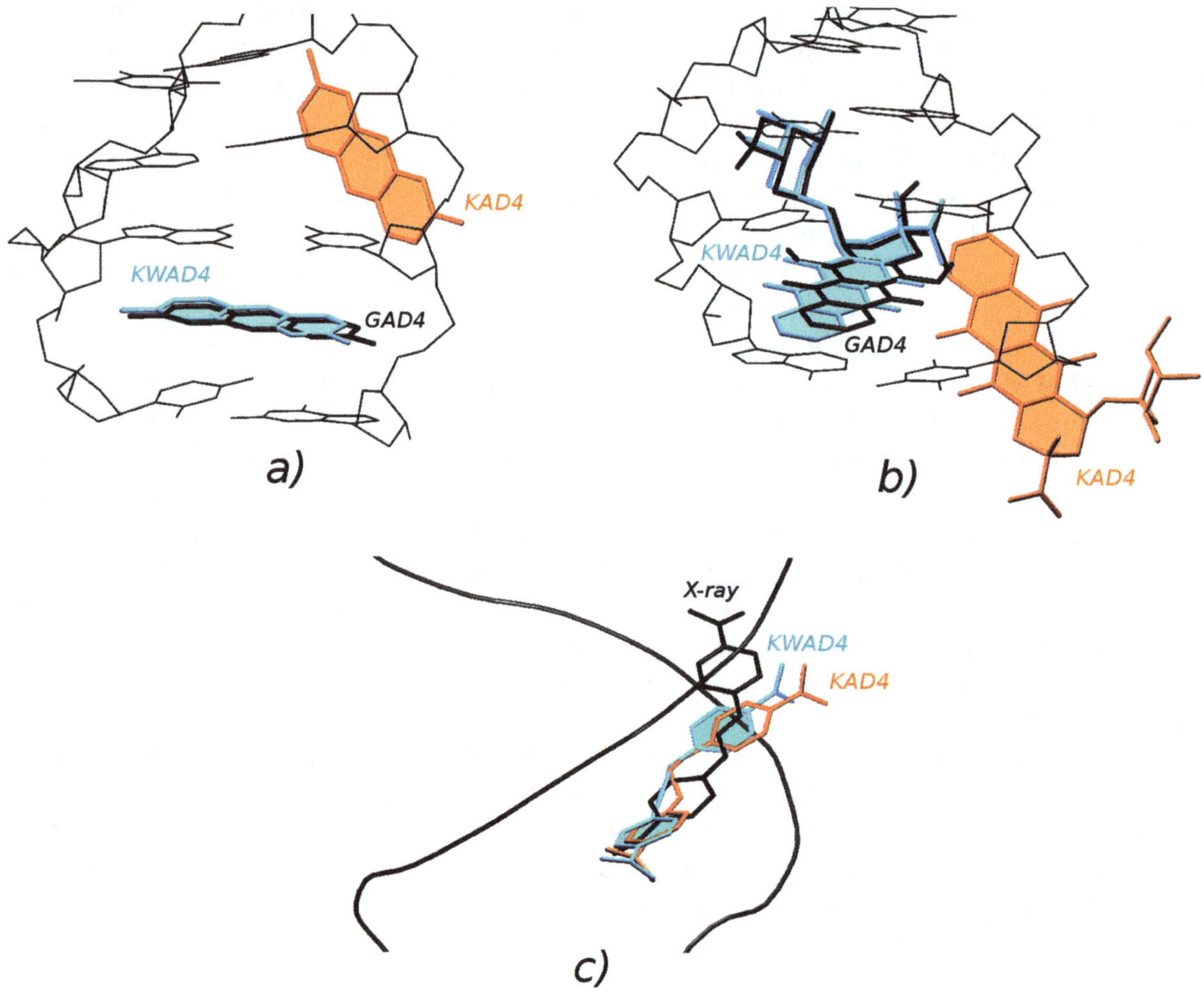

THE INFLUENCE OF THE DNA-TARGET CONFORMATION ON THE DOCKING RESULTS

Upon binding, the structures of macromolecule-target and ligand adapt to each other. While many programs account for the flexibility of the small ligand, the modeling of the flexibility of the target molecule remains a challenging task. Most docking programs can only partially take into account the conformational flexibility of macromolecule-target (changing some torsion angles in the macromolecule-target or allowing an overlap between the atoms of ligand and target in the docking complex). At the same time, ligands-intercalators greatly disturb the structure of the target-DNA helix: it lengthens and unwinds upon ligand binding. Therefore, in order to construct an intercalation complex of DNA and ligand using docking method, one needs first to create an artificial intercalation cavity in the DNA-target – the site for ligand insertion.

Intercalation Cavity Formation

The procedure of intercalation cavity formation is not standardized, while at the same time, it can affect the final structure of the resulting docking complex. Usually, the DNA helix is "cleaved" in the step of ligand insertion and lengthened by 3-4 Å and also unwound by the certain angle. Each intercalator has its own specific unwinding angle that can be determined in experiments with closed circular DNA. The analysis of NDB structures of intercalation complexes indicates that unwinding of the DNA-target can be localized not only in the intercalation step but the adjacent steps could also be involved. It is known that an insertion of ethidium into the DNA helix results in the significant unwinding of helix in the intercalation step, whereas, upon the daunomycin intercalation, there is no unwinding in the ligand insertion step but instead the adjacent helical steps are unwound. In this study, the examples of DNA intercalation targets formation protocols for ligands ethidium and daunomycin published earlier by the authors will be given (Miroshnychenko & Shestopalova, 2010).

The B-DNA decamer d(CGTACGTACG)$_2$ built in program FIBER of package 3DNA was used as DNA-target. There were three stages in intercalation site formation. First, in order to create an intercalation cavity, the DNA decamer was "cleaved" in the central CG-step, and one of the parts was shifted from the other on 3.4 Å along the helical axis. Then the unwinding was applied to the intercalation step or adjacent helical steps depending on the structural features of ligand X-ray complexes. And finally, the structure of sugar phosphate backbone of the deformed helical steps was optimized in module SANDER of AMBER 9 package with positional restraints on heavy base pair atoms. The optimization consisted of 2000 cycles: the first 500 of them were done with the steepest descent method and the rest – with the conjugate gradient method. The parmbsc0 force field was used upon optimization. The solvent was treated implicitly with Generalized Born model. The obtained DNA intercalation target was used for docking with the corresponding ligand. Docking was done in AD3 with the following parameters for Lamarckian genetic algorithm: maximum number of energy evaluations=10^7, maximum number of generations =27000, population size=100 individuals. The AutoDock grid of 70x70x40 points was centered on the geometric center of intercalation site base pairs. For each target and ligand 100 dockings were performed and the complexes obtained were clustered using rmstol=1 Å. Only the structure of highly populated clusters (with more than 10 conformations) was analyzed.

Intercalation Target for Ethidium

The unwinding angle of 26° determined for ethidium experimentally corresponds well to the observed reduction of twist angle in the intercalation site of the X-ray structure. Therefore, in the DNA intercalation target built for ethidium, all unwinding was localized in the ligand insertion step: the DNA helix was unwound there by 26°. The alternating sugar conformations C4'exo (3'-5') C1'exo were obtained for bases in the intercalation site after the optimization of sugar-phosphate backbone of DNA-target was performed. Docking of ethidium to this target resulted in complexes that well corresponded to the X-ray data: ligand chromophore was intercalated parallel to the long base pair axis with its benzene ring lying in the DNA minor groove. In order to calculate the RMSD with respect to the ethidium crystal pose, the base pairs of the intercalation site of DNA-target and those of the X-ray complex were superimposed.

The symmetry of the intercalation site was taken into account upon RMSD calculation. The obtained best correspondence RMSD value for the most highly populated cluster with respect to the crystal pose was equal to 1.0 Å (Figure 15).

Intercalation Targets for Daunomycin

In the X-ray structures of daunomycin complexes, the unwinding was observed in the helical step adjacent to the intercalation step. As ligand was inserted into the DNA terminal steps, the unwinding of helix could be observed only from one side of the ligand insertion step. Therefore it was unclear where there would be a symmetrical unwinding of DNA helix from both sides of the intercalation step or an asymmetrical unwinding only from one side, if ligand intercalates into the central step of DNA fragment. Two types of DNA intercalation targets were considered for daunomycin: the first target was built by symmetrical unwinding of two helical steps neighboring to the intercalation cavity (by 5.5° each), in the second intercalation target, the DNA helix was asymmetrically unwound by 11° only from one side of the intercalation step.

The docking complexes of daunomycin with both types of DNA intercalation targets were, in general, consistent with the X-ray structures: the chromophore of ligand was intercalated perpendicularly to the long base pair axis and its amino sugar was in the DNA minor groove. At the same time, the RMSD with respect to the crystal pose was lower for the daunomycin complex with asymmetrically unwound DNA-target (RMSD = 1.5 Å) than for the daunomycin docking complex with symmetrically unwound DNA-target (RMSD = 2.3 Å). In the former complex the angle between the long axis of daunomycin chromophore and the mean C1'C1' vector of the intercalation site was the same as in the crystal structure (Figure 16). Thus, the protocol with the asymmetrical unwinding of the DNA-target should be recommended for the construction of intercalation targets for daunomycin.

The examples of successful docking of ethidium and daunomycin to artificially constructed DNA intercalation targets indicate that such protocol of intercalation cavity formation can be applied on a regular basis for the construction of DNA-intercalation targets for various ligands. However, the peculiarities of each intercalator, the unwinding angle specific to it and the way of unwinding (in the intercalation step or adjacent step) should be taken into account upon target formation.

Figure 15. The top-ranked ethidium docking pose obtained for artificial intercalation target. Docking pose is shown in cyan (gray) with filled rings, crystal pose is shown in black.

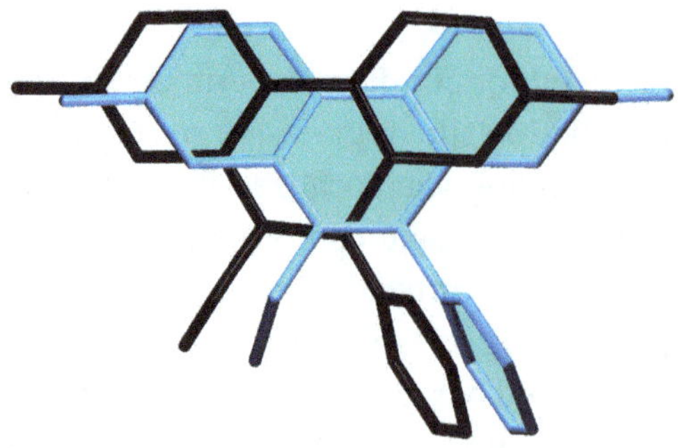

Figure 16. The docking poses of daunomycin obtained for symmetrically unwound DNA-target (a) and for asymmetrically unwound DNA-target (b). Docking pose is in cyan (gray), crystal pose is in black. The mean C1'C1' vector of the intercalation site is shown.

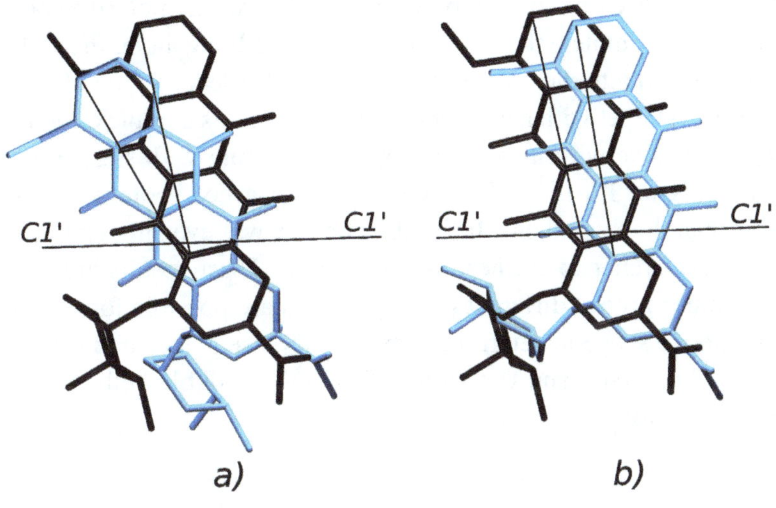

The Influence of the Structure of DNA-Target on the Minor-Groove Complex

During binding of ligands in the DNA minor groove, the structure of target usually changes insignificantly; therefore one can use structures of free DNA fragments deposited in the NDB as targets for minor groove binders. Besides, many programs (such as NAB of AmberTools or FIBER of 3DNA (Lu & Olson, 2008)) are able to build any DNA sequence in the canonical B or A-form. Minor groove binders prefer to bind to the DNA in the B-form due to the narrowness of its minor groove compared to the A-form. There is experimental evidence that binding of ligand in the DNA minor groove stabilizes the B-form of helix and prevents the B-A conformational transition of DNA (Song et al., 1997). But even in the B-DNA, the minor groove parameters can vary within a wide range for different DNA sequences. The AT-rich sequences are characterized by a narrower minor groove than the GC ones, which explains the specificity of most minor groove binders for the AT-regions.

In this study, the examples of PF minor-groove complexes with B-DNA duplexes of different sequence will be given. PF is an intercalator but it can also form a strong complex in the DNA minor groove (Sasikala & Mukherjee, 2013). The binding of PF to DNA occurs in two steps: first the strong minor-groove complex is formed and then – the intercalation complex (Li & Crothers, 1969). Two B-DNA-targets of 30 bp length were built in the module FIBER of 3DNA package: an alternating poly(dCG) oligonucleotide (structure id 5 in FIBER) with two-step helical twist of 72° and two-step helical rise of 6.72 Å and a poly(dA)·poly(dT) oligonucleotide (structure id 18 in FIBER) with helical twist of 36° and helical rise of 3.23 Å. The shape of the minor groove in these targets was quite different. The width of the groove was estimated as the minimal distance between the backbone atoms of opposite DNA strands. The poly(dCG) target had a very wide minor groove with width being equal to 7.9 Å. The minor groove of the poly(dA)·poly(dT) target was almost two times narrower: its width was equal to 4.4 Å. The docking of PF to these targets was done in AD3 with following parameters: maximum number of energy evaluations = 10^7, maximum number of generations = 27000, population size = 150 individuals. The grid of 70x70x70 points (gridspacing = 0.375 Å) centered in the middle step of the DNA-target was used.

As a result of docking of PF to these targets, two different minor-groove complexes were obtained. The first minor-groove complex was formed between the PF and the poly(dA)·poly(dT) target with a narrow minor groove. In this complex, the PF chromophore was lying along the DNA minor groove and its amino groups were directed into the groove (Figure 17a). The second type of minor-groove complex was obtained for poly(dCG) target with a wide minor groove. In this complex, the PF chromophore was located across the minor groove with its amino groups interacting with sugar phosphate backbone of one of the DNA chains (Figure 17b). The first type PF minor-groove complex was 1.5 kcal/mol more energetically favorable than the second one. Its binding energy was even higher than that of the PF intercalation complexes (Miroshnychenko & Shestopalova, 2013). These results indicate that depending on the shape of the DNA minor groove different binding modes are possible for one and the same ligand. At the same time, the groove shape is determined by the DNA sequence and the environmental conditions. The conformational diversity and flexibility of the DNA double helix appears to play a crucial role in the drug-DNA recognition.

CONCLUSION

In this study, the comparative analysis of the DNA-ligand docking with programs AD3, AD4 and Vina was performed. The old version of AutoDock program (AD3) was shown to be the most successful in re-docking both intercalation complexes and minor-groove complexes. This program can be recommended for a general docking of ligands to DNA-targets. At the same time, it should be taken in mind, that the scoring function of AD3 was calibrated for the set of protein-ligand complexes. Therefore the

Figure 17. Two different types of PF minor-groove complexes: for DNA-target with narrow minor groove (a) and for DNA-target with wide minor groove (b). PF is shown in cyan (gray).

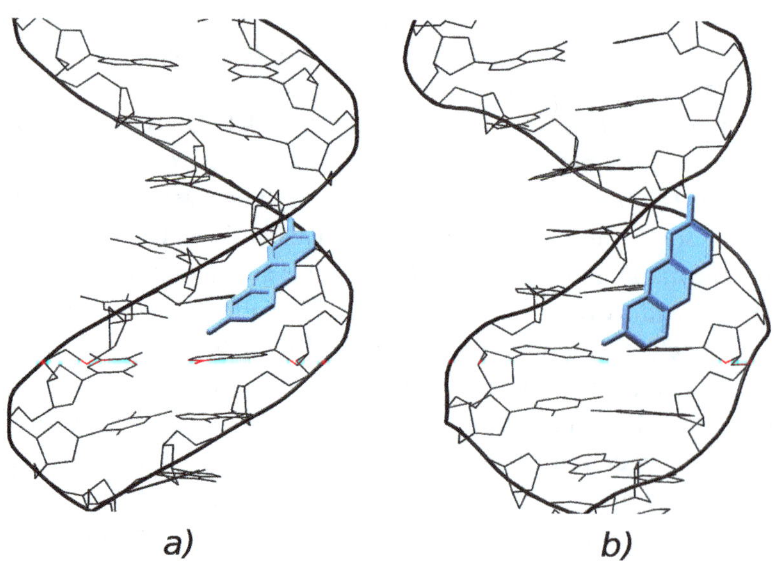

obtained values of DNA-ligand binding energies can hardly be compared to the experimental values. In general, to obtain the meaningful binding energies, the recalibration of AD3 scoring function is needed with respect to the set of DNA-ligand complexes.

The program AD4 showed good results in re-docking of DNA-ligand intercalation complexes, and as AD3, it can also be used for docking of ligands-intercalators to appropriately prepared DNA intercalation targets. In this study, the protocols for the construction of the DNA intercalation targets were given.

While the program Vina is not recommended for the construction of DNA-ligand intercalation complexes, it can be used with caution for docking of DNA-targets and minor groove binders if small and appropriately centered search space will be defined.

The proper account of electrostatic interactions in DNA-ligand docking is of great importance. It was shown that the deviation from the charge model used upon the docking scoring function calibration could give unpredictable docking results.

REFERENCES

Allen, W. J., & Rizzo, R. C. (2014). Implementation of the Hungarian algorithm to account for ligand symmetry and similarity in structure-based design. *Journal of Chemical Information and Modeling*, *54*(2), 518–529. doi:10.1021/ci400534h PMID:24410429

Baraldi, P. G., Bovero, A., Fruttarolo, F., Preti, D., Tabrizi, M. A., Pavani, M. G., & Romagnol, R. (2004). DNA minor groove binders as potential antitumor and antimicrobial agents. *Medicinal Research Reviews*, *24*(4), 475–528. doi:10.1002/med.20000 PMID:15170593

Berman, H. M., Olson, W. K., Beveridge, D. L., Westbrook, J., Gelbin, A., Demeny, T., & Schneider, B. et al. (1992). The Nucleic Acid Database: A comprehensive relational database of three-dimensional structures of nucleic acids. *Biophysical Journal*, *63*(3), 751–759. doi:10.1016/S0006-3495(92)81649-1 PMID:1384741

Bikadi, Z., & Hazai, E. (2009). Application of the PM6 semi-empirical method to modeling proteins enhances docking accuracy of AutoDock. *Journal of Cheminformatics*, *1*(15). doi:10.1186/1758-2946-1-15 PMID:20150996

Brana, M. F., Cacho, M., Gradillas, A., de Pascual-Teresa, B., & Ramos, A. (2001). Intercalators as anticancer drugs. *Current Pharmaceutical Design*, *7*(17), 1745–1780. doi:10.2174/1381612013397113 PMID:11562309

Case, D. A., Babin, V., Berryman, J. T., Betz, R. M., Cai, Q., Cerutti, D. S. … Kollman, P. A. (2014). AMBER 14. San Francisco, CA: University of California.

Chang, M. W., Ayeni, C., Breuer, S., & Torbett, B. E. (2010). Virtual screening for HIV protease inhibitors: A comparison of AutoDock 4 and Vina. *PLoS ONE*, *5*(8), e11955. doi:10.1371/journal.pone.0011955 PMID:20694138

Chen, Q., Shafer, R. H., & Kuntz, I. D. (1997). Structure-based discovery of ligands targeted to the RNA double helix. *Biochemistry*, *36*(38), 11402–11407. doi:10.1021/bi970756j PMID:9298959

Chen, Y.-C. (2015). Beware of docking! *Trends in Pharmacological Sciences*, *36*(2), 78–95. doi:10.1016/j.tips.2014.12.001 PMID:25543280

Cornell, W. D., Cieplak, P., Bayly, C. I., Gould, I. R., Merz, K. M. Jr, Ferguson, D. M., & Kollman, P. A. et al. (1995). A second generation force field for the simulation of proteins, nucleic acids, and organic molecules. *Journal of the American Chemical Society*, *117*(19), 5179–5197. doi:10.1021/ja00124a002

Dailey, M. M., Hait, C., Holt, P. A., Maguire, J. M., Meier, J. B., Miller, M. C., & Trent, J. O. et al. (2009). Structure-based drug design: From nucleic acid to membrane protein targets. *Experimental and Molecular Pathology*, *86*(3), 141–150. doi:10.1016/j.yexmp.2009.01.011 PMID:19454265

Deligkaris, C., Ascone, A. T., Sweeney, K. J., & Greene, A. J. Q. (2014). Validation of a computational docking methodology to identify the non-covalent binding site of ligands to DNA. *Molecular BioSystems*, *10*(8), 2106–2125. doi:10.1039/C4MB00239C PMID:24853173

Detering, C., & Varani, G. (2004). Validation of automated docking programs for docking and database screening against RNA drug targets. *Journal of Medicinal Chemistry*, *47*(17), 4188–4201. doi:10.1021/jm030650o PMID:15293991

Dhanik, A., McMurray, J. S., & Kavraki, L. E. (2013). DINC: A new AutoDock-based protocol for docking large ligands. *BMC Structural Biology*, *13*(Suppl 1), S11. doi:10.1186/1472-6807-13-S1-S11 PMID:24564952

Evans, D. A., & Neidle, S. (2006). Virtual screening of DNA minor groove bonders. *Journal of Medicinal Chemistry*, *49*(14), 4232–4238. doi:10.1021/jm0601957 PMID:16821782

Friesner, R. A., Banks, J. L., Murphy, R. B., Halgren, T. A., Klicic, J. J., Mainz, D. T., & Shenkin, P. S. et al. (2004). Glide: A new approach for rapid, accurate docking and scoring. 1. Method and assessment of docking accuracy. *Journal of Medicinal Chemistry*, *47*(7), 1739–1749. doi:10.1021/jm0306430 PMID:15027865

Gasteiger, J., & Marsili, M. (1980). Iterative partial equalization of orbital electronegativity—a rapid access to atomic charges. *Tetrahedron*, *36*(22), 3219–3228. doi:10.1016/0040-4020(80)80168-2

Gilad, Y., & Senderowitz, H. (2014). Docking studies on DNA intercalators. *Journal of Chemical Information and Modeling*, *54*(1), 96–107. doi:10.1021/ci400352t PMID:24303988

Guilbert, C., & James, T. L. (2008). Docking to RNA via Root-Mean-Square-Deviation-Driven Energy Minimization with Flexible Ligands and Flexible Targets. *Journal of Chemical Information and Modeling*, *48*(6), 1257–1268. doi:10.1021/ci8000327 PMID:18510306

Herman, T., & Tor, Y. (2005) RNA as a target for small-molecule therapeutics. *Expert Opinion on Therapeutic Patents, 15*(1), 49-62. doi:10.1517/13543776.15.1.49

Holt, P. A., Buscaglia, R., Trent, J. O., & Chaires, J. B. (2011). A discovery funnel for nucleic acid binding drug candidates. *Drug Development Research*, *72*(2), 178–186. doi:10.1002/ddr.20414 PMID:21566705

Holt, P. A., Chaires, J. B., & Trent, J. O. (2008). Molecular docking of intercalators and groove-binders to nucleic acids using Autodock and Surflex. *Journal of Chemical Information and Modeling*, *48*(8), 1602–1615. doi:10.1021/ci800063v PMID:18642866

Huey, R., Morris, G. M., Olson, A. J., & Goodsell, D. S. (2007). A semiempirical free energy force field with charge-based desolvation. *Journal of Computational Chemistry*, *28*(6), 1145–1152. doi:10.1002/jcc.20634 PMID:17274016

Humphrey, W., Dalke, A., & Schulten, K. (1996). VMD - Visual Molecular Dynamics. *Journal of Molecular Graphics*, *14*(1), 33–38. doi:10.1016/0263-7855(96)00018-5 PMID:8744570

Jain, A. N. (2003). Surflex: Fully automatic flexible molecular docking using a molecular similarity-based search engine. *Journal of Medicinal Chemistry*, *46*(4), 499–511. doi:10.1021/jm020406h PMID:12570372

Lang, P. T., Brozell, S. R., Mukherjee, S., Pettersen, E. F., Meng, E. C., Thomas, V., & Kuntz, I. D. et al. (2009). DOCK6: Combining techniques to model RNA-small molecule complexes. *RNA (New York, N.Y.)*, *15*(6), 1219–1230. doi:10.1261/rna.1563609 PMID:19369428

Li, H. J., & Crothers, D. M. (1969). Relaxation studies of the proflavine DNA complex: The kinetics of an intercalation reaction. *Journal of Molecular Biology*, *39*(3), 461–477. doi:10.1016/0022-2836(69)90138-7 PMID:5390557

Li, Y., Shen, J., Sun, X., Li, W., Liu, G., & Tang, Y. (2010). Accuracy assessment of protein-based docking protocols against RNA targets. *Journal of Chemical Information and Modeling*, *50*(6), 1134–1146. doi:10.1021/ci9004157 PMID:20481574

Lisgarten, J. N., Coll, M., Portugal, J., Wright, C. W., & Aymami, J. (2002). The antimalarial and cytotoxic drug cryptolepine intercalates into DNA at cytosine-cytosine sites.[Letter to the editor]. *Nature Structural Biology*, *9*(1), 57–60. doi:10.1038/nsb729 PMID:11731803

Lu, X.-J., & Olson, W. K. (2008). 3DNA: A versatile, integrated software system for the analysis, rebuilding and visualization of three-dimensional nucleic-acid structures. *Nature Protocols*, *3*(7), 1213–1227. doi:10.1038/nprot.2008.104 PMID:18600227

Mihasan, M. (2012). What *in silico* molecular docking can do for the 'bench-working biologists'. *Journal of Biosciences*, *37*(6), 1089–1095. doi:10.1007/s12038-012-9273-8 PMID:23151798

Miroshnychenko, K. V., & Shestopalova, A. V. (2010). The effect of drug-DNA interactions on the intercalation site formation. *International Journal of Quantum Chemistry*, *110*(1), 161–176. doi:10.1002/qua.22290

Miroshnychenko, K. V., & Shestopalova, A. V. (2013). The study of different binding modes of proflavine with DNA and RNA sequences by molecular docking method: evidence for a proflavine as a minor groove binder. In Book of Abstracts of the 5th International Symposium. Kharkiv, Ukraine: Methods and Applications of Computational Chemistry. Retrieved from http://macc.org.ua/MACC-5/MACC5colored.pdf

Morley, S. D., & Afshar, M. (2004). Validation of an empirical RNA-ligand scoring function for fast flexible docking using RiboDock. *Journal of Computer-Aided Molecular Design*, *18*(3), 189–208. doi:10.1023/B:JCAM.0000035199.48747.1e PMID:15368919

Morris, G. M., Goodsell, D. S., Halliday, R. S., Huey, R., Hart, W. E., Belew, R. K., & Olson, A. J. (1998). Automated docking using a Lamarckian genetic algorithm and an empirical binding free energy function. *Journal of Computational Chemistry*, *19*(14), 1639–1662. doi:10.1002/(SICI)1096-987X(19981115)19:14<1639::AID-JCC10>3.0.CO;2-B

Morris, G. M., Huey, R., Lindstrom, W., Sanner, M. F., Belew, R. K., Goodsell, D. S., & Olson, A. J. (2009). Autodock4 and AutoDockTools4: Automated docking with selective receptor flexiblity. *Journal of Computational Chemistry*, *30*(16), 2785–2791. doi:10.1002/jcc.21256 PMID:19399780

Neese, F. (2012). The ORCA program system. *Wiley Interdisciplinary Reviews: Computational Molecular Science*, *2*(1), 73–78. doi:10.1002/wcms.81

Neidle, S. (2001). DNA minor-groove recognition by small molecules. *Natural Product Reports*, *18*(3), 291–309. doi:10.1039/a705982e PMID:11476483

Norgan, A. P., Coffman, P. K., Kocher, J.-P. A., Katzmann, D. J., & Sosa, C. P. (2011). Multilevel parallelization of Autodock 4.2. *Journal of Cheminformatics*, *3*(12). doi:10.1186/1758-2946-3-12 PMID:21527034

O'Boyle, N. M., Banck, M., James, C. A., Morley, C., Vandermeersch, T., & Hutchison, G. R. (2011). Open Babel: An open chemical toolbox. *Journal of Cheminformatics*, *3*(33). doi:10.1186/1758-2946-3-33 PMID:21982300

Pfeffer, P., & Gohlke, H. (2007). DrugScore^{RNA} – knowledge-based scoring function to predict RNA-ligand interactions. *Journal of Chemical Information and Modeling*, *47*(5), 1868–1876. doi:10.1021/ci700134p PMID:17705464

Ricci, C. G., & Netz, P. A. (2009). Docking studies on DNA-ligand interactions: Building and application of a protocol to identify the binding mode. *Journal of Chemical Information and Modeling*, *49*(8), 1925–1935. doi:10.1021/ci9001537 PMID:19655805

Sasikala, W., & Mukherjee, A. (2013). Intercalation and de-intercalation pathway of proflavine through the minor and major grooves of DNA: Roles of water and entropy. *Physical Chemistry Chemical Physics*, *15*(17), 6446–6455. doi:10.1039/c3cp50501d PMID:23525232

Song, Z., Antzutkin, O. N., Lee, Y. K., Shekar, S. C., Rupprecht, A., & Levitt, M. H. (1997). Conformational transitions of the phosphodiester backbone in native DNA: Two-dimensional magic-angle-spinning 31P-NMR of DNA fibers. *Biophysical Journal*, *73*(3), 1539–1552. doi:10.1016/S0006-3495(97)78186-4 PMID:9284321

Srivastava, H. K., Chourasia, M., Kumar, D., & Sastry, G. N. (2011). Comparison of computational methods to model DNA minor groove binders. *Journal of Chemical Information and Modeling*, *51*(3), 558–571. doi:10.1021/ci100474n PMID:21375336

Trott, O., & Olson, A. J. (2010). AutoDock Vina: Improving the speed and accuracy of docking with a new scoring function, efficient optimization and multithreading. *Journal of Computational Chemistry*, *31*(2), 455–461. doi:10.1002/jcc.21334 PMID:19499576

Verdonk, M. L., Cole, J. C., Hartshorn, M. J., Murray, C. W., & Taylor, R. D. (2003). Improved protein-ligand docking using GOLD. *Proteins*, *52*(4), 609–623. doi:10.1002/prot.10465 PMID:12910460

Wang, J.-C., Lin, J.-H., Chen, C.-M., Perryman, A. L., & Olson, A. J. (2011). Robust scoring functions for protein-ligand interactions with quantum chemical charge models. *Journal of Chemical Information and Modeling*, *51*(10), 2528–2537. doi:10.1021/ci200220v PMID:21932857

Wu, G., Robertson, D. H., Brooks, C. L. III, & Vieth, M. (2003). Detailed analysis of grid-based molecular docking: A case study of CDOCKER – a CHARMm-based MD docking algorithm. *Journal of Computational Chemistry*, *24*(13), 1549–1562. doi:10.1002/jcc.10306 PMID:12925999

Chapter 6
Molecular-Docking-Based Drug Design and Discovery:
Rational Drug Design for the Subtype Selective GPCR Ligands

Soo-Kyung Kim
California Institute of Technology, USA

William A. Goddard III
California Institute of Technology, USA

ABSTRACT

Currently 30-50% of drug targets are G Protein-Coupled Receptors (GPCRs). However, the clinical useful drugs for targeting GPCR have been limited by the lack of subtype selectivity or efficacy, leading to undesirable side effects. To develop subtype-selective GPCR ligands with desired molecular properties, better understanding is needed of the pharmacophore elements and of the binding mechanism required for subtype selectivity. To illustrate these issues, we describe here three successful applications to understand the binding mechanism associated with subtype selectivity: 5-HT2B (5-Hydroxytryptamine, 5-HT) serotonin receptor ($HT_{2B}R$), H_3 histamine receptor (H_3HR) and A_3 adenosine receptor (A_3AR). The understanding of structure-function relationships among individual types and subtypes of GPCRs gained from such computational predictions combined with experimental validation and testing is expected the development of new highly selective and effective ligands to address such diseases while minimizing side-effects.

INTRODUCTION

G protein-coupled receptors (GPCRs) with >800 including 340 non-olfactory receptors are the largest superfamily in the human genome. These are customarily partitioned into 5 families: glutamate, rhodopsin, adhesion, frizzled, and secretin (Fredriksson, Lagerstrom, Lundin, & Schioth, 2003). GPCRs regulate essential physiological processes (e.g. cellular metabolism, cell growth, secretion, immune

DOI: 10.4018/978-1-5225-0362-0.ch006

defense, neurotransmission, and differentiation) through various endogenous ligands which include biogenic amines, peptides, lipids, nucleotides, and proteins, modulate (Lefkowitz, Pierce, & Luttrell, 2002). GPCRs also involves in various important cell recognition and communication processes (Ellis, 2004). Thus, GPCRs are important drug targets for all major disease areas, including neurodegenerative, psychiatric, metabolic, cardiovascular, cancer, and infectious diseases (Tang & Insel, 2005). Indeed, currently 30-50% of drug targets are GPCRs (Hopkins & Groom, 2002). The currently marked GPCR-targeting drugs with ~80 account for ~$50 billion in annual sales, and many have annual sales > $2 billion (Goddard III & Abrol, 2007). Identifying GPCR subtypes with specific cell and tissue has been accelerated for target evaluation, lead identification, and optimization of GPCRs. However, the clinical use as drugs for targeting GPCR have been limited by the lack of subtype selectivity or efficacy, leading to undesirable side effects.

BACKGROUND

Endogenous ligands regulate multiple GPCR subtypes. Serotonin activates 15 serotonin (5-HT) receptors. With the exception of the 5-HT_3 receptor, a ligand-gated ion channel, all other 14 serotonin receptors are GPCRs. The 5-HT_1 (1A- 1F) and 5-HT_5 (5A, 5B) receptors decrease cellular level of cAMP through coupling with Gi/ Go protein, while 5HT_4, 5-HT_6, and 5-HT_7 receptors increase cellular level of cAMP through coupling with Gs protein. 5-HT_2 (2A– 2C) receptors increase cellular level of inositol triphosphate (IP_3) and diacylglycerol (DAG) through coupling with G_q/ G_{11} protein (Nichols & Nichols, 2008). Various biological and neurological processes were regulated through the serotonin receptors such as aggression, anxiety, appetite, cognition, learning, memory, mood, nausea, sleep, and thermoregulation. Thus, the serotonin receptors are the target of a variety of pharmaceutical drugs, including many antidepressants, antipsychotics, anorectics, antiemetics, gastroprokinetic agents, antimigraine agents, hallucinogens, and entactogens (Nichols & Nichols, 2008).

Histamine acts via four histamine receptors (HRs); H_1, H_2, H_3 and H_4. Histamine has a critical role in immumomodulation and allergic diseases. Other biological activities include cell proliferation, differentiation, hematopoiesis, embryonic development, regeneration, wound healing, aminergic neurotransmission, secretion of pituitary hormones and regulation of gastrointestinal and circulatory functions (Jutel, Blaser, & Akdis, 2005). The H_3 and H_4 HRs inhibit the cellular level of cAMP through coupling with Gi/ Go protein, while the H_2HR activates the cellular level of cAMP through coupling with Gs protein. The H_1HR increases the cellular level of IP_3 and DAG through coupling with G_q/ G_{11} protein.

Adenosine binds all 4 subtypes of adenosine receptors (ARs), denoted A_1, A_{2A}, A_{2B}, and A_3, which coupled to G proteins. Activation of the A_1 and A_3 ARs inhibit the cellular level of cAMP via Gi/ Go protein, while the A_{2A} and A_{2B} ARs activate the cellular level of cAMP via Gs protein (Olah & Stiles, 1995). ARs are involved in many of the body's cytoprotective functions. Thus, ARs are important pharmacological targets in the treatment of a variety of diseases because of their key roles in controlling numerous physiological processes. For example, many therapeutic agents under development for treatment of central nervous system disorders, inflammatory diseases, asthma, kidney failure and ischemic injuries exert their effects via interactions with ARs.

To achieve high selectivity for specific GPCR subtype, an alternative approach is the development of selective allosteric modulators of the specific receptor subtypes. There are two marketed GPCR allosteric modulators, a positive allosteric modulator of the calcium sensing receptor used to treat hyperparathyroidism and Maraviroc a negative allosteric modulator of chemokine receptor 5 that inhibits HIV entry into cells and is used to treat HIV infections (Conn, Christopoulos, & Lindsley, 2009). However, in this review we will focus on the subtype selective orthosteric ligands. Several groups studied the subtype selectivity. The 3D models of all four subtypes using the crystal structure of the $hAA_{2A}R$ as a template were generated by homology modeling and applied the methodology of ligand-guided receptor optimization for refinement (Katritch, Kufareva, & Abagyan, 2011). The crystal structures of 5-HT_{1B} and 5-HT_{2B} receptors were compared (Wang, Jiang, Ma, & Xu, 2013). The 5-HT_{2A} selectivity was discussed based on the steered molecular dynamics (MD) approach of the 5-HT_{2A} model (Isberg, Balle, Sander, Jørgensen, & Gloriam, 2011). The development of subtype selective nicotinic acetylcholine receptor ligands was discussed (Astles et al., 2002). Fine-tune ligand selectivity for the angiotensin 2 receptors AT2R subtypes through electronic control of ligand aromatic-prolyl interactions results in an AT2R high affinity (Ki = 3 nM) agonist analogue with 18,000-fold higher selectivity over AT1R(Magnani et al., 2014). More recently, the binding pocket and subtype specificity in the X-ray structures of opioid receptors (OR) were discussed (Manglik et al., 2012b). In the binding pocket of the δ-OR within 4 Å of the ligand, 11 out of the 14 residues, are identical between μ-OR and δ -OR. The three differences are at μ-OR positions $E229^{EC2}$, $K303^{6.58}$ and $W318^{7.35}$, which are Asp, Trp and Leu in the δ-OR, respectively. In particular, the point mutation $W318L^{7.35}$ markedly increases the affinity of the δ-OR selective ligands at the μ-OR. But the qualitative analysis is not done for which residues are involved in ligand subtype selectivity according to energy. The interaction energy includes the non-bonding van der Waals (vdW) and electrostatic Coulombic energy between the ligand and protein to see how much or which interaction the subtype selective residues contribute in the binding site.

MAIN FOCUS OF THE CHAPTER

Issues, Controversies, Problems

To develop subtype-selective GPCR ligands with desired molecular properties, we need much better understanding of the binding sites for ligands to GPCRs and of how the mechanism for subsequent activation is affected by the structural parameters. Particularly difficulty here is that there are often many subtypes for the same endogenous ligand (4 for histamine and adenosine, 13 for serotonin), which we can expect might have someone similar binding sites. Moreover, there is clear evidence that the GPCR structure suitable for recognizing antagonists can be quite different than those for agonists. Thus even with a crystal structure of a GPCR bound to a specific antagonist or agonist, the best GPCR conformation for binding a different agonist or antagonist might be very different. Thus to predict the proper binding site to a particular GPCR, we need to first predict the ensemble of stable apo-protein structures which might be suitable for the various ligands.

To predict the ensemble of low energy GPCR conformations, we developed the GEnSeMBLE (GPCR Ensemble of Structures in Membrane BiLayer Environment) Monte Carlo Method. GEnSeMBLE compares the sequences of any one target to those of all other GPCRs to predict the hydrophobic regions likely to span the membrane and then predicts the optimum packing of the 7 transmembrane (TM) domains into 7-helix bundles. To ensure sampling of all possible packings, GEnSeMBLE starts with templates for GPCRs obtained previously from experiment (or theory) and samples up to ~12 trillion tilts and rotations of these helices, which it reduces down to the best ~10 to 20.

Then the DarwinDock Monte Carlo Technique was performed to predict the best binding pose for each ligand to each of the 7-helix bundles in this 10-20 bundle ensemble.

These procedures, which have been validated against several X-ray crystal structures, are applied here to three systems which were compared with the X-ray studies and validated by other experiments. We consider here three successful applications of computations to predict the binding mechanism for subtype selectivity:

1. Human 5-HT_{2B} (5-Hydroxytryptamine, 5-HT) serotonin receptor ($hHT_{2B}R$), (S.-K. Kim, Li, Abrol, Heo, & Goddard III, 2011)
2. Human H_3 Histamine receptor (hH_3HR) (S. K. Kim, Fristrup, Abrol, & Goddard III, 2011) and
3. Human A_3 adenosine receptor (hA_3AR) (S. K. Kim, Riley, Abrol, Jacobson, & Goddard III, 2011)

These calculations improved understanding of how these subtype selective ligands interact with each subtype at the molecular level and would be useful to diminish undesired side effects through the cross-selectivity with other subtypes and to design novel classes of subtype selective drugs.

1. METHODS

1.1. GEnSeMBLE Monte Carlo Protocol (Abrol, Bray, & Goddard III, 2011b)

The MembEnsemb (later version of the MembStruck) techniques which were used to predict the 3D structure for the hHT2BR and hHT2CR is the earlier version of the GEnSeMBLE Monte Carlo method. For each of the 7-helix conformations selected from the BiHelix analysis we built full 7-helix bundle, scream all 7 helixes simultaneously and evaluated the energy, ScreamTot. In addition, we evaluated the membrane solvation free energy (MembSolE, Es) using a multi-dielectric (80|7|2|7|80) implicit membrane model of the total solvation free energy change associated with transfer of protein from implicit bulk water. In the GEnSemBLE method, the scoring energies are ranked 4 different ways the charged total energy, the neutralized total energy, the charged interhelical energy, and the neutralized interhelical energy. Although we did not apply the same methods for the three successful applications, we validated our previous result using our new methods.

The GEnSeMBLE method (Abrol, et al., 2011b) was used to generate the 3D structures for the ensemble conformations of GPCRs, as summarized in Figure 1. GEnSeMBLE generates a complete sampling (millions to trillions) of rotations and tilts of the helices for the ensemble of low lying structures expected to be energetically accessible for binding of ligands. This method replaced our earlier MembStruk method which sampled helix rotations helix by helix, rather than combinatorial (Vaidehi et al., 2002).

The structure prediction methodology described previously (Abrol, Bray, & Goddard III, 2011a) is summarized as follows:

1. **PredicTM:** Uses multiple sequence alignments of over 2,000 GPCR sequences using the MAFFT (Katoh, Kuma, Toh, & Miyata, 2005) program combined with the hydrophobicity scales from White and von Heijne(Wimley, Creamer, & White, 1996) to predict the TM domains for membrane protein.

Figure 1. Flowchart of GEnSeMBLE (GPCR Ensemble of Structures in Membrane BiLayer Environment) Monte Carlo method

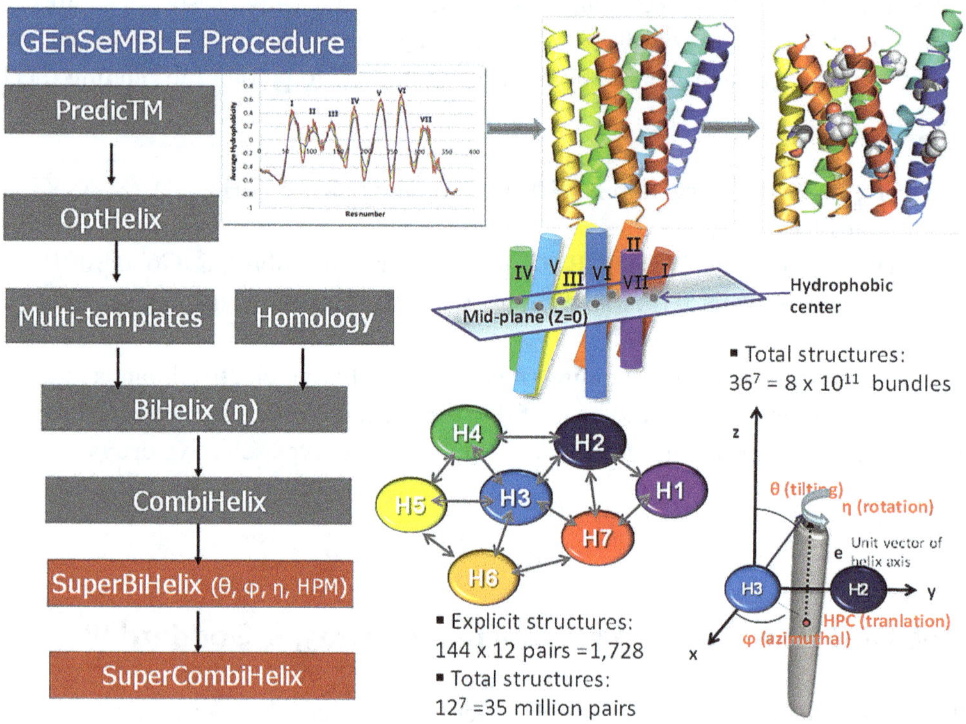

These TM domains were extended by capping rules and the secondary structure prediction using PORTER(Pollastri & McLysaght, 2005) and SSPro(Pollastri, Przybylski, Rost, & Baldi, 2002).

2. **OptHelix:** OptHelix uses molecular dynamics to predict the structure for each TM domain, which is usually helical except for kinks that may be caused by Proline residues.
3. **Homologize Helices:** When closely related X-ray structures with higher than 30% of sequence identity are available, we also use homology models to determine the helix shape. We consider the currently available X-ray structures; human nociceptin/ orphanin FQ receptor,(Thompson et al., 2012) mOPRD mouse delta opioid receptor,(Granier et al., 2012) mouse mu opioid receptor,(Manglik et al., 2012a) kappa opioid receptor,(H. Wu et al., 2012) human D_3 dopamine receptor,(Chien et al., 2010) turkey β_1 adrenergic receptor,(T. Warne, Serrano-Vega, M.J., Baker, J.G., Moukhametzianov, R., Edwards, P.C., Henderson, R., Leslie, A.G.W., Tate, C.G., Schertler, G.F.X., 2008) human adenosine A_{2A} receptor,(Jaakola et al., 2008) human β_2 adrenergic receptor,(V. Cherezov, Rosenbaum, D.M., Hanson, M.A., Rasmussen, S.G.F., Thian, F.S., Kobilka, T.S., Choi, H-J., Kuhn, P., Weis, W.I., Kobilka, B.K., Stevens, R.C., 2007) human sphingosine 1-phosphate,(Hanson et al., 2012) human Chemokine CXCR4,(B. Wu et al., 2010) human H_1 histamine receptor,(Shimamura et al., 2011) rat M_3 muscarinic receptor,(Kruse et al., 2012) human M_2 muscarinic receptor,(Haga et al., 2012) and bovine rhodopsin (Li, 2004; T. Okada, Fujiyoshi, Y., Silow, M., Navarro, J., Landau, E.M., Shichida, Y., 2002; T. Okada, Sugihara, M., Bondar, A.N., Elstner, M., Entel, P., Buss, V., 2004; Palczewski, 2000; Teller, 2001) Based on the sequence identities of the TM regions for these systems and the diversity of structures, we chose a starting template.

4. **BiHelix:** To cover all possible rotations of all 7 helices, consider 30° increments in the rotation about each TM axis (eta η angle), leading to $(12)^7$ ~ 35 million packings. The energies for all 35 million packing are estimated in by evaluating the energies of all $(12)^2$=144 interactions for each of the 12 pairs of helices in direct contact. For each of the 1,728 pairwise interactions, SCREAM(V.W.T. Kam & W.A. Goddard III, 2008) was used to optimize the side-chains. The BiHelix mean field energies for all 35,000,000 packings are used to select the best 2000. We then build the full 7-helix bundle for each of these 2,000 structures, optimize the side-chains for each using SCREAM, and neutralize the charged residues for more accurate energy scoring. The Dreiding D3 force field (D3FF)(Mayo, Olafson, & Goddard III, **1990**) was used throughout wherever energies were evaluated.
5. **SuperBiHelix(Bray, Abrol, Goddard III, Trzaskowski, & Scott, 2014):** Starting with the X-ray tilt angles we validated(Abrol, et al., 2011a) that the BiHelix analysis always identifies the experimental set of rotation angles correctly (this validates the procedure and the scoring function). However we have shown that even for closely related GPCRs, we must optimize the tilt angles: theta θ (tilting away from the z-axis) and phi φ (the azimuthal angle of the tilting from the xy plane) angle in order to get an accurate structure for a structure based on a different template (Abrol, Griffith, Bray, & Goddard III, 2012) Here we have found that starting with the best angles from the BiHelix optimization, it is sufficient to sample ±10° for θ tilt angle while simultaneously sampling by ±30° for both φ and η angles, leading to a total of $(3x5x5)^7$ ~ 10 trillion combinations. Generally the hydrophobic center (HPC) for each TM domain from the PredicTM analysis is placed at z=0, but we can use SuperBiHelix to optimize translation along the TM axis simultaneously with optimizing rotations.

In SuperBiHelix, The BiHelix mean field energies for all 10 trillion combinations packings are used to select the best 2000, which are then examined much more accurately by building the 7 TM bundles and optimizing side chains. Our experience is that the mean field energies are sufficiently accurate that the best 7-helix bundle energies are all in the first 700 of the 2,000, indicating that analyzing the best 2,000 is sufficient.

6. **Loop Generation:** The loops and the N/C terminus including helix 8 for a new structure are modeled through homology with the template structure, and then relaxed through 10 cycles of simulated annealing between 50K and 600K each for 0.1 ps, followed by minimization. The disulfide bridge between C3.25 and Cys in the second extracellular loop (EC2), which is conserved among Class A GPCRs, is constructed. Then keeping the 7 TM bundle fixed, we first optimize the loop structures (up to1000 steps or down to 0.5 RMS force) followed by quench annealing between 50 K and 600K for 10 cycles.

The best examples of using our procedures to successfully predict mechanisms relating to activation or inactive character of GPCRs are our studies on various mutants of CB1 that were subsequently confirmed by experiment. We were able to show that mutations that strengthening the coupling of TM6 to TM2 and TM3 make the CB1 receptor totally inactive, while mutations that break this interaction are essentially totally active (Ahn, Scott, Abrol, Goddard III, & Kendall, 2013) Moreover we predicted mutations that would switch either from totally inactive to constitutively active or from totally active to constitutively active that were confirmed by subsequent experiments.

1.2. DarwinDock

The lowest energy predicted structures from BiHelix/ SuperBiHelix were used for docking several ligands, using DarwinDock. The starting structure and charges of the ligands are calculated using quantum mechanics (B3LYP with the 6-311G** basis set). Then we develop a set of conformations by LigCluster and select 10 to 20 for docking.

1. **Scanning the receptor for potential binding regions:** Using the predicted 7-helix bundle structure, we predicted putative ligands binding regions as follows. The 6 hydrophobic residues, I, L, V, F, Y, and W, were alanized for the whole protein and scanned for potential binding regions. The overall molecular surface of the protein structure was mapped with spheres representing the empty volume of the protein (currently using the Sphgen procedure in DOCK4.0 suite). The entire set of protein spheres is partitioned into ~30 to 50 overlapping cubes of 10 to 14 Å sides. For each of these 30 - 50 regions, 1,000 poses were generated.
2. **DarwinDock:** The GenMSCDock is the later version of the GenMSCDock techniques to predict the binding site for HT2B/ HT2CR agonists. The GenMSCDock procedures sampled a complete set of configurations for each conformation of the ligand. This leads to a hierarchy of families with a specified range of diversities, allowing the best to be selected at a coarse level by evaluating energies of family heads. This also leads to an ensemble of the best diversity families that are used for higher-level calculations. The GenMSCDock method does two rounds, one to ensure completeness and the other to enrich the better families. To achieve an average of 6 children in each family, we resort to obtain 0.6 Å families in the enrichment step. The steps in GenMSCDock include bulky residue alanination, diversity finder, Voronoi reclustering, side-chain refinement and neutralization. The DarwinDock method which is the later version of GenMSCDock increases the sampling number for all binding configurations from 2500 to 5000 with a larger diversity parameter of 1.2 – 1.4 Å in the completeness step. In addition, we include the scoring energies of ligand strain and solvation.

DarwinDock has replaced our earlier methods, the HierDock and GenMSCDock. For each ligand conformation, DarwinDock iteratively produces a total of ~50,000 poses into the putative binding regions of the alanized protein. The poses partitions into families based on CRMS and the energy scoring of family heads and select the top 10% ordered by total energy. The 100 conformations are scored for the children of these top families for further optimization. For 100 conformations, we dealanize the protein side-chains to optimize the side chains using SCREAM. Then we neutralize the protein and ligand by transferring protons within salt bridges and protonating or deprotonating exterior ligands, followed by further full geometry minimization.

We validated DarwinDock using 3 crystal structures of ligand/ GPCR complexes: human β_2-adrenergic receptor (0.4 Å RMSD),(V. Cherezov et al., 2007) human $AA_{2A}R$ (0.8 Å RMSD),(Jaakola, et al., 2008) and turkey β_1-adrenergic receptor (0.1 Å RMSD) (T. Warne et al., 2008) This supports that we can accurately identify ligand binding sites in proteins and optimize the ligands with desirable properties.

3. **Neutralization for scoring energy:** Quantum mechanics (QM) calculations show that if an effective dielectric constant is < 8Å, the extra proton on a Lys or Arg transfers back to the negative carboxylate of an Asp or Glu.[22] Thus buried salt bridges is expected to have neutral residues. We

find that neutralizing these exposed residues removes the sensitivity to details of the distances of charged residues (and counter ions) remote from the binding cavity. This neutralization leads to differential binding energies that are dominated by the local cavity interactions and leads to much smaller solvation energies.

We used the unified cavity energy as the scoring function for docking. It was calculated by the total non-bonding energy in 4 Å of ligand which is the sum of van der Waals (vdW), Coulomb, and H-bond energy (kcal/mol) in the unified cavity.

An excellent example validating our DarwinDock is the predicted binding site for the Maraviroc anti HIV drug from a selection of 8 diverse packings for the chemokine CCR5 GPCR (Abrol et al., 2014) Then we predicted 12 mutations designed to increase or decrease binding significantly. These mutations, carried out after the predictions, led to a mean error of <1 kcal/mol in the predicted correlation of relative binding energies. Interestingly, the optimum protein packing for binding Maraviroc was different than for the original wild type binding site for each of the 12 mutations.

2. APPLICATIONS

2.1. Subtype Selectivity of Antagonists for $hHT_{2B}R$ (Kim et al., 2011)

Three 3 subtypes of 5-HT_2 GPCRs are important drug targets for schizophrenia, depression, migraines, anxiety, hallucinogens, feeding disorders, perception, hypertension, and gastrointestinal dysfunctions. In particular, $hHT_{2B}R$ antagonists have a possible therapeutic treatment of pulmonary hypertension (Fredriksson, et al., 2003).

The sequence identities of overall and TM region between $hHT_{2B}R$ and $hHT_{2C}R$ were 57.38 and 72.51%, respectively. The sequence identities of each TM region have 65.22% for TM1, 75.00% for TM2, 87.50% for TM3, 55.56% for TM4, 62.50% for TM5, 81.82% for TM6, and 80.80% for TM7. The predicted structure of $hHT_{2B}R$ has an average RMSD of 0.39 Å relative to the $hHT_{2c}R$ TM bundle (RMSD of TM1: 0.63 Å, TM2: 0.39 Å, TM3: 0.26 Å, TM4: 0.52 Å, TM5: 0.45 Å, TM6: 0.21 Å, TM7: 0.30 Å).

We compare here the binding sites and energies for two cases

- A non-selective antagonist (SB-206533 **1**:5 -methyl-1-3,5-dihydro-2H-pyrrolo[2,3-f]indole-1-carboxylic acid pyridine-3-ylamide) bound to hHT_{2B} and $hHT_{2C}R$ (serotonin) GPCR
- A selective antagonist **2** (1-methyl-1-1,6,7,8-tetrahydro-pyrrolo[2,3-g]quinoline-5-carboxylic acid pyridine-3-ylamide) bound to hHT_{2B} (serotonin) GPCR

Figure 1 shows the predicted structures(S.-K. Kim, et al., 2011) for non-selective SB-206533 **1** in cyan, to both of the $hHT_{2B}R$ and $hHT_{2C}R$. For residue numbering, we used the Ballesteros-Weinstein number; the most conserved residue in each of the 7 TM domains is taken as the reference and numbered as 50. This residue is designated x.50 where x is the number of the TM helix (Ballesteros & Weinstein, 1995) From the cavity analysis, we find similar interaction energies for this antagonist to both $hHT_{2B}R$ and the $hHT_{2C}R$. For example the hydrogen bonding (HB) interactions between the ureido-CO group and the Ser-OH side chain at $S139^{3.36}$ is conserved among 5-HT2 family receptors, 2A, 2B, and 2C. The complex is stabilized through additional hydrophobic interactions at conserved hydrophobic residues, $V136^{3.33}$,

F274$^{6.51}$ (2B)/ F272$^{6.51}$ (2C), and F299$^{7.38}$/ F298$^{7.38}$. The N-methyl group in the cyclopenta-indole ring forms the hydrophobic interactions with the subtype variable residues, Non-bond energies of L132$^{3.29}$, I186$^{4.56}$, and V190$^{4.60}$ in the hHT$_{2B}$R are -2.48, -0.10, and -0.76 kcal/mol, respectively. The corresponding amino acids, I132$^{3.29}$, V186$^{4.56}$, and I190$^{4.60}$, at the hHT$_{2C}$R also have favorable non-bond energies. Thus, similar total cavity energies of SB-206533 **1** at the hHT$_{2B}$R and hHT$_{2C}$R are -42.48 and -42.28 kcal/mol, respectively. In addition, subtype-selective residues also reveal similar interactions (Table 1).

Table 1. Cavity analysis of SB-206533 **1** *and* **2** *at human 5-HT2B and 2C receptors. The residues are ordered by the non-bonding energy difference (2C -2B)*

Ballesteros-Weinstein #	2 (pKi, 2B: 7.3)			2 (pKi, 2C: 5.4)			Diff.	1 (2B: 7.6)	1 (2C: 7.9)
	Res	#	NonBond	Res	#	NonBond	2C-2B	NonBond	NonBond
3.29	**LEU**	**132**	**-2.35**	**ILE**	**131**	**6.12**	**8.46**	**-2.48**	**-0.84**
6.55	ASN	344	-2.40	ASN	331	-1.54	0.86	-2.19	-0.47
5.47	PHE	226	-2.05	PHE	223	-1.38	0.68	-1.02	-2.39
6.54	THR	343	-1.33	THR	330	-0.66	0.67	-1.17	0.00
3.33	VAL	136	-5.34	VAL	135	-4.91	0.43	-3.78	-5.44
5.46	ALA	225	-1.09	ALA	222	-0.68	0.41	-0.48	-1.12
6.47	MET	336	-0.40	MET	323	0.00	0.40	-0.06	0.00
6.52	PHE	341	-0.38	PHE	328	0.00	0.38	0.00	0.00
3.34	LEU	137	-0.38	LEU	136	0.00	0.38	0.00	0.00
6.58	**LEU**	**347**	**-0.56**	**SER**	**334**	**-0.23**	**0.33**	**-0.67**	**0.00**
5.39	**MET**	**218**	**-0.32**	**VAL**	**215**	0.00	**0.32**	**-0.35**	**-0.28**
5.40	LEU	219	-0.28	LEU	216	0.00	0.28	-0.27	0.00
5.44	**LEU**	**223**	**-0.26**	**PHE**	**220**	0.00	**0.26**	**-0.20**	**-0.30**
3.40	ILE	143	-0.51	ILE	142	-0.26	0.25	-0.28	-0.31
5.43	SER	222	-1.70	SER	219	-1.49	0.21	-2.03	-2.92
6.44	PHE	333	-0.59	PHE	320	-0.42	0.17	-0.52	-0.09
6.53	ILE	342	-0.16	ILE	327	0.00	0.16	0.00	-0.10
7.45	SER	372	-0.15	CYS	360	0.00	0.15	0.00	0.00
7.41	ILE	368	-0.12	ILE	356	0.00	0.12	-0.10	0.00
5.41	**PHE**	**220**	**-0.07**	**ILE**	**217**	**0.00**	**0.07**	**0.00**	**-0.08**
3.36	*SER*	*139*	*-6.34*	*SER*	*138*	-6.28	*0.06*	*-5.67*	*-6.95*
5.45	**ALA**	**224**	**-0.05**	**VAL**	**221**	0.00	**0.05**	**0.00**	**-0.14**
6.49	CYS	338	-0.03	CYS	325	0.00	0.03	-0.03	0.00
6.46	**LEU**	**335**	**-0.03**	**ILE**	**321**	**0.00**	**0.03**	**-0.03**	**0.00**
6.51	PHE	340	-5.95	PHE	327	-5.93	0.02	-4.43	-5.80
5.32	**LYS**	**211**	**-0.01**	**VAL**	**208**	0.00	**0.01**	**-0.06**	0.00
6.45	**ILE**	**334**	**0.00**	**LEU**	**321**	0.00	**0.00**	**-0.02**	0.00
3.28	TRP	131	-0.32	TRP	130	-0.32	**0.00**	-0.37	-0.44
4.56	**ILE**	**186**	**0.00**	**VAL**	**185**	**0.00**	**0.00**	**-0.10**	**-1.27**

continued on following page

Table 1. Continued

Ballesteros-Weinstein #	2 (pKi, 2B: 7.3)			2 (pKi, 2C: 5.4)			Diff.	1 (2B: 7.6)	1 (2C: 7.9)
	Res	#	NonBond	Res	#	NonBond	2C-2B	NonBond	NonBond
4.57	**ALA**	**187**	**0.00**	**SER**	**186**	**0.00**	**0.00**	**-0.12**	**-0.58**
5.37	ASP	216	0.00	ASP	213	0.00	0.00	-0.38	0.00
7.46	SER	373	-0.85	SER	361	-0.86	0.00	-0.77	0.00
7.33	**MET**	**360**	**0.01**	**LYS**	**348**	**0.00**	**-0.01**	0.00	0.00
2.57	VAL	107	-0.08	VAL	106	-0.10	-0.02	-0.10	0.00
7.34	LEU	361	-0.23	LEU	349	-0.26	-0.03	-0.26	0.00
6.42	PHE	331	0.05	PHE	318	0.00	-0.05	0.00	0.00
3.35	PHE	138	-0.70	PHE	137	-0.76	-0.06	-0.63	0.00
7.43	TYR	370	-1.59	TYR	358	-1.65	-0.06	-1.35	-0.53
3.37	THR	140	-1.16	THR	139	-1.32	-0.17	-0.96	-2.12
4.60	**VAL**	**190**	**-0.51**	**ILE**	**189**	-0.72	**-0.20**	**-0.76**	**-1.24**
7.42	GLY	369	-0.87	GLY	357	-1.10	-0.23	-0.73	-0.31
6.48	TRP	337	-0.84	TRP	269	-1.14	-0.30	-1.42	-1.77
7.39	VAL	366	-1.72	VAL	354	-2.19	-0.48	-1.83	-0.09
7.35	LEU	362	-1.05	LEU	350	-1.59	-0.54	-1.21	-0.69
3.32	ASP	135	-2.87	ASP	134	-3.63	-0.75	-3.81	-3.50
7.38	PHE	365	-1.95	PHE	353	-2.76	-0.82	-1.87	-1.54
	SUM		**-47.53**	**SUM**		**-36.06**	**11.47**	**-42.48**	**-42.28**

S.-K. Kim, et al. (2011)

Then we studied the conformational restricted analogue **2** (1-methyl-1-1,6,7,8-tetrahydro-pyrrolo[2,3-g] quinoline-5-carboxylic acid pyridine-3-ylamide) known to bind much more strongly to 5-HT_{2B}R affinity (pA2, 5-HT2B: 7.27 vs. pKi, 5-HT2C: 5.39) (Forbes et al., 1995)

Indeed we find that arises because the introduction of the six membered ring is clearly detrimental for hHT_{2C}R. The major hydrogen bond interaction is at the ureido-CO group with Ser-OH at conserved S139$^{3.36}$. Conserved hydrophobic residues, V136$^{3.33}$, F274$^{6.51}$ / F272$^{6.51}$ (2C), and F299$^{7.38}$/ F298$^{7.38}$ (2C) also stabilize the complex through hydrophobic interactions. We observed that bad contacts at I132$^{3.29}$ with unfavorable vdW interaction (+6.24 kcal/mol) result in reduced interaction at the hHT_{2C}R, compared with the favorable vdW interaction at the hHT_{2B}R (-2.32 kcal/mol).

In addition, similar binding energies were shown at other variable residues, V190$^{4.60}$ / I190$^{4.60}$ in 2C, L281$^{6.58}$/ S279$^{6.58}$ in 2C. We find favorable interaction at hHT_{2B}R (-47.53 kcal/mol) but unfavorable interaction at the hHT_{2C}R (+11.47 kcal/mol), consistent with experiment (Table 1). However, A225$^{5.46}$ in 2B and A223$^{5.46}$ in 2C which were reported as specific interaction for 5-HT_{2A} selectivity(Isberg, et al., 2011) shows the similar interaction at both receptors.

The superimposition of **1** in cyan and **2** in blue in Figure 2 suggest that the *N*-methyl of the cyclohexa-indole ring of 2B selective compound **2** pointing toward the upper TM 3 leads to unfavorable interactions with the bulkier Ile side chain in hHT_{2C}R (I132$^{3.29}$), while the *N*-methyl of the cyclopenta-indole ring in the

*Figure 2. Superimposition of the predicted structures for the nonselective 5-HT2B/2C receptor antagonist (SB-206533 1) and the selective 5-HT2B receptor antagonist (***2***) at the human 5-HT2B receptor. The N-methyl of the cyclohexa-indole ring of* **2** *is closer to the upper part of TM3 leading to unfavorable interactions with bulkier Ile side chains in the 5-HT2C receptors (I132$^{3.29}$), while the methyl of the cyclopenta-indole ring in SB-206533* **1** *is pointing toward the upper TM 4.*
S.-K. Kim, et al. (2011)

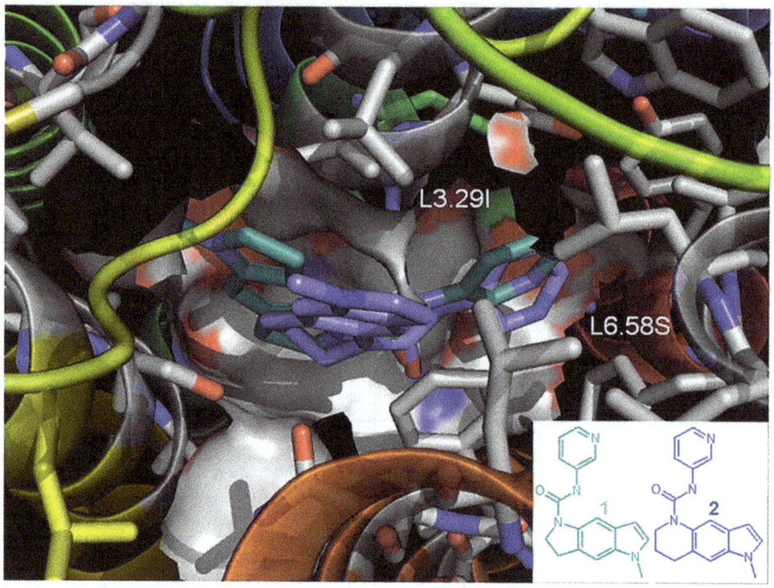

non-selective compound **1** directs toward the upper TM 4. In addition, the cyclohexa-indole ring is near the hydrophilic residue S279$^{6.58}$ in the hHT$_{2C}$R, while it interacts with hydrophobic L281$^{6.58}$ in hHT$_{2B}$R.

In the binding cavity of hHT$_{2B}$R, 13 out of 46 residues are variable compared with the binding cavity of hHT$_{2C}$R with 72% sequence identity. Among them, L3.29I is the major contribution to the hHT$_{2B}$R selectivity. L6.58S and M5.39V are the next important residues for the hHT$_{2B}$R selectivity.

When compared with the recently published crystal structures of the partial agonist ergotamine-bound hHT$_{2B}$R (4IB4),(Wacker et al., 2013) the predicted hHT$_{2B}$R has a Ca RMSD of 2.49 Å relative to the hHT$_{2B}$R crystal TM bundle. We observed the same TMs 1-2-7 networks among N1.50-D2.50-N7.49 and TMs 2-3-4 networks among S2.45-H3.42-W4.50 which is observed in the crystal structure. The H-bonding between D3.32 and Y7.43 which is conserved in most biogenic amine receptor was also observed near the binding site.

2.2. Subtype Selectivity of hH$_3$HR (Kim, Fristrup, et al., 2011)

HRs are major drug targets for the treatment of various diseases, such as schizophrenia, psychosis, depression, migraine, allergies, asthma ulcers, and hypertension. In particular, hH$_3$HR antagonists have been reported as specific therapeutic applications for the treatment of Alzheimer's disease, attention deficit hyperactivity disorder (ADHD), epilepsy, and obesity (Lorenzi et al., 2005) We report here the predicted lowest energy structure of all four human HR subtypes (H$_1$, H$_2$, H$_3$, and H$_4$), for each of which we have predicted the binding sites for H_3 selective antagonist: clobenpropit **3**, (*N*'-[(4-chlorophenyl) methyl]-1-[3-(3H-imidazol-4-yl)propylsulfanyl]formamidine).

The hH_3HR shared 54.74% sequence identity with the hH_4HR, 32.15% sequence identity with the hH_1HR, and 33.08% sequence identity with the hH_2HR in the TM regions. The predicted structure of hH_3HR has an average RMSD of 0.04 Å for the hH_1HR, 1.03 Å for the hH_2HR, 1.01 Å for the hH_4HR, and 1.33 Å for the crystal structure of hH_1HR (3RZE) (Shimamura, et al., 2011). The predicted hH_1HR also has the same RMSD of 1.33 Å for the crystal structure of hH_1HR (Shimamura, et al., 2011).

We docked hH_3HR selective clobenpropit, to the lowest energy conformations of all four subtypes. From the cavity analysis of antagonist clobenpropit at hH_3HR, the major contributing amino acids are $Y115^{3.33}$ (-4.96 kcal/mol), $W402^{7.43}$ (-4.71 kcal/mol), and $D114^{3.32}$ (-3.73 kcal/mol) in Table 2. $D114^{3.32}$ and $Y374^{6.51}$ form the major H-bonding with the isothiourea group. The terminal imidazole ring forms additional H-bonding with at $E206^{5.46}$. Hydrophobic residues, $L401^{7.42}$ and $W402^{7.43}$ surround the *para*-chloro-benzyl group of clobenpropit (Figure 3).

To study subtype selectivity of HRs, the predicted best binding pose of the highly H_3-selective ligand clobenpropit at hH_3HR was matched into other three subtypes (H_1, H_2, H_4) of hHRs, and then SCREAM(V.W.T. Kam & W.A. Goddard III, 2008) was used to optimize the side chain position of residues in the binding cavity, followed by minimization and neutralization of the final ligand/ protein complex.

Table 2 displays predicted variable residues among four subtypes in the binding site (H_4: 29%, 8 out of 28, H_1: 57%, 16 out of 28, H_2: 68%, 19 out of 28). The corresponding amino acids of the same E5.46/ S5.43 in both of hH_3HR and hH_4HR are $N198^{5.46}$/ $A195^{5.43}$ in hH_1HR and $T190^{5.46}$/ $G187^{5.43}$ in hH_2HR (Figure 4) with weakened interactions in the cavity, resulting in a 2.2 to 2.4 kcal/mol favorable interaction in hH_3HR in Table 2. But, a similar interaction is observed at these two conserved residues E5.46/ S5.43 of hH_4HR. Thus, the final cavity energy leads to a substantial decrease in binding affinity (cavity sum = -34.11 kcal/mol in hH_1HR, -33.77 kcal/mol in hH_2HR) in agreement with the dramatically decreased experimental binding affinity at hH_1HR (PKi: 5.6) and hH_2HR (PKi: 5.2).

The final cavity energy shows a weakened binding affinity (-35.12 kcal/mol) for hH_4HR (PKi: 7.4) is a compared with the cavity energy (-36.94 kcal/mol) of hH_3HR (PKi: 9.4), Thus, this predicted binding energy is well correlated with the experimental binding affinity of H_3 subtype selective clobenpropit (Esbenshade et al., 2003).

In the binding cavity of hH_3HR, 7 (25%), 16 (57%), and 18 (72%) out of 28 residues are variable for hH_4HR, hH_1HR, and hH_2HR, respectively. The important residues for the hH_3HR selectivity are E5.46/ S5.43 in both of hH_3HR and hH_4HR which are $N198^{5.46}$/ $A195^{5.43}$ in hH_1HR and $T190^{5.46}$/ $G187^{5.43}$ in hH_2HR, respectively. When compared with the hH_4HR, M6.55T is the most important residue to overcome the hH_3HR selectivity.

Table 2 shows the unified cavity energy (UCav E) in all subtypes of hHRs, which are ordered by experimental binding affinity including. We find that the predicted binding energies for all four HRs are consistent with their experimental binding affinities with the r^2 values (correlation coefficients) ranging from 0.67 to 0.99.

2.3. Subtype Selectivity of $hAA_{2A}R$ and hAA_3R (Kim, Riley, et al., 2011)

ARs are targets for treatments of central nervous system disorders, inflammatory diseases, asthma, kidney failure, and ischemic injuries. Each agonist for 4 subtypes of ARs has various therapeutic applications. For example, the hAA_1R activation involves in cardioprotection, antilipolysis, type II diabetes,

Table 2. Cavity energy of the H_3 selective antagonist clobenpropit **3** *to four human histamine receptors (HRs). Residues are ordered by total NonBond energy (H_3), which is the sum of van der Waals (vdW), Coulomb, and H-bond energy (kcal/mol) in the unified cavity. Predicted subtype residues that vary among four subtypes (H4: 29%, 8/28, H1: 57%, 16/28, H2: 68%, 19/28) are displayed in italic font. The color coding for contributions of each residue to binding of the adenosine ligand is: Dark grey: > 3 kcal/mol, Grey: 1 - 3 kcal/mol, Light grey: 0.5 - 1.0 kcal/mol. In the Ballesteros-Weinstein numbering, the most conserved residue in each of the 7 TM domains is taken as the reference and numbered as 50. This residue is designated x.50 where x is the number of the TM helix.*

Ballesteros-Weinstein #	hH_3HR (pKi: 9.4)		hH_4HR (pKi: 7.4)		hH_1HR (pKi: 5.6)		hH_2HR (pKi: 5.2)	
	Res. #	NonBond	Res. #	NonBond	Res. #	NonBond	Res. #	NonBond
3.33	Y115	-4.96	Y95	-5.70	Y108	-5.41	*V99*	*-3.45*
7.43	W402	-4.71	W348	-5.20	*Y458*	*-4.02*	*Y278*	*-2.78*
3.32	D114	-3.73	D94	-3.55	D107	-2.73	D98	-5.19
6.51	Y374	-2.54	Y319	-2.51	Y431	-3.65	Y250	-3.19
5.43	S203	-2.43	S179	-1.79	*A195*	*-0.53*	*G187*	*-1.21*
5.46	E206	-2.22	E182	-3.39	*N198*	*-1.92*	*T190*	*-1.07*
3.28	W110	-1.68	W90	0.00	W103	-1.60	*Y94*	*0.00*
3.36	C118	-1.66	C98	-3.11	*S111*	*-3.63*	C102	-2.80
2.57	C87	-1.58	*S68*	*-2.02*	*V80*	*-1.47*	*V71*	*-2.24*
2.53	V83	-1.33	V64	-1.26	V76	2.25	*L67*	*0.15*
6.55	M378	-1.28	*T323*	*-0.41*	*F435*	*-1.66*	*F254*	*-2.40*
2.51	Y81	-1.27	Y72	-1.35	*N84*	*-0.84*	*S75*	*-0.67*
7.39	F398	-1.20	F344	-0.12	*I454*	*-1.12*	*L274*	*-0.98*
6.48	W371	-1.14	W316	-0.58	W428	-1.51	W247	-2.83
2.58	I88	-1.03	I69	-0.78	*M81*	*0.14*	*L72*	*-1.25*
5.47	F207	-0.77	F183	-1.07	F199	-0.48	F191	-0.83
3.35	L117	-0.75	L97	-0.95	*A110*	*-0.65*	L101	-0.73
3.37	T119	-0.74	T99	-0.50	T112	-0.74	T103	-0.39
6.52	T375	-0.60	*S320*	*-0.24*	*F432*	*-1.06*	*F251*	*-0.81*
5.39	L199	-0.57	L175	-0.45	*K191*	*-0.22*	*G183*	*-0.22*
5.42	L401	-0.39	*Q347*	*0.03*	*G457*	*-0.22*	*G457*	*0.00*
3.31	V113	-0.34	*T93*	*-0.43*	*M106*	*-0.41*	*L97*	*0.00*
5.38	F198	-0.15	*I174*	*-0.14*	F190	0.00	*Y182*	*0.00*
4.57	Y167	0.00	*N147*	*0.15*	V159	-0.64	*S150*	*0.00*
4.61	I171	0.00	I151	0.00	*L163*	*-0.31*	I154	0.00
7.40	W399	0.00	W345	0.00	W455	0.00	W275	-0.75
7.46	S405	0.00	S351	0.00	S461	-0.23	S281	0.00
5.42	A202	0.13	*T178*	*0.23*	*T194*	*-1.47*	*D186*	*-0.13*
	SUM	**-36.94**	**SUM**	**-35.12**	**SUM**	**-34.11**	**SUM**	**-33.77**

S. K. Kim, P. Fristrup, et al. (2011)

*Figure 3. Predicted best structures of the H_3 selective antagonist clobenpropit **3** bound to the human histamine H_3 receptor (hH_3HR). The H-bonding is represented by the arrows between the donor and the acceptor. The residues are numbered according to the order of unified cavity energy in Table 2. A schematic structure of the predicted binding site is displayed on the bottom figure.*
S. K. Kim, P. Fristrup, et al. (2011)

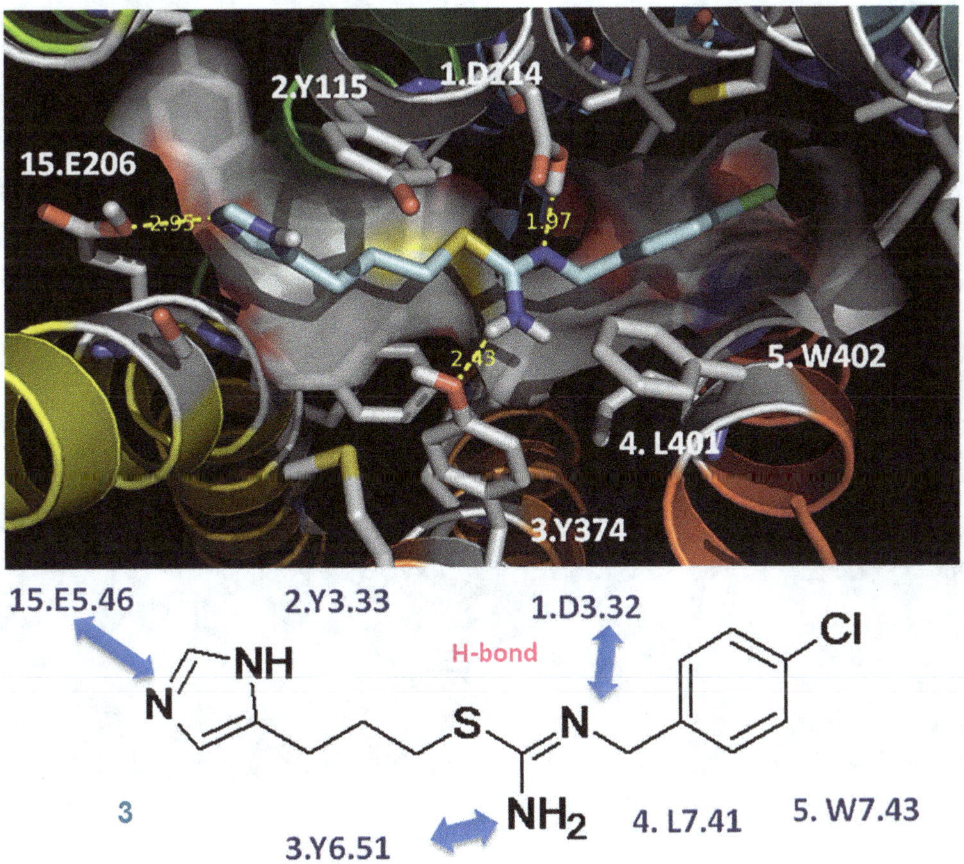

anti-arrhythmic, antiepileptic, analgesic/antinociceptive, ischemia, tachycardia, obesity, cachexia, while the hAA_3R activation involves in cardioprotection, cerebrovascular protection, prevention of stroke, anti-inflammatory, anti-asthmatic, neurological disease, ischemia, immune disorder, anticancer, and anti-arthritic. In addition, the $hAA_{2A}R$ agonists are useful for the target of antihypotensive, antipsychotic, and anti-inflammatory, while the $hAA_{2B}R$ agonists are useful for the target of anti-asthma cystic fibrosis, anti-inflammatory, septic shock, cardiac disease, diarrhea, impotence, and angiogenesis. The only other adenosine agonist currently in clinical use is non-selective adenosine itself, for the treatment of supraventricular tachycardia and as an aid in cardiac imaging. But the $hAA_{2A}R$-selective agonist Lexiscan (regadenoson, CV Therapeutics, Palo Alto, CA, USA), CVT-3146, recently approved for cardiac imaging in patients (K.A. Jacobson, A.M. Klutz, D.K. Tosh, A.A. Ivanov, D. Preti, & P.G. Baraldi, 2009). Selective hAA_3R antagonists are especially interesting as potential anti-glaucoma agents (Yang et al., 2005). Moreover, such selective hAA_3R agonists as IB-MECA (*N*[6]-(3-Iodobenzyl)-5'-*N*-methylcarboxamidoadenosine) and Cl-IB-MECA (2-Chloro-*N*[6]-(3-iodobenzyl)-5'-*N*-methylcarboxamidoadenosine),

Figure 4. Predicted best structures of the H_3 selective antagonist clobenpropit **3** *bound to four human histamine receptors (hH_1HR, hH_2HR, hH_3HR, hH_4HR). H-bonding is indicated by red dots, and subtype selective residues are shown in red.*
S. K. Kim, P. Fristrup, et al. (2011)

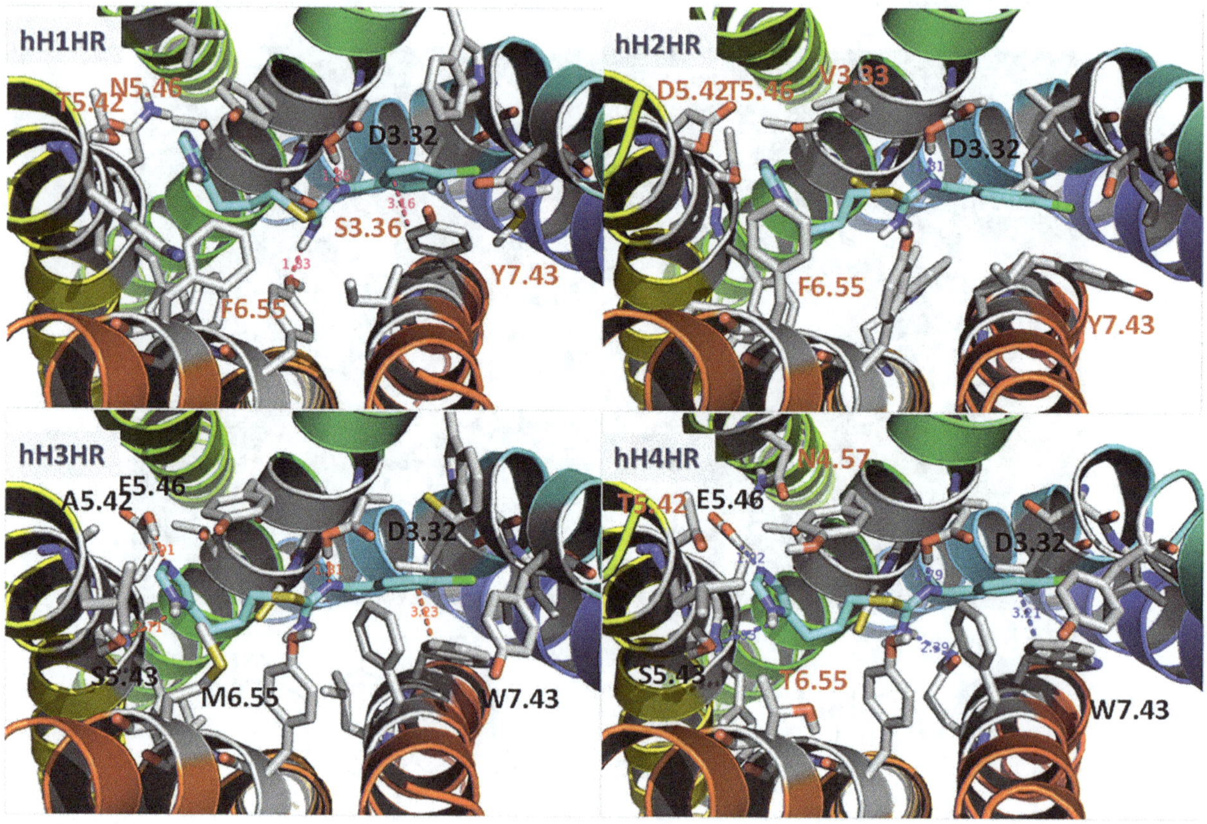

are now in Phase II clinical trials for treatment of liver cancer (K.A. Jacobson, A.M. Klutz, D.K. Tosh, A.A. Ivanov, D. Preti, & P.G. Baraldi, 2009). Thus, the development of the subtype selective ligand is critical for the specific treatment of human diseases.

To map out the binding site of A_3 subtype selective agonist, we predicted the ensemble of the 20 most stable structures for each of the four ARs. Here we used the high-resolution X-ray structure of $hAA_{2A}R$(Jaakola, et al., 2008) as the template. Because this crystal structure has $hAA_{2A}R$ bound to an antagonist, it provides little information about the structural changes involved in activation by agonist binding.

Then we predicted the binding of the above two A_3 selective agonists all 20 structures from this ensemble for each of the human adenosine subtypes (A_1, A_{2A}, A_{2B}, and A_3). In addition, we predicted the high hAA_3R selectivity for the agonist, Cl-IB-MECA and the high $hAA_{2A}R$ selectively for the endogenous agonist adenosine, validating the models for understanding the subtype selectivity. In addition we predicted binding sites for several A_3 selective antagonists to these structures for hA_3 and hA_{2A}.

The sequence identities of overall and TM region between hA_3AR and $hA_{2A}AR$ were 31.31 and 51.01%, respectively. The sequence identities of each TM region have 44.44% for TM1, 56.14% for TM2, 47.37% for TM3, 48.00% for TM4, 57.14% for TM5, 52.94% for TM6, and 51.04% for TM7. The predicted structure of hA_3AR has an average RMSD of 0.49 Å relative to the $hA_{2A}AR$ TM bundle (RMSD of TM1: 0.50 Å, TM2: 0.32 Å, TM3: 0.50 Å, TM4: 0.56 Å, TM5: 0.55 Å, TM6: 0.53 Å, TM7: 0.48 Å).

1. **$hAA_{2A}R$ selective adenosine:** We selected adenosine because of ~ 30 fold higher selectivity to $hAA_{2A}R$ than hAA_3R. The endogenous agonist adenosine was docked to the ensemble of 25 low-lying structures of apo-$hAA_{2A}R$.

All adenosine agonists show similar ribose moieties, except for the amide substitution at the 5'-position. Indeed an intact furanose moiety is present in most potent adenosine agonists previously reported (Jacobson et al., 2000). Table 3 shows the cavity analysis of the adenosine bound to $hAA_{2A}R$. Three major contributing amino acids are $N253^{6.55}$ (-7.22 kcal/mol), $F168^{EC2}$ (-6.33 kcal/mol), and $E169^{EC2}$ (-5.13 kcal/mol), based on non-bonding energies as the sum of vdW, electrostatic Coulomb with 2.5 dielectric constant and H-bond energies.

In the binding site of adenosine-$hAA_{2A}R$, the exo-cyclic amino group of the adenine ring interacts with $N253^{6.55}$ and $E169^{EC2}$ as the X-ray structure of $hAA_{2A}R$ also interacts with the amino group in the tri-cyclic ring of ZM241385 (Doré et al., 2011). We also observed an additional H-bonding interaction (-2.82 kcal/mol) between the 2' and 3'-OH group of the nucleoside ring and the side chains of the double protonated basic $H278^{7.43}$, which forms the H-bonding with $E13^{1.39}$. The mutation of $N253^{6.55}$ to alanine reports loss of binding for agonist and antagonist, while the $H278E^{7.43}$ mutation displays substantial decrease of binding for agonist and antagonist except in the case of N^6-substituted agonists (S.-K. Kim et al., 2003)

Later this adenosine-bound $hA_{2A}AR$ was compared with the crystal structures of the agonist UK43207-bound structure (3QAK)(Xu et al., 2011) and the agonist adenosine-bound structure (2YDO) (Lebon et al., 2011) and has a Ca RMSD of 1.61 and 1.69 Å relative to the $hAA_{2A}R$ crystal TM bundle, respectively. After the TM alignment, the adenosine conformation has a RMSD of 1.33 Å relative to the crystal structure.

The predicted ligand binding structure from $hAA_{2A}R$ to hAA_3R was then matched, leading to common interactions at $N250^{6.55}$ (-5.82 kcal/mol), $F168^{EC2}$ (-4.52 kcal/mol), and $H272^{7.43}$ (-4.03 kcal/mol). However, the $V169^{EC2}$ in hAA_3R (corresponding amino acid of $E169^{EC2}$ in $hAA_{2A}R$) results in unfavorable interactions in hAA_3R because of the different loop geometry of EC2 as well as loss of the interaction with the terminal amino group (Table 3 and Figure 5A and 4B). The result is a dramatic reduced binding affinity (cavity sum = -37.56 for $hAA_{2A}R$ vs. -28.92 for hAA_3R), which agrees with the dramatically decreased experimental binding affinity of endogenous adenosine at both rat ARs ($rAA_{2A}R$ estimated K_i: 30 nM vs. rAA_3R estimated K_i, 1,000 nM) (Jacobson, et al., 2000).

2. **hAA_3R selective Cl-IB-MECA agonist:** We docked A_3 subtype selective Cl-IB-MECA to the lowest energy conformation of hAA_3R. From the cavity analysis of Cl-IB-MECA at hAA_3R, the major contributing amino acids are $F168^{EC2}$ (-9.79 kcal/mol), $N250^{6.55}$ (-4.44 kcal/mol), $L246^{6.51}$ (-4.28 kcal/mol), and $H272^{7.43}$ (-3.35 kcal/mol) in Table 4. The hydrophobic residues, $V169^{EC2}$ and $I253^{6.58}$, surround the terminal benzyl group of Cl-IB-MECA (Figure 5D). The N^6-bulky benzyl group forms an additional interaction between EC2 and upper TM6, stabilizing the complex.

To study the origin of subtype selectivity, the predicted best binding pose of the highly A_3-selective ligand Cl-IB-MECA was matched to the $hAA_{2A}R$ predicted structure, and then SCREAM(V.W.T. Kam &

Table 3. Cavity energy of the endogenous agonist adenosine 4 bound to the human adenosine A_3 (AA_3R) and A_{2A} receptors ($AA_{2A}R$) () Residues are ordered by total NonBond energy (A_{2A}), which is the sum of van der Waals (vdW), Coulomb, and H-bond energy (kcal/mol) in the unified cavity. Subtype residues that are predicted to vary between A_{2A} and A_3 are displayed in bold face. The main interactions leading to subtype selectivity (underlined) in the predicted structure are E169 in A_{2A} vs. V169 in A_3, and V84 in A_{2A} vs. L90 in A_3. The color coding for contributions of each residue to binding of the adenosine ligand is: Dark grey: > 3 kcal/mol, Grey: 1 - 3 kcal/mol, Light grey: 0.5 - 1.0 kcal/mol.

Ballesteros-Weinstein #	**$AA_{2A}R$ Ki: estimated 30 nM**			**AA_3R Ki: estimated 1,000 nM**		
	Res	**#**	**NonBond**	**Res**	**#**	**NonBond**
6.55	ASN	253	-7.22	ASN	250	-5.82
EC2	PHE	168	-6.33	PHE	168	-4.52
EC2	**GLU**	**169**	**-5.13**	**VAL**	**169**	**0.00**
7.43	HSE	278	-2.82	HSE	272	-4.03
7.39	ILE	274	-2.72	ILE	268	-2.02
3.32	**VAL**	**84**	**-2.39**	**LEU**	**90**	**1.07**
6.51	LEU	249	-2.19	LEU	246	-3.39
7.42	SER	277	-1.78	SER	271	-1.77
3.33	LEU	85	-1.49	LEU	91	-0.96
5.38	MET	177	-1.39	MET	177	-1.34
3.36	THR	88	-1.29	THR	94	-0.80
7.35	**MET**	**270**	**-1.05**	**LEU**	**264**	**-1.27**
2.61	ALA	63	-0.49	ALA	69	-0.37
6.52	**HIS**	**250**	**-0.30**	**SER**	**247**	**0.00**
2.64	**ILE**	**66**	**-0.25**	**VAL**	**72**	**-0.18**
3.29	**ALA**	**81**	**-0.24**	**THR**	**87**	**-0.67**
6.54	ILE	252	-0.17	ILE	249	-0.12
5.35	MET	174	-0.17	MET	174	0.00
1.39	GLU	13	-0.15	GLU	19	0.00
6.48	TRP	246	0.00	TRP	243	-1.94
EC2	**LEU**	**167**	**0.00**	**GLN**	**167**	**-0.44**
2.57	**ALA**	**59**	**0.00**	**VAL**	**65**	**-0.34**
6.58	**THR**	**256**	**0.00**	**ILE**	**253**	**-0.04**
	SUM		**-37.56**	**SUM**		**-28.92**

W.A Goddard III, 2008) was used to optimize the side chains in the binding pocket, after which the final ligand/protein complex was neutralized and minimized. In $hAA_{2A}R$, the interactions between protein and ligands were hampered by the conformationally restricted loop between $E169^{EC2}$ and $H264^{EL3}$ through a salt-bridge (Figure 5C). The corresponding amino acids of $I253^{6.58}$, $V169^{EC2}$, and $Q167^{EC2}$ in hAA_3R are $T256^{6.58}$, $E169^{EC2}$, and $L167^{EC2}$ in $hAA_{2A}R$ with weakened interactions, resulting in a 7.31 kcal/mol favorable interaction in hAA_3R (Table 4). This predicted binding mode is in parallel with the experi-

Figure 5. Predicted binding sites (A) adenosine 4 to the adenosine A_{2A} receptor ($AA_{2A}R$), (B) adenosine to the adenosine A_3 receptor (AA_3R), (C) A_3 selective agonist Cl-IB-MECA **5** *at $AA_{2A}R$, (D) A_3 selective agonist Cl-IB-MECA at AA_3Rs. The high selectivity of Cl-IB-MECA for AA_3R relative to $AA_{2A}R$ is due to the salt-bridge between E169 and H264 (red circle in (C)) in AA_3R. In $AA_{2A}R$ we have V169 but there is no residue corresponding to H264 because of the gap in $AA_{2A}R$. H-bonding is indicated by red dots, and subtype selective residues are shown in red.*
S. K. Kim, L. Riley, et al. (2011)

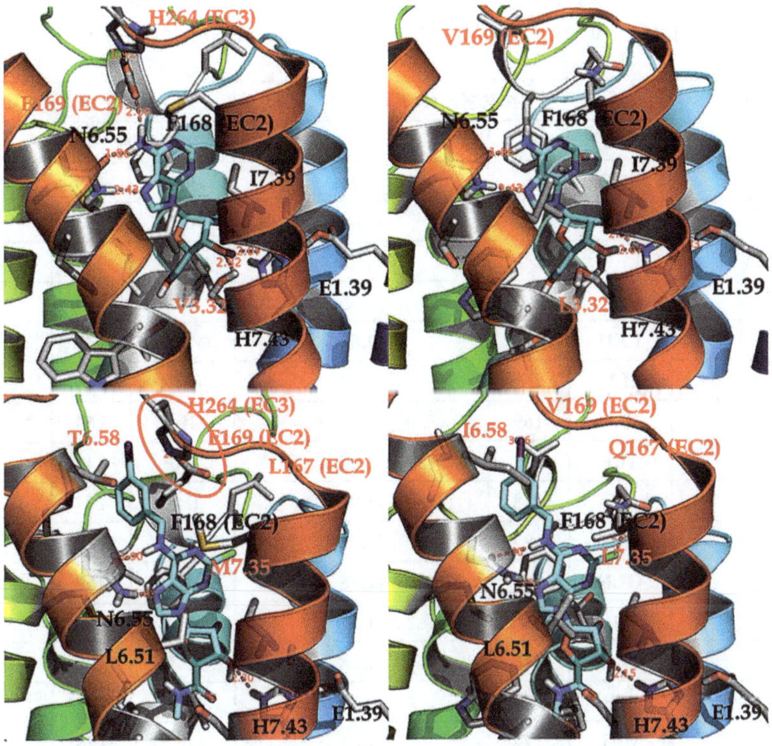

mental binding affinity of A_3 subtype selective Cl-IB-MECA (AA_3R Ki: 1.4 nM vs. $AA_{2A}R$ Ki, 5,360 nM) (Jeong et al., 2007). The docking result also agrees with reported observations, where modification of the N^6-amine moiety (where large hydrophobic groups tend to produce AA_1R and AA_3R selectivity) and the C2 position (where large hydrophobic groups tend to produce $AA_{2A}R$ selectivity) increase the subtype selectivity of adenosine derivatives as AR agonists for each subtype (K. A. Jacobson et al., 2009).

From the docking studies of the subtype selective antagonists, $I253^{6.58}$ and $V169^{EC2}$ in hAA_3R are involved in hydrophobic interactions with the N^6-bulky group in the A_3 selective ligands, while $E169^{EC2}$ in $hAA_{2A}R$ (which is the corresponding amino acid of $V169^{EC2}$ in hAA_3R) stabilizes the exocyclic amino group of the A_{2A} selective ligands through additional H-bonding. Thus this predicted structure explains the increase of A_3 selectivity for the additional N^6-bulky substituent at hAA_3R. However, $Q167^{EC2}$ in hAA_3R is located in the proximity of the C2 substituent which is oriented toward the extracellular loop 2 (EC2) in the docking model. The corresponding amino acid $L167^{EC2}$ in $hAA_{2A}R$ could explain the $hA_{2A}AR$ selectivity for large hydrophobic C2 substituents which are observed in most A_{2A} selective antagonists (K. A. Jacobson, et al., 2009).

Table 4. Cavity energy of the A_3 selective Cl-IB-MECA **5** *bound to the human adenosine A_3 (AA_3R) and A_{2A} receptors ($AA_{2A}R$). The residues are ordered by the total NonBond energy (A_3), which is the sum of van der Waals (vdW), Coulomb, and H-bond energy (kcal/mol) in the unified cavity. Subtype residues that vary between A_{2A} and A_3 are displayed in bold face. The main interactions leading to subtype selectivity are I253 in A_3 vs. T256 in A_{2A}, V169 in A_3 vs. E169 in $A_{2A,}$ and Q167 in A_3 vs. L167 in A_{2A}, leading to 3.85 of the predicted 7.31 differential binding. The other major differences, involving conserved residues are for F168, S273, M174 favoring A_3 and N250, I268 favoring A_{2A}, leading to a differential of 2.16 of the predicted 7.31 differential binding. The color coding for contributions of each residue to binding of the adenosine ligand is: Dark grey: > 3 kcal/mol, Grey: 1 - 3 kcal/mol, Light grey: 0.5 - 1.0 kcal/mol*

Ballesteros-Weinstein #	AA_3R Ki: 1.4 nM			$AA_{2A}R$ Ki: 5,360 nM		
	Res	#	NonBond	Res	#	NonBond
EC2	PHE	168	-9.79	PHE	168	-7.99
6.50	ASN	250	-4.44	ASN	253	-5.86
6.51	LEU	246	-4.28	LEU	249	-4.19
7.43	HSE	272	-3.35	HSE	278	-2.55
7.35	**LEU**	**264**	**-2.87**	**MET**	**270**	**-3.49**
6.58	**ILE**	**253**	**-2.87**	**THR**	**256**	**-1.29**
EC2	**VAL**	**169**	**-2.82**	**GLU**	**169**	**-1.93**
6.48	TRP	243	-2.02	TRP	246	-1.20
7.39	ILE	268	-1.91	ILE	274	-3.10
5.38	MET	177	-1.72	MET	177	-1.73
EC2	**GLN**	**167**	**-1.38**	**LEU**	**167**	**0.00**
3.29	**THR**	**87**	**-1.37**	**ALA**	**81**	**-0.75**
3.32	**LEU**	**90**	**-1.28**	**VAL**	**84**	**-1.74**
7.42	SER	271	-1.05	SER	277	0.17
5.35	MET	174	-1.05	MET	174	0.72
7.36	TYR	265	-0.67	TYR	271	-0.63
EC2	**MET**	**172**	**-0.65**	**VAL**	**172**	**-0.66**
3.36	THR	94	-0.61	THR	88	0.01
6.54	ILE	249	-0.53	ILE	252	-0.57
3.33	LEU	91	-0.48	LEU	85	-0.92
2.64	**VAL**	**72**	**-0.40**	**ILE**	**66**	**-0.42**
2.61	ALA	69	-0.38	ALA	63	-0.62
2.57	**VAL**	**65**	**-0.25**	**ALA**	**59**	**0.01**
6.47	**SER**	**242**	**-0.05**	**CYS**	**245**	**-0.15**
EC2	**ARN**	**173**	**0.00**	**PRO**	**173**	**-0.24**
6.62	**GLY**	**257**	**0.00**	**PRO**	**260**	**-0.40**
EC3				**HIS**	**264**	**0.64**
	SUM		**-46.17**	**SUM**		**-38.86**

S. K. Kim, L. Riley, et al. (2011)

From 27 residues in the binding site of hA_3AR, 14 residues are conserved with 52% sequence identity with the $hA_{2A}AR$. For the $hA_{2A}AR$ selectivity, E169V in EC2 (Val in hA_3AR) is the most important residue which could form an extra H-bonding interaction. V3.32L is also important for the $hA_{2A}AR$ selectivity. However, for the hA_3AR selectivity, Q167L (Leu in $hA_{2A}AR$) can increase the hA_3AR binding. T3.29A can also add a hydrophilic interaction compared with the $hA_{2A}AR$.

3. EXECUTIVE SUMMARY

The GEnSeMBLE Monte Carlo Technique was used to generate an ensemble of the 10 to 20 low energy packing of the 7-helix bundles for all subtypes of hHT_2Rs, hHRs and hARs to understand the origin of subtype selectivity for several subtype selective antagonists of $hHT_{2B}R$, hH_3HR and hA_3AR. Then the DarwinDock Monte Carlo Technique was performed to predict the best binding pose to each of the structures of this ensemble the

1. We found that the antagonist **2** is selective to HT_{2B} because of favorable vdW interaction at $L132^{3.29}$ and $L281^{6.58}$ in $hHT_{2B}R$, while become unfavorable at $I132^{3.29}$ and $S279^{6.58}$in $hHT_{2C}R$.
2. We found that clobenpropit **3** binds selective to both hH_3HR and hH_4HR because of the additional H-bonding to E5.46/ S5.43, whereas the corresponding amino acids $N198^{5.46}$/ $A195^{5.43}$ in hH_1HR and $T190^{5.46}$/ $G187^{5.43}$ in hH_2HR lead to a loss in these hydrophilic interactions.
3. We found that the A_3 selective ligand, Cl-IB-MECA **5**, has favorable vdW interactions of the N^6-bulky group with $I253^{6.58}$ and $V169^{EC2}$ in hAA_3R while $E169^{EC2}$ in $hAA_{2A}R$ (the amino acid corresponding to $V169^{EC2}$ in hAA_3R) instead interacts with the free amino group of the A_{2A} selective endogenous ligand adenosine **4** through additional H-bonding.

SOLUTIONS AND RECOMMENDATIONS

Although GPCRs are the most fruitful targets for marketed drugs, the success of developments of highly selective subtype drug candidates have been hampered by difficulty in achieving high selectivity for individual GPCR subtypes. After lead or target discovery, management of drug cross-reactivity is a crucial part in the lead-optimization process. Since a major problem of drug is cross-reactivity across members of a single GPCR subfamily because of high conservation of the orthosteric binding sites, development of subtype selective GPCR agonists or antagonists through the aid of molecular modelling is the first step of understanding for the structural requirements for subtype selectivity and the molecular properties required for activation or inactivation.

FUTURE RESEARCH DIRECTIONS

While high resolution 3D structures of $hAA_{2A}R$, hH_1HR, and $hHT_{2B}R$ represent an excellent template for homology modeling for other subtypes, the developments of subtype selective ligands has yet to be shown. Direct comparison of binding modes of subtype selective ligands at other subtypes was applied to rational design of novel subtype-selective ligands without side-effects. The non-conserved residues in the binding cavity which serve as determinants of ligand selectivity will affect overall protein conformation, leading to the different local environments in the binding cavity. Analysis of the subtype-specific ligand receptor interactions allowed identification of the major determinants of ligand selectivity, which may facilitate discovery of more efficient drug candidates. Application of these computational approaches is advancing the understanding of structure-function relationships among individual types, subtypes, and other family of GPCRs. Site directed mutagenesis experiments are also an effective tool to investigate the amino acid properties of specific residues for determining the subtype-selectivity. However, the issue in

Figure 6. Diagram of rational drug design for the subtype selective GPCR ligands

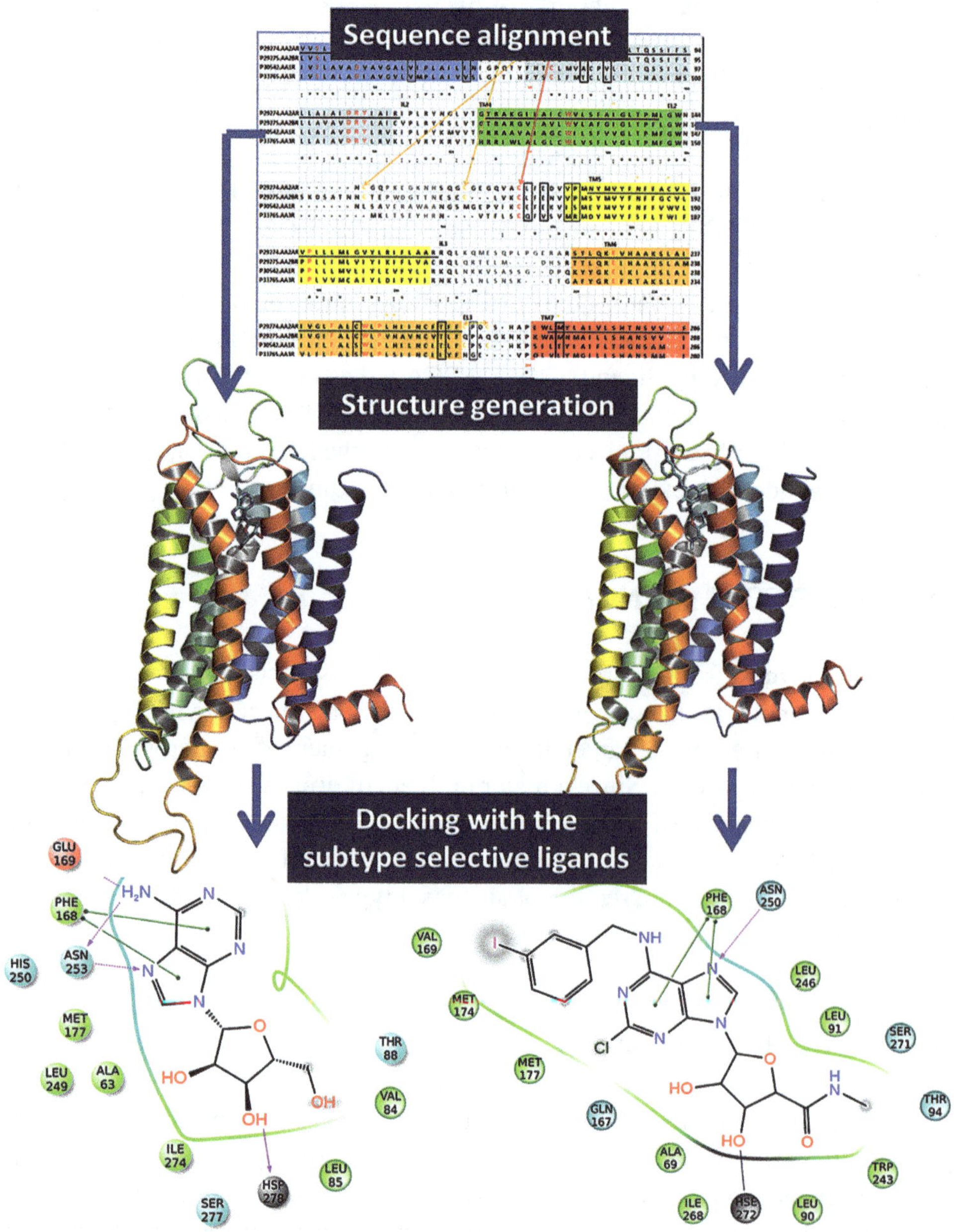

GPCR targeted drug discovery is the presence of multiple conformations. GPCRs exhibit several distinct low-energy conformations, each of which might favor binding to different ligands and/or lead to different downstream functions, understanding their function requires the various ensembles of low-energy configurations that might play a role in this pleiotropic functionality. Thus take-home message in this review is the activation conformation might be various among the subtypes which were regulated by different G protein, in which might be the one of the major causes of attrition in drug discovery. Accounting for 28% of cases was the lack of systemic drug efficacy because of lack of information of GPCR activation. In addition, the existence of functional dimers or larger oligomeric complexes through homo- or heterodimerization makes it more difficult to interpret the experimental data. Despite all of problems, we expect that the improved understanding of the microscopic interactions at the various binding site of the various low energy conformations of the GPCRs will guide dramatically in the development of novel subtype selective ligands. Since GPCRs are important as physiological and pharmacologic targets, such results from computational predictions validated by experiment should likely provide useful new insights of broad biological and medical importance.

CONCLUSION

The availability of high resolution X-ray structures has provided new insights into the structures of GPCRs, however for distinguishing between the GPCR conformations favored by the various ligands it is essential to determine the low energy packings of the GPCR that might be favored by different ligands for different subtypes. The GEnSeMBLE methodology provides the means to predict these low lying structures, but the accuracy of GEnSeMBLE has been dramatically improved by the availability of the many new X-ray structures. We studied the subtype selectivity of GPCRs using molecular modeling. As summarized in Figure 6, we generated the 3D-structures from the sequence alignment and docked the subtype-selective ligands. Finally, the comparison of binding site at each subtype gives the information which residues are directly involved in the subtype selectivity and how much subtype selective residues are energetically affected by ligands through the cavity analysis. This information will be the best tools for understanding the binding mechanism of each subtypes and further developing the subtype selective ligands through rational drug design to minimize the undesirable side effects because of non-specificity. Thus the computation and experiment working together can provide the information required for determining the optimum packing and binding pose for various ligands to various subtypes.

REFERENCES

Abrol, R., Bray, J. K., & Goddard, W. A. III. (2011). Bihelix: Towards de novo structure prediction of an ensemble of G-protein coupled receptor conformations. *Proteins*, *80*(2), 505–518. doi:10.1002/prot.23216 PMID:22173949

Abrol, R., Griffith, A. R., Bray, J. K., & Goddard, W. A. III. (2012). Structure prediction of G protein-coupled receptors and their ensemble of functionally important conformations. *Methods in Molecular Biology (Clifton, N.J.)*, *914*, 237–254. PMID:22976032

Abrol, R., Trzaskowski, B., Goddard, W. A., Nesterovb, A., Olave, I., & Irons, C. (2014). Ligand- and mutation-induced conformational selection in the CCR5 chemokine G protein-coupled receptor. *Proceedings of the National Academy of Sciences of the United States of America*, *111*(36), 13040–13045. doi:10.1073/pnas.1413216111 PMID:25157173

Ahn, K. H., Scott, C. E., Abrol, R., Goddard, W. A. III, & Kendall, D. A. (2013). Computationally-predicted CB1 cannabinoid receptor mutants show distinct patterns of salt-bridges that correlate with their level of constitutive activity reflected in G protein coupling levels, thermal stability, and ligand binding. *Proteinsx*, *81*(8), 1304–1317. doi:10.1002/prot.24264 PMID:23408552

Astles, P. C., Baker, S. R., Boot, J. R., Broad, L. M., Dell, C. P., & Keenan, M. (2002). Recent progress in the development of subtype selective nicotinic acetylcholine receptor ligands. *Current Drug Targets. CNS and Neurological Disorders*, *1*(4), 337–348. doi:10.2174/1568007023339256 PMID:12769608

Ballesteros, J. A., & Weinstein, H. (1995). Integrated methods for the construction of three-dimensional models and computational probing of structure-function relations in G protein coupled receptors. *Methods Neurosci*, *25*, 366–428. doi:10.1016/S1043-9471(05)80049-7

Bray, J. K., Abrol, R., Goddard, W. A. III, Trzaskowski, B., & Scott, C. E. (2014). SuperBiHelix method for predicting the pleiotropic ensemble of G-protein-coupled receptor conformations. *Proceedings of the National Academy of Sciences of the United States of America*, *111*(1), E72–E78. doi:10.1073/pnas.1321233111 PMID:24344284

Cherezov, V., Rosenbaum, D. M., Hanson, M. A., Rasmussen, S. G. F., Thian, F. S., Kobilka, T. S., & Stevens, R. C. et al. (2007). High-resolution crystal structure of an engineered human beta2-adrenergic G protein-coupled receptor. *Sci.*, *318*(5854), 1258–1265. doi:10.1126/science.1150577 PMID:17962520

Chien, E. Y., Liu, W., Zhao, Q., Katritch, V., Han, G. W., Hanson, M. A., & Stevens, R. C. et al. (2010). Structure of the human dopamine D3 receptor in complex with a D2/D3 selective antagonist. *Science*, *330*(6007), 1091–1095. doi:10.1126/science.1197410 PMID:21097933

Conn, P. J., Christopoulos, A., & Lindsley, C. W. (2009). Allosteric modulators of GPCRs: A novel approach for the treatment of CNS disorders. *Nature Reviews. Drug Discovery*, *8*(1), 41–54. doi:10.1038/nrd2760 PMID:19116626

Doré, A. S., Robertson, N., Errey, J. C., Ng, I., Hollenstein, K., Tehan, B., & Marshall, F. H. et al. (2011). Structure of the adenosine A(2A) receptor in complex with ZM241385 and the xanthines XAC and caffeine. *Structure (London, England)*, *19*(9), 1283–1293. doi:10.1016/j.str.2011.06.014 PMID:21885291

Ellis, C. (2004). The state of GPCR research in 2004. *Nature Reviews. Drug Discovery*, *3*(7), 577–626. doi:10.1038/nrd1458 PMID:15272499

Esbenshade, T. A., Krueger, K. M., Miller, T. R., Kang, C. H., Denny, L. I., Witte, D. G., & Hancock, A. A. et al. (2003). Two novel and selective nonimidazole histamine H3 receptor antagonists A-304121 and A-317920: I. In vitro pharmacological effects. *The Journal of Pharmacology and Experimental Therapeutics*, *305*(3), 887–896. doi:10.1124/jpet.102.047183 PMID:12606603

Forbes, I. T., Ham, P., Booth, D. H., Martin, R. T., Thompson, M., Baxter, G. S., & Wood, M. D. et al. (1995). 5-Methyl-1-(3-pyridiiylcarbamoyl)-1,2,3,5-tetrahydropyrrolo[2,3-f]indole: A novel 5-HT2C/5-HT2B receptor antagonist with improved affinity, selectivity, and oral activity. *Journal of Medicinal Chemistry*, *38*(14), 2524–2530. doi:10.1021/jm00014a004 PMID:7629791

Fredriksson, R., Lagerstrom, M. C., Lundin, L. G., & Schioth, H. B. (2003). The G-protein-coupled receptors in the human genome form five main families. Phylogenetic analysis, paralogon groups, and fingerprints. *Molecular Pharmacology*, *63*(6), 1256–1272. doi:10.1124/mol.63.6.1256 PMID:12761335

Goddard, W. A. III, & Abrol, R. (2007). 3-Dimensional structures of G protein-coupled receptors and binding sites of agonists and antagonists. *The Journal of Nutrition*, *137*, 1528S–1538S. PMID:17513420

Granier, S., Manglik, A., Kruse, A. C., Kobilka, T. S., Thian, F. S., Weis, W. I., & Kobilka, B. K. (2012). Structure of the delta-opioid receptor bound to naltrindole. *Nature*, *485*(7398), 400–404. doi:10.1038/nature11111 PMID:22596164

Haga, K., Kruse, A. C., Asada, H., Yurugi-Kobayashi, T., Shiroishi, M., Zhang, C., & Kobayashi, T. et al. (2012). Structure of the human M2 muscarinic acetylcholine receptor bound to an antagonist. *Nature*, *482*(7386), 547–551. doi:10.1038/nature10753 PMID:22278061

Hanson, M. A., Roth, C. B., Jo, E., Griffith, M. T., Scott, F. L., Reinhart, G., & Stevens, R. C. et al. (2012). Crystal structure of a lipid G protein-coupled receptor. *Science*, *335*(6070), 851–855. doi:10.1126/science.1215904 PMID:22344443

Hopkins, A. L., & Groom, C. R. (2002). The druggable genome. *Nature Reviews. Drug Discovery*, *1*(9), 727–730. doi:10.1038/nrd892 PMID:12209152

Isberg, V., Balle, T., Sander, T., Jørgensen, F. S., & Gloriam, D. E. (2011). G protein- and agonist-bound serotonin 5-HT2A receptor model activated by steered molecular dynamics simulations. *Journal of Chemical Information and Modeling*, *51*(2), 315–325. doi:10.1021/ci100402f PMID:21261291

Jaakola, V. P., Griffith, M. T., Hanson, M. A., Cherezov, V., Chien, E. Y., Lane, J. R., & Stevens, R. C. et al. (2008). The 2.6 angstrom crystal structure of a human A2A adenosine receptor bound to an antagonist. *Sci.*, *322*(5905), 1211–1217. doi:10.1126/science.1164772 PMID:18832607

Jacobson, K. A., Ji, X., Li, A. H., Melman, N., Siddiqui, M. A., Shin, K. J., & Ravi, R. G. et al. (2000). Methanocarba analogues of purine nucleosides as potent and selective adenosine receptor agonists. *Journal of Medicinal Chemistry*, *43*(11), 2196–2203. doi:10.1021/jm9905965 PMID:10841798

Jacobson, K. A., Klutz, A. M., Tosh, D. K., Ivanov, A. A., Preti, D., & Baraldi, P. G. (2009). Medicinal chemistry of the A3 adenosine receptor: Agonists, antagonists, and receptor engineering. *Handbook of Experimental Pharmacology*, *193*, 123–159. doi:10.1007/978-3-540-89615-9_5 PMID:19639281

Jeong, L. S., Choe, S. A., Gunaga, P., Kim, H. O., Lee, H. W., Lee, S. K., & Moon, H. R. et al. (2007). Discovery of a New Nucleoside Template for Human A3 Adenosine Receptor Ligands: D-4'-Thioadenosine Derivatives without 4'-Hydroxymethyl Group as Highly Potent and Selective Antagonists. *Journal of Medicinal Chemistry*, *50*(14), 3159–3162. doi:10.1021/jm070259t PMID:17555308

Jutel, M., Blaser, K., & Akdis, C. A. (2005). Histamine in chronic allergic responses. *Journal of Investigational Allergology & Clinical Immunology*, *15*, 1–8. PMID:15864876

Kam, V. W. T., & Goddard, W. A. III. (2008). Flat-Bottom Strategy for Improved Accuracy in Protein Side-Chain Placements. *Journal of Chemical Theory and Computation*, *4*(12), 2160–2169. doi:10.1021/ct800196k PMID:26620487

Katoh, K., Kuma, K., Toh, H., & Miyata, T. (2005). MAFFT version 5: Improvement in accuracy of multiple sequence alignment. *Nucleic Acids Research*, *33*(2), 511–518. doi:10.1093/nar/gki198 PMID:15661851

Katritch, V., Kufareva, I., & Abagyan, R. (2011). Structure based prediction of subtype-selectivity for adenosine receptor antagonists. *Neuropharmacology*, *60*(1), 108–115. doi:10.1016/j.neuropharm.2010.07.009 PMID:20637786

Kim, S. K., Fristrup, P., Abrol, R., & Goddard, W. A. III. (2011). Structure-based prediction of subtype selectivity of histamine H3 receptor selective antagonists in clinical trials. *Journal of Chemical Information and Modeling*, *51*(12), 3262–3274. doi:10.1021/ci200435b PMID:22035233

Kim, S.-K., Gao, Z.-G., Van Rompaey, P., Gross, A. S., Chen, A., Van Calenbergh, S., & Jacobson, K. A. (2003). Modeling the adenosine receptors: Comparison of the binding domains of A2A agonists and antagonists. *Journal of Medicinal Chemistry*, *46*(23), 4847–4859. doi:10.1021/jm0300431 PMID:14584936

Kim, S.-K., Li, Y., Abrol, R., Heo, J., & Goddard, W. A. III. (2011). Predicted structures and dynamics for agonists and antagonists bound to serotonin 5-HT2B and 5-HT2C receptors. *Journal of Chemical Information and Modeling*, *51*(2), 420–433. doi:10.1021/ci100375b PMID:21299232

Kim, S. K., Riley, L., Abrol, R., Jacobson, K. A., & Goddard, W. A. III. (2011). Predicted structures of agonist and antagonist bound complexes of adenosine A3 receptor. *Proteins*, *79*(6), 1878–1897. doi:10.1002/prot.23012 PMID:21488099

Kruse, A. C., Hu, J., Pan, A. C., Arlow, D. H., Rosenbaum, D. M., Rosemond, E., & Kobilka, B. K. et al. (2012). Structure and dynamics of the M3 muscarinic acetylcholine receptor. *Nature*, *482*(7386), 552–556. doi:10.1038/nature10867 PMID:22358844

Lebon, G., Warne, T., Edwards, P. C., Bennett, K., Langmead, C. J., Leslie, A. G. W., & Tate, C. G. (2011). Agonist-bound adenosine A2A receptor structures reveal common features of GPCR activation. *Nature*, *474*(7352), 521–525. doi:10.1038/nature10136 PMID:21593763

Lefkowitz, R. J., Pierce, K. L., & Luttrell, L. M. (2002). Dancing with different partners: Protein kinase A phosphorylation of seven membrane-spanning receptors regulates their G protein-coupling specificity. *Molecular Pharmacology*, *62*(5), 971–974. doi:10.1124/mol.62.5.971 PMID:12391258

Li, J., Edwards, P. C., Burghammer, M., Villa, C., & Schertler, G. F. X. (2004). Structure of bovine rhodopsin in a trigonal crystal form. *Journal of Molecular Biology*, *343*(5), 1409–1438. doi:10.1016/j.jmb.2004.08.090 PMID:15491621

Lorenzi, S., Mor, M., Bordi, F., Rivara, S., Rivara, M., Morini, G., & Plazzi, P. V. et al. (2005). Validation of a histamine H3 receptor model through structure-activity relationships for classical H3 antagonists. *Bioorganic & Medicinal Chemistry*, *13*(19), 5647–5657. doi:10.1016/j.bmc.2005.05.072 PMID:16085419

Magnani, F., Pappas, C. G., Crook, T., Magafa, V., Cordopatis, P., Ishiguro, S., & Tzakos, A. G. et al. (2014). Electronic sculpting of ligand-GPCR subtype selectivity: The case of angiotensin II. *ACS Chemical Biology*, *9*(7), 1420–1425. doi:10.1021/cb500063y PMID:24787922

Manglik, A., Kruse, A. C., Kobilka, T. S., Thian, F. S., Mathiesen, J. M., Sunahara, R. K., & Granier, S. et al. (2012). Crystal structure of the μ-opioid receptor bound to a morphinan antagonist. *Nature*, *485*(7398), 321–326. doi:10.1038/nature10954 PMID:22437502

Mayo, S. L., Olafson, B. D., & Goddard, W. A. III. (1990). DREIDING - a generic force field for molecular simulations. *Journal of Physical Chemistry*, *94*(26), 8897–8909. doi:10.1021/j100389a010

Nichols, D. E., & Nichols, C. D. (2008). Serotonin receptors. *Chemical Reviews*, *108*(5), 1614–1641. doi:10.1021/cr078224o PMID:18476671

Okada, T., Fujiyoshi, Y., Silow, M., Navarro, J., Landau, E. M., & Shichida, Y. (2002). Fundamental role of internal water molecules in rhodopsin revealed by Xray crystallography. *Proceedings of the National Academy of Sciences of the United States of America*, *99*(9), 5982–5987. doi:10.1073/pnas.082666399 PMID:11972040

Okada, T., Sugihara, M., Bondar, A. N., Elstner, M., Entel, P., & Buss, V. (2004). The retinal conformation and its environment in rhodopsin in light of a new 2.2 Å crystal structure. *Journal of Molecular Biology*, *342*(2), 571–583. doi:10.1016/j.jmb.2004.07.044 PMID:15327956

Olah, M. E., & Stiles, G. L. (1995). Adenosine receptor subtypes: Characterization and therapeutic regulation *Annu. Rev. Phamloeol. Toxieol.*, *35*, 581–606. PMID:7598508

Palczewski, K., Kumasaka, T., Hori, T., Behnke, C.A., Motoshima, H. (2000). Crystal structure of rhodopsin: a G protein-coupled receptor. *Sci.*, 739-745.

Pollastri, G., & McLysaght, A. (2005). Porter: A new, accurate server for protein secondary structure prediction. *Bioinformatics (Oxford, England)*, *21*(8), 1719–1720. doi:10.1093/bioinformatics/bti203 PMID:15585524

Pollastri, G., Przybylski, D., Rost, P., & Baldi, B. (2002). Improving the Prediction of Protein Secondary Structure in Three and Eight Classes Using Recurrent Neural Networks and Profiles. *Proteins*, *47*(2), 228–235. doi:10.1002/prot.10082 PMID:11933069

Shimamura, T., Shiroishi, M., Weyand, S., Tsujimoto, H., Winter, G., Katritch, V., & Iwata, S. et al. (2011). Structure of the human histamine H1 receptor complex with doxepin. *Nature*, *475*(7354), 65–70. doi:10.1038/nature10236 PMID:21697825

Tang, C. M., & Insel, P. A. (2005). Genetic variation in G-protein-coupled receptorsconsequences for G-protein-coupled receptors as drug targets. *Expert Opinion on Therapeutic Targets*, *9*(6), 12472–12465. doi:10.1517/14728222.9.6.1247 PMID:16300474

Teller, D. C., Okada, T., Behnke, C. A., Palczewski, K., & Stenkamp, R. E. (2001). Advances in determination of a high-resolution three-dimensional structure of rhodopsin, a model of G-protein-coupled receptors (GPCRs). *Biochemical*, *40*(26), 7761–7762. doi:10.1021/bi0155091 PMID:11425302

Thompson, A. A., Liu, W., Chun, E., Katritch, V., Wu, H., Vardy, E., & Stevens, R. C. et al. (2012). Structure of the nociceptin/orphanin FQ receptor in complex with a peptide mimetic. *Nature*, *485*(7398), 395–399. doi:10.1038/nature11085 PMID:22596163

Vaidehi, N., Floriano, W. B., Trabanino, R., Hall, S. E., Freddolino, P., Choi, E. J., & Goddard, W. A. III. (2002). Structure and Function of GPCRs. *Proceedings of the National Academy of Sciences of the United States of America*, *99*, 12622–12627. doi:10.1073/pnas.122357199 PMID:12351677

Wacker, D., Wang, C., Katritch, V., Han, G. W., Huang, X. P., Vardy, E., & Stevens, R. C. et al. (2013). Structural features for functional selectivity at serotonin receptors. *Science*, *340*(6132), 615–619. doi:10.1126/science.1232808 PMID:23519215

Wang, C., Jiang, Y., Ma, J., & Xu, H. E. (2013). Structural Basis for Molecular Recognition at Serotonin Receptors. *Science*, *340*(6132), 610–614. doi:10.1126/science.1232807 PMID:23519210

Warne, T., Serrano-Vega, M. J., Baker, J. G., Moukhametzianov, R., Edwards, P. C., Henderson, R., & Schertler, G. F. X. et al. (2008). Structure of a β1-adrenergic G-protein coupled receptor. *Nature*, *454*(7203), 486–491. doi:10.1038/nature07101 PMID:18594507

Wimley, W. C., Creamer, T. P., & White, S. H. (1996). Solvation energies of amino acid side chains and backbone in a family of host-guest pentapeptides. *Biochem*, *35*(16), 5109–5124. doi:10.1021/bi9600153 PMID:8611495

Wu, B., Chien, E. Y., Mol, C. D., Fenalti, G., Liu, W., Katritch, V., & Stevens, R. C. et al. (2010). Structures of the CXCR4 chemokine GPCR with small-molecule and cyclic peptide antagonists. *Sci.*, *330*(6007), 1066–1071. doi:10.1126/science.1194396 PMID:20929726

Wu, H., Wacker, D., Mileni, M., Katritch, V., Han, G. W., Vardy, E., & Stevens, R. C. et al. (2012). Structure of the human k-opioid receptor in complex with JDTic. *Nature*, *485*(7398), 327–332. doi:10.1038/nature10939 PMID:22437504

Xu, F., Wu, H., Katritch, V., Han, G. W., Jacobson, K. A., Gao, Z. G., & Stevens, R. C. et al. (2011). Structure of an agonist-bound human A2A adenosine receptor. *Science*, *332*(6027), 322–327. doi:10.1126/science.1202793 PMID:21393508

Yang, H., Avila, M. Y., Peterson-Yantorno, K., Coca-Prados, M., Stone, R. A., Jacobson, K. A., & Civan, M. M. (2005). The cross-species A3 adenosine-receptor antagonist MRS1292 inhibits adenosine-triggered human nonpigmented ciliary epithelial cell fluid release and reduces mouse intraocular pressure. *Current Eye Research*, *30*(9), 747–754. doi:10.1080/02713680590953147 PMID:16146920

KEY TERMS AND DEFINITIONS

BiHelix: To cover all possible rotations of all 7 helices, consider 30° increments in the rotation about each TM axis (eta η angle), leading to $(12)^7$ ~ 35 million packings.

GEnSeMBLE: GPCR Ensemble of Structures in Membrane BiLayer Environment) Monte Carlo method.

GPCR: G protein-coupled receptors.

Subtype Selectivity: The requirement to gain control of receptor selectivity.

SuperBihelix: To cover all possible rotations of all 7 helices, optimize the tilt angles: theta θ (tilting away from the z-axis) and phi φ (the azimuthal angle of the tilting from the xy plane) angle in addition to the rotation about each TM axis (eta η angle) leading to ~12 trillion packings.

Chapter 7
Molecular Modelling, Dynamics, and Docking of Membrane Proteins:
Still a Challenge

Nanda Kumar Yellapu
Vector Control Research Centre, India

ABSTRACT

Computational tools and techniques are now most popular and promising to progress the research at rapid rate. Molecular modelling studies contribute their maximum role in wide variety of disciplines especially in proteomics and drug discovery strategies. Molecular dynamics and molecular docking algorithms are now became an essential part in daily research activities of every laboratory throughout the world. These strategies are now well established and standardised to study any specific protein of interest and drug molecule. But still there exist considerable drawbacks in a special concern with membrane proteins as the presently available tools and methods cannot be applied directly to them. Modelling, dynamics and docking studies of membrane proteins need a special care and attention as several challenges are to be crossed with an intensive care to produce a reliable result. This chapter is aimed to discuss such challenges and solutions to handle membrane proteins.

INTRODUCTION

Membrane Proteins

Cell membranes are the most indispensable attributes of any living organisms that safeguard the integrity of a cell by wrapping around. It acts as a potential barrier to perform several physiological functions by offering complex molecular machineries. These multifarious functions and mechanisms of membranes are facilitated and regulated by an inimitable group of proteins called as membrane proteins which are of both surface and integral types (Figure 1). They play a vital role as channels for the transport of ions and molecule to meet the regular demands of biochemical functions (Cooper, 2000).

DOI: 10.4018/978-1-5225-0362-0.ch007

Figure 1. Membrane and membrane proteins; their significant role as transporters and signal transducers

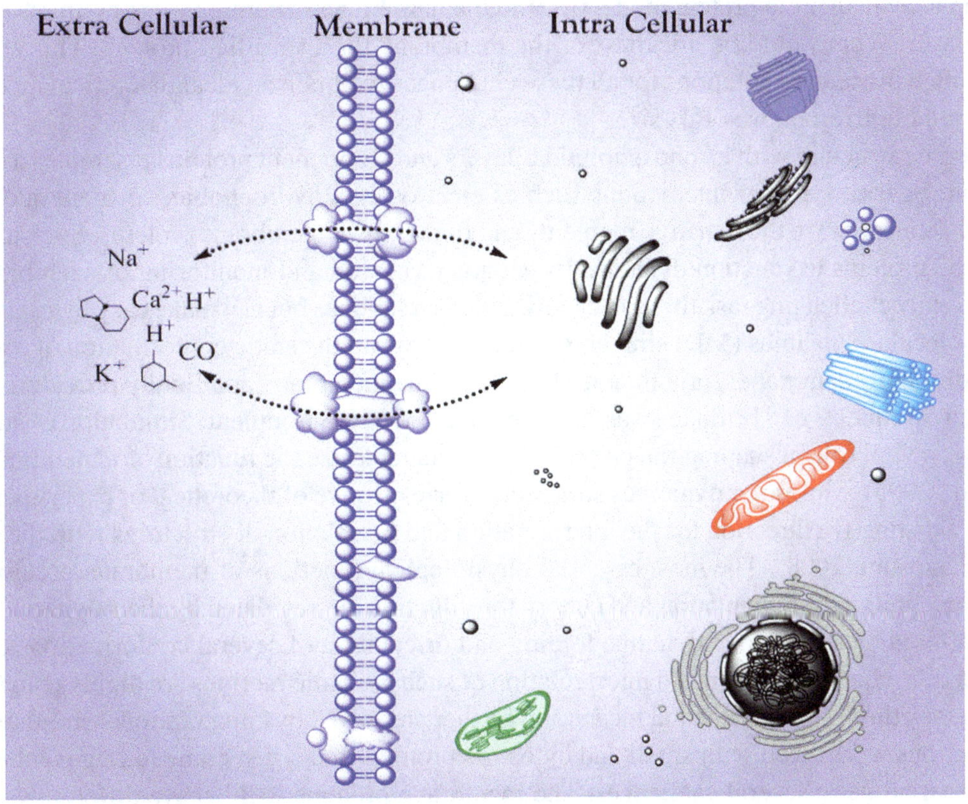

The structural and functional characterization of membrane proteins is typically a thorny task as the difficulties are associated with providing membrane environment that limits the purification and crystallization. Recent advancement in the technologies has led to the structural and functional characterization of several membrane proteins by providing their experimental structural information (Tatulian, 2000). But still it is not meeting the required demand as presently only 2% of membrane protein structures were solved. Such demanding situations paved a spectacular way for the construction of membrane protein models through computational modelling methods (Arinaminpathy et al., 2009). There exist very narrow traditions to construct the protein models but providing an immense support to the structural biologists (Lacapère et al., 2007). Such ways even relies on the experimental data of pre-existing proteins (Ash et al., 2004). Such methods are incredibly known as homology modelling methods; nevertheless homology modelling also fails many times with membrane models as the availability of experimental data for them is very limited. Construction of such models remains as a challenge when the subject of protein is very demanding and necessitous. This chapter is going to describe how such situations will overcome. Stepwise recurrent methodologies with unequivocal and particularized algorithms can be followed where the propensities of amino acids can be utilized to construct the membrane protein models where *de novo* modelling is one of the methods comes under this technology. Even, this also never solve the problem entirely, as this idea works out well with the small proteins and still leavening the challenge unsolved as the membrane proteins usually bigger in size. This chapter will further deals, how the larger size membrane protein models are constructed. The constructed models must represent or resemble a natural conformation so as to mimic the experimental model. How it can be made? Still another challenge remnants, whether the protein model fits into the membrane structure in line or aligns properly in the lipid bilayer. What would be thickness of the membrane for a specified protein? This chapter will explain detailed procedural solutions for all these challenges to obtain an ideal theoretical model for any membrane protein irrespective of its size.

A membrane structure with its phospholipid bilayers and component proteins resembles a disordered environment. Several types of interactions such as electrostatic, hydrophobic, covalent and hydrogen bonding etc., contribute to the dynamic moments and functions of membrane proteins. Such interactions will drive the proteins to function dynamically. Characterization and monitoring of such behaviour of channels is really challenging task through experimental techniques but can make easy by *in silico* means applying molecular dynamics (MD) simulation studies. Beyond the static conformation of a membrane protein, addition of membranes, orientation of model in the membrane and finally providing the membrane like environments can be done so as to mimic the *in vivo* environment. Simulation of such atomic models plays a vital role in shaping the perfect view to characterize the function of a membrane protein (Chang et al., 2005). Molecular dynamics simulations, irrespective of theoretical or experimental structures, offer a distinctive direction for the interpretation and correlation of structures with their function (Lindahl & Sansom, 2008). The most essential physiological functions of membrane proteins such as molecular recognition, ion regulation and energy transduction are regulated by their dynamic moments which are offered by the internal dynamic folding and orientation of several conformations of the proteins. Inspection, characterization and interpretation of such dynamic motions are highly convoluted and challenging task through experimental means and further are solved by computational simulations. Such characterizations works well with small and cytosolic proteins and when come to large and membrane proteins it will prone to several challenges. The membrane proteins will behave differently when they are present in lipid environment as they are embedded in lipid bilayer their folding is majorly influenced by the membrane. The dynamic motions of the membrane can be monitored in presence of simulated peripheral forces that duplicate some aspect of the environment, such as its voltage and surface tension. This chapter deals with the challenges associated with the membrane protein dynamics especially how the simulation of ions/molecules conductance works out which is the most challenging exertion in computational studies of membrane proteins because of the complexity of the potential energy surface, alternate hydrophilic and hydrophobic environments that finally makes the passage of ions/molecules more specific.

Membrane Proteins as Transporters/Ion Channels

The small molecules and ions possessing charge cannot pass through the cell membrane on their own and they require transport proteins that are available in the membranes. The transporter proteins in the transmembrane regions are of two types (1) Channel proteins that functions as selective pores that opens and closes in response to an electro-chemical gradient and (2) Carrier proteins that use energy processes for the translocation of a substrate against concentration gradient (Saier, 2000). Human cell membranes contain several kinds of transporters for transporting glucose, amino acids, ions, neuro transmitters etc. Some transporters are also involved in the multi drug resistance. Such multi drug resistance is the particular limitation in the cancer chemo therapy and antibiotic treatments. Hence the drug molecules that inhibit drug efflux transporters are required to use as a complement to the existing therapy in order to overcome multidrug resistance (Dantzig et al., 2003). Usually drugs show their effect on transport system by binding to transporter proteins and either inhibits their activity so as preventing the passage of ions/molecules or by acting as false substrate for the transport process.

So far only a few transporter proteins from some bacterial species have been crystallized to determine their structure. There is only less structural data is available on the membrane transporter proteins that limits the development of new leads that could interfere with the structure and functions of transporter proteins. Such situations demand the molecular modelling studies for the bio computation of alternative methods for the structural studies of membrane transporters (Dahl et al., 2004). However, these methods depend both on the computational studies and experimental interference for the perfect molecular modelling studies.

Membrane Proteins as Potential Drug Targets to Discover New Drugs

The three dimensional information of a membrane structure and its interaction with a specific drug reveals several critical factors about the topology of the complex and their interacting surfaces between the membrane proteins and drug molecules (Klebe, 2006). Such information provides several possible ways for the virtual screening experiments and helps to develop novel potential lead molecules that specifically bind with a defined target membrane protein (Blundell et al., 2002). Several docking softwares can be utilised to obtain the orientation and binding affinities of the drugs against a membrane protein. This information will help a chemist to optimize the potential lead molecule by incorporating several functional groups in different orientations finally getting an improved molecule with high potency and selectivity (Hajduk & Greer, 2007).

Several drugs came into existence into the market in the present days which are developed based on the concept of structure based drug design. Some of them include HIV drugs Agenerase and Viracept developed based on the structure of HIV proteinase (Tomasselli & Heinrikson, 2000; Stoffler et al., 2002), Flu drug Zanamivir developed based on the structure of neuraminidase (Monto et al., 1999) and the inhibitors of angiotensin-converting enzyme (Cushman et al., 1987; Fournie et al., 1996). The structural information obtained from molecular and biophysical studies when combined with bioinformatics and computational approaches followed by the amalgamation of knowledge from synthetic chemistry will paves a pivotal role in the discovery of potential drug molecules which could be effective and safer.

The structural information of the membrane proteins which are involved in the cellular networking is very essential to understand the cellular machineries. The information about the structure of membrane proteins, their regulation and interacting partners forms an important basis to discover new therapeutic targets. Membrane proteins play a crucial role as transporters and also acts as co-players in the cellular metabolic activities. Such membrane transporter proteins can be targeted for the development of novel drugs.

Most of the protein targets where the drugs are supposed to bind are classified into four different types such as (1) Receptors, (2) Enzymes, (3) Ion channels and (4) Transporters. Majority of these targets includes the membrane integral proteins only rather than the cellular components. Such membrane integral proteins are involved in wide variety of metabolic networks and processes and represent themselves as plethora of molecular targets for pharmacological interference. The structural information of such membrane proteins with high resolution in the cell membranes is highly significant to develop novel drugs with high potential and specificity. Such information also provides an insight into the understanding of molecular mechanisms involved in cellular networks and functions.

Therefore the structural biology concepts such as theoretical predictions and three-dimensional structural information are becoming more significant criteria in present days.

Drugs and Membrane Proteins

Several transporter membrane proteins are targeted so far and the drugs were discovered based on their structural information and the most important target among them are the ABC transporters (Leslie et al., 2005). There are justifications why the drugs are specifically to be targeted on membrane proteins and primary reason is that, several diseases are caused by mutations in the membrane proteins such as neurological disorders; schizophrenia and depression are due to mutations in the iron transport channels. Mutations in the ABC transporter protein system cause cystic fibrosis. The next important reason is more than 50% of known drugs binds to the membrane proteins on the surface of the membrane. If better drugs are to be discovered by aiming a disease, actual molecular structure of the drug molecule and its intermolecular interactions are required. Such structural information assists to develop better drug molecules with improved activity. Hence membrane proteins could be the major determinants of pharmacokinetic, safety and efficacy profiles of drugs (Giacomini et al., 2010).

Drugs will have their own specific way in the membrane proteins and according to their solubilities they distribute in the hydrophobic regions of the structure. Not only this, several questions also to be justified for a successful drug discovery such as selection of specific membrane transporter protein that should be clinically important in drug absorption and disposition. This will also clarify what kind of *in vitro* mechanisms are to be chosen. Specific attention has to be paid for the transporters that are confined to the epithelia of the intestine, liver and kidney, and in the endothelium of the blood–brain barrier.

MODELLING OF MEMBRANE PROTEIN STRUCTURES

Molecular modelling is not only a robust method for generating the theoretical models of proteins it is also an integral part of crystallography and Nuclear Magnetic Resonance (NMR) structure determination techniques. Experimental determination of membrane protein structures and their complexes is extremely a difficult task and more over the problem hinders with the size of the membrane protein also. The Membrane proteins represents approximately one third of the total encoded proteins of human and other genomes, but the currently available crystallographic structural information of membrane proteins is very limited (Frishman & Mewes, 1997; Wallin & Von Heijne, 1998). Many important puzzles in biology, chemistry and medicine remain unsolved because of limited understanding of the membrane structures, behaviour and their molecular interactions.

Usually the integral membrane proteins have two types of conformations such as (1) α−helices and (2) β−barrels. Till now it has been observed that the plasma membrane and reticulum membranes of eukaryotes contain α−helices whereas the β−barrel conformations are observed mainly in the proteins of outer membrane of gram-negative bacteria, mitochondria and in chloroplast membranes (Basyn et al., 2003). The α−helical conformations contain usually long hydrophobic helical structures and densely packed together. β−barrel conformations are very huge proteins possessing anti parallel β−sheets those folds into a barrel shape and closed by the first and last β−strands (Oberai et al., 2006; Elofsson & Heijne, 2007). In a protein with known amino acid sequence and unknown structure and function, the long run hydrophobic residues that represent spanning regions of membrane can be recognised easily in the sequence than the less hydrophobic trans-membrane β-strands. *In silico* approaches works best for the prediction and identification of such membrane spanning α–helical bundles and β–sheet barrels. As it is difficult to determine the 3D-structure of an integral membrane protein experimentally, prediction of such secondary conformational elements from the sequence information helps best for the annotation of protein sequences to membrane protein families. Such information when combined with the recognition of structural motifs by bioinformatics approaches provides valuable structural information that will also helps to determine the function (Amico et al., 2006; Granseth et al., 2006). *In silico* methods offers robust methods and algorithms for the accurate construction of protein models.

The molecular structure of a membrane protein contains very essential information such as architecture of active sites on the surface of membrane, possible binding sites for ligands and antigens and evolutionary determinants. Such information from membrane protein structure is utilised for several proteomic approaches and designing of successful experiments. The electrostatic properties and the orientation of the membrane protein structure are essential to understand its interaction with its interacting partners which are all involved in the regulation of metabolic activities. Hence it is very essential to acquire a three dimensional structural information of a membrane protein.

Experimentally the membrane structures are determined and solved by either X-ray crystallography or by NMR method. Computationally, there are two main popular ways to construct the membrane models and even applicable universally for all the proteins. The first one is the homology or comparative modelling and the other is *Ab initio* or *de novo* modelling. Homology modelling is strictly based on the scaffold of the existing structure with homology, nothing but a template and the *de novo* modelling is based only on the sequence information where the propensities of each amino acid is taken into consideration that tends to form a specific kind of conformation (Figure 2). There will be other consideration

Figure 2. Criteria and methods involved in the protein structure determination

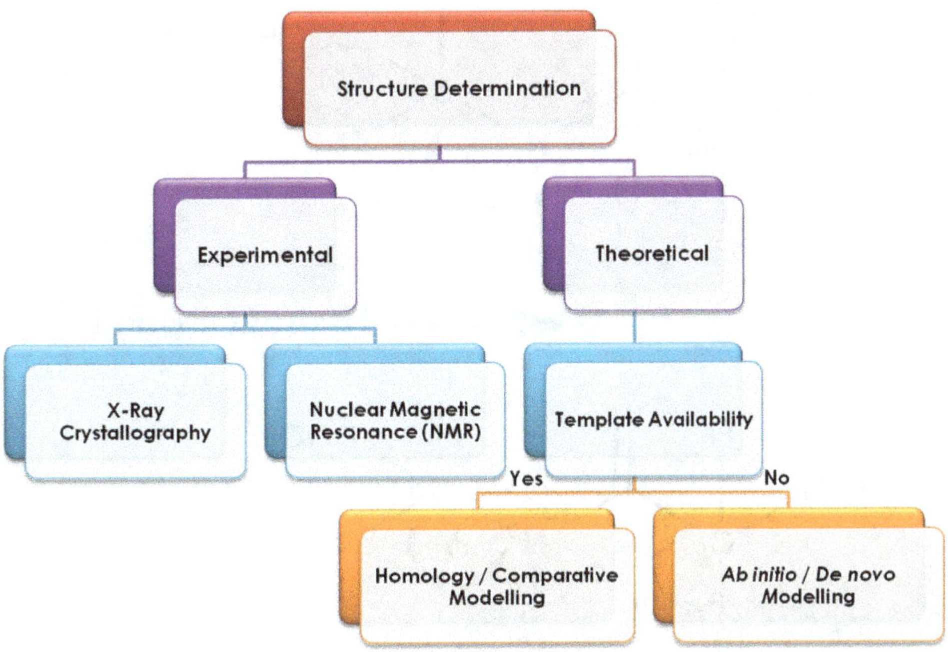

such as prediction of secondary structure and membrane spanning regions before constructing the final three dimensional model.

Irrespective of the model chosen for the prediction of theoretical models for a specific protein, certain preliminary analysis is required to be done to acquire a reliable and accurate model without any obligations. Such criteria are discussed in detailed as follows.

Secondary Structure Prediction

Before proceeding for the construction of a three dimensional conformation of a protein, it is mandatory to determine their secondary conformations. These secondary conformations only further folds into a three dimensional conformations producing a functional protein. Hence the secondary conformational elements such as helices, sheets, turns and coils are the basements to build a perfect conformation of any protein. The multiple conformations of a protein are possible due to the rotation of CΨ and CΦ angles of peptide bond. Each amino acid with its specific R-group has its own role to form a specific kind of secondary conformation. The secondary structure is popularly said to be a local conformation of a polypeptide chain that is strictly formed and stabilised by hydrogen bonding. A hydrogen bond always forms in between the carbonyl oxygen of one peptide bond and amide hydrogen of another peptide bond by means of a hydrogen atom (Figure 3). These are the responsible factors for the stability of any secondary conformation of a protein.

A α-helical conformation forms when n^{th} peptide bond forms H-bond with $n+4^{th}$ peptide bond whereas a β-conformations forms when H-bond forms between a normal of 5–10 sequential amino acids in one

Figure 3. Formation of Hydrogen bond between two peptide bonds

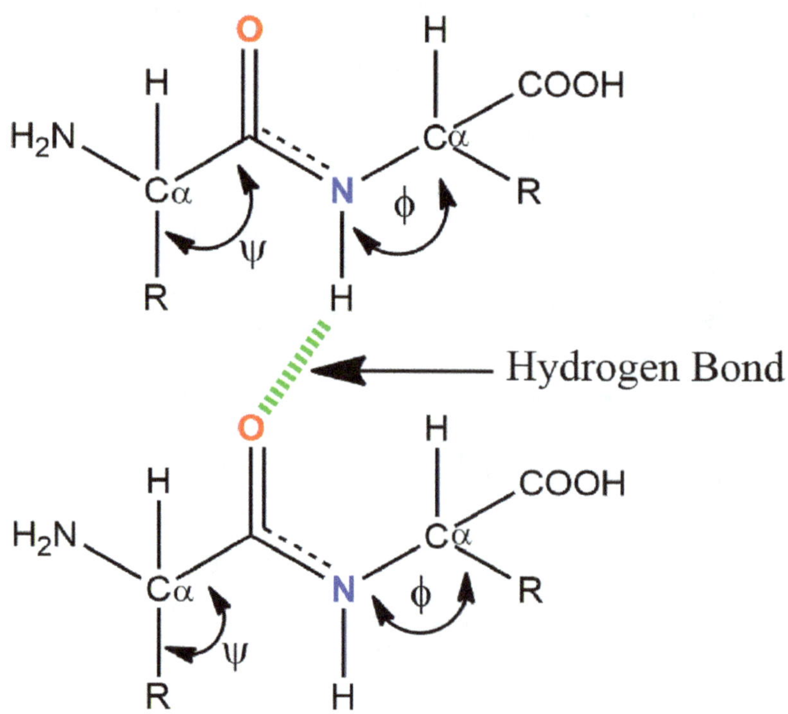

bit of the chain with another 5–10 more remote down the chain. Loops are the locales of a protein chain that are between α helices and β sheets, of different lengths and three-dimensional arrangements, and on the surface of the structure. Hairpin loops that cause a complete turn in the polypeptide chain joining two anti parallel β strands may be as short as two amino acids long. A locality of auxiliary structure that is not a α-helix, a β-sheet, or a conspicuous turn is ordinarily eluded as a coil. All these types of conformational elements are taken into consideration while prediction of a full pledge secondary conformation from a given amino acid sequence. This will paves a reliable way for the next step to predict accurate three dimensional conformation of a protein.

Several secondary structure prediction methods were introduced in the early 1960 itself (Guzzo, 1935; Kotelchuck & Scheraga, 1969; Lewis et al., 1970; Prothero, 1966; Schiffer & Edmundson, 1967) relying on the recognition of likely α-helices and were constructed principally in light of helix-coil conversion models. More precised expectations that included β-sheets were presented in the 1970s and depended on measurable evaluations in view of likelihood parameters got from known comprehended structures. The transformative conservation of secondary structures can be abused numerous homologous sequences in a diverse sequence alignments, by computing the net optional structure inclination of an adjusted segment of amino acids. Working together with bigger databases of known protein structures and cutting edge machine learning routines, for example, neural networks and support vector machines, these strategies can accomplish up to 80% accuracy rate (Dor et al., 2006).

Algorithms in Secondary Structure Prediction

Chou-Fasman Method

The Chou–Fasman system is an experimental method for the prediction of secondary conformations in proteins, initially grew in the 1970s by Peter Y. Chou and Gerald D. Fasm (Chou & Fasman, 1974, 1978). The technique is in light of examinations of the relative frequencies of every amino acid in alpha helices, beta sheets and turns taking into account known crystal structures of proteins. From these frequencies an arrangement of likelihood parameters were inferred for the presence of every amino acid in every secondary structure, and these parameters are utilized to anticipate the likelihood that a given succession of amino acids would shape a helix, a beta strand, or a turn in a protein. The parameters discovered some solid propensities among individual amino acids to incline toward a specific kind of structural conformation over others.

GOR Method

The GOR method (Garnier-Osguthorpe-Robson) is an information theory-based method developed in the late 1970s (Garnier et al., 1996). This system is in light of likelihood parameters got from exact investigations of known protein tertiary structures from X-ray crystallography. Nonetheless, not like Chou-Fasman, the GOR technique considers not just the propensities of individual amino acids to frame specific secondary structures, additionally the contingent likelihood of the amino acids to shape a specific secondary structure given that its quick neighbours have effectively framed that structure. This technique dissects successions to anticipate alpha helix, beta sheet, turn, or arbitrary coils at every position in view of 17-amino-acid sequence windows. The unique explanation of this method is, it includes four scoring matrices with a size of 17×20, where the columns correspond to the log-odds score that reflects the likelihood of judging a certain amino acid at each position in the 17-residue sequence. The four matrices imitate the probabilities of the vital, ninth amino acid tends to be in a helical, sheet, turn, or coil conformation. This method was considered as finest requiring at least four neighbouring amino acids to attain alpha helices to organize the region as α-helical, and at least two adjoining residues for a β-sheet.

Machine Learning

Such methods popularly include Artificial Neural Networks (ANNs) and Support Vector Machines (SVMs). ANNs are a group of statistical learning models enlivened by biological neural networks and are utilized to estimate functions that can rely upon large quantity of inputs and are usually anonymous. They use the training set formed from experimental structures and investigate the similar sequence motifs connected with certain arrangements of secondary elements. SVMs have been demonstrated especially helpful for predicting the areas of turns, which are hard to identify using statistical methods. All machine learning strategies endeavour to predict the more fine-grained nearby properties of proteins, for example, backbone dihedral angles in unassigned areas (Kuang et al., 2004; Zimmermann et al., 2006).

Servers for Secondary Structure Prediction

A wide variety of servers are available for the prediction of secondary structures from a given protein sequence among which some are listed as in Table 1.

Transmembrane Prediction

The most important task in the membrane protein modelling is the identification of transmembrane regions. Several factors are to be considered for predicting membrane spanning regions. The free energies are summed over residues of 10-15 and the regions with high hydrophobicity are considered to be in the membrane. This idea works well with helical structures as all the residues present in the helix will have same kind of hydropathy scales. But this type of prediction is unable to work out to predict the membrane spans beta conformations. Utilising the evolutionary information from multiple sequence alignments and using machine learning approaches helps to predict transmembrane regions accurately. Several tools and servers exist based on these approaches that can predict the perfect topology of the membrane protein. The perfect orientation of membrane protein i.e. external regions, spanning regions and inside regions will be predicted based on the propensities, artificial neural networks (ANN), hidden markov models (HMM) and support vector machines (SVM).

Some of the online tools available for the prediction of transmembrane regions are given in Table 2.

Table 1. Secondary structure prediction tools

Tool	Web Link
APSSP	http://imtech.res.in/raghava/apssp/_
CFSSP	http://cho-fas.sourceforge.net/_
GOR	https://npsa-prabi.ibcp.fr/cgi-bin/npsa_automat.pl?page=npsa_gor4.html_
Jpred	http://www.compbio.dundee.ac.uk/jpred/_
PROF	http://www.aber.ac.uk/~phiwww/prof/_
SOPMA	https://npsa-prabi.ibcp.fr/cgi-bin/npsa_automat.pl?page=npsa_sopma.html_

Table 2. Transmembrane prediction tools

Tool	Web Link
TMpred	http://www.ch.embnet.org/software/TMPRED_form.html_
HMMTOP	http://www.enzim.hu/hmmtop/_
Phobius	http://phobius.sbc.su.se/_
TMHMM	http://www.cbs.dtu.dk/services/TMHMM-2.0/_
DAS	http://www.sbc.su.se/~miklos/DAS/_
TOPCONS	http://topcons.cbr.su.se/_

Once the secondary structure and transmembrane spanning regions are predicted from a given sequence, the next step is to move towards the prediction of tertiary structure. It can be done in two methods i.e. homology modelling and de novo modelling.

Homology Modelling

Homology modelling otherwise called as comparative modelling is a most widely used popular method to construct the protein models based on the template similarity. This method is very important in case of membrane proteins as the experimental determination of their structures is prohibitively time-consuming and tough task (Wieman et al., 2004). Homology modelling found very important for membrane proteins as many of them such as ion channels are the primary drug targets for wide variety of diseases.

Unfortunately there are very less reports on the homology modelling of the membrane proteins which is due to several factors such as lacking of proper template, unambiguous prediction methods, poor resolutions, high rate of errors in model validation etc. Recently, there is an extensive increase in the number of structures in the PDB database that helps to analyse the hydrophilic surface areas and hydrophobic interiors based on the structural information but rather than the sequence information alone. Although the number of crystal structures is being increased in the PDB, membrane proteins are found to have poor resolution where 10% of available membrane structures have a resolution ranging from 3-4 Å. This situation made the homology modelling as an exceptionally significant task to construct their 3D models and also for the refinement of existing crystal structures with poor resolution (Sanchez et al., 2000). There are several steps for the prediction of membrane protein structures which are explained as follows.

Template Selection and Alignment

A template is the mandatory element to construct a homology model of a protein. The template should have high similarity to the target so that high quality model will be produced. Sequence coverage is also an important point under consideration during template selection. A template assuring a minimum of 98% coverage is responsible to produce the overall reliable quality of the model. A template with a similarity of 70% produces a model with an RMSD of 1-2 Å where as the low similarities of templates ranging around 25% will produce a model with an RMSD of 3-4 Å.

The quality of sequence alignment between the target and template sequence will decide the quality of the final model where the quality of the alignment is improved by BLOSUM and PAM matrices. But for the membrane protein alignment there is specific matrix for the improvements of alignments such as PHAT (Ng et al., 2000) that utilises the membrane protein specific substitutions and PRALINETM (Pirovano et al., 2008) is an example that utilises such matrix. JSUBST (Shi et al., 2001) is another membrane specific substitution matrix which is amalgamated with MP-T (Hill & Deane, 2013) tool that is a sequence-structure alignment tool. This tool also considers other environment factors such as secondary structure, membrane spanning region and solvent accessibility regions such as intra cellular and extra cellular regions.

In many instances membrane proteins does not have any suitable templates for the entire sequence and in such cases discontinuous templates may be available. Discontinuous templates have similarity with the target protein at multiple locations rather being at a confined location and such templates may be from different sources. So such cases demand the use of multiple templates to construct the final model. Such methods are called as threading methods that utilises the alignment between target and multiple templates. Further sequence-structure alignment followed by scoring functions which are based on the factors such as secondary structure, residue pairing and mutatability produces a quality model.

Model Construction

Once the improved alignment has been generated, it can be used for the construction of homology model of the target using any homology modelling tools. Several tools are available for the homology modelling and some of them are given as follows.

- **RosettaMembrane:** This is the most reliable and oldest tool available for the membrane protein homology modelling. This tool is also able to model the open and closed conformations of the ion channels (Yarov et al., 2005).
- **MODELLER:** This is the most popular tool in the present days both in the modelling and experimental areas. This tool constructs the models by satisfying the spatial constraints derived from sequence alignment (Eswar et al., 2006).
- **MEDELLER:** This is a coordinate generation tool especially used for the membrane proteins. This tool works in conjunction with other tools like MEMOIR, HHPred and SWISS Model.

The overall outline of the homology modelling is represented in Figure 4.

Figure 4. Outline of homology modelling of membrane proteins

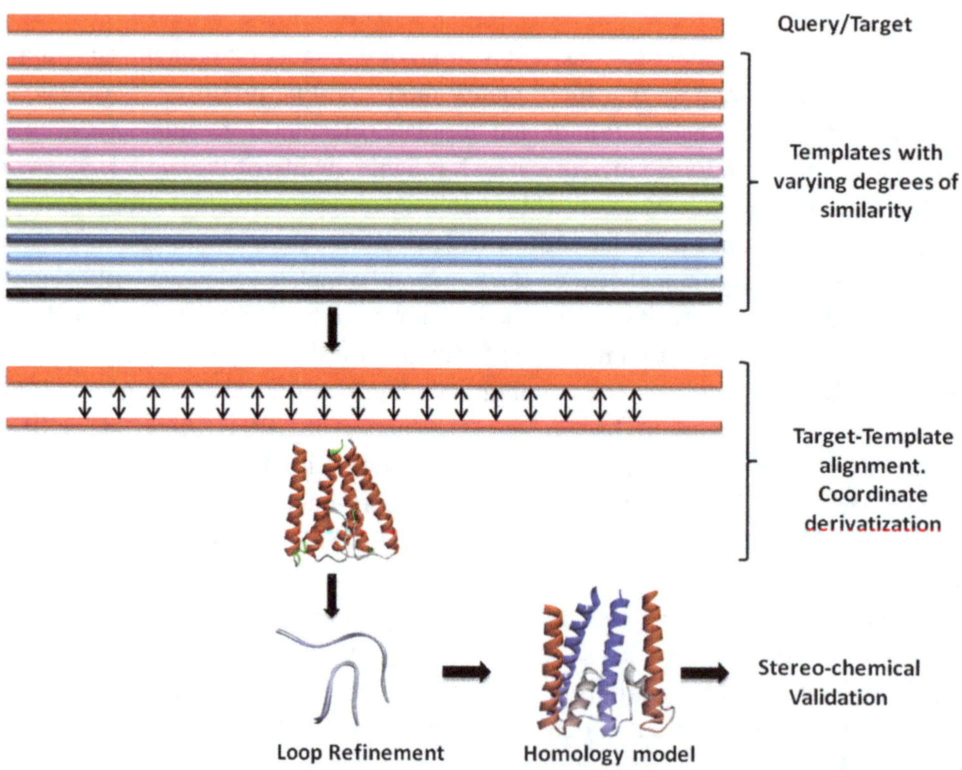

De Novo Modelling

De novo modelling is an alternate to homology method where it is preferred when no suitable template is available for the construction of reliable model. De novo modelling works with the propensity of each amino acid in a protein sequence and build a specific kind of secondary structural conformation and produce the model. De novo methods able to predict the tertiary structure of a protein based on the sequence principles which are basis for protein folding energetics and statistical tendencies of conformational characteristics that experimental structures possess and never rely on template information. Amino acid sequence is the best and sole starting material for the de novo predictions.

De novo prediction algorithms usually predicts the structure initially by producing the multiple conformations of the protein and chose the best amongst based on the thermodynamic stability and energy levels. Most of the de novo prediction servers considers the following factors, (1) Accurate energy function that filters thermodynamically stable conformation and resembles to native structure (2) A potential search method for the quick identification of low energy conformation (3) Ability to select most near native conformation among multiple conformations.

The outline of de novo modelling method is represented in Figure 5.

De novo prediction uses the strategies such as,

- **Ab Initio:** Initially the secondary structural propensities such as helices, sheets, coils and turns are elucidated from the primary sequence based on the physico chemical properties and neural net algorithms. Based on this information the tertiary folding will be determined.

Figure 5. Outline of de novo modelling

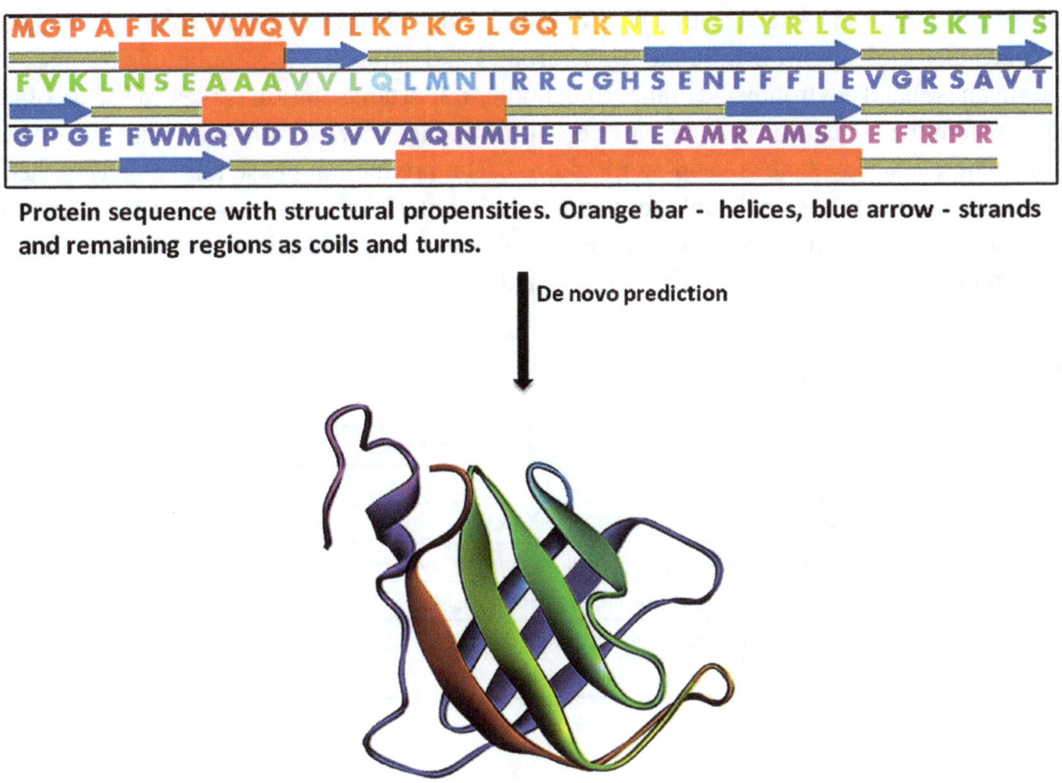

- **Fold Prediction:** This method also initially predicts the secondary structure information and compare with the existing protein fold libraries such as CATH and SCOP databases.
- **Threading:** The fold recognition process is progressed to threading where the interaction of residue pairs is used based on the empirical energy functions and placed onto a putative backbone of the protein utilising the gaps and gap penalties.

The main disadvantage of de novo prediction methods is they require huge amount of efforts and computational time for the successful prediction of near native conformation of the protein. Few examples of de novo modelling servers are given in Table 3.

MOLECULAR DYNAMICS OF MEMBRANE PROTEIN MODELS

The majority and comprehensive information about the theoretical models of ion channels well provides by Molecular Dynamics (MD) simulations. The present available MD methodologies are capable to generate the trajectories that are very close to the realistic atomic models of typical biological membrane channel systems. MD simulations can be used to discover detailed atomic information about the molecular mechanisms behind the function of a specific membrane protein such as ion transport, selective entry and voltage gating of channel proteins. Although there is a vast improvement in the MD simulations of proteins of different sizes for the past two decades, still it remains as a challenge with the membrane protein simulations. The heavy size of the membrane proteins limiting the success rates of the simulations and several approaches are being put forward to solve the critical hindrance arising with membrane proteins. One such a solutions is limiting the size of protein to the short possible fragments and proceed for simulations. The complex membrane structure may be broken down into smaller fragments of specific importance such as function, domain, and structural elements so on. Such significant fragments are then added with membrane layers around and subjected to simulations to observe their behaviour in an isolated environment (Figure 6). Another important solution is the time scale of production phase, where the simulations can be carried out for a longer period rather than the traditional time boundaries where a steady state is keep on monitored. The production phase may be terminated when a steady state is attained. It is applicable for the simulation of all proteins irrespective of their size, but the limitation could be the time factor needed for the computation. This situation will also demand the high end computing systems for the successful simulations.

Table 3. De novo modelling servers

Server	Web Link
Robetta	http://robetta.bakerlab.org/_
CABS	http://biocomp.chem.uw.edu.pl/CABSfold/index.php_
Phyre2	http://www.sbg.bio.ic.ac.uk/phyre2/html/page.cgi?id=index_
RaptorX	http://raptorx.uchicago.edu/StructurePrediction/predict/_
Pep-Fold	http://bioserv.rpbs.univ-paris-diderot.fr/services/PEP-FOLD/_
I-TASSER	http://zhanglab.ccmb.med.umich.edu/I-TASSER/_

Figure 6. Alignment and orientation of a membrane protein into the lipid bilayer

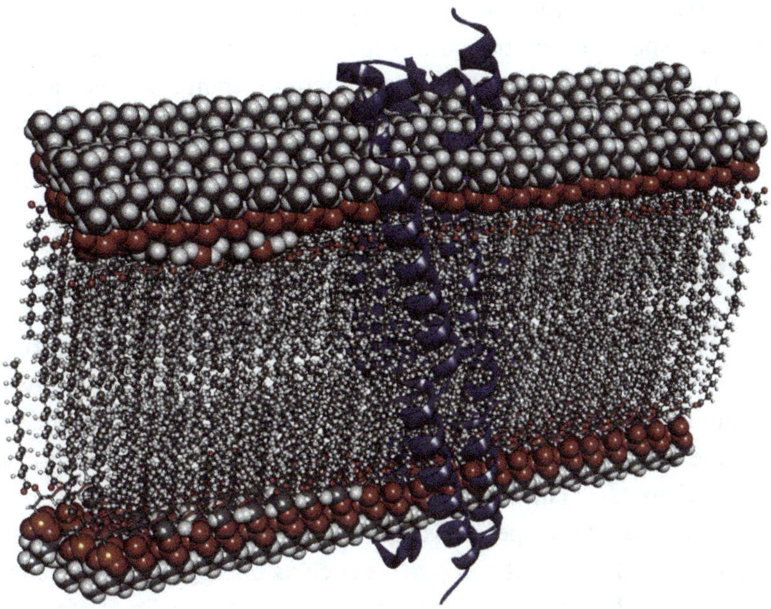

Force Fields and Softwares

The folding of membrane proteins at the atomic levels in explicit membrane proteins proves challenges and basic problems with the specific force fields and simulation algorithms (Nymeyer et al., 2005; Ulmschneider & Ulmschneider, 2009). Employing new lipid force fields by the concern of the detailed atomic partitioning and folding of individual transmembrane regions into explicit lipid regions helps to achieve successful unbiased long time MD runs (Ulmschneider et al., 2011). Coarse-grained MARTINI force field is the most acceptable method developed specifically to model explicit lipid bilayers (Marrink et al., 2004). This force filed is used extensively to study the membrane bilayer parameters such as raft formation, membrane joining and membrane interactions. This force filed is more popular to study the large scale MD systems especially drug delivery strategies through lipid bilayers. To study the molecular mechanism of membrane folding, partitioning and it's inter molecular interactions requires detailed representation of hydrogen bonds at atomic level. This force filed will uncover the mechanisms behind ion transport, specificity, selectivity and gating properties of membranes. Several most common force fields are also available such as AMBER (Wang et al., 2004), OPLS (Jorgensen & Tirado, 1988) and CHARMM (Karplus, 1983) for the past few years. Highly accurate lipid parameters are added for these force fields in the recent years that made them to utilise for the simulation of membrane proteins with a variety of atomic models. But all these force fields are supplemented with MARTINI force field that has coarse-grained representations of proteins and lipids. Finally all the kinds of force fields are available with several kinds of software packages. The best example of such package is GROMACS (Van et al., 2005), fastest and most popularly used package. Other best examples count with NAMD (Phillips et al., 2005) and Desmond (Shaw, 2005) which are also fastest but not used as popularly as GROMACS.

Several methods are in progress for the acceleration of MD simulations at faster rates in addition to the general approaches discussed above. One such popular method is replica exchange sampling or parallel tempering that creates indistinguishable replicas of an MD system which are simulated in parallel manner with an exponential ladder of temperature. The replicas are swapped at regular intervals using Monte Carlo approach and speed up the transitions towards near native state. Another related method to accelerate the MD process is using the elevated temperature ranges. This method is preferred during the unbiased partitioning of peptides into membranes. But the experimental confirmation of high thermo stability of the peptide is required. Equilibrium MD method is best suitable to accelerate the MD simulations of low energy conformations.

Conformational Analysis

Although the structure of a membrane protein is determined, it is not sufficient to predict its function. A single protein structure represents only a discrete conformational state and it never represent itself as a sole maker of its function, interaction with other molecules and its regulation by other elements in the bilayer. To understand the function of a membrane protein, different conformational states are needed which is hard to capture (Figure 7). The structure of a particular conformation can be characterized by altering the variable interactions and a specific snap shot of interest could be trapped. The function of membrane protein can be modulated by its properties such as energy levels and charge distribution. In addition the spring constant of the lipid bilayer also induce the changes in conformation and properties of a membrane protein. These specific considerations help to trace the required conformation of the protein that can be used for the interpretation of the function.

These computational studies of membrane protein modelling even though a challenging effort, a prospective and massive progress has been made in the field of computational modelling. So far the major efforts were put on the homology models whereas de novo modelling with a special concern with membrane proteins has yet to be progressed to a maximum extent. As of now we are able to model the membrane proteins in a coarse grained process further research is likely to focus on predicting the function of a membrane protein from multiple functional states of the protein. In addition the structures are needed to represent at high resolution both for protein and lipid bilayers.

Figure 7. Membrane proteins with multiple conformations represented in multiple colours

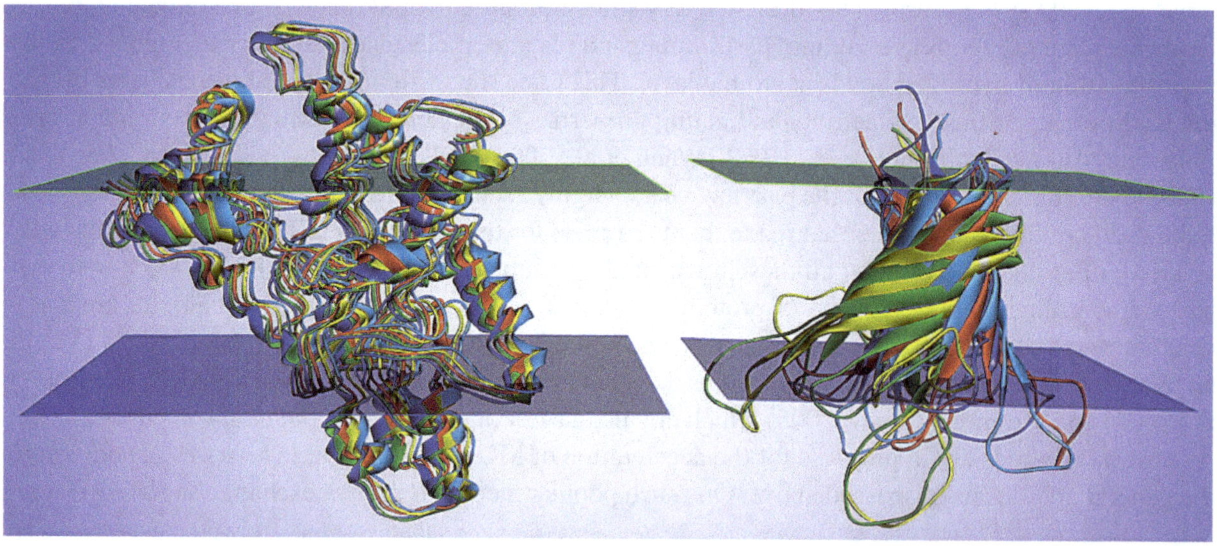

PREDICTION OF BINDING SITES

The next challenging task with the membrane proteins is accurate prediction of binding sites that offers sites of interactions and molecular passages. Prediction and blocking of specific binding sites makes them efficient targets for several drugs especially the membrane proteins of potential pathogenic organisms. The ideal membrane protein models with explicit water molecules both inside and outside and with added phospholipid bilayer are the only perfect starting materials to find out such potential sites (Bordner, 2009). Like cytosolic proteins, it is not an easy task to predict the binding sites based on the cavities information obtained from the three dimensional models. Membrane proteins have a peculiar compartmentalized composition and orientation. The initial challenge is associated in solving the different regions of the membrane proteins, where such region/segment of the membrane protein has its own function. Then those regions with functional significance can be used to find out the binding sites. The actual challenges are concurrent with the accurate prediction of transmembrane, extra and intra cellular regions. Extra cellular regions are associated with protein-protein interactions especially involved in signal transduction and also offer specific sites as pores where the ions/molecule come and sits then enters the pore channels. Membrane regions are to be monitored to find out such pore regions where the channels are formed especially by beta barrel structures. Further the intracellular regions are to be concentrated where the pores opened into the cytosolic regions and also offers sites for the interactions with intracellular components that may be either proteins or small molecules. All these criteria are to be considered for the perfect prediction of binding sites (Figure 8).

Membrane proteins forms complexes with other molecules to perform their functions where the structural information about such interactions is helpful to generate experimentally acceptable hypotheses which also further helps for structure based drug designing methods. The computational methods can address to predict the appropriate residues on the surface of membrane proteins which are involved in the interactions. Such predicted residues can also be used as constraints for the molecular docking

Figure 8. Predicted binding sites in a membrane protein highlighted as surface clouds

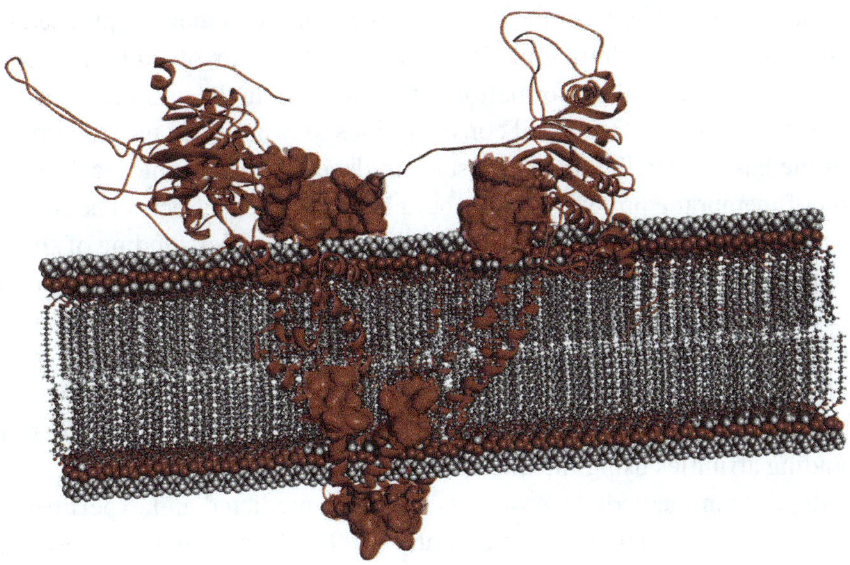

processes such as protein-ligand or protein-proteins interactions. Presently available computational tools are able to predict the binding sites on the membrane proteins based on two approaches; (1) sequence based, and (2) structure based.

Sequence based approaches are applied only when no structural information is available and even no template structure is available. Structure based methods provide additional information such as solvent accessibility and residue proximity in the three dimensional orientation of the molecules. Structure based methods relies on the scoring functions, ANNs and SVMs that were trained on various properties for the efficient prediction of binding sites. But the problem lies with the membrane proteins especially with the intra membrane regions. The above said methods are well works with non-membranous proteins as they are trained on more experimental structures available. This situation now can demand the separate predictors that are trained on the specific class of membrane proteins. Random forest prediction methods came into existence that made for individual surface residues which then are averaged above a local surface region and a final prediction set is prepared.

MOLECULAR DOCKING OF MEMBRANE PROTEINS

The final great challenge is, finding the perfect intermolecular interactions with protein partners and small molecules. If the above steps are followed stringently, then this task will be facile and the result will be veracious (Deepshikha, 2002). There will be bit variations in the parameters to be chosen while setting a docking process for membrane protein. First, the site of interaction, nothing but the binding site and its location whether it is present in water or lipid environment. The interaction will be completely different as the hydrophillicity and hydrophobicity greatly alter the interaction mechanisms. Some sites will be spanning both membrane-intracellular and membrane-extracellular regions. Such situations are crucial to specify what kind of environment or force-fields. Membrane protein docking is more trivial as the length of transmembrane regions is not conserved and structural elements occasionally contain bulges and kinks that will challenge the perfect molecular interactions. Another challenge exists with numerous and diverse conformations with near similar energies and RMSD values (Nygaard et al., 2013). Choosing a correct one from thousands of likely conformations is of greatest challenge to proceed for docking. The next provoking challenge exists with multiple loops which are also involved in the generation of kinks in the structure and disturb the local conformations and interrupting the ensemble stabilization (Klco et al., 2005). These investigations and technical considerations may probably provide immense opportunities to understand the basic signalling that are useful to develop pharmacological tools and drugs that modulate function of membrane proteins which could be therapeutic effects. The molecular models of docking complexes provide vast opportunities for the fundamental understanding of signalling processes and help to develop pharmacological agents and drugs that modulate function of membrane proteins.

The docking methods such as DOCK, LibDock, CDOCKER and GOLD with their different scoring functions are used to predict small molecule interactions in the binding sites of membrane proteins both in intra and extra cellular regions. The binding affinity estimations relies on the scoring functions whereas the binding orientations on the docking methods. Regression-based scoring functions are used to estimate the binding affinities using the interaction terms derived from experimental complex structural parameters. Regression methods give weights by fitting predicted and experimentally determined affinities to the training set complexes (Bohm & Stahl, 1999). The size and composition of the training set also a major concern to derive the weights. But the drawback is any new molecular scaffolds will be ignored that were absent in the training set complexes (Tame, 1999). Principle based approaches also calculate binding free energies based on the individual contribution of intermolecular interactions between the protein and ligands that are derived from physico-chemical theory and are not based on experimental data. The implicit solvent methods like Possion-Boltzmann and surface generalized born methods are used to calculate the salvation energies (Shoichet et al., 1999).

The flexibility of the integral membrane regions of the protein and their inter molecular interfaces influences the binding affinities and specificities of the ligands and their recognitions. The stabilities and flexibilities of the membrane structures during the molecular recognitions interfere with the binding energies and contact surface areas that could be linked with several functions such as specificity and permissiveness in recognition. All such points complicate the correct orientations of the protein-ligand complexes especially in the instance where the flexible binding sites in the membrane are unknown. Hierarchical approaches solve such problems to an extent by the consideration of the flexibilities of both the ligand and membrane proteins. Such approaches include multistage docking methods and hierarchical energy functions.

CONCLUSION

Membrane proteins have major problems especially construction of theoretical models, binding site prediction and molecular docking that are complicated by the flexible moments and multiple conformations of the huge membrane structures. Application of computational methods to solve the membrane protein structures and their conformational analyses were discussed in detail in a technical manner so as to give a clear glance to the readers. The methods were explained to construct the complete structure of huge membrane protein models in fragment based methods, combining and analysing them and choosing the possible reliable model. A range of molecular dynamics simulations were explained specific to membrane proteins to study the relevant moments and multiple conformations in the membrane regions. Further coarse grained methods are discussed to model, simulate, analyze and to predict probable binding sites. Such a specific analysis helps to predict the function of several membrane receptors in a less time.

REFERENCES

Amico, M., Finelli, M., Rossi, I., Zauli, A., Elofsson, A., Viklund, H., & Casadio, R. et al. (2006). PONGO: A web server for multiple predictions of all-alpha transmembrane proteins. *Nucleic Acids Research*, *34*(suppl 2), W169–W172. doi:10.1093/nar/gkl208 PMID:16844984

Arinaminpathy, Y., Khurana, E., Engelman, D. M., & Gerstein, M. B. (2009). Computational analysis of membrane proteins: The largest class of drug targets. *Drug Discovery Today*, *14*(23), 1130–1135. doi:10.1016/j.drudis.2009.08.006 PMID:19733256

Ash, W. L., Zlomislic, M. R., Oloo, E. O., & Tieleman, D. P. (2004). Computer simulations of membrane proteins. Biochimica et Biophysica Acta (BBA)-. *Biomembranes*, *1666*(1), 158–189. doi:10.1016/j.bbamem.2004.04.012

Basyn, F., Spies, B., Bouffioux, O., Thomas, A., & Brasseur, R. (2003). Insertion of X-ray structures of proteins in membranes. *Journal of Molecular Graphics & Modelling*, *22*(1), 11–21. doi:10.1016/S1093-3263(03)00122-0 PMID:12798387

Blundell, T. L., Jhoti, H., & Abell, C. (2002). High-throughput crystallography for lead discovery in drug design. *Nature Reviews. Drug Discovery*, *1*(1), 45–54. doi:10.1038/nrd706 PMID:12119609

Bohm, H. J., & Stahl, M. (1999). Part 1--Docking and Scoring: Methods Development-rapid empirical scoring functions in virtual screening applications. *Medicinal Chemistry Research*, *9*(7-8), 445–462.

Bordner, A. J. (2009). Predicting protein-protein binding sites in membrane proteins. *BMC Bioinformatics*, *10*(1), 312. doi:10.1186/1471-2105-10-312 PMID:19778442

Chang, C., Ray, A., & Swaan, P. (2005). In silico strategies for modeling membrane transporter function. *Drug Discovery Today*, *10*(9), 663–671. doi:10.1016/S1359-6446(05)03429-X PMID:15894231

Chou, P. Y., & Fasman, G. D. (1974). Prediction of protein conformation. *Biochemistry*, *13*(2), 222–245. doi:10.1021/bi00699a002 PMID:4358940

Chou, P. Y., & Fasman, G. D. (1978). Empirical predictions of protein conformation. *Annual Review of Biochemistry*, *47*(1), 251–276. doi:10.1146/annurev.bi.47.070178.001343 PMID:354496

Cooper, G. M. (2000). *The Cell: A Molecular Approach*. Sunderland: Sinauer Associates.

Cushman, D. W., Ondetti, M. A., Gordon, E. M., Natarajan, S., Karanewsky, D. S., Krapcho, J., & Petrillo, E. W. Jr. (1987). Rational design and biochemical utility of specific inhibitors of angiotensin-converting enzyme. *Journal of Cardiovascular Pharmacology*, *10*, 17–30. doi:10.1097/00005344-198706107-00004 PMID:2485059

Dahl, S. G., Sylte, I., & Ravna, A. W. (2004). Structures and models of transporter proteins. *The Journal of Pharmacology and Experimental Therapeutics*, *309*(3), 853–860. doi:10.1124/jpet.103.059972 PMID:14988415

Dantzig, A. H., de Alwis, D. P., & Burgess, M. (2003). Considerations in the design and development of transport inhibitors as adjuncts to drug therapy. *Advanced Drug Delivery Reviews*, *55*(1), 133–150. doi:10.1016/S0169-409X(02)00175-8 PMID:12535578

Deepshikha, D. (2002). *Protein-Ligand Interactions: Docking, Design and Protein Conformational Change*. (Thesis). California Institute of Technology, Pasadena, CA.

Dor, O., & Zhou, Y. (2007). Achieving 80% ten-fold cross-validated accuracy for secondary structure prediction by large-scale training. Proteins. *Structure, Function, and Bioinformatics*, *66*(4), 838–845. doi:10.1002/prot.21298 PMID:17177203

Elofsson, A., & Heijne, G. V. (2007). Membrane protein structure: Prediction versus reality. *Annual Review of Biochemistry*, *76*(1), 125–140. doi:10.1146/annurev.biochem.76.052705.163539 PMID:17579561

Eswar, N., Webb, B., Marti-Renom, M. A., Madhusudhan, M. S., Eramian, D., Shen, M. Y., & Sali, A. et al. (2006). Comparative protein structure modeling using Modeller. *Current Protocols in Bioinformatics*, 5–6. PMID:18428767

Fournie-Zaluski, M. C., Coric, P., Thery, V., Gonzalez, W., Meudal, H., Turcaud, S., & Roques, B. P. et al. (1996). Design of orally active dual inhibitors of neutral endopeptidase and angiotensin-converting enzyme with long duration of action. *Journal of Medicinal Chemistry*, *39*(13), 2594–2608. doi:10.1021/jm950783c PMID:8691458

Frishman, D., & Mewes, H. W. (1997). Protein structural classes in five complete genomes. *Nature Structural Biology*, *4*(8), 626–628. doi:10.1038/nsb0897-626 PMID:9253410

Garnier, J., Gibrat, J., & Robson, B. (1996). GOR method for predicting protein secondary structure from amino acid sequence. *Methods in Enzymology Computer Methods for Macromolecular Sequence Analysis*, 540-553.

Giacomini, K. M., Huang, S. M., Tweedie, D. J., Benet, L. Z., Brouwer, K. L., Chu, X., & Zhang, L. et al. (2010). Membrane transporters in drug development. *Nature Reviews. Drug Discovery*, *9*(3), 215–236. doi:10.1038/nrd3028 PMID:20190787

Granseth, E., Viklund, H., & Elofsson, A. (2006). ZPRED: Predicting the distance to the membrane center for residues in α-helical membrane proteins. *Bioinformatics (Oxford, England)*, *22*(14), e191–e196. doi:10.1093/bioinformatics/btl206 PMID:16873471

Guzzo, A. V. (1965). The influence of amino acid sequence on protein structure. *Biophysical Journal*, *5*(6), 809–822. doi:10.1016/S0006-3495(65)86753-4 PMID:5884309

Hajduk, P. J., & Greer, J. (2007). A decade of fragment-based drug design: Strategic advances and lessons learned. *Nature Reviews. Drug Discovery*, *6*(3), 211–219. doi:10.1038/nrd2220 PMID:17290284

Hill, J. R., & Deane, C. M. (2013). MP-T: Improving membrane protein alignment for structure prediction. *Bioinformatics (Oxford, England)*, *29*(1), 54–61. doi:10.1093/bioinformatics/bts640 PMID:23110969

Jorgensen, W. L., & Tirado-Rives, J. (1988). The OPLS [optimized potentials for liquid simulations] potential functions for proteins, energy minimizations for crystals of cyclic peptides and crambin. *Journal of the American Chemical Society*, *110*(6), 1657–1666. doi:10.1021/ja00214a001

Karplus, M. (1983). CHARMM: A program for macromolecular energy, minimization, and dynamics calculations. *Journal of Computational Chemistry*, *4*, 187217.

Klco, J. M., Wiegand, C. B., Narzinski, K., & Baranski, T. J. (2005). Essential role for the second extracellular loop in C5a receptor activation. *Nature Structural & Molecular Biology*, *12*(4), 320–326. doi:10.1038/nsmb913 PMID:15768031

Klebe, G. (2006). Virtual ligand screening: Strategies, perspectives and limitations. *Drug Discovery Today*, *11*(13), 580–594. doi:10.1016/j.drudis.2006.05.012 PMID:16793526

Kuang, R., Leslie, C. S., & Yang, A. S. (2004). Protein backbone angle prediction with machine learning approaches. *Bioinformatics (Oxford, England)*, *20*(10), 1612–1621. doi:10.1093/bioinformatics/bth136 PMID:14988121

Lacapere, J. J., Pebay-Peyroula, E., Neumann, J. M., & Etchebest, C. (2007). Determining membrane protein structures: Still a challenge! *Trends in Biochemical Sciences*, *32*(6), 259–270. doi:10.1016/j.tibs.2007.04.001 PMID:17481903

Leslie, E. M., Deeley, R. G., & Cole, S. P. (2005). Multidrug resistance proteins: Role of P-glycoprotein, MRP1, MRP2, and BCRP (ABCG2) in tissue defense. *Toxicology and Applied Pharmacology*, *204*(3), 216–237. doi:10.1016/j.taap.2004.10.012 PMID:15845415

Lewis, P. N., Kotelchuck, D., & Scheraga, H. A. (1970). Helix probability profiles of denatured proteins and their correlation with native structures. *Proceedings of the National Academy of Sciences of the United States of America*, *65*(4), 810–815. doi:10.1073/pnas.65.4.810 PMID:5266152

Lindahl, E., & Sansom, M. S. (2008). Membrane proteins: Molecular dynamics simulations. *Current Opinion in Structural Biology*, *18*(4), 425–431. doi:10.1016/j.sbi.2008.02.003 PMID:18406600

Marrink, S. J., De Vries, A. H., & Mark, A. E. (2004). Coarse grained model for semiquantitative lipid simulations. *The Journal of Physical Chemistry B*, *108*(2), 750–760. doi:10.1021/jp036508g

Monto, A. S., Fleming, D. M., Henry, D., De Groot, R., Makela, M., Klein, T., & Man, C. Y. et al. (1999). Efficacy and safety of the neuraminidase inhibitor zanamivir in the treatment of influenza A and B virus infections. *The Journal of Infectious Diseases*, *180*(2), 254–261. doi:10.1086/314904 PMID:10395837

Ng, P. C., Henikoff, J. G., & Henikoff, S. (2000). PHAT: A transmembrane-specific substitution matrix. *Bioinformatics (Oxford, England)*, *16*(9), 760–766. doi:10.1093/bioinformatics/16.9.760 PMID:11108698

Nygaard, R., Zou, Y., Dror, R. O., Mildorf, T. J., Arlow, D. H., Manglik, A., & Kobilka, B. K. et al. (2013). The dynamic process of β 2-adrenergic receptor activation. *Cell*, *152*(3), 532–542. doi:10.1016/j.cell.2013.01.008 PMID:23374348

Nymeyer, H., Woolf, T. B., & Garcia, A. E. (2005). Folding is not required for bilayer insertion: Replica exchange simulations of an α-helical peptide with an explicit lipid bilayer. Proteins. *Structure, Function, and Bioinformatics*, *59*(4), 783–790. doi:10.1002/prot.20460

Oberai, A., Ihm, Y., Kim, S., & Bowie, J. U. (2006). A limited universe of membrane protein families and folds. *Protein Science*, *15*(7), 1723–1734. doi:10.1110/ps.062109706 PMID:16815920

Phillips, J. C., Braun, R., Wang, W., Gumbart, J., Tajkhorshid, E., Villa, E., & Schulten, K. et al. (2005). Scalable molecular dynamics with NAMD. *Journal of Computational Chemistry*, *26*(16), 1781–1802. doi:10.1002/jcc.20289 PMID:16222654

Pirovano, W., Feenstra, K. A., & Heringa, J. (2008). PRALINE™: A strategy for improved multiple alignment of transmembrane proteins. *Bioinformatics (Oxford, England)*, *24*(4), 492–497. doi:10.1093/bioinformatics/btm636 PMID:18174178

Prothero, J. W. (1966). Correlation between the distribution of amino acids and alpha helices. *Biophysical Journal*, *6*(3), 367–370. doi:10.1016/S0006-3495(66)86662-6 PMID:5962284

Saier, M. H. (2000). A functional-phylogenetic classification system for transmembrane solute transporters. *Microbiology and Molecular Biology Reviews*, *64*(2), 354–411. doi:10.1128/MMBR.64.2.354-411.2000 PMID:10839820

Sánchez, R., Pieper, U., Melo, F., Eswar, N., Martí-Renom, M. A., Madhusudhan, M. S., & Šali, A. et al. (2000). Protein structure modeling for structural genomics. *Nature Structural & Molecular Biology*, *7*, 986–990. doi:10.1038/80776 PMID:11104007

Schiffer, M., & Edmundson, A. B. (1967). Use of helical wheels to represent the structures of proteins and to identify segments with helical potential. *Biophysical Journal*, *7*(2), 121–135. doi:10.1016/S0006-3495(67)86579-2 PMID:6048867

Shaw, D. E. (2005). A fast, scalable method for the parallel evaluation of distance-limited pairwise particle interactions. *Journal of Computational Chemistry*, *26*(13), 1318–1328. doi:10.1002/jcc.20267 PMID:16013057

Shi, J., Blundell, T. L., & Mizuguchi, K. (2001). FUGUE: Sequence-structure homology recognition using environment-specific substitution tables and structure-dependent gap penalties. *Journal of Molecular Biology*, *310*(1), 243–257. doi:10.1006/jmbi.2001.4762 PMID:11419950

Shoichet, B. K., Leach, A. R., & Kuntz, I. D. (1999). Ligand solvation in molecular docking. Proteins. *Structure, Function, and Bioinformatics*, *34*(1), 4–16. doi:10.1002/(SICI)1097-0134(19990101)34:1<4::AID-PROT2>3.0.CO;2-6

Stoffler, D., Sanner, M. F., Morris, G. M., Olson, A. J., & Goodsell, D. S. (2002). Evolutionary analysis of HIV-1 protease inhibitors: Methods for design of inhibitors that evade resistance. Proteins. *Structure, Function, and Bioinformatics*, *48*(1), 63–74. doi:10.1002/prot.10130 PMID:12012338

Tame, J. R. (1999). Scoring functions: A view from the bench. *Journal of Computer-Aided Molecular Design*, *13*(2), 99–108. doi:10.1023/A:1008068903544 PMID:10091117

Tatulian, S. A. (2013). Structural characterization of membrane proteins and peptides by FTIR and ATR-FTIR spectroscopy. In *Lipid-Protein Interactions* (pp. 177–218). Humana Press. doi:10.1007/978-1-62703-275-9_9

Tomasselli, A. G., & Heinrikson, R. L. (2000). Targeting the HIV-protease in AIDS therapy: A current clinical perspective. Biochimica et Biophysica Acta (BBA)-. *Protein Structure and Molecular Enzymology*, *1477*(1), 189–214. doi:10.1016/S0167-4838(99)00273-3

Ulmschneider, J. P., Smith, J. C., White, S. H., & Ulmschneider, M. B. (2011). In silico partitioning and transmembrane insertion of hydrophobic peptides under equilibrium conditions. *Journal of the American Chemical Society*, *133*(39), 15487–15495. doi:10.1021/ja204042f PMID:21861483

Ulmschneider, J. P., & Ulmschneider, M. B. (2009). Sampling efficiency in explicit and implicit membrane environments studied by peptide folding simulations. Proteins. *Structure, Function, and Bioinformatics*, *75*(3), 586–597. doi:10.1002/prot.22270 PMID:19003985

Van Der Spoel, D., Lindahl, E., Hess, B., Groenhof, G., Mark, A. E., & Berendsen, H. J. (2005). GROMACS: Fast, flexible, and free. *Journal of Computational Chemistry*, *26*(16), 1701–1718. doi:10.1002/jcc.20291 PMID:16211538

Wallin, E., & Von Heijne, G. (1998). Genome-wide analysis of integral membrane proteins from eubacterial, archaean, and eukaryotic organisms. *Protein Science*, *7*(4), 1029.

Wang, J., Wolf, R. M., Caldwell, J. W., Kollman, P. A., & Case, D. A. (2004). Development and testing of a general amber force field. *Journal of Computational Chemistry*, *25*(9), 1157–1174. doi:10.1002/jcc.20035 PMID:15116359

Wieman, H., Anderssen, E., & Drablos, F. (2004). Homology-based modelling of targets for rational drug design. *Mini Reviews in Medicinal Chemistry*, *4*(7), 793–804. PMID:15379646

Yarov-Yarovoy, V., Schonbrun, J., & Baker, D. (2006). Multipass membrane protein structure prediction using Rosetta. Proteins. *Structure, Function, and Bioinformatics*, *62*(4), 1010–1025. doi:10.1002/prot.20817

Zimmermann, O., & Hansmann, U. H. (2006). Support vector machines for prediction of dihedral angle regions. *Bioinformatics (Oxford, England)*, *22*(24), 3009–3015. doi:10.1093/bioinformatics/btl489 PMID:17005536

Chapter 8
In Silico Perspective into Interactions and Mutations in Human TLR4 and Ebola Glycoprotein

Sujay Ray
University of Kalyani, India

Arundhati Banerjee
National Institute of Technology, India

ABSTRACT

Toll-Like Receptor-4 (TLR4) senses life-threatening Ebola virus Glycoprotein (GP) and produces pro-inflammatory cytokines, resulting in lethal Ebola virus infections. GP2-subunit of Ebola promotes viral entry via membrane fusion. The present study models, optimizes and demonstrates the 3D monomer of the responsible human protein. The essential residue (studied from wet-laboratory research) was observed to be functionally conserved from multiple-sequence alignment. Thus, after performing point-mutation, the mutant protein was satisfactorily re-modelled; keeping its functionality preserved. Comparable residual participation in GP2 and each of the proteins was examined, individually. Stability of the proteins and protein-GP2 complexes on mutation; were discerned via energy calculations, solvent-accessibility area and conformational switching, with supportive statistical significances. Therefore, this probe paves a pathway to examine the weaker interaction of the stable mutated human protein with Ebola GP2 protein, thereby defending the Ebola viral entry.

INTRODUCTION

Functioning of the immune system depends greatly upon the Toll-Like Receptors (TLRs), which are distinct, membrane-spanning, non-catalytic receptors generally expressed in sentinel cells including not only dendritic cells but also macrophages. Among them, Toll-Like Receptor 4 (TLR4) protein, which is encoded by *tlr4* gene (Rock et.al., 1998), senses the presence of Ebola virus glycoprotein (GP).

DOI: 10.4018/978-1-5225-0362-0.ch008

Initially, Ebola virus targets the monocytes and macrophages (Yong-chen, Wen-chen & Pamela, 2008). This leads to a stimulus to generate the pro-inflammatory cytokines and chemokines (Yong-chen, Wen-chen & Pamela, 2008). These inflammatory cytokines generally results further to the development of a shock in the circulatory system (Okumura, Pitha, Yashimura & Harty, 2010). This shock is thus observed in the fatal infections by Ebola virus, which has turned out to serve as world's alarming danger (Okumura, Pitha, Yashimura & Harty, 2010). TLR4 identifies the Ebola virus glycoprotein (GP) upon particles which are specifically virus-like in nature (VLPs) (Okumura, Pitha, Yashimura & Harty, 2010). The resultant TLR4 signalling pathways further aid to release the pro-inflammatory cytokines and suppressor of cytokine signalling 1 (SOCS1) not only in a human monocytic cell line (Okumura, Pitha, Yashimura & Harty, 2010) but also in HEK293 associated TLR4/MD2 cells (Yamamoto, Sato, Hemmi et. al., 2003; Re & Strominger, 2002). This therefore, steadily and firmly expresses the TLR4/MD2 complex (Yamamoto, Sato, Hemmi et. al., 2003; Re & Strominger, 2002). Certain experimental studies involving immune-precipitation/Western blot analyses (Okumura, Pitha, Yashimura & Harty, 2010) were undergone previously in this regard to reveal that Ebola virus and GP interacts with the TIR (Toll/Interleukin-1 Receptor) domain of TLR4 and thereby excites the expression of NF-kappaB in a TLR4-dependent manner (Yamamoto, Sato, Hemmi et. al., 2003).

The Ebola glycoprotein subunit; GP0 undergoes a cleavage with the performance of a cellular protease, also known as furin convertase, into two subunits (Balaji, Jizhen, Haiqing & Lijun, 2005). One of the subunit is GP1 having approximately 130kDa, and the other one is GP2 (~24 kDa), which resides mainly in the transmembrane. GP1 is correlated with GP2, possibly via both non-covalent interactions as well as disulfide linkages (Jeffers, Sanders & Sanchez, 2002; Volchkov, Feldmann, Volchkova & Klenk, 1998) For receptor binding purpose, the GP1 subunit serves to be responsible one (Takada et.al., 1997; Wool-Lewis & Bates, 1998). Then further cell membrane fusion of the virus and its entry is mediated by GP2 subunit (Takada et.al., 1997; Wool-Lewis & Bates, 1998).

In the TLR4 protein (TIR domain), mutational studies (Ota, Monika & Roman, 2012) involving point mutation of proline to histidine is documented to prevent the signalling mechanism when Ebola attacks and thereby the TIR domain of TLR protein expresses a hypo-responsiveness towards the Ebola glycoprotein (Ota, Monika & Roman, 2012).

So far no such investigation was done at the computational level involving the explorations of the interactions in the life-threatening Ebola virus glycoprotein and TLR4 protein complex, along with their interactive patterns; to examine the importance of the residues involved in the membrane fusion. No studies previously, aimed to investigate the molecular level changes including the interactions, due to point mutations in the TIR domain of TLR4 protein in humans.

Therefore, in this present study, the structure of the TIR domain from human TLR4 protein was modelling via homology modelling, analyzed and demonstrated. The X-ray crystal structure was analysed for the extraction of GP2 protein. Further, multiple sequences of the TIR domains from different related species having TLR4 protein were obtained via extraction from the parent TLR4 protein and were aligned to examine the conserved residues. The conserved residues serve to be the important residues for the functionality of the protein. Moreover, based on the previous wet laboratory research disclosures, point mutation was performed in the sequence of the parent protein of TIR domain from human TLR4 after the analysis of the residue to be point mutated being functionally vital from multiple sequence alignment. The mutated protein (MT) was further remodelled by homology or comparative modelling after the satisfaction of its varied stereo-chemical properties. The functionality and the fold of the mutated protein was observed and compared with the wild type (WT) protein. On the whole, energy minimization of each of the homology modelled tertiary structure was executed for obtaining a stable protein conformation. The wild-Type (WT) and MT protein monomers underwent protein-protein docking phenomena to interact with GP2 protein individually. Further again energy optimization followed by Molecular Dynamics (MD) simulation was performed for the docked complexes in either of the cases. The calculation of several stability determining energy parameters, net solvent accessibility area values were estimated for the examination of the stability and the strengthening impact upon their binding. Residual involvement from the relevant proteins to interact with GP2 protein strongly was discerned and investigated. Earlier investigation (Simanti, Amit, Semanti, Rakhi & Angshuman, 2014; Angshuman, 2015) reveals the molecular computational level studies for many such diseases but a lethal life-threatening disease from Ebola virus was dealt in none.

In a digest, this present computational exploration focused into the basic source for the important participation of the two indispensably important proteins in the cell signalling phenomena for the membrane fusion technique by Ebola glycoprotein subunit 2 (GP2). The study further extended to discern the effect on point mutation in the human TLR4 protein in association with its altered interaction with GP2 protein of Ebola and change in the stability of the simulated mutated complex with supportive statistical significances. The examination of residual dependencies of the relevant proteins in the respective WT and MT cases, from this probe would be additionally favourable for upcoming therapeutic research in future. So, this study might provide a limelight in the zone of defensive mechanism for the Ebola virus at its molecular level analysis.

MATERIALS AND METHODS

Sequence Analysis and Template Search for TLR4_TIR Protein Domain

For homology modelling of the protein, first the sequence of the amino acids in TLR4 protein of *Homo sapiens* was obtained from GenBank (Accession No.: O00206.2, GI No.: 20140413). The sequence result was verified from UniProtKB too. NCBI CD-BLAST aided to identify the conserved domains (extremely conserved protein segments that are specifically responsible for protein functionality) (Marchlar-Bauer et.al, 2015). The results were further validated by Pfam (Punta et.al., 2011). Domains often serve as an important interaction site for proteins too (George & Heringa, 2002; Jones et.al., 1998). As documentation (Ota, Monika & Roman, 2012) states the TIR domain from TLR4 to participate in the interactions, so, the amino acid sequence of this domain was extracted from EMBOSS. The domain length was observed to be 139 amino acid residues long. For homology modelling, also familiar as comparative modelling; prerequisite is the determination of the suitable template for the protein to be modelled. So for the purpose, the amino acid sequence of this TIR protein domain was utilized to search the Brookhaven Protein Data Bank (PDB) (Berman et.al., 2000) via the PBLAST software tool (Altschul et.al., 1990). A comprehensive result was observed for the purpose of template search from HH-Pred (Johannes et.al, 2005). The template search results for the domain selected the X-Ray crystal structure of TIR domain of human TLR2 protein from *Homo sapiens* (PDB Code: 1O77, chain A) posing 41% sequence identity and a query coverage of 100%.

Comparative (Homology) Modelling of Monomer Wild-Type TLR4_TIR Domain

Studies from earlier documentations state that for homology modelling, the most essential necessity lies in the sequence identity of templates with respect to the target protein to be more than 30% (Xiang, 2006; Sander and Schneider, 1991). So, as the template sequence identity is found to be 41%, homology modelling was executed here, using the MODELLER 9.14 software suite and thus the Wild-Type (WT) protein was obtained. Without varying the coordinate system of atomic positions in the obtained template (A chain of PDB ID: 1O77); the modelled structure was then superimposed separately on the available crystal templates. The root-mean-square-deviations (RMSD) by PyMOL (DeLano, 2002) for the backbone superimpositions were investigated to be 0.182Å for the modelled protein, on superimposing upon its existing template. DaliLite server from European Bioinformatics Packages generated a validating result for the RMSD values (Analysis Tool Web Services from the EMBL-EBI, 2013). The pictorial view of the superimposition of the crystal template structure upon the modelled protein in subject is well depicted through Figure 1.

Loop Optimization of the Protein Domain

Deformations in model-target overlap are often caused because of the structural differences involving loops and errors in side-chain packing (Krieger et.al, 2003; Marti-Renom et.al, 2000). Many-a-times, certain residues reside in the disallowed zones in the Ramachandran map. For the proper conformation of ψ-φ angles, the modelled protein needs to be re-modelled further, after optimization of its loops using ModLoop (Fiser and Sali, 2003). Generally the automated modelling of the disordered loop regions is

Figure 1. Backbone superimposition of modelled wild type (WT) human TLR4_TIR domain (pink shade) upon its crystal template in cyan shade (PDB Code: 1O77_A)

performed by ModLoop using MODELLER. Moreover, in addition, the loop conformations are predicted by contentment of the spatial restraints (Fiser and Sali, 2003). The modelled protein structure was thus subjected for loop optimization.

Refinement of the Protein Structure

The loop optimised structures were refined further by subjecting to ModRefiner (Xu and Zhang, 2011). ModRefiner is not only an algorithm for atomic-level but also performs refinement of the protein structure at a high resolution. Here, exploration into the conformational stability is performed via the utilization of two force-fields; one being physics based in nature while the other being template information dependant in nature. The main focus remains to draw the modelled proteins near to their most native and interactive state with the utilization of a high resolution algorithm (Xu and Zhang, 2011). The benefit is that at the native states of the individual proteins, maximum interaction between residues exists.

Validation of the Modelled Structure

To estimate Z scores, ProSA was utilized (Weiderstein and Sippl, 2007). This server provides an assessment for the deviation in the net total energy of the modelled structure with reference to an energy distribution obtained from abrupt conformations and therefore, evaluated the Z-scores as a measure of overall model quality (Weiderstein and Sippl, 2007). The negative Z-scores for all models hold for the appropriateness of the modelled protein. The result disclosed that the predicted homology models contented the scope of typical native structures (Wiederstein, 2004). Main chain properties of the modelled monomer of the protein domain structure were verified by SAVeS server. To explore and analyze the specific contour for each residue of amino acid from the 3D model, VERIFY3D was used (Eisenberg et.al., 1997). Analyses with PROCHECK (Laskowski et.al., 1993) were executed for the purpose to calculate the stereo chemical qualities of the models. WHAT_CHECK examinations supplied a net total outline of the nature of each of the structures in comparison to a set of reliable structures (Hooft et.al., 1996). Ramachandran plots (Ramachandran & Sashisekharan, 1968) were drawn. It finally evaluated and presented that there were no such amino acid residues that occupied the disallowed regions.

Sequence and Structure Analysis of Ebola GP2

Search results for the Ebola GP2 protein for the purpose of fusion and viral entry selected its X-ray crystal structure from Ebola virus having PDB ID: 1EBO (Weisssenhorn, Carfi, Lee, Skehel & Wiley, 1998). The structure has six chains (A, B, C, D, E and F) and is total 670 amino acids long.

Search for Conserved Regions: Multiple Sequence Alignment

Amino acid sequences of the TLR4 protein from the most closely related primates before the evolution of *Homo sapiens* were extracted from GenBank followed by the identification of the range of their TIR domain region within each of the TLR4 proteins, individually, with the help of pfam (Punta et.al., 2011). The most closely related primates include gorilla (*Gorilla gorilla* with Accession No.: Q8SPE8.1), gibbon (*Hylobates lar* with Accession No.: ACC68078.1) and chimpanzee (*Pan troglodytes* with Accession No.: JAA42756.1) (Grove, 2001; Cartmil & Smith, 2011).Furthermore, as the genomes of human and mouse are documented to be closely related (Richard, Leo, Eitan & Chris, 2003), so the TIR domain from TLR4 of *Mus musculus* (Accession No.: EDL31078.1) was also aligned along with the other protein sequences. The parent TLR4 proteins, in each case were provided to EMBOSS software tool (McWilliam et.al, 2013) to extract the respective TIR domains individually. For each of the TLR4 proteins, this entire procedure followed the similar protocol, as performed for the sequence analysis of only human TLR4_TIR domain. The set of these collected sequence data of TIR domains, from TLR4 proteins of different organisms were utilised to perform the Multiple Sequence Alignment (MSA) with the aid of PRALINE server (Simmosis & Heringa, 2005). Comprehensive results were observed from CLUSTAL-W (Thompson, Higgins & Gibson, 1994). The conserved residues that are essential for the functionality of the proteins were identified and studied further.

Mutation and Re-Modelling of the TLR4_TIR Protein Domain

Mutational Analysis

Earlier studies investigated that a point mutation of a proline residue from human TLR4 to histidine led to the less effective or weakened sensing of the viral glycoprotein (Ota, Monika & Roman, 2012). This specific proline residue was investigated in the human TLR4_TIR protein domain via the MSA results. So, the point mutation of Pro to His residue was thereby performed at the sequence level of the TIR domain of human TLR4 and was further modelled.

Homology Modelling of the Mutated Monomer of TLR4_TIR Domain

The mutated human TLR4_TIR domain sequence was further re-modelled following the same protocol as previously described in "Comparative *(Homology) Modelling of Monomer TLR4_TIR domain*". The modelled mutated structure (MT) was again *authenticated* for the approval of its stereo-chemical properties.

Conservation of the Function via Fold Recognition

To observe the conserved functionality, TM-Align (Zhang & Skolnick, 2005) was used to obtain the TM-Score value. The TM-Score >0.5 designates that the two proteins (wild-type and mutated) share a common SCOP/CATH fold and thus, preserve their functionality (Zhang & Skolnick, 2005). Generation of protein residue-to-residue alignment based on structural similarity was performed by TM-Align (a highly optimized algorithm for protein structure comparison and alignment) (Zhang & Skolnick, 2005).

Docking Studies with the Ebola GP2

Scoring function holds one of the paramount factors to be considered for protein-protein docking simulations. Cluspro2.0 (Comeau et.al, 2004; Kozakov et.al, 2013) performs the docking operations by clustering and minimization of the protein complex, at each time. It additionally opts for the best electrostatic as well as desolvation free energies before clustering. ClusPro 2.0 was primarily used to dock the modelled structure of both WT and MT protein with the Ebola virus GP2 separately. GP2 having 5 chains, being longer, was uploaded as receptor whereas; TIR protein was uploaded as ligand in each of the cases. Using advanced option for structure optimization of ClusPro 2.0, unstructured residues were removed. For both the cases, the finest protein complex model was selected and examined. This is because among all the 10 docked structures available, the selected protein complex had the superior cluster size for either of the cases (WT and MT). Furthermore, GRAMM-X (Vakser, 1995) and ZDOCK (Chen et.al, 2003) performed protein-protein docking to accomplish an inclusive result. The best complex model was opted. Both Z-DOCK and GRAMM-X operates on fast Fourier transformation to explore all plausible modes for the protein-protein binding. Z-DOCK performs the starting stage of the unbound type docking phenomena by utilizing the scoring function that is energy-based. It further estimates the shape complementarity, electrostatic potential as well as the desolvation energy. Z-DOCK performs docking simulations by taking into account the rotational as well as the translational gap amongst the two proteins in subject. On the other hand, GRAMM-X utilizes smooth and even surface potentials and a scoring program that is knowledge-based and thereby performs the protein docking simulations efficiently.

Energy Minimization of Modelled Structures and the Complexes

The modelled TLR4_TIR domains as well as the docked GP2_TIR protein complexes were energy minimized using CHARMM force field. Utilizing the Chiron Energy Minimization tool, the overall energy of the modelled WT GP2 and MT GP2 complex structure was minimized (Ramachandran, Kota, Ding & Dokholyan, 2011). It follows a specific force field; mainly CHARMM to minimize the gross and aggregate energy of the docked complex (Brooks et.al., 1983). The tool Chiron utilizes a high heat exchange rate of the individual proteins with the bath in simulations by Molecular Dynamics (MD). Therefore, Chiron not only removes the steric clashes rapidly, but also produces minimal disconcertion of the protein backbone.

Comparable Study for the Alterations in the Interacting Sites

Protein Interaction Calculator (P.I.C) helped to perceive varied interactions; such as disulphide bonds, hydrophobic interactions, ionic interactions, hydrogen bonds, aromatic-aromatic interactions, aromatic-sulphur interactions and cation-π interactions in each of the docked TIR-GP2 complexes (Tina et.al., 2007). So, 3D molecular file of the docked ebola-human complexes were uploaded for the interaction calculation separately. Operation of Discovery Studio (DS) with the assistance of CHARMM force field (Brooks et.al., 1983) was performed to examine the protein-protein complexes. Binding residues and thus binding regions were detected with Analyze Binding Site tool from DS package for the docked complex for the validation of the ebola-TLR4 docked complexes.

Stability Examination of Wild-Type Protein, Mutated Protein, and Docked Complexes

Few Parameters for Energy Calculation

To investigate the stability of the mutant modelled structure in comparison to the wild-type modelled protein, several parameters for energy calculation were taken into account. At first, the DFire energy value was evaluated to observe the most stable complex. DFire evaluates the atomic interactions (in protein model) that are more specifically, non-bonded in nature (Yuedong & Yaoqi, 2008). The complex with minimum DFire energy value is considered to be closer to its native conformation and thus to be more stable (Yuedong & Yaoqi, 2008). Free Energy of Folding in kcal/mol for the WT and MT proteins were analyzed from VADAR 2.0 (Tartaglia et.al, 2008). Free energy of folding is well-recognized to be indirectly proportional to the foldability of any protein (Tartaglia et.al, 2008). Furthermore, along with DFire energy, free energy of folding also offered supportive results for the complex having mutated protein.

Moreover, SDM (Site Directed Mutator) (Topham, Srinivasan & Blundell, 1997) was utilized as a function for the calculation of statistical potential energy as well as mutated residual solvent accessible percent. It was used to forecast the consequence that will result upon the protein stability due to the point or due to Single Nucleotide Polymorphisms (SNPs) on the stability of proteins. For the folded as well as the unfolded state of the wild-type and mutant protein structures, the algorithm estimates the alteration in the stability scores (Topham, Srinivasan & Blundell, 1997). In comparison to other published methods, SDM was seen to perform comparably better than the other methods for classifying mutations as stabilizing or destabilizing (Topham, Srinivasan & Blundell, 1997). For the purpose it was tested utilizing a set of mutants with thermodynamic measurements conducted under physiological conditions (Topham, Srinivasan & Blundell, 1997). SDM is well acknowledged to however provide an increased and more balanced sensitivity upon mutations (Topham, Srinivasan & Blundell, 1997).

Another energy calculation parameter to identify with the roles of mutations in disease, their impacts was evaluated not only on protein stability but also on the protein-protein interaction and the complex stability. For the purpose, mCSM value (Douglas, David & Tom, 2013) was calculated which relies on graph-based signatures by predetermining the distance prototypes between atoms. Experimental evidences presents that mCSM can calculate stability alterations of an extensive range of mutations occurring in p53 (a tumor suppressor protein), thereby exhibiting the applicability of the recommended technique in a challenging disease scenario (Douglas, David & Tom, 2013).

Calculation of the Net Solvent Accessibility Values

The net solvent accessibility value for the Ebola GP2 protein after interaction with WT and MT human proteins individually was inspected for identifying the best interactive complex with a curiosity to justify the results from energy calculations. Lower solvent accessibility value points out a stronger interactive complex.

Exploration into the Conformational Switching

The conformational alteration in the mutated protein was studied in contrast to the wild-type protein. So, the individual secondary structure distribution was studied using DSSP method (Kabsch & Sander, 1983) and also from Discovery Studio packages from Accelyrs. Increment in the interaction through helical structures and β-sheets lead to stronger and better interaction (Paul, Thomas & Ken, 1993). The alteration and switching of the secondary structures for both of the complexes (the mutated and wild-type) were examined and analyzed.

Analysis via Statistical Significance

Further analysis through statistical significance was executed for the variations in the secondary structure conformations, net solvent accessibility values as well as the comparable studies involving energy parameters for the structures, pre and post mutation. So for the purpose, the paired T-test was performed. It is generally performed when the two separate measurements in each of the cases are observed before and after a treatment, in present study the treatment being- mutational alterations. An assumption, usually accepted for this T-Test is that, the standard deviations (SDs) are not approximately same in both the measurements in the individual cases. It is examined to get a confirmation that the difference between the two means is statistically significant ($P<0.05$).

RESULTS

Structural Portrayal of Wild-Type (WT) TLR4_TIR Domain

The homology modelled structure of WT TLR4_TIR protein domain is 139 amino acid residues long. The modelled structure is very analogous to its potential template from the X-Ray crystal structure of TLR2 protein from Ebola virus (PDB Code: 1O77, chain A). The protein is mainly composed of α-helices, 4 parallel β-sheets interconnected with coil regions. The structure begins with 4 residues in the N-terminal, forming β-sheets and further, including this set of β-sheets, four parallel sets of β-sheets are observed to be present (amino acid residues: 1-4, 29-32 and 57-62) in the entire structure. The structure ends with 1 residue (residue: Ala139) forming coil at the C-terminal end. The structure forms helical structures with amino acid residue positions as 7-23, 43-54, 64-77, 79-81, 108-116 and 126-138 respectively. In total, the protein exhibited 43.9%, 12.2% and 43.9% residues forming helical, β-sheet and coil like conformation, respectively. The structure is well depicted in Figure 2 with α-helices and β-sheets in cyan and red shades, respectively having interspersing yellow shaded coils. The essential Pro39 residue in the coil region was represented with green stick like projection.

Analysis of Conserved Residue

The sequence of all the TIR domains from the most closely related species to humans were aligned sequentially for the execution of multiple sequence alignment (MSA). The results from MSA were presented in Figure 3. The most conserved residues are functionally vital. Accumulating evidences documents the weakened sensitivity of human TLR4_TIR domain towards the Ebola virus, due to the point mutation of a particular proline residue to histidine (Ota, Monika & Roman, 2012). From the MSA results, that proline residue was observed in 39th position of the TLR4_TIR protein domain in humans. The 39th position residue was also observed to be conserved throughout (Figure 3), suggesting it to be the most functional residue additionally. So, to examine the mutational effects, this proline residue was point mutated to histidine in the sequence of the TLR4_TIR protein domain in humans.

Figure 2. Structural demonstration of the modelled monomer of Wild-Type (WT) TIR protein having prominent secondary structures as helices in cyan shade and red shaded four parallel β-sheets having yellow shaded interspersing coil regions. The Pro39 residue to be mutated is represented in green stick like representation.

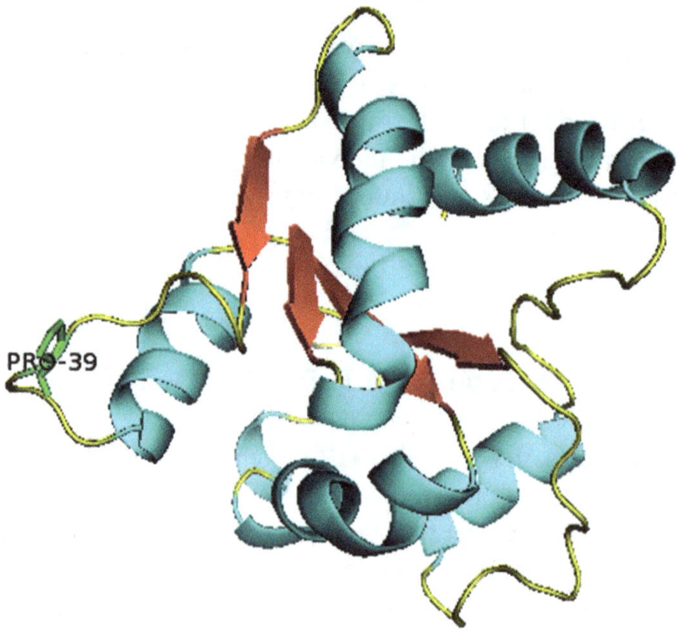

Figure 3. Conserved residues (deep red) were examined after multiple sequence alignment (MSA) for all the different TLR4_TIR domains

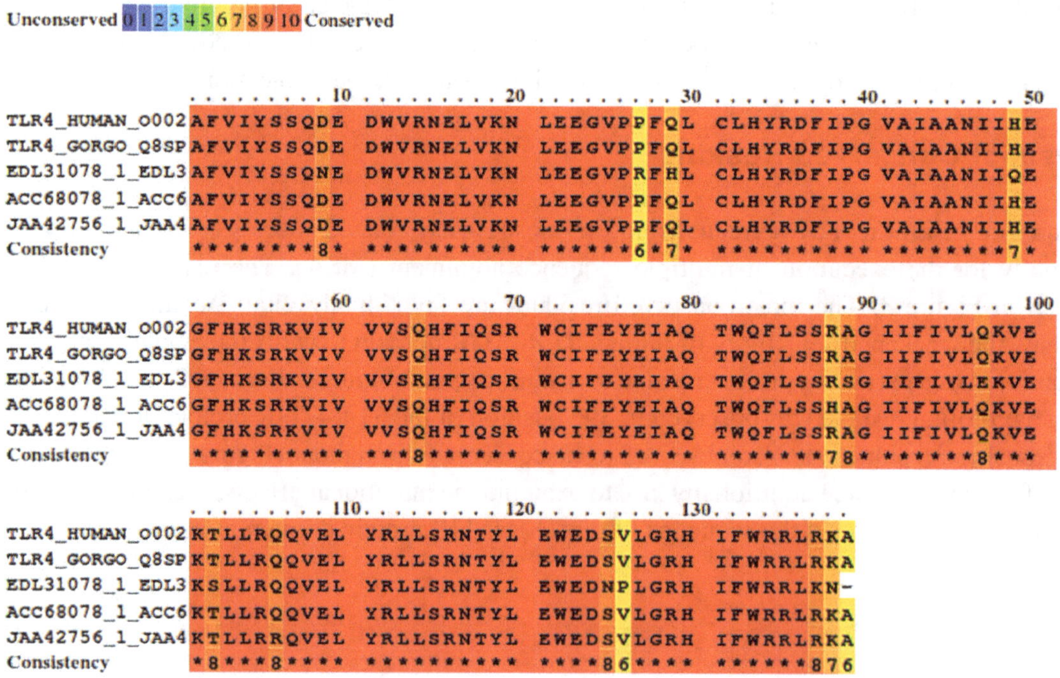

Structural Portrayal of Mutant-Type (MT) TLR4_TIR Domain

The P39H mutated TLR4_TIR domain was also observed to have 139 amino residues beginning with four residues at N-terminal, adopting β-sheets (amino acid residues: 1-4). Reduced number of residues adopted 4 β-sheet conformations. They occupied positions as 1-4, 28-31, 57-62 and 92-95 as shown in yellow shades in Figure 4. The structure again ends with alanine139 forming coil region after several helical regions that were interspersed with coils (blue shaded coil regions in Figure 4). The helices (in red shade in Figure 4) were observed to be occupying the positions as 7-23, 43-53, 64-77, 79-81, 101-103, 108-116 and 126-138 respectively. Upon P39H mutation, several residues from β-sheet and helical conformation switched over to coil-like conformation, leading to the poorer conformation of the mutated protein. Therefore, the percentage of residues adopting helices and β-sheet conformation reduced to 41% and 10.8%, respectively with an increase in the coil like conformation to 48.2%. The entire functional tertiary structure of the protein is well presented in Figure 4, with the mutated residue; His39 represented in green stick like projection.

TM-Score for Fold Prediction

TM-score (Zhang & Skolnick, 2005) was estimated by residue-to-residue alignment based on structural resemblance for WT over MT. Generally, TM-score > 0.5 specifies that the structures exhibit same SCOP/ CATH fold (Zhang & Skolnick, 2005). In the present study, the TM-score yielded to 0.7, implying that even after mutational and conformational switches, the MT exhibited same SCOP/ CATH fold with the WT.

Figure 4. Structural demonstration of the modelled mutated monomer (MT) of TIR protein having prominent secondary structures as helices in red shade and yellow shaded four parallel β-sheets having marine shaded interspersing coil regions. The mutated residue, His39 is represented in green stick like representation.

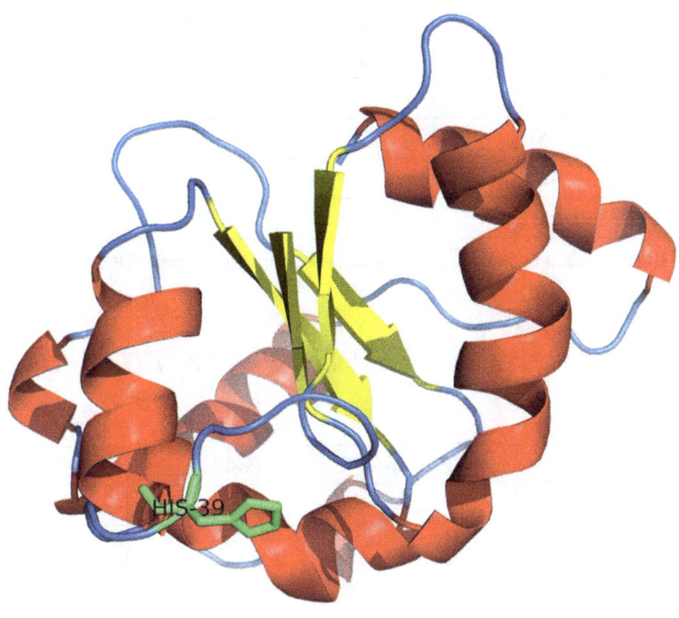

Inter Protein-Protein Interactions for WT and MT Complexes

After performing the protein-protein docking of WT and MT with GP2 protein, the interaction studies examined mainly side chain-side chain inter-hydrogen bonds and ionic-ionic interactions to occur in each of the cases. Ionic interactions are usually known to be more paramount for an interaction to be strengthened (Baldwin, 1996). The number of ionic-ionic bonds was observed to get reduced from 6 in WT_GP2 (Table 1) to 2 for the MT_GP2 (Table 2). This leads to make the MT protein to interact less firmly with the Ebola GP2. Chain D and chain F from Ebola GP2 was observed to participate with TLR4 protein, whereas, only chain F was observed to interact forming ionic bonds with TLR4 protein. Figure 5 and Figure 6 clearly depicts few of the interacting ionic residues labelled and ionic bonds represented in pink and blue dashed lines for WT_GP2 and MT_GP2 complex respectively.

Stability Outcomes on Mutation

Few Parameters for Energy Calculations

To investigate the stability of the individual modelled proteins as well as the protein-protein complex, several energy calculating parameters were studied. For the purpose, VADAR 2.0 (Tartaglia et.al, 2008) was operated to analyze the free energy of folding of the individual protein first. It revealed an increase in the free energy of folding values for the sole MT protein. Free energy of folding value of -127.89kcal/mol and -138.46kcal/mol was estimated for MT and WT models respectively (Table 3). So, decline in the free energy of folding indicates better and improved folds in the WT protein.

Table 1. The ionic-ionic interacting residues for Wild-Type (WT) TLR4_TIR protein and Ebola glycoprotein (GP2) protein

Position	Residue	Protein	Position	Residue	Protein
388	ARG	D	780	GLU	G
428	ASP	D	783	ARG	G
435	ASP	D	787	ARG	G
438	ASP	D	787	ARG	G
644	ASP	F	776	ARG	G
646	ARG	F	748	GLU	G

G represents the WT_TLR4_TIR protein and D and F are the interacting chains from Ebola GP2 protein

Table 2. The ionic-ionic interacting residues for Mutant-Type (MT) TLR4_TIR protein and Ebola glycoprotein (GP2) protein

Position	Residue	Protein	Position	Residue	Protein
633	ARG	F	746	GLU	G
633	ARG	F	780	GLU	G

G represents the MT_TLR4_TIR protein and F is the interacting chain from Ebola GP2 protein

Figure 5. The warm-pink dashed lines represent the H-bonds in the complex interface of WT with three parallel β-sheets and Ebola glycoprotein subunit 2 (GP2) in cyan shade and yellow shade respectively

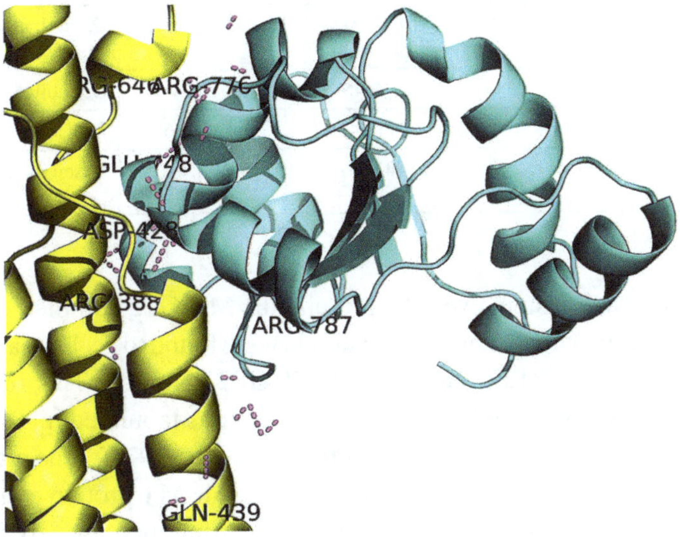

Figure 6. The reduced number of blue dashed lines represent the H-bonds in the complex interface of MT with four parallel β-sheets and Ebola glycoprotein subunit 2 (GP2) in pink shade and yellow shade respectively

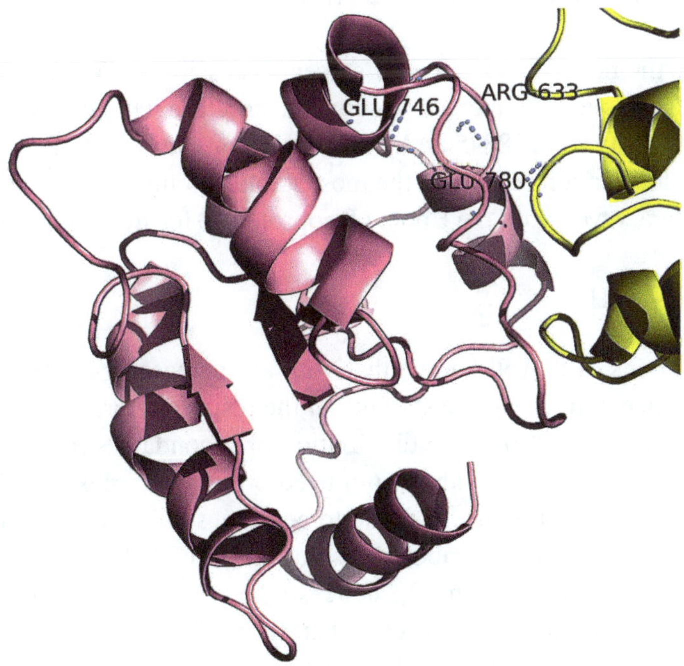

Table 3. Few stability calculating parameters for Wild-Type and Mutant-Type human TLR4_TIR protein and their respective complexes formed with Ebola glycoprotein-2 (GP2)

Stability Calculating Parameters	WT Protein	MT Protein	Whether Stabilizing or Destabilizing the MT
Free Energy of Folding of Protein (kcal/mol)	-138.46	-127.89	Destabilizing
DFire Energy of Protein *Complex* (kcal/mol)	-2211.41	-2181.86	Destabilizing
Net Area for Solvent Accessibility (square Å) for Protein *Complex*	38845.84Å^2	39187.3 Å^2	Destabilizing with Weaker Interaction

Like, the stability of the WT protein over MT protein, the MT-GP2 complex was less stable than the complex having the WT protein. DFire energy values again were found to get increased from -2211.41kcal/mol (in WT-GP2 complex) to -2181.86kcal/mol for MT-GP2 complex. This further deduces that the MT_GP2 complex was less stable than the WT_GP2 complex as tabulated in Table 3.

SDM generated energy value (Topham, Srinivasan & Blundell, 1997) for the prediction of the stabilizing or destabilizing effect on point mutation showed that the sole MT protein exhibited an energy value ($\Delta\Delta G$) of -0.45kcal/mol with residual solvent accessibility getting increased from 84.1% (for Pro39) to 107.1% (for His39). This therefore, states it to be unstable. Again to evaluate the protein stability, the mCSM value (Douglas, David & Tom, 2013) generated a $\Delta\Delta G$ energy value of -0.05kcal/mol and stated it to be an unstable upon mutation.

Net Area of Solvent Accessibility for the Complexes

A comparative study on the net available area for solvent accessibility WT and MT protein complexes individually, apprehends the strongly interactive complex structure. Herein, the increase in the value of net area of solvent accessibility from 38845.84Å^2 (in the WT complex) to 39187.3 Å^2 for complex having MT protein, depicts the former complex to be the most stable and interactive one, again. Therefore, the MT protein does not interact well with the Ebola glycoprotein-2 for membrane fusion and viral entry.

Conformational Switching

From earlier documentations, it is well known that with an increase in the % of residues undergoing helical structures and β-sheet structures, the stability of the respective protein complex increases (Paul, Thomas & Ken, 1993). For the purpose, the distribution of secondary structures was studied for WT and MT sole proteins as well as the individual optimized protein complexes with GP2. From Table 4, it was observed that the sole WT protein was found to be more stable conformationally as on its mutation the individual protein had reduced number of residues forming helices and β-sheets. Moreover, the WT protein exhibited few residues forming 3_{10} helices which accompanied the pure α-helices. Such accompanying 3_{10} helices were seen to be absent when the WT protein was mutated (P39H). As per documentation suggests that protein structures having α-helices combined with few 3_{10} helices exhibits a stronger interaction than those having only pure α-helices (Toniolo & Benedetti, 1991), so the WT protein is further affirmed to be stronger in its conformation. Figure 7 depicts the conformational variations through comparable column representation.

Table 4. Categorization of secondary structure for Wild-Type (WT) and Mutant-Type (MT) human TLR4_TIR protein before interaction with Ebola glycoprotein-2 (GP2)

% Categorization of Secondary Structure	WT Protein	MT Protein	Whether Stabilizing or Destabilizing the MT
% net α-helices	41.7%	41%	Destabilizing with Weaker Interaction
% net 3-ten helices	2.2%	0%	Destabilizing with Weaker Interaction
% β-sheets	12.2%	10.8%	Destabilizing with Weaker Interaction
% coils	43.9%	48.2%	Destabilizing with Weaker Interaction

Figure 7. Conformational variations (with respect to % of net pure α-helices, % of 3_{10} helices, % of β-sheets and % of coils) in the sole WT and MT protein before interaction with Ebola GP2

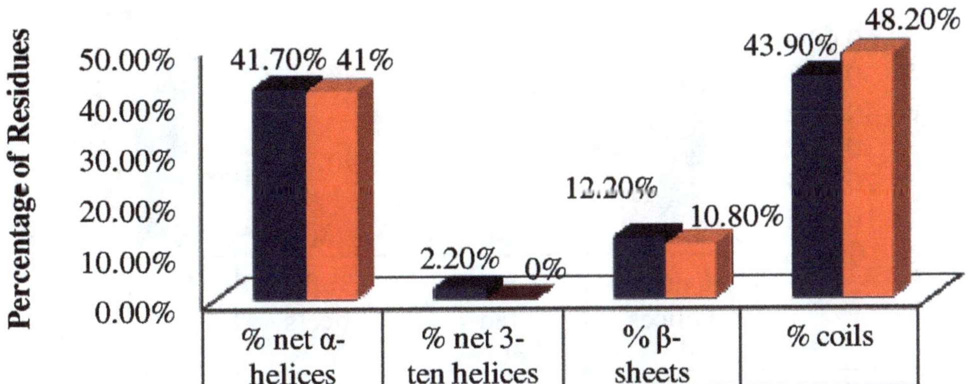

	% net α-helices	% net 3-ten helices	% β-sheets	% coils
WT protein	41.70%	2.20%	12.20%	43.90%
MT protein	41%	0%	10.80%	48.20%

Similar to the case for the sole proteins (before the interaction of Ebola GP2 protein), the WT protein even after interaction with Ebola GP2 complex had more residues forming helices and β-sheets in comparison to the MT protein after its Ebola GP2 interaction (Table 5). Fascinatingly this time, the MT protein also had few residues forming 3_{10} helices accompanied by pure α-helices but it was comparatively lesser than that in the WT protein after Ebola GP2 interaction (Table 5). So, this further deduces that the MT protein even after interaction with Ebola GP2 protein turned out to be feeble in its interaction when compared to the WT protein on interaction with the Ebola GP2,, thereby, preventing the membrane fusion and viral entry of Ebola virus. Figure 8 presents the conformational variations in the WT and MT protein upon interaction with GP2 of Ebola.

Examination of Statistical Significance

A statistical analysis was primarily observed for the change in the energy values for WT and MT proteins and respective protein-GP2 complexes. The requisite for statistical significance is $P< 0.05$. The lesser stability of the single MT modelled protein structure was statistically significant over that of WT with a P value of 0.012 (that is, <0.05) from T-test. The stability of the WT_GP2 complex structure from

Table 5. Categorization of secondary structure for Wild-Type (WT) and Mutant-Type (MT) human TLR4_TIR protein upon interaction with Ebola glycoprotein-2 (GP2)

% Categorization of Secondary Structure	WT	MT	Whether Stabilizing or Destabilizing the MT
% net α-helices	35.3%	34.5%	Destabilizing with Weaker Interaction
% net 3-ten helices	4.3%	2.2%	Destabilizing with Weaker Interaction
% β-sheets	12.9%	10.1%	Destabilizing with Weaker Interaction
% coils	47.5%	53.2%	Destabilizing with Weaker Interaction

Figure 8. Conformational variations (with respect to % of net pure α-helices, % of 3_{10} helices, % of β-sheets and % of coils) in the WT and MT proteins on interaction with Ebola GP2 individually

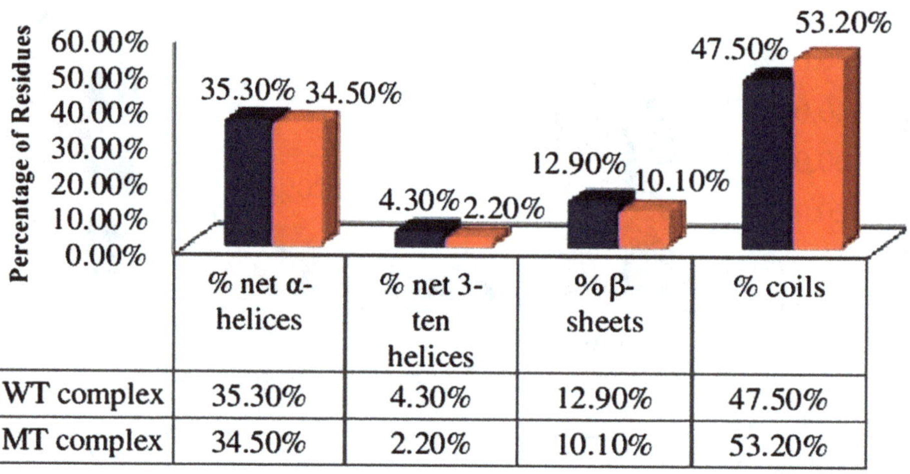

its energy calculations were also statistically significant over the MT_GP2 complex with P=0.01522. The increase in net area for solvent for the complex structure possessing the MT protein were statistically significant over those possessing the WT protein having a P-value of 0.02134. Further statistical evaluations for the individual WT and MT proteins, for % of total helices and % of β-sheets with the assistance of paired T-test, before and after mutational alteration supported the reduction in the % of pure α-helices, 3_{10} helices and % β-sheets, post mutation. In the current study, statistical calculation using T-test revealed a P-value of P=0.01521, P=0.02163 and P=0.01425 for α-helices, 3_{10} helices and β-sheets, respectively. This would imply the MT protein to be feeble and unstable one having weaker folds. Again, the increment in the % of 3_{10} helices accompanied by the pure α-helices as well as the % of β-sheets in WT protein after the Ebola GP2 complex formation over the MT protein after its interaction with the Ebola GP2 protein were statistically significant exhibiting a P-value of 0.01623, 0.02415 and 0.02613, respectively. Therefore, it also verifies significantly that the MT protein before and after interaction with Ebola GP2 protein was weakly interactive with poorer conformation when comparable with the WT protein. This affirms further that the MT protein does not interact firmly with the Ebola virus GP2 protein, thus preventing the virus to enter the human physique.

DISCUSSION

The scenario of the present optimized computational study investigates the participation of human TLR4 (Toll-Like Receptor-4) during the attack of Ebola virus. Ebola virus glycoprotein is known to have a glycoprotein subunit-2 (GP2) that helps in the membrane fusion and viral entry by interacting with the TIR domain of TLR4 protein in humans. After the interaction with the human protein, pro-inflammatory cytokines and suppressor of cytokine signaling 1 is generated. Therefore, it enters the human anatomy and affects the normal functioning of the immune system leading to a lethal impact.

So, for the investigation at the basic molecular level, homology modelling of the TIR domain from human TLR4 was performed and the obtained structure was demonstrated too. Well stated documentations (Ota, Monika & Roman, 2012) help to acquire the fact that point mutation in the TIR domain of TLR4 protein, at a specific proline residue to histidine, leads to a hypo-responsiveness towards the glycoprotein. From the multiple sequence alignment of the TIR domains from the TLR4 protein of related species, the position of the functionally essential proline residue in the TIR domain of the human TLR4 protein was recognized. The mutated protein was further remodelled using the same protocol as that for the wild-type (WT) protein. All the stereo-chemical properties for the individual modelled proteins were satisfied and the functionality of the protein was found to remain conserved even after mutation (through the TM-score value of 0.7). The protein structures were refined and optimized to attain a stable conformation. For a comparable study, the evaluation of the stability calculating parameters disclosed the mutant-type (MT) protein to be more stable in nature with supportive statistical significances. The results from the estimation of free energy of folding value deduced the MT protein to be solely unstable when compared with the sole WT protein. Furthermore, SDM generated energy value predicted an energy value ($\Delta\Delta G$) of -0.45kcal/mol for the WT upon mutation from P39H, stating the WT protein to be more stable. Additionally the increment in the solvent accessibility for only Pro39 and His39 discloses the MT protein to be destabilising one. The mCSM value, that is calculated to examine the stability of the mutated protein on protein-protein interactions, generated a $\Delta\Delta G$ energy value of -0.05kcal/mol and stated the MT protein to exhibit a destabilizing effect having weaker interaction with the partner protein.

Protein-Protein docking studies were thereby performed with the individual modelled proteins to interact with the x-ray crystallography obtained GP2 structure, separately. After the overall energy minimizations followed by Molecular Dynamics Simulation studies for the individual docked complexes, the stability of the complexes were evaluated. Astonishingly, it was perceived from the evaluations with the relevant supportive statistical significances that the MT structure on interacting with the GP2 protein exhibited poorer interaction turning the complex into an unstable one. Fascinatingly, the DFire energy and the net area for solvent accessibility for the complex protein in either case also revealed the decreased stability for the complex having MT protein. DFire energy value again was observed to get changed from -2211.41kcal/mol (in WT-GP2 complex) to -2181.86kcal/mol for MT-GP2 complex (Table 3). Net available solvent area for the interacting GP2 protein was found to be increased on the interaction of the MT complex with GP2 protein than that in the WT-GP2 complex. This also infers that the WT protein binds more efficiently with the GP2 protein allowing the membrane fusion and viral entry. On studying the alterations due to conformational variations, it was discerned that the MT protein turned out to share a poor unstable interaction with the GP2 protein (Table 4-5). According to documentations (Paul, Thomas & Ken, 1993; Toniolo & Benedetti, 1991), proteins with more β-sheets, net α-helices accompanied by few 3_{10} helices are more stable and experience a strengthened interaction than those with few β-sheets and purely net α-helices. The strengthened interaction was thus shown by the WT, on interacting with GP2 protein (Table 4-5).

To study the residual participation from either of the two proteins, it was revealed that the WT_GP2 protein had 6 ionic-ionic interactions to strengthen the complex (Table 1). These interactions reduced to 2 ionic-ionic interactions with one chain involvement from Ebola GP2, when the WT protein was mutated and interacted with GP2 protein. In the presence of WT protein, remarkably, 2 ionic-ionic interactions were accomplished by Arg787 from the WT protein with polar negatively charged Asp from 435th and 438th positions from D chain of the Ebola GP2. Furthermore, polar negatively charged Glu780 and Arg783 from the WT protein with Arg388 and Asp428 from the chain D of Ebola GP2 protein. Two more ionic interactions were accomplished by the F chain of GP2 protein (Asp644 and Arg646) with Arg776 and Glu748 from the WT protein. Documentation documents ionic interactions to offer the complex to be more strongly interactive one (Baldwin, 1996). On mutation, only Arg633 from the F chain of Ebola GP2 formed only two ionic interactions with Glu residues from 746th and 780th position of the MT protein (Table 2). These altogether provides a strong interaction by formation of a cavity to accommodate GP2 and aid in the mediation of the membrane fusion. On mutation, no such aromatic interactions or ionic interactions were observed (Table 2).

Consequently, this present study provides an acquaintance in the interactive examination and residue-level disclosure between TIR domain of human TLR4 and Ebola glycoprotein. This residual computational study for the basis of the interaction involving the life-threatening Ebola virus is one of the most essential zones to be explored into. Prior to this, several molecular level studies (Simanti, Amit, Semanti, Rakhi & Angshuman, 2014; Angshuman, 2015) were documented for other diseases but none dealt with the detailed molecular level signalling mechanism for the entry of Ebola virus by membrane fusion. This *in silico* analysis therefore, indulges the disclosure of the residual participation, binding demonstration and analysis of the wild-type human TIR and Ebola GP2 protein complex. Moreover, it also explores into the multiple sequence alignment for the relevance of the functionally paramount conserved residues. Further, investigation into the residual point mutation in the TIR protein domain of TLR4 in humans and the residual participation of the remodelled mutated protein and the GP2 protein was performed. Several stability calculating parameters and interaction evaluations, aided to infer that after the point mutation, the MT (Mutant-Type) protein alone becomes less steady than the Wild-Type protein. Fascinatingly upon interaction with GP2, it was perceived that the MT protein showed hypo-responsiveness towards the Ebola GP2, making the complex very unstable. It thus, obstructs the viral entry through the membrane. It provides with an avenue for the future therapeutic and pharmaceutical research for avoiding the fatal Ebola entry.

CONCLUSION AND FUTURE SCOPE

In a nutshell, the findings from the current study, demonstrated a central regulatory role of human TLR4 during Ebola virus attack, revealing the paramount residues involved. Furthermore, the effects on point mutation was also studied and analyzed with the involvement of several energy calculating parameters to study the extent of stability (whether stabilize or destabilize) of the complexes due to mutations. Net area of solvent accessibility was one more additionally important parameters that were dealt with, to justify the strength in the interactive complexes. Finally, it shows the way to the disclosure that the Wild-Type (WT) human protein, when mutated (P39H) led to a weaker and les stable

Mutant-Type (MT) protein. This MT further turned out to be hypo-responsive towards the Ebola GP2 protein with a less stable MT_GP2 complex having very fewer unstable interactions in comparison to the WT_GP2 complex. Therefore, this study aims at the immune-pathogenesis of the dreadful Ebola virus infection involving the TLR4 protein in humans. Involvement in the therapeutic zones can also be studied in future. Moreover, future scope remains in investigating the binding of any other protein that might strengthen the interaction pattern and create a protective shield against the Ebola virus.

List of Abbreviations: MT= Mutant Type, WT=Wild Type, SD=Standard Deviation, GP2=Glyco-Protein subunit 2, TLR=Toll-Like Receptor, TIR=Toll/Interleukin-1 Receptor, P.I.C= Protein Interaction Calculator, DS-Discovery Studio and MD=Molecular Dynamics.

Conflicts of Interest: No conflicts of interest.

ACKNOWLEDGMENT

Authors are deeply indebted for the immense help, paramount suggestions and continuous encouragement rendered by Dr. Angshuman Bagchi, Assistant Professor, Department of Biochemistry and Biophysics, University of Kalyani, Kalyani, Nadia, India. Without his support and cooperation, it would not have been possible to complete the manuscript in a successful manner. Deep gratitude is extended to Bioinformatics Resources and Applications Facility (BRAF), C-DAC, Pune for the necessary support.

REFERENCES

Altschul, S. F., Gish, W., Miller, W., Myers, E. W., & Lipman, D. J. (1990). Basic local alignment search tool. *Journal of Molecular Biology*, *25*(3), 403–410. doi:10.1016/S0022-2836(05)80360-2 PMID:2231712

Analysis Tool Web Services from the EMBL-EBI. (2013). *Nucleic Acids Research, 41*, 597–600.

Angshuman, B. (2015). Structural Characterizations of Metal Ion Binding Transcriptional Regulator CueR from Opportunistic Pathogen *Pseudomonasaeruginosa* to Identify Its Possible Involvements in Virulence. *Applied Biochemistry and Biotechnology*, *175*(2), 649–656. doi:10.1007/s12010-014-1304-5 PMID:25342258

Balaji, M., Jizhen, W., Haiqing, J., & Lijun, R. (2005). Comprehensive Analysis of Ebola Virus GP1 in Viral Entry. *Journal of Virology*, *79*(8), 4793–4805. doi:10.1128/JVI.79.8.4793-4805.2005 PMID:15795265

Baldwin, R. L. (1996). How Hofmeister ion interactions affect protein stability. *Biophysical Journal*, *71*(4), 2056–2063. doi:10.1016/S0006-3495(96)79404-3 PMID:8889180

Berman, M. H. et al.. (2000). The protein data bank. *Nucleic Acids Research*, *28*(1), 235–242. doi:10.1093/nar/28.1.235 PMID:10592235

Bhattacharya, S., Das, A., Ghosh, S., Dasgupta, R., & Bagchi, A. (2014). Hypoglycosylation of dystroglycan due to T192M mutation: A molecular insight behind the fact. *Gene*, *537*, 108–114. PMID:24361964

Brooks, B. R., Bruccoleri, R. E., Olafson, B. D., States, D. J., Swaminathan, S., & Karplus, M. (1983). CHARMM: A program for macromolecular energy, minimization, and dynamics calculations. *Journal of Computational Chemistry*, *4*(2), 187–217. doi:10.1002/jcc.540040211

Cartmill, M., & Smith, F. H. (2011). *The Human Lineage*. John Wiley & Sons.

Chen, R., Li, L., & Weng, Z. (2003). ZDOCK: An initial-stage protein docking algorithms. *Proteins*, *51*, 82–87.

Comeau, S. R., Gatchel, D. W., Vajda, S., & Camacho, C. J. (2004). ClusPro: An automated docking and discrimination method for the prediction of protein complexes. *Bioinformatics (Oxford, England)*, *20*(1), 45–50. doi:10.1093/bioinformatics/btg371 PMID:14693807

DeLano, W. L. (2002). *The PyMOL Molecular Graphics System DeLano Scientific*. San Carlos, CA, USA; doi:10.1093/nar/gki408

Douglas, E. V. Pires, David, B. Ascher, & Tom, L. Blundell. (2013). mCSM: Predicting the effects of mutations in proteins using graph-based signatures. *Bioinformatics (Oxford, England)*. doi:10.1093/bioinformatics/btt691

Eisenberg, D., Luthy, R., & Bowie, J. U. (1997). VERIFY3D: Assessment of protein models with three-dimensional profiles. *Methods in Enzymology*, *277*, 396–404. doi:10.1016/S0076-6879(97)77022-8 PMID:9379925

Fiser, A., & Sali, A. (2003). ModLoop: Automated modeling of loops in protein structures. *Bioinformatics (Oxford, England)*, *19*(18), 2500–2501. doi:10.1093/bioinformatics/btg362 PMID:14668246

George, R. A., & Heringa, J. (2002). An analysis of protein domain linkers: Their classification and role in protein folding. *Protein Engineering*, *15*(11), 871–879. doi:10.1093/protein/15.11.871 PMID:12538906

Groves, C. P. (2001). *Primate Taxonomy*. Smithsonian Institution Press.

Hansson, G. K., & Edfeldt, K. (2005). Toll to be paid at the gateway to the vessel wall. *Arteriosclerosis, Thrombosis, and Vascular Biology*, *25*(6), 1085–1087. doi:10.1161/01.ATV.0000168894.43759.47 PMID:15923538

Hooft, R. W., Vriend, G., Sander, C., & Abola, E. E. (1996). Errors in protein structures. *Nature*, *381*(6580), 272. doi:10.1038/381272a0 PMID:8692262

Jeffers, S. A., Sanders, D. A., & Sanchez, A. (2002). Covalent modifications of the Ebola virus glycoprotein. *Journal of Virology*, *76*(24), 12463–12472. doi:10.1128/JVI.76.24.12463-12472.2002 PMID:12438572

Johannes, S¨., Andreas, B., & Andrei, N. L. (2005). The HHpred interactive server for protein homology detection and structure prediction. *Nucleic Acids Research*, *33*(Web Server), W244–W248. doi:10.1093/nar/gki408 PMID:15980461

Jones, S., Stewart, M., Michie, A., Swindells, MB., Orengo, C., & Thornton, J. M., (1998). Domain assignment for protein structures using a consensus approach: characterization and analysis. *Protein Sci*, *7*(2), 233–42. doi:10.1002/pro.5560070202

Kabsch, W., & Sander, C. (1983). Dictionary of protein secondary structure: Pattern recognition of hydrogen-bonded and geometrical features. *Biopolymes*, *22*(12), 2577–2637. doi:10.1002/bip.360221211 PMID:6667333

Kozakov, D., Beglov, D., Bohnuud, T., Mottarella, S., Xia, B., Hall, D. R., & Vajda, S. (2013). How good is automated protein docking? *Proteins: Structure, Function, and Bioinformatics*, *81*(12), 2159–2166. doi:10.1002/prot.24403 PMID:23996272

Krieger, E., Nabuurs, S. B., & Vriend, G. (2003). Homology modeling. In Structural Bioinformatics. Wiley. doi:10.1002/0471721204.ch25

Kuhn, J. H., Becker, S., Ebihara, H., Geisbert, T. W., Johnson, K. M., Kawaoka, Y., … Ana, I. (2010). Proposal for a revised taxonomy of the family Filoviridae: Classification, names of taxa and viruses, and virus abbreviations. *Archives of Virology, 155*(12), 2083–103. doi:10.1007/s00705-010-0814-x

Laskowski, R. A., MacArthur, M. W., Moss, D. S., & Thornton, J. M. (1993). PROCHECK: A program to check the stereochemistry of protein structures. *Journal of Applied Crystallography*, *26*(2), 283–291. doi:10.1107/S0021889892009944

Marchler-Bauer, A., Derbyshire, M. K., Gonzalez, N. R., Lu, S., Chitsaz, F., Geer, L. Y., & Bryant, S. H. et al. (2015). CDD: NCBI's conserved domain database. *Nucleic Acids Research*, *43*(D), 222–226. doi:10.1093/nar/gku1221 PMID:25414356

Marti-Renom, M. A., Stuart, A. C., Fiser, A., Sanchez, R., Melo, F., & Sali, A. (2000). Comparative protein structure modeling of genes and genomes. *Annual Review of Biophysics and Biomolecular Structure*, *29*(1), 291–325. doi:10.1146/annurev.biophys.29.1.291 PMID:10940251

McWilliam, H., Li, W., Uludag, M., Squizzato, S., Park, Y.M., Buso, N., Cowley, A.P., & Lopez, R. (2013). Analysis Tool Web Services from the EMBL-EBI. *Nucleic Acids Research, 41*, 597-600. DOI: 10.1093/nar/gkt376

Okumura, A., Pitha, P. M., Yoshimura, A., & Harty, R. N. (2010). Interaction between Ebola virus glycoprotein and host toll-like receptor 4 leads to induction of proinflammatory cytokines and SOCS1. *Journal of Virology*, *1984*(1), 27–33. doi:10.1128/JVI.01462-09 PMID:19846529

Ota, F., Monika, A., & Roman, J. (2012). Suppression of TLR Signaling by Targeting TIR domain-Containing Proteins. *Current Protein & Peptide Science*, *13*(8), 776–788. doi:10.2174/138920312804871148 PMID:23305364

Paul, D. (1993). Local and nonlocal interactions in globular proteins and mechanisms of alcohol denaturation. *Protein Science*, *2*(12), 2050–2065. doi:10.1002/pro.5560021206 PMID:8298455

Punta, M., Coggill, P. C., Eberhardt, R. Y., Mistry, J., Tate, J., Boursnell, C., Pang, N., & Forslun (2011). The Pfam protein families database. *Nucleic Acids Research, 40*(D1), D290–D301.. PMC 3245129. PMID 22127870.10.1093/nar/gkr1065

Ramachandran, G. N., & Sashisekharan, V. (1968). Conformation of polypeptides and proteins. *Advances in Protein Chemistry*, *23*, 283–438. doi:10.1016/S0065-3233(08)60402-7 PMID:4882249

Ramachandran, S., Kota, P., Ding, F., & Dokholyan, N. V. (2011). PROTEINS: Structure. *Function and Bioinformatics*, *79*, 261–270. doi:10.1002/prot.22879

Re, F., & Strominger, J. L. (2002). Monomeric recombinant MD-2 binds toll-like receptor 4 tightly and confers lipopolysaccharide responsiveness. *The Journal of Biological Chemistry*, *277*(26), 23427–23432. doi:10.1074/jbc.M202554200 PMID:11976338

Richard, D. E., Leo, G., Eitan, E. W., & Chris, P. P. (2003). Comparison of the genomes of human and mouse lays the foundation of genome zoology. *Human Molecular Genetics*, *12*(7), 701–709. doi:10.1093/hmg/ddg078 PMID:12651866

Rock, F. L., Hardiman, G, Timans, J. C., Kastelein, R. A., Bazan, J. F. (1998). A family of human receptors structurally related to Drosophila Toll. *Proc Natl Acad Sci U S A., 95*(2), 588–93. doi:10.1073/pnas.95.2.588

Sander, C., & Schneider, R. (1991). Database of homology-derived protein structures and the structural meaning of sequence alignment. *Proteins*, *9*(1), 56–68. doi:10.1002/prot.340090107 PMID:2017436

Simmosis, V. A., & Heringa, J. (2005). PRALINE: A multiple sequence alignment toolbox that integrates homology-extended and secondary structure information. *Nucleic Acids Research*, *33*(Web Server), W289–W294. doi:10.1093/nar/gki390 PMID:15980472

Takada, A., Robison, C., Goto, H., Sanchez, A., Murti, K. G., Whitt, M. A., & Kawaoka, Y. (1997). A system for functional analysis of Ebola virus glycoprotein. *Proceedings of the National Academy of Sciences of the United States of America*, *94*(26), 14764–14769. doi:10.1073/pnas.94.26.14764 PMID:9405687

Tartaglia, G. G., Pawar, A. P., Campioni, S., Dobson, C. M., Chiti, F., & Vendruscolo, M. (2008). Prediction of aggregation-prone regions in structured proteins. *Journal of Molecular Biology*, *380*(2), 425–436. doi:10.1016/j.jmb.2008.05.013 PMID:18514226

Thompson, J. D., Higgins, D. G., & Gibson, T. J. (1994). CLUSTAL W: Improving the sensitivity of progressive multiple sequence alignment through sequence weighting, position-specific gap penalties and weight matrix choice. *Nucleic Acids Research*, *22*(22), 4673–4680. doi:10.1093/nar/22.22.4673 PMID:7984417

Tina, K. G., Bhadra, R., & Srinivasan, N. (2007). PIC: Protein Interactions Calculator. *Nucleic Acids Research*, *35*, W473–W476. doi:10.1093/nar/gkm423 PMID:17584791

Toniolo, C., & Benedetti, E. (1991). Article. *Trends in Biochemical Sciences*, *16*, 350–353. doi:10.1016/0968-0004(91)90142-I PMID:1949158

Topham, C. M., Srinivasan, N., & Blundell, T. L. (1997). Prediction of protein mutants based on structural environment-dependent amino acid substitution and propensity tables. *Protein Engineering*, *10*(1), 7–21. doi:10.1093/protein/10.1.7 PMID:9051729

Vakser, I. A. (1995). Protein docking for low-resolution structures. *Protein Engineering*, *8*(4), 371–377. doi:10.1093/protein/8.4.371 PMID:7567922

Volchkov, V. E., Feldmann, H., Volchkova, V. A., & Klenk, H. D. (1998). Processing of the Ebola virus glycoprotein by the proprotein convertase furin. *Proceedings of the National Academy of Sciences of the United States of America*, *95*(10), 5762–5767. doi:10.1073/pnas.95.10.5762 PMID:9576958

Weiderstein, & Sippl. (2007). ProSA-web: Interactive web service for the recognition of errors in three-dimensional structures of proteins. *Nucleic Acid Residues, 35*, W407-W410.

Weisssenhorn, W., Carfi, A., Lee, K. H., Skehel, J. J., & Wiley, D. C. (1998). Crystal Structure of the Ebola virus membrane fusion subunit, GP2, from the envelope glycoprotein ectodomain. *Molecular Cell*, *2*(5), 605–616. doi:10.1016/S1097-2765(00)80159-8 PMID:9844633

Wiederstein, M. (2004). *Evolutionary methods in biotechnology*. Weinheim: Wiley-VCH.

Wool-Lewis, R. J., & Bates, P. (1998). Characterization of Ebola virus entry by using pseudotyped viruses: Identification of receptor-deficient cell lines. *Journal of Virology*, *72*(4), 3155–3160. PMID:9525641

Xiang, Z. (2006). Advances in Homology Protein Structure Modeling. *Current Protein & Peptide Science*, *7*(3), 217–227. doi:10.2174/138920306777452312 PMID:16787261

Xu, D., & Zhang, Y. (2001). Improving the Physical Realism and Structural Accuracy of Protein Models by a Two-step Atomic-level Energy Minimization. *Biophysical Journal*, *101*(10), 2525–2534. doi:10.1016/j.bpj.2011.10.024 PMID:22098752

Yamamoto, M, Sato, S, Hemmi, H, Uematsu, S., Hoshino, K., Kaisho, T., Takeuchi, O., Takeda, K. & Akira, S. (2003). TRAM is specifically involved in the Toll-like receptor 4-mediated MyD88-independent signaling pathway. *Nat. Immunol.*, *4*(11), 1144–50. doi: 14556004.10.1038/ni986.PMID

Yong-Chen, L., Wen-Chen, Y., & Pamela, S. O. (2008). LPS/TLR4 signal transduction pathway. *Cytokine*, *42*(2), 145–151. doi:10.1016/j.cyto.2008.01.006 PMID:18304834

Yuedong, Y., & Yaoqi, Z. (2008). Specific interactions for ab initio folding of protein terminal regions with secondary structures. *Proteins*, *72*(2), 793–803. doi:10.1002/prot.21968 PMID:18260109

Zhang, Y., & Skolnick, J. (2005). TM-Align: A protein structure alignment algorithm based on TM-score. *Nucleic Acids Research*, *33*(7), 2302–2309. doi:10.1093/nar/gki524 PMID:15849316

Chapter 9
Molecular-Docking-Based Anti-Allergic Drug Design

Anamika Basu
Gurudas College, India

Piyali Basak
Jadavpur University, India

Anasua Sarkar
Jadavpur University, India

ABSTRACT

Allergens are foreign proteins that when come in contact of part(s) of human body stimulate the production of immunoglobulin types of proteins (antibodies). These allergens react with antibodies (immunoglobulin type E or IgE) produces allergic reactions, also known as immediate-type hypersensitivity reactions. As much as 20% of the general population may be affected by grass pollen as a major cause of allergic disease. EXPB class of proteins are known in the immunological literature as group-1 grass pollen allergens Molecular docking method can be used to identify the predicated the interaction of pollen allergen EXPB1 (Zea m 1), a beta-expansin and group-1 pollen allergen from maize with IgE molecules of human. The World Health Organization recognised allergen immunotherapy, as therapeutics for allergic diseases. RNA Interference (RNAi) is a biological process in which RNA molecules e.g. Small Interfering RNAs (siRNAs) inhibit gene expression, by cleavage and destruction of specific mRNA molecules. Use of Small Interfering RNA (siRNA) is a novel method in the induction of RNA Interference (RNAi), which is a potent method for therapeutics of allergic reactions. Due to various effects of STAT 6 proteins during hypersensitivity reactions caused by pollen allergens, mRNA of STAT6 gene is selected as target gene for allergy therapeutics via Post-Transcriptional Gene Silencing (PTGS). Using molecular docking study a specific sense siRNA is identified as anti allergic drug to treat allergic asthma during immediate type of hypersensitivity reaction, caused by Zea m 1 pollen allergen.

DOI: 10.4018/978-1-5225-0362-0.ch009

INTRODUCTION

Allergens are foreign proteins that when come in contact of part(s) of human body stimulate the production of immunoglobulin types of proteins (antibodies). These allergens react with antibodies (immunoglobulin type E or IgE) produces allergic reactions, also known as immediate-type hypersensitivity reactions (Kay, 2008). As much as 20% of the general population may be affected by grass pollen as a major cause of allergic disease. An evolutionary conserved protein in plant developmental biology, expansins play an important role in cell wall expansion by slippage or rearrangement of matrix polymers during plant cell growth (Basu, 2013). But among the four classes of expansin protein family, EXPB class of proteins are known in the immunological literature as group-1 grass pollen allergens (Anderson, 2003). These EXPBs proteins cause hay fever and seasonal asthma in humans (Ball, 2005). Allergy symptoms are caused by these proteins when these allergens come into contact with the moist surface of the human respiratory tract (Knox, 1996). Molecular docking method can be used to identify the predicated the interaction of pollen allergen EXPB1 (Zea m 1), a beta-expansin and group-1 pollen allergen from maize with IgE molecules of human. Antibodies, expressed on the surface of B –cell of human immune system, recognize antigenic determinants, also called epitopes on their antigen. The interacting part of the antibody involved in the antigen-antibody interaction, is called the paratope (Stave, 2013). Paratope is formed in combination of different amino acids in the complementarity Determining Regions (CDRs) of antibody or immunoglobulin (Collis, 2003).

At present drugs such as antihistamines, leukotriene receptor antagonists, and corticosteroids are used for symptomatic treatment, but they do not prevent the allergic response (Natt, 2011). The World Health Organization recognised allergen immunotherapy, as therapeutic vaccines for allergic diseases, also known as desensitisation or hyposensitisation. The use of chemically altered allergens, allergoids, recombinant allergens, and relevant T-cell epitope peptides are some common approaches for immunotherapy.RNA interference (RNAi) is a biological process in which RNA molecules e.g. small interfering RNAs (siRNAs) and microRNAs (miRNAs) inhibit gene expression, by cleavage and destruction of specific mRNA molecules. Small interfering RNAs (siRNAs) are 21-25 nucleotide single-stranded RNAs containing 2 nucleotide 3′ overhangs produced by cleavage of longer double-stranded RNAs by the enzyme Dicer. After cleavage by Dicer the 21-25 nucleotide double-stranded product is loaded into an Argonuate protein (humans contain 4 Argonautes) and rendered single-stranded.

Small interfering RNAs (siRNAs) can regulate eukaryotic gene expression when transcribed endogenously or preformed, synthetic siRNA introduced into cells. Among two strands of siRNA, the guide strand pairs with a complementary sequence in a messenger RNA molecule and induces cleavage of mRNA molecule with the help of catalytic component of the RNA-induced silencing complex (RISC) complex. The siRNA component guides RISC to mRNA molecules containing a homologous antisense sequence, resulting in cleavage and degradation of that mRNA. RISC is composed of dicer protein, the double-stranded RNA binding protein TAR RNA Binding Protein (TRBP), and Argonaute2 (AGO2) and does not required ATP for its activity. Dicer protein converts long double-stranded RNAs into siRNAs by the endonuclease activity and Argonaute protein performs the cleavage of the mRNA. At first the process of RNA interference is initiated through Dicer protein by converting double-stranded RNA into small interfering RNA (siRNA). Then the siRNA guide strand after binding with the Argonaute protein in RISC recognizes its complementary sequence in mRNA and targets mRNA for its cleavage.

RNA interference or RNA silencing occurs due to endonucleolytic cleavage of the mRNA specifically by AGO2 protein, when RISC complex binds with a perfectly complementary mRNA (Kandeel, 2013).

The 'minimal RISC' complex composed of AGO2 bound to a short interfering RNA (siRNA). Among two domains of AGO2, The PIWI domain may perform RNA cleavage by a mechanism similar to that of RNase H and the PAZ domain is an RNA-binding module (Song, 2003). It has been already established that PAZ domain of AGO2 protein in RISC anchors the 2-nucleotide 3'-overhang of the siRNA duplex causing two types of effects (Lingel, 2004). The first effect is the binding the 7-nucleotide phosphodiester backbone of the overhang-containing strand and the second one is the capping the 5'-terminal residue of the complementary strand in a sequence-independent manner. In RISC, PAZ acts as an anchoring site for the 3'-end of guide RNA within silencing effector complexes. Similarly the PAZ domain of AGO2 serves as a siRNA-end-binding module for siRNA transfer in the RNA interference method.

Use of small interfering RNA (siRNA) is a novel method in the induction of RNA interference (RNAi), which is a potent method for therapeutics of allergic reactions (Bantounas, 2004). Effect on human immune system can be modulated by siRNAs in immunotherapy of pollen allergen Zea m1. Structures of Zea m1 allergen (PDB ID 2HCZ), human immunoglobulin E (PDB ID 4J4P) and human Argonaute2 protein (PDB ID 4OLA) have been extracted from RCSB PDB web server. Protein- protein molecular docking between pollen allergen Zea m 1 and human immunoglobulin E was done using PatchDock protein- protein docking server. In our present work we propose a strategy for vaccine preparation for Plant pollen allergen Zea m 1 by RNA Interference by using In Silico approaches such as molecular docking and designing siRNA. Analyzing the docking results for pollen allergen Zea m1 with human immunoglobulin E protein, we can identify the paratopes present in that specific antibody, which is essential for preparation of vaccine for that pollen allergen. Findings from the molecular docking study of AGO2 protein with siRNAs would help us to guide the future design of modified siRNA analogues as anti allergic drug in immunotherapy of pollen allergen Zea m 1.

BACKGROUND

Zea m 1 as Pollen Allergen

Allergens from various sources (e.g. pollen, house dust mites, pets, moulds, food and insects) and are mainly proteins or glycoproteins (Valenta, 2008). When allergen comes in contact to human body the production of allergen-specific IgE antibodies increases and that bind to their receptors on immune cells .That binding causes allergic inflammatory response. Cross-linking of IgE antibodies that are bound to the high-affinity receptors for IgE (i.e. FcεRI) on mast cells and basophils causes degranulation within a few minutes .The release of inflammatory mediators, proteases and pro-inflammatory cytokines after degranulation is known as immediate hypersensitivity reaction and that is the cause for the majority of allergic symptoms.

Inhaled pollen allergens encounter antigen presenting cells (APC) that are present in the lining of respiratory tract. After recognition as antigen allergenic proteins are presented by APC, leads to cell differentiation of naive T cells into TH2 cells. Activated TH2 stimulate the formation of IgE by B cells. IgE molecules bind to IgE receptors located on mast cells. The cross linking of mast-cell-bound IgE by allergens leads to the release of biologically active mediators (histamine, leukotrienes) by means of degranulation and, so, to the immediate symptoms of allergy as shown in Figure 1. Mast cells also release chemotactic factors that contribute to the recruitment of inflammatory cells, particularly eosinophils, whose proliferation and differentiation from bone marrow progenitors is promoted by IL-5.There are

Figure 1. Immediate-type hypersensitivity reactions. A. Recognition of specific epitope of allergen by paratope of Ig E on B cell, B. Complementary binding of epitope and paratope between allergen and Ig E, C. Histamine release after Ig E binding with the receptor on mast cell, causes allergic symptoms.

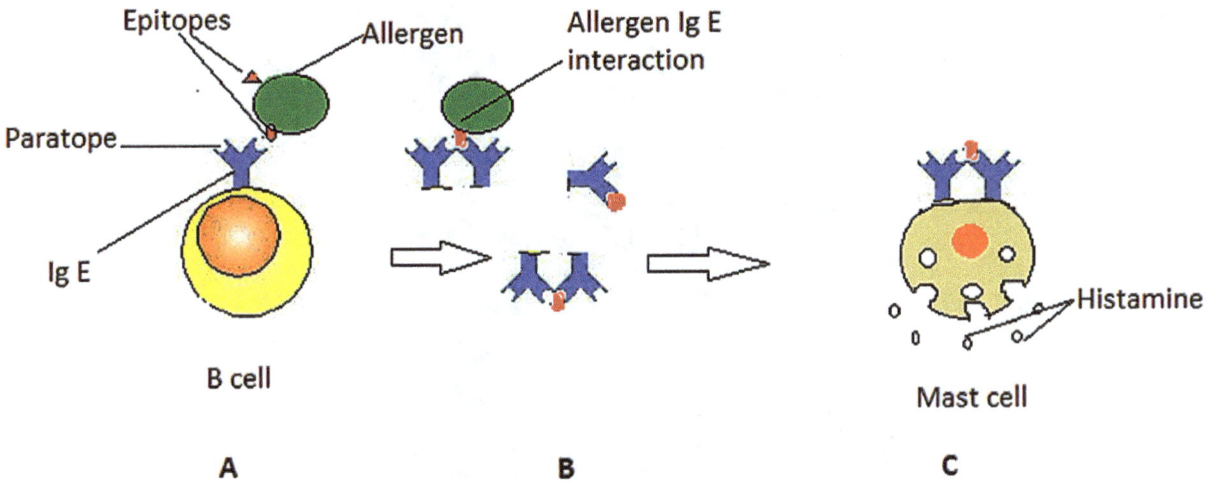

several other protein factors are involved in airflow obstruction, bronchial hyperresponsiveness, and airway inflammation .These are some important symptoms of immediate type of hypersensitivity reaction, caused by pollen allergens.

Molecular docking study with Zea m1 allergen (PDB ID 2HCZ) and human immunoglobulin E (PDB ID 4J4P) using PatchDock web server (Duhovny 2002; Schneidman-Duhovny, 2005) with clustering RMSD 4.0 and score 10504 is shown in Figure 3. PatchDock is an algorithm for molecular docking. The algorithm has three major stages:

Figure 2. KEGG pathway for allergic response due to allergen in human
© Kanahisa Laboratories

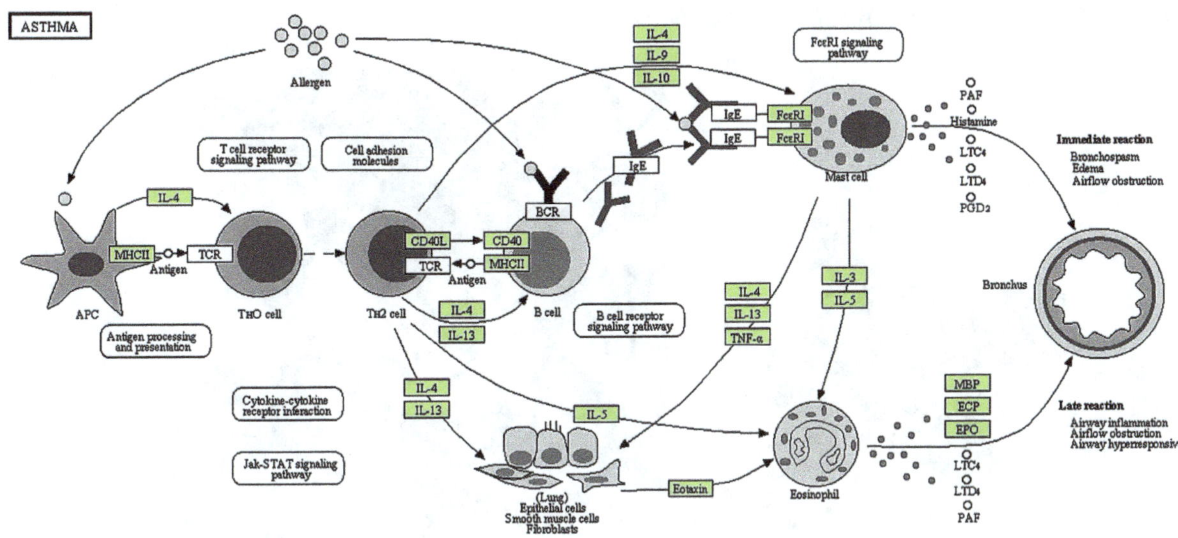

Figure 3. Molecular docking structure of Zea m1 with human IgE

- Molecular Shape Representation
- Surface Patch Matching
- Filtering and Scoring

After refinement of docking results with FireDock web server (Andrusier, 2007; Mashiach, 2008) the following figure is obtained as shown in figure 4. The FireDock web server solves the refinement problem of protein-protein docking solutions. The method simultaneously targets the problem of flex-

Figure 4. Molecular docking result from FireDock

ibility and scoring of solutions produced by fast rigid-body docking algorithms. Here the candidate solutions for FireDock are generated by rigid-body docking method, such as PatchDock as shown in table 2. Each result from PatchDock is subsequently refined by restricted interface side-chain rearrangement and by soft rigid-body optimization. The side-chain flexibility is modelled by rotamers and the obtained combinatorial optimization problem is solved by integer linear programming. Following rearrangement of the side-chains, the relative position of the docking partners is refined by Monte Carlo minimization of the binding score function. The refined candidates are ranked by the binding score. The binding free energy is the change in the free energy of the system, which occurs upon complex formation, which combines the free energy of the receptor-ligand complex and the free energies of the uncomplexed receptor and ligand, respectively. The desolvation free energy is estimated by the ACE potential. The electrostatic contribution is approximately calculated by the atomic pair-wise partial Coulomb energy. The torsion energy, contributions of the hydrogen bonds, disulfide bonds, van der Waals interactions and Π-Stacking and aliphatic interactions are also considered to calculate binding score (Andrusier,2007). Interacting amino acid Thr 12 of Zea m1 with heavy and light chains of human immunoglobulin E are shown in Figure 4. Molecular docking result proves that Zea m 1 allergen binds with human IgE and causes immediate-type hypersensitivity reaction and allergic symptoms.

Where,

- **Global Energy:** The binding energy of the solution.
- **Attractive and Repulsive VdW:** The contribution of the van der Waals forces to the global binding energy.
- **ACE:** The contribution of the atomic contact energy (ACE) to the global binding energy. The ACE (Atomic Contact Energy) function is a statistical contact potential for all (non-hydrogen) atom pairs less than 6 Å apart,
- **HB:** The contribution of the hydrogen bonds to the global binding energy.

Antiallergic Drug Design for Pollen Allergen Zea m 1

Compounds, Histamine, Prostaglandins and Leukotrienes are involved in allergic reactions in human as shown in Figure 2. Nowadays drugs such as antihistamines, leukotriene receptor antagonists, and corticosteroids are used for symptomatic treatment, but they do not prevent the allergic response. In order to eradicate allergic reaction due to any allergen, allergen specific immunotherapy can be applied. Allergen-specific immunotherapy is based on the repeated lower dose administration of disease-causing allergens to modify the allergen-specific immune response in patients so that higher doses of the allergen can be tolerated during hyper sensitivity reaction. (Valenta, 2012) reviewed several approaches for allergen specific immunotherapy using T-cell epitope-containing peptides, Recombinant hypoallergens and chemically modified allergens. In first method allergen-derived synthetic peptides containing T-cell

Table 1. Result from FireDock

Rank	Solution Number		Attractive VdW	Repulsive VdW	ACE	HB
1	5	-21.47	-28.30	15.74	10.06	-2.56

epitopes are used to treat sensitivity reactions. Because these peptides which are synthesized from linear sequences of small allergen fragments, that are epitopes of T cells. These peptides can bind with the receptor of allergen-specific T cells and show no reactivity with IgE antibodies of B cells. The T-cell epitope-containing peptides can be used to reduce the IgE-mediated allergic reactions for that allergen. In second method recombinant hypoallergenic allergens, developed in Escherichia coli, by introducing mutations into the allergen sequence, are used to reduce IgE reactivity. Combination of allergens with DNA sequences containing CpG motifs as immunomodulatory components are used to activate the innate immune system through toll-like receptors (TLRs).This method is applied to inhibit Th2 immune responses and finally later stages of hypersensitivity reaction.

RNA interference (RNAi) technologies (Ball, 2004) are very new approach in the development of allergy therapeutics via post-transcriptional gene silencing (PTGS). From one mRNA molecule about 5,000 copies of that protein can be produced. So, it is a wise decision to target mRNA molecules related to allergic reaction rather than blocking different proteins involved in hypersensitivity reactions. (Popescu, 2005) identified different types of protein molecules as target for potential anti-mRNA drugs in asthma and other allergic disorders. They are shown in Table 2.

Miller et al (1998) suggested that the regulation of class switching to IgE by cytokines is mediated by STAT transcription factors, by analyzing the cytokine signalling in patients with extrinsic asthma and hyper immunoglobulin E. IgE production in asthma and hyper immunoglobulin E usually is associated with elevated levels of IL-4in the peripheral blood. The protein encoded by STAT6 is a member of the STAT family of transcription factors. In response to cytokines and growth factors, STAT family members are phosphorylated by the receptor associated kinases, and then form homo- or heterodimers that translocate to the cell nucleus where they act as transcription activators. This protein plays a central role in exerting IL4 mediated biological responses. Walford and Doherty (2013) elaborately discussed about the various functions of STAT6 protein, which play crucial roles in the pathogenesis of allergen-induced airway inflammation, mucus production, and airway hyper responsiveness (AHR). In asthma patients different structural cells (airway epithelium, smooth muscle, and fibroblasts) and immune cells (B and T cells, macrophages, dendritic cells, and innate lymphoid cells) are involved in many pathological conditions e.g. Th2 cell differentiation and production of cytokines (IL-4, IL-5, IL-9, and IL-13),

Table 2. Different types of protein molecules as target for potential anti-mRNA drugs in asthma and other allergic disorders

Serial Number	Type of Protein Molecule	Examples
1.	Cell surface receptors	adenosine A1 receptor, high-affinity receptor Fc-ε RI-α, cytokine receptors
2.	Adhesion molecules and ligands	ICAM-1, VLA-4
3.	Ion channels	calcium-dependent chloride channel-1
4.	Cytokines and related factors	IL-4, IL-5, IL-13, SCF, TNF-α, TGF-β1
6.	Intracellular signal transduction molecules	tyrosine-protein kinases (Syk, Lyn, Btk),
7.	Serine/threonine-protein kinases	p38 α MAP kinase, Raf-1
8.	Non-kinase signaling proteins	RasGRP4
9.	Transcription factors involved in Th2 differentiation and allergic inflammation	STAT-6, GATA-3, NF-κB

IgE production from B cells, airway eosinophilia, epithelial mucus production, change in contractility of smooth muscle. Effects of STAT6 protein on different types of cell are summarized in Figure 5.

Due to various effects of STAT 6 proteins during hypersensitivity reactions caused by pollen allergens, mRNA of STAT6 gene is selected as target gene for allergy therapeutics via post-transcriptional gene silencing (PTGS). For drug designing using siRNA for STAT6 mRNA, following steps are performed:

- Search for siRNA for signal transducer and activator of transcription 6, interleukin-4 induced (STAT6) for human.
- Comparative modelling of siRNA 3D structures.
- Secondary structure determination of siRNAs.

Search result for siRNA for signal transducer and activator of transcription 6, interleukin-4 induced (STAT6) for human in NCBI, shows six siRNA sequences as shown in Table 3.

Using ModeRNA (Rother, 2011) a program for comparative modelling of RNA 3D structures, three dimensional structure for all six siRNAs can be elucidated. But this program requires a pairwise sequence alignment and a structural template to generate a 3D structural model of the target RNA sequence. But template finding results for those six siRNAs show that only fifth and sixth sequences have structural templates with PDB Ids (shown in Table 4) and their models were built using ModeRNA server (Rother, 2011).

Secondary structure determination of Probe|16741853| sense and antisense siRNAs was carried out using RNAstructure, Version 5.7 (Mathews, 2014) which is a complete package for RNA secondary structure prediction and analysis including facility to predict base pairing probabilities. Results from

Figure 5. Role of STAT6 protein on various types of cells tissues during allergic reaction

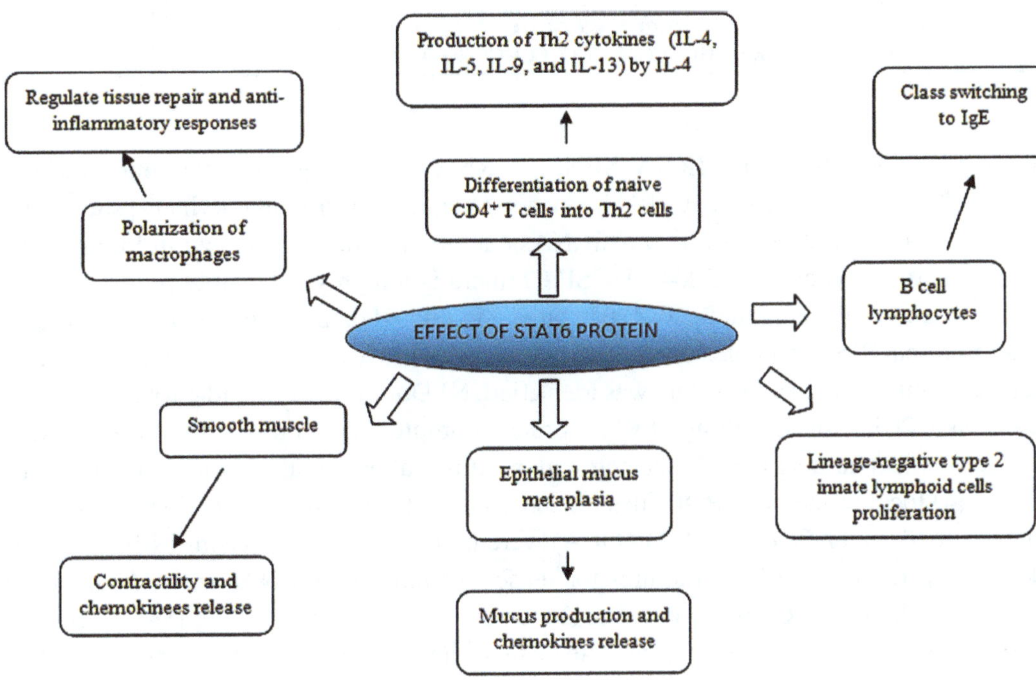

Table 3. siRNA sequences for STAT6 genes in human from NCBI

NAME OF siRNA PROBES	SEQUENCE OF siRNA	POSITION OF siRNA FEATURE OF SEQUENCE
>Probe|16741855|SIRNA_SENSE siRNA sense sequence (21b)	AGACCUGUCCAUUCGCUCATT	1..19 sense sequence [rna] 20..21 sirna overhang [dna]
>Probe|16741855|SIRNA_ ANTISENSE siRNA antisense sequence (21b)	UGAGCGAAUGGACAGGUCUTT	1..19 antisense sequence [rna] 20..21 sirna overhang [dna]
>Probe|16741854|SIRNA_SENSE siRNA sense sequence (21b)	GGUGCCUUCUUAUGACCUUTT	1..19 sense sequence [rna] 20..21 sirna overhang [dna]
>Probe|16741854|SIRNA_ ANTISENSE siRNA antisense sequence (21b)	AAGGUCAUAAGAAGGCACCAT	1..19 antisense sequence [rna] 20..21 sirna overhang [dna]
>Probe|16741853|SIRNA_SENSE siRNA sense sequence (21b)	AGAACCUGCUUCUCAAGAATT	1..19 sense sequence [rna] 20..21 sirna overhang [dna]
>Probe|16741853|SIRNA_ ANTISENSE siRNA antisense sequence (21b)	UUCUUGAGAAGCAGGUUCUTG	1..19 antisense sequence [rna] 20..21 sirna overhang [dna]

Table 4. Template finding results for Probe\|16741853\| sense and antisense siRNAs respectively

Rfam Family	PDB ID	Alignment	Identity	Similarity
RF00005	1UOB_A	>Target - - - - - - - - - - - A - - - - - - - - - G – A - - - - - - - - - - - - - - - - - - A C - - - - - - - CU GCUUC - - - UCAA - - - - GAA - - - - - - - - - - - - - - > 1 U0B A GGCGCGUUAACAAAGCGGUUAUGUAGCGGAUUGCAAAUC CGUCUAGUCCGGUUCGACUCCGGAACGCGCCUCCA	0.162	0.189
RF00005	2DXI_D	>Target - - - - - - - - - - - UUCUUG - - - - - - -AGAAG - - - - - - - - - - - -C- - - - - - - - - AGG- - - - - - - - -UUC- - -U- >2DXI D GGCCCCAUCGUCUAGCGGUUAGGACGCGGCCCUCUCAA GGCCGAAACGGGGGUUCGAUUCCCCCUGGGGUCACCA	0.16	0.18

RNAstructure for Probe|16741853|SIRNA_SENSE siRNA sense sequence (without overhang) and Probe|16741853|SIRNA_ANTISENSE siRNA antisense sequence (without overhang) shown in Figure 6. Predicted 3D secondary structures of two siRNAs can be visualized as shown in Figure 7 by using UCSF Chimera software (Petersen, 2004). UCSF Chimera is a highly extensible program for interactive visualization and analysis of molecular structures and related data e.g. molecular docking results.

Since the 'minimal RISC' complex composed of AGO2 bound to a short interfering RNA (siRNA), a docking server for nucleic acid and protein was identified. NPDock (Nucleic acid-Protein Dock) is a web server (Tuszyńska, 2015) for modelling of RNA-protein complex structures. It combines (1) GRAMM for global macromolecular docking, (2) scoring with a statistical potential, (3) clustering of best-scored structures, and (4) local refinement. PDB files for sense and antisense siRNA obtained from RNAstructure (Mathews, 2014) and AGO2 (PDB ID 4OLA) were used as input files. Using NPDock web server (Tuszyńska, 2015) 'minimal RISC' structures for sense and antisense siRNAs with AGO2 proteins are built and visualized by UCSF Chimera software (Petersen, 2004) as shown in Figure 8 where sense and antisense siRNA were shown in green colour. The distance and orientation-dependent knowledge-based

Figure 6. Predicted secondary structures of two siRNAs

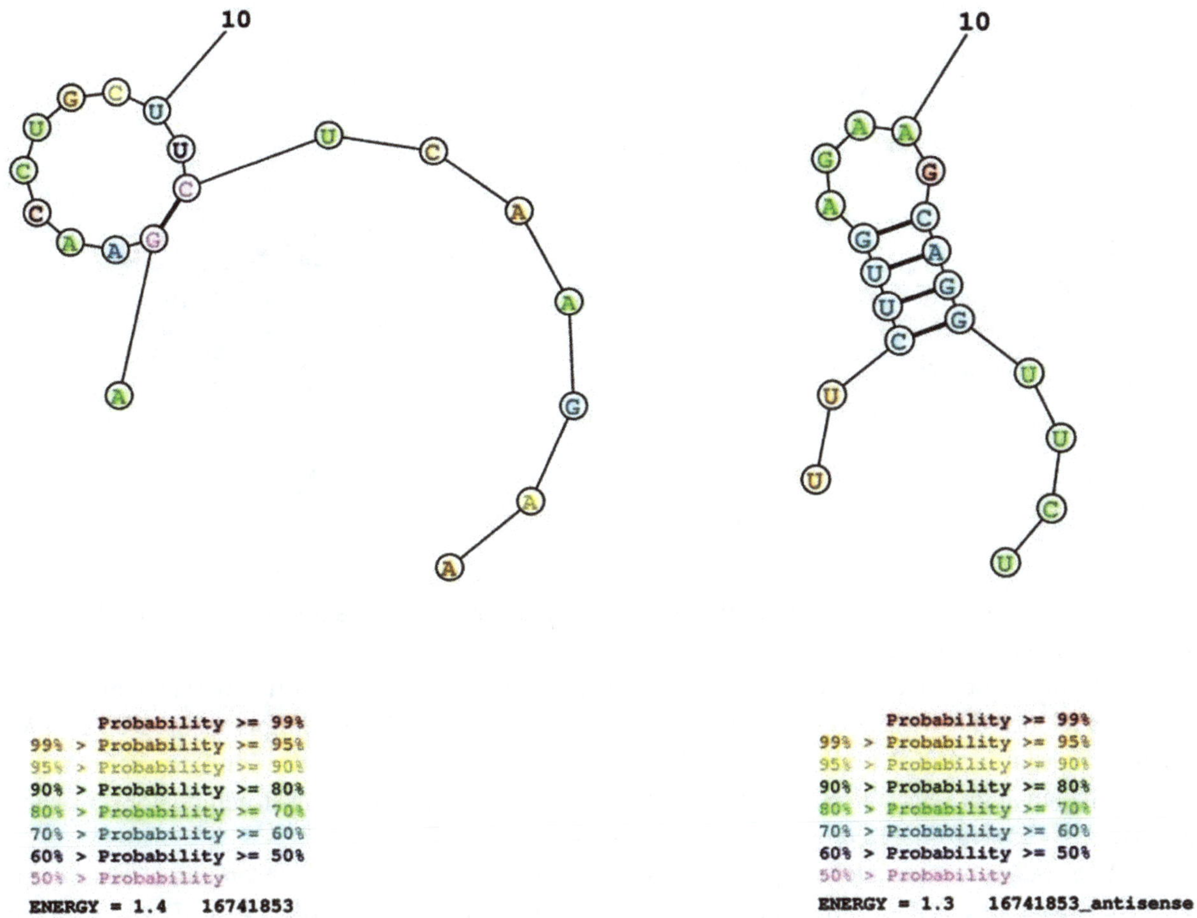

Figure 7. 3D structures of sense and antisense siRNAs

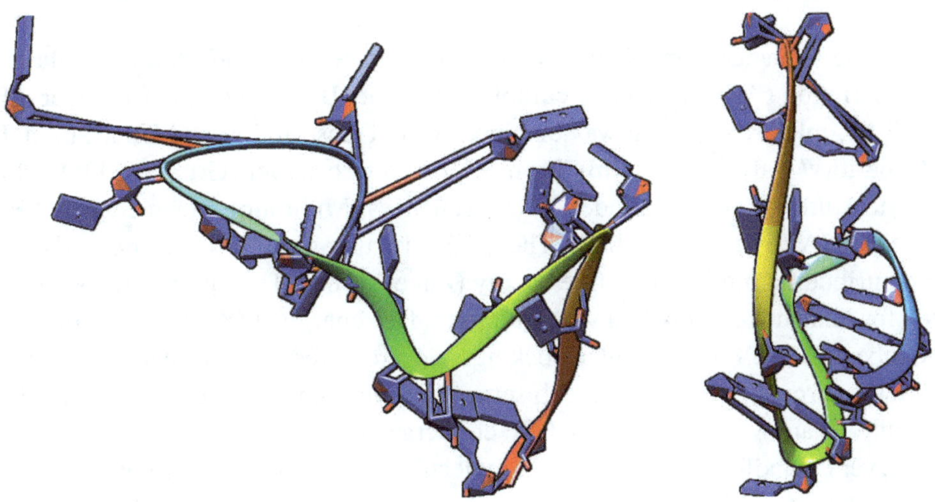

Figure 8. Binding structures of siRNAs with AGO2. a. For sense strand b. For antisense strand.

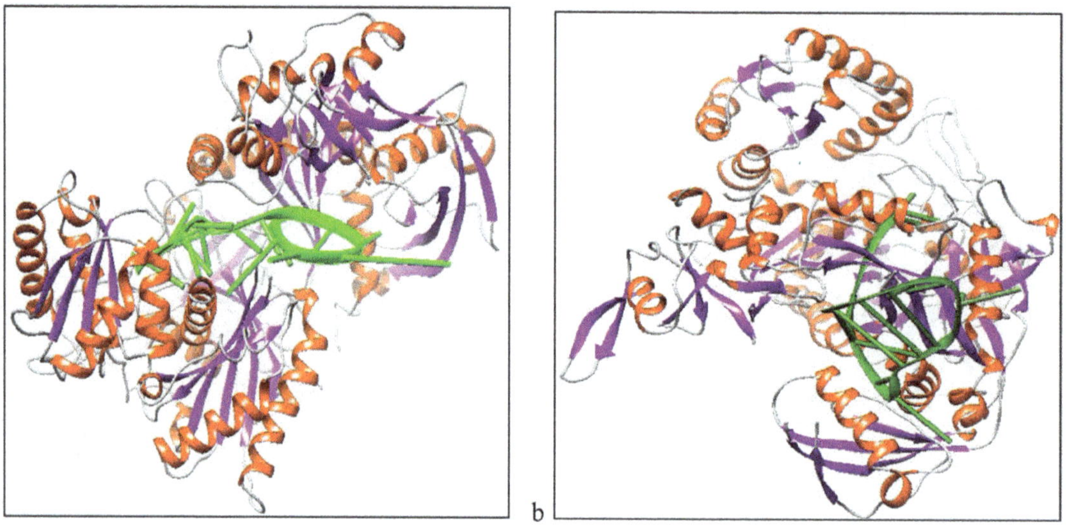

a b

potentials (DARS-RNP and QUASI-RNP) were generated for both sense and antisense RNAs when bound with AGO2 protein using reverse Boltzmann statistics.

From the best scored docking results, after visualizing with UCSF chimera software (Petersen, 2004), it can be concluded that both sense and antisense siRNA strands bind with PIWI domain of AGO2. PIWI domain is present in AGO2 protein from 517-818 amino acids sequence. Since binding energy score was more negative for sense siRNA, so it has been selected among two binding structures for sense and antisense siRNAs with AGO2 protein.

In RISC, PAZ acts as an anchoring site for the 3'-end of guide RNA within silencing effecter complexes. For sense siRNA the 'minimal RISC' complex composed of AGO2 bound to a siRNA as shown in Figure 9 where PAZ domain of AGO2 protein shown in red and sense siRNA in green.

STAT6 Gene as Target Gene for Allergy Therapeutics

After retrieving gene sequence from NCBI for Homo sapiens STAT6 mRNA for signal transducer and activator of transcription 6 nirs variant 2, complete cds (GenBank Accession number AB103089.1) containing 2022 bp, mRNA sequence was obtained from Transcription and Translation Tool (www.attotron.com/cybertory/analysis/trans.htm). From RNAfold web server (Gruber, 2008), which can predict secondary structures of single stranded RNA sequences, Minimum free energy (MFE) secondary structure of that mRNA molecule was built. The MFE structure of an RNA sequence is the secondary structure that contributes a minimum of free energy (Zuker, 1981). The optimal secondary structure of STAT6 mRNA for human was modelled with minimum free energy -1584.20 kcal.mol using RNAfold web server (Gruber, 2008). For molecular docking study required three dimensional structure of the same molecule was formed from 3dRNA web server (http://biophy.hust.edu.cn/3dRNA/3dRNA.html) as shown in Figure 10 and visualized with UCSF chimera software (Petersen, 2004).

Using web server LigandRNA (Philips, 2013) molecular docking was performed with 'minimal RISC ' complex (formed between AGO2 protein and sense siRNA, shown in Figure 4a) and mRNA of STAT6

Figure 9. Interaction of siRNA with PAZ domain of ARO2 protein

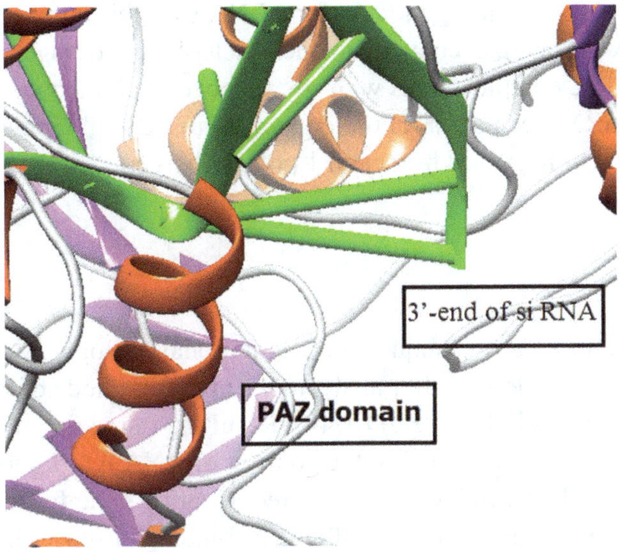

Figure 10. 3D structure of mRNA of STAT6 transcript2 gene

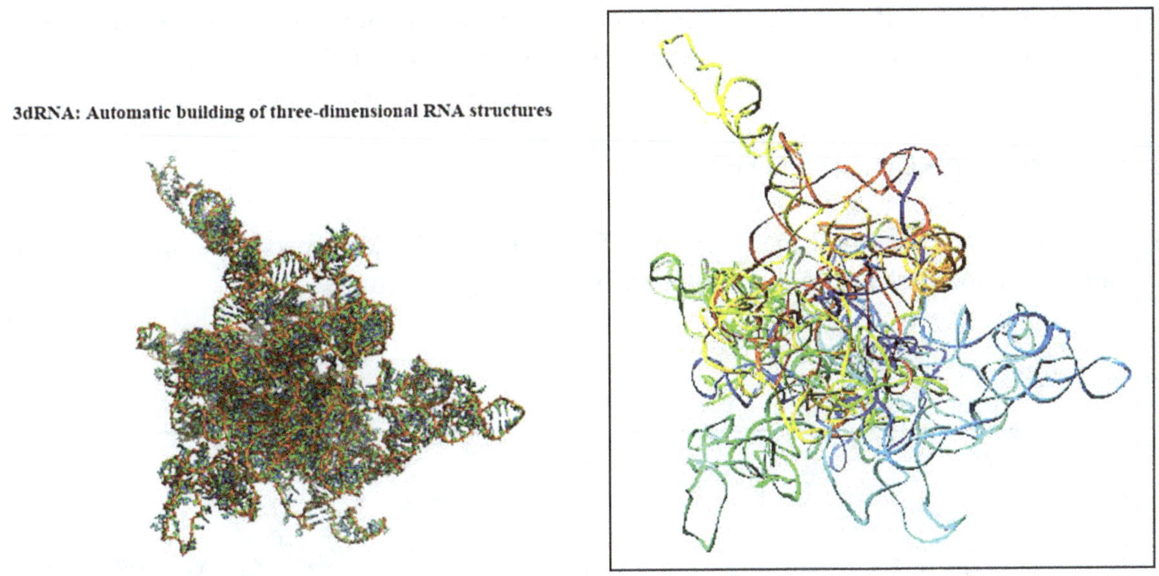

transcript2 gene (shown in Figure 10). Since this mRNA is 2022 nt long, its 3' end was selected (red coloured molecule in Figure 11 A). The molecular docking study results in following conformer with LigandRNA score -61.34, shown in Figure 11 B. In this figure mRNA molecule shown in black, different domains of AGO2 protein in red, blue, cyan, yellow colour and sense siRNA bound with AGO2 protein in dark blue colour. From the Figure 11 B it is clear that minimal RISC complex (AGO2 and sense siRNA) interacts with STAT6 mRNA through PIWI domain of AGO2 protein (red in colour) and may perform RNA cleavage by a mechanism similar to that of RNase H. Similarly in RISC, PAZ acts as an anchoring site for the 3'-end of guide RNA within silencing effecter complexes. Similarly the PAZ domain of AGO2 serves as a siRNA-end-binding module for siRNA transfer in the RNA interference method. RNA interference or RNA silencing occurs due to endonucleolytic cleavage of the mRNA specifically by AGO2 protein, when RISC complex binds with a perfectly complementary mRNA. Finally it can be concluded that the sense siRNA can act as antiallergic drug to treat allergic asthma during immediate type of hypersensitivity reaction, caused by Zea m 1 pollen allergen.

CONCLUSION

Molecular docking is an important technique in computational biology. By using this technique we first of all identified Zea m 1 as allergen after analyzing the binding mode of Zea m1 allergen (PDB ID 2HCZ) and human immunoglobulin E (PDB ID 4J4P). Resulting immediate-type hypersensitivity reaction and allergic symptoms can be treated with the concept of RNA interference (RNAi) technologies. In this work Homo sapiens STAT6 mRNA was selected as target gene for allergy therapeutics. Drug designing for allergy was done by selecting, modelling and binding of sense siRNA with STAT6 mRNA in presence of AGO2 protein. Molecular docking study reveals that in RISC, PAZ acts as an anchoring site for the 3'-end of guide RNA within silencing effecter complexes and RNA silencing occurs due to endonucleolytic cleavage of the STAT6 mRNA specifically by AGO2 protein. Thus crucial roles of STAT6 proteins in the pathogenesis of allergen-induced airway inflammation, mucus production, and airway hyper responsiveness (AHR) can be avoided by using siRNA of STAT6 mRNA.

FUTURE RESEARCH DIRECTIONS

For future work plan we propose the study on 'off target binding' of that specific sense siRNA, which is essential for rational therapeutic use of this siRNA in allergic reaction.

Figure 11. RISC complex

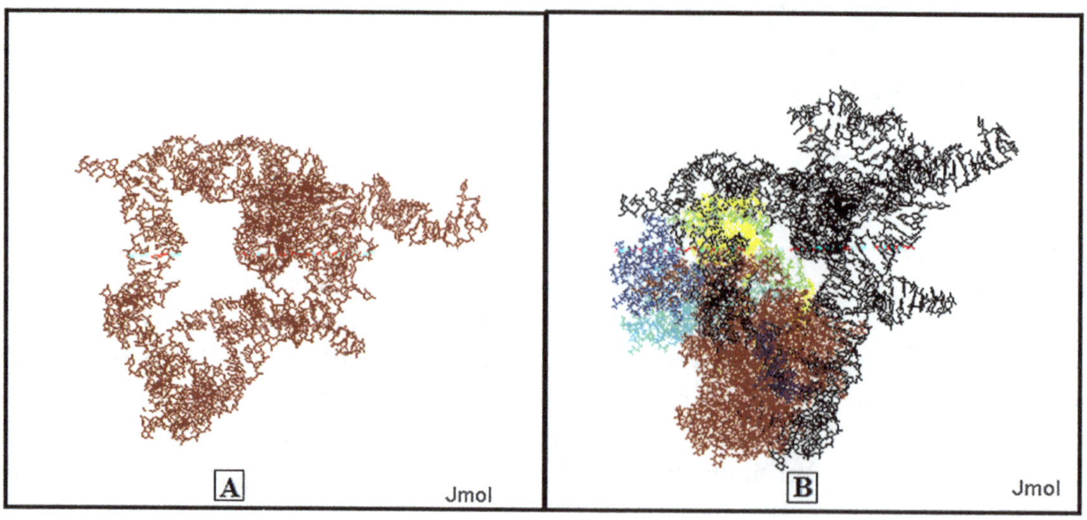

REFERENCES

Andersson, K., & Lidholm, J. (2003). Characteristics and immunobiology of grass pollen allergens. *International Archives of Allergy and Immunology*, *130*(2), 87–107. doi:10.1159/000069013 PMID:12673063

4. Andrusier, N., Nussinov, R., & Wolfson, H. J. (2007). FireDock: Fast Interaction Refinement in Molecular Docking. *Proteins*, *69*(1), 139–159. doi:10.1002/prot.21495 PMID:17598144

Ball, H. A., van Scott, M. R., & Robinson, C. B. (2004). Sense and antisense: Therapeutic potential of oligonucleotides and interference RNA in asthma and allergic disorders. *Clinical Reviews in Allergy & Immunology*, *27*(3), 207–217. doi:10.1385/CRIAI:27:3:207 PMID:15630157

Ball, T., Edstrom, W., Mauch, L., Schmitt, J., Leistler, B., Fiebig, H., & Valenta, R. et al. (2005). Gain of structure and IgE epitopes by eukaryotic expression of the major Timothy grass pollen allergen, Phl p 1. *The FEBS Journal*, *272*(1), 217–227. doi:10.1111/j.1432-1033.2004.04403.x PMID:15634345

Bantounas, I., Phylactou, L. A., & Uney, J. B. (2004). RNA interference and the use of small interfering RNA to study gene function in mammalian systems. *Journal of Molecular Endocrinology*, *33*(3), 545–557. doi:10.1677/jme.1.01582 PMID:15591019

Basu, A., & Sarkar, A. (2013). Emphasizing the of role of proteins in construction of the developmental genetic toolkit in plants. In Y. Pan & A. Y. Zomaya (Eds.), *Algorithmic and AI Methods for Protein Bioinformatics* (pp. 3–40). John Wiley and Sons Inc. doi:10.1002/9781118567869.ch1

Collis, A. V., Brouwer, A. P., & Martin, A. C. (2003). Analysis of the antigen combining site: Correlations between length and sequence composition of the hypervariable loops and the nature of the antigen. *Journal of Molecular Biology*, *325*(2), 337–354. doi:10.1016/S0022-2836(02)01222-6 PMID:12488099

Duhovny, D., Nussinov, R., & Wolfson, H. J. (2002) Efficient Unbound Docking of Rigid Molecules. In *Proceedings of the 2nd Workshop on Algorithms in Bioinformatics(WABI)* (LNCS), (vol. 2452, pp. 185-200). Springer Verlag.

Gruber, A. R., Lorenz, R., Bernhart, S. H., Neuböck, R., & Hofacker, I. L. (2008). The Vienna RNA Websuite. *Nucleic Acids Research*, *36*, W70–W74. doi:10.1093/nar/gkn188 PMID:18424795

Kandeel M., Kitade Y. (2013) Computational Analysis of siRNA Recognition by the Ago2 PAZ Domain and Identification of the Determinants of RNA-Induced Gene Silencing. *PLoS ONE, 8*(2). doi:10.1371/journal.pone.0057140

Kay, A. B. (2008). *Allergy and Allergic Diseases*. Oxford, UK: Blackwell Publishing Ltd. doi:10.1002/9781444300918

Knox, R. B., & Suphioglu, C. (1996). Pollen allergens: Development and function. *Sexual Plant Reproduction*, *9*(6), 318–323. doi:10.1007/BF02441949

Lingel, A., Simon, B., Izaurralde, E., & Sattler, M. (2004). Nucleic acid 3'-end recognition by the Argonaute2 PAZ domain. *Nature Structural & Molecular Biology*, *11*(6), 576–577. doi:10.1038/nsmb777 PMID:15156196

Mashiach E., Schneidman-Duhovny D., Andrusier N., Nussinov, R. Wolfson H. J. (2008). FireDock: a web server for fast interaction refinement in molecular docking. *Nucleic Acids Res., 36*, W229-32.

Mathews, D. H. (2014). Article. *Current Protocols in Bioinformatics, 46*, 12.6.1-12.6.25.

Miller, R. L., Eppinger, T. M., McConnell, D., Cunningham-Rundles, C., & Rothman, P. (1998). Analysis of cytokine signaling in patients with extrinsic asthma and hyperimmunoglobulin E. *The Journal of Allergy and Clinical Immunology, 102*(3), 503–511. doi:10.1016/S0091-6749(98)70141-1 PMID:9768594

Natt. (2011). *BMC Ear, Nose and Throat Disorders*. Retrieved from http://www.biomedcentral.com/1472-6815/11/3

Petersen, E. F., Goddard, T. D., Huang, C. C., Couch, G. S., Greenblatt, D. M., Meng, E. C., & Ferrin, T. E. (2004). UCSF Chimera--a visualization system for exploratory research and analysis. *Journal of Computational Chemistry, 25*(13), 1605–1612. doi:10.1002/jcc.20084 PMID:15264254

Philips, A., Milanowska, K., Lach, G., & Bujnicki, J. M. (2013). LigandRNA: Computational predictor of RNA-ligand interactions. *RNA (New York, N.Y.), 19*(12), 1605–1616. doi:10.1261/rna.039834.113 PMID:24145824

Popescu, F. D. (2005). Antisense- and RNA interference-based therapeutic strategies in allergy. *Journal of Cellular and Molecular Medicine, 9*(4), 840–853. doi:10.1111/j.1582-4934.2005.tb00383.x PMID:16364194

Rother, M., Rother, K., Puton, T., & Bujnicki, J. M. (2011). Mode RNA: A tool for comparative modelling of RNA 3D structure. *Nucleic Acids Research, 39*(10), 4007–4022. doi:10.1093/nar/gkq1320 PMID:21300639

Schneidman-Duhovny, D., Inbar, Y., Nussinov, R., & Wolfson, H. J. (2005). PatchDock and SymmDock: Servers for rigid and symmetric docking. *Nucleic Acids Research, 33*, W363-367. doi:10.1093/nar/gki481 PMID:15980490

Song, J. J., Liu, J., Tolia, N. H., Schneiderman, J., Smith, S. K., Martienssen, R. A., & Joshua-Tor, L. et al. (2003). The crystal structure of the Argonaute2 PAZ domain reveals an RNA binding motif in RNAi effector complexes. *Nature Structural Biology, 10*(12), 1026–1032. doi:10.1038/nsb1016 PMID:14625589

Stave, J. W., & Lindpaintner, K. (2013). Antibody and Antigen Contact Residues Define Epitope and Paratope Size and Structure. *Journal of Immunology (Baltimore, MD.: 1950), 191*(3), 1428–1435. doi:10.4049/jimmunol.1203198 PMID:23797669

Tuszynska, I., & Bujnicki, J. M. (2011). DARS-RNP and QUASI-RNP: New statistical potentials for protein-RNA docking. *BMC Bioinformatics, 12*(1), 348. doi:10.1186/1471-2105-12-348 PMID:21851628

Tuszyńska, I., Magnus, M., Jonak, K., Dawson, W., & Bujnicki, J.M. (2015) NPDock – a web server for protein-nucleic acid docking *Nucleic Acid Research, 43*(W1), W425-30.

Valenta, R. (2008). Biochemistry of allergens and recombinant allergens. In A. B. Kay (Ed.), *Allergy and Allergic Diseases* (pp. 895–912). Oxford, UK: Blackwell Publishing Ltd. doi:10.1002/9781444300918.ch42

Valenta, R., Campana, R., Marth, K., & van Hage, M. (2012). Allergen-specific immunotherapy: From therapeutic vaccines to prophylactic approaches. *Journal of Internal Medicine*, *272*(2), 144–157. doi:10.1111/j.1365-2796.2012.02556.x PMID:22640224

Walford HH, Doherty T A. (2013) STAT6 and lung inflammation. *JAKSTAT, 2*(4).

Zuker, M., & Stiegler, P. (1981). Optimal computer folding of large RNA sequences using thermodynamics and auxiliary information. *Nucleic Acids Research*, *9*(1), 133–148. doi:10.1093/nar/9.1.133 PMID:6163133

ADDITIONAL READING

Nakanishi, K., Ascano, M., Gogakos, T., Ishibi-Murakami, S., Serganov, A. A., Briskin, D., & Patel, D. J. et al. (2013). Eukaryote-specific insertion elements control human ARGONAUTE slicer activity. *Cell Reports*, *3*(6), 1893–1900. doi:10.1016/j.celrep.2013.06.010 PMID:23809764

Patel, D. J., Ma, J.-B., Yuan, Y.-R., Ye, K., Pei, Y., Kuryavyi, V., & Tuschl, T. et al. (2007). Structural biology of RNA silencing and its functional implications.*Cold Spring Harbor Laboratory Symposium on Regulatory RNAs* 71, 81-93.

Swarts, D. C., Makarova, K., Wang, Y., Nakanishi, K., Ketting, R. F., Koonin, E. V., & van der Oost, J. et al. (2014). The evolutionary journey of Argonaute proteins. *Nature Structural & Molecular Biology*, *21*(9), 743–753. doi:10.1038/nsmb.2879 PMID:25192263

KEY TERMS AND DEFINITIONS

Argonaute Proteins: Argonaute proteins (AGO) play a central role in RNA silencing processes, as essential catalytic components of the RNA-induced silencing complex (RISC). Argonaute proteins bind different classes of small non-coding RNAs, including microRNAs (miRNAs), small interfering RNAs (siRNAs) and Piwi-interacting RNAs (piRNAs). Small RNAs guide Argonaute proteins to their specific targets through sequence complementarity (base pairing), which then leads to mRNA cleavage or translation inhibition.

RISC: The RNA-induced silencing complex, or RISC, is a multiprotein complex, specifically a ribonucleoprotein. RISC complex is responsible for the gene silencing phenomenon known as RNA interference (RNAi), which incorporates one strand of a double-stranded RNA (dsRNA) fragment, such as small interfering RNA (siRNA) or microRNA (miRNA).

Small Interfering RNA (siRNA): siRNA is a class of double-stranded RNA molecules, 20-25 base pairs in length. siRNA interferes with the expression of specific genes with complementary nucleotide sequences in RNAi or RNA interference technique. siRNA functions by cutting mRNA into pieces after transcription, resulting in no translation.

STAT Protein: Signal Transducers and Activators of Transcription (STAT) family proteins contains transcription factors that are specifically activated to regulate gene transcription when cells encounter cytokines and growth factors; hence they act as signal transducers in the cytoplasm and transcription activators in the nucleus. STAT proteins play a critical role in regulating innate and acquired host immune responses.

Chapter 10
Protein-Protein Interactions (PPIs) as an Alternative to Targeting the ATP Binding Site of Kinase:
In Silico Approach to Identify PPI Inhibitors

Sailu Sarvagalla
Pondicherry University, India

Mohane Selvaraj Coumar
Pondicherry University, India

ABSTRACT

Most of the developed kinase inhibitor drugs are ATP competitive and suffer from drawbacks such as off-target kinase activity, development of resistance due to mutation in the ATP binding pocket and unfavorable intellectual property situations. Besides the ATP binding pocket, protein kinases have binding sites that are involved in Protein-Protein Interactions (PPIs); these PPIs directly or indirectly regulate the protein kinase activity. Of recent, small molecule inhibitors of PPIs are emerging as an alternative to ATP competitive agents. Rational design of inhibitors for kinase PPIs could be carried out using molecular modeling techniques. In silico tools available for the prediction of hot spot residues and cavities at the PPI sites and the means to utilize this information for the identification of inhibitors are discussed. Moreover, in silico studies to target the Aurora B-INCENP PPI sites are discussed in context. Overall, this chapter provides detailed in silico strategies that are available to the researchers for carrying out structure-based drug design of PPI inhibitors.

INTRODUCTION

Protein kinase enzymes, also known as phosphotransferases, are the most extensively pursued class of drug targets in current pharmaceutical research. Several kinase inhibitors are registered for pre-clinical & clinical trial evaluation for different ailments, including cancer (O'Brien & Fallah Moghaddam, 2013; Rask-Andersen, Zhang, Fabbro, & Schioth, 2014; J. Zhang, Yang, & Gray, 2009). These enzymes transfers

DOI: 10.4018/978-1-5225-0362-0.ch010

ATP terminal phosphate group to the substrate protein, and thereby regulate various activities including cell proliferation, differentiation, survival, transcription, apoptosis, metabolism, and a wide array of other signal transduction process (Adams, 2001; P. Wu, Nielsen, & Clausen, 2015). The involvement of kinases in various pathological conditions, makes them an attractive drug target for therapeutic intervention in diseases such as cancer (Fabbro, Cowan-Jacob, Mobitz, & Martiny-Baron, 2012), vascular (Abeyrathna & Su, 2015; Kikuchi et al., 2014) & central nervous system (CNS) disorders (Chico, Van Eldik, & Watterson, 2009), inflammatory disease conditions (Barnes, 2013; Rommel, 2010), and diabetes (Banks et al., 2015; Y. Wu & Chakrabarti, 2015). Consequently, for the past one and half decades, pharmaceutical companies and academic researchers are mounting intense efforts to develop small molecule kinase inhibitors. As a result of this, the first successful kinase inhibitor imatinib was approved in 2001 by FDA for the treatment of chronic myeloid leukemia (Gambacorti-Passerini & Piazza, 2015; Wisniewski et al., 2002). Since then, understanding of kinase structural and functional mechanism and their involvement in pathological conditions has significantly improved by the advancement of cell and molecular biology, structural biology, genetics, and associated fields. Concomitantly, the kinase inhibitors approval has also increased, and presently around 28 small molecule kinase inhibitors are approved by the US FDA for therapeutic usage in various cancers (Fabbro, 2015; P. Wu et al., 2015; Z. Zhao et al., 2014). The structures of approved kinase inhibitor drugs are represented in Figure 1.

The entire human genome encodes approximately 518 protein kinases (Manning, Whyte, Martinez, Hunter, & Sudarsanam, 2002), and have conserved kinase domain fold. The 3D structure of kinase contains a helix-dominated conserved C-terminal region, and a β sheet-dominated N-terminal region, which varies in sequence length and amino acid composition. A flexible hinge region connects the N- and C-terminal lobes, and in between these two lobes there exist a conserved ATP binding cleft. Glycine rich loop contains a conserved GXGXXG motif, and acts like a clamp to stabilize the binding of ATP at this cleft. Binding of ATP to the active site is controlled by the activation loop N-terminal starting residues Asp-Phe-Gly, also called DFG motif which adapts "DFG-in" and "DFG-out" conformation according to protein active and inactive state, respectively. The DFG motif residue Asp together with metal ion Mg^{2+} regulates kinase transactivation mechanism during phosphorylation process (Eswaran & Knapp, 2010; Thaimattam, Banerjee, Miglani, & Iqbal, 2007).

KINASE INHIBITORS, THEIR BINDING MECHANISM AND DRAWBACKS

The small molecule kinase inhibitors are useful as therapeutic agents, as well as to understand various cellular functions in normal and disease conditions. To date, most of the discovered kinases inhibitors are ATP competitive in nature; the heterocyclic core group in the inhibitor forms one to three hydrogen bond interactions with the conserved hinge region residues of the kinase (Garuti, Roberti, & Bottegoni, 2010; J. Zhang et al., 2009). Thus, they mimic the typical hydrogen bond interactions of adenine ring of ATP with the kinase. Based on the inhibition mechanism, kinase inhibitors are classified as either reversible or irreversible (P. Wu et al., 2015). The reversible inhibitors are further categorized as Type-I, Type-II, Type-III and Type-IV inhibitors, based on their binding mechanism as well as DFG motif conformation (Z. Zhao et al., 2014). The type-I inhibitors binds to active state of the protein where the DFG motif adapts into "DFG-in" conformation; in this state, the DFG motif residues Asp and Phe are placed inside and outside of the ATP binding pocket, respectively. The type-II inhibitors bind to the inactive state of the protein, where the DFG motif adapts into "DFG-out" conformation; in this conformation, the DFG

Figure 1. FDA-approved kinase inhibitors with binding modes.
Reproduced from reference Z. Zhao et al. (2014).

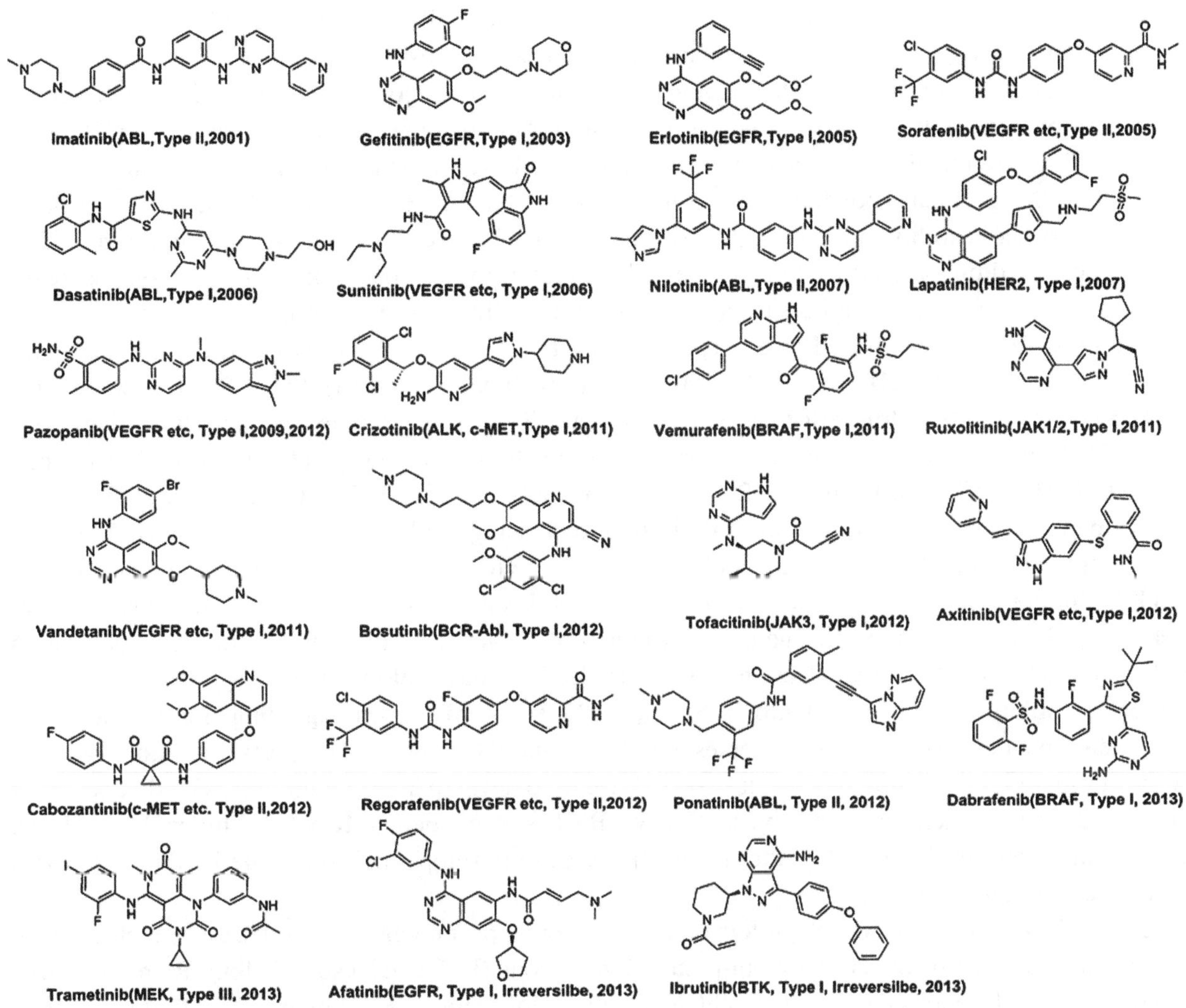

motif residues Asp and Phe are placed outside of the ATP binding (under αC-helix) pocket. Moreover, in this conformation, the gate keeper residue permits the inhibitors to access the back pocket region. The type-III inhibitors also called allosteric inhibitors which binds exclusively to the ATP adjacent pocket (back pocket), and modulates the ATP binding by allosteric mechanism. While, type-IV inhibitors binds to a distant place from ATP binding site. Both type-III and type-IV inhibitors show highest degree of kinase selectivity because of their allosteric binding mode and regulatory mechanism that are unique to particular kinase enzymes (P. Wu et al., 2015). The irreversible inhibitors of kinases have tendency to bind covalently with ATP binding site residue, most frequently by interacting with nucleophilic Cys residue, which results in irreversible inhibition (Garuti, Roberti, & Bottegoni, 2011).

So far, most of the developed kinase inhibitors are ATP competitive in nature, and have serious drawbacks because of the shared conserved ATP binding site among the human kinome. The primary concerns regarding ATP competitive inhibitors are: off- and on-target toxicity issues (Burtness, 2014;

Ho, Yeo, Kang, & Chua, 2014; Liu & Kurzrock, 2014), development of drug resistance (Barouch-Bentov & Sauer, 2011; Murray & Miller, 2015) and unfavorable intellectual property situation. Because of the conserved ATP binding site, development of selective kinase inhibitors is very challenging, and most of the approved and developed kinase inhibitors are also having reasonable multi-kinase activity which ultimately causes serious on- and off-target toxicity. The second major concern about kinase inhibitors is the development of drug resistance because of mutations in the gate keeper residue of the ATP binding site, which eventually causes loss of drug binding and drug efficacy to a particular kinase target. Because, at this particular position (gate keeper residue), the size of amino acid side chain regulates the inhibitors accessibility to the back pocket region or ATP adjacent pocket. As the kinase inhibitors make crucial hydrophobic interactions at the ATP adjacent pocket, loss of these interactions decrease the binding affinity of the inhibitors to the target kinase (J. Zhang et al., 2009).

The most frequently reported gate keeper residue mutations are observed in different kinases, including BCR-ABL1 (T315I), EGFR (T790M), FLT3 (G697R), KIT (T670I), PDGFR (T674I) and BTK (C481S) which results in drug resistance (J. Zhang et al., 2009). For example, gate keeper residue mutation (T315I) of BCR-ABL1 confers drug resistance to imatinib (Gleevec) in chronic myeloid leukemia patients (O'Hare, Eide, & Deininger, 2007). Further, mutation of EGFR gate keeper residue T790M results in drug resistance to geftinib and erlotinib by decreasing inhibitors accessibility to the ATP back pocket region (Godin-Heymann et al., 2008). Moreover, the recently approved Bruton's tyrosine kinase (BTK) inhibitor ibrutinib also showed resistance to gate keeper residue mutation (C481S) (D'Cruz & Uckun, 2013). In addition to these gate keeper mutations, other aberration such as target amplification and up-regulation of alternative kinase pathways also leads to the development of drug resistance to ATP competitive inhibitors (Barouch-Bentov & Sauer, 2011). The third concern regarding kinase inhibitors is the development of diverse set of molecules with favorable IP (intellectual property) situation. Because of the conserved ATP binding pocket, it is very difficult to develop kinase inhibitors with diverse set of functional groups which are protectable by IP laws (Backes, Zech, Felber, Klebl, & Muller, 2008). Thus, there is an urgent need to develop alternative strategies to develop small molecule inhibitors towards these indispensable drug targets.

Beside the ATP binding pocket, protein kinases have other prominent structural features including the binding sites that are involved in protein-protein interactions (PPIs) (Dickson, Wilbur, Reinke, Young, & Vojtek, 2015; Pearson et al., 2001; Rechfeld, Gruber, Hofmann, & Kirchmair, 2011; Rudolph, 2007; White, Westwell, & Brahemi, 2008). These PPIs directly or indirectly monitor the protein kinase activity, and interruption of interaction at these sites (PPI sites) have emerged as an attractive and alternative strategy to develop selective kinase inhibitors with favorable IP situation (Cierpicki & Grembecka, 2015; Jin, Wang, & Fang, 2014). As the activity of different kinases are controlled by numerous protein-protein interactions, and their binding interface significantly varies from one kinase family to other kinase family member proteins, targeting these PPI sites have advantage over targeting the conserved ATP binding site. This might allow higher level of selectivity as compared to the ATP competitive inhibitors. Moreover, development of kinase PPI inhibitors permits for combinatorial treatment with ATP competitive inhibitors which could synergistically block kinase activity as compared to ATP competitive inhibitors alone. However, targeting kinases PPI site is very challenging, because of their flat and featureless large hydrophobic interface region; further, these interactions are transient and are exposed to solvent regions, causing difficulties in PPI inhibitor development. Nevertheless, growing evidence indicates that targeting PPI sites of kinases are emerging as an important therapeutic strategy in drug discovery and development (Corbel et al., 2011; Dickson et al., 2015; Han, Kim, & Yang, 2012; Kaidanovich-Beilin & Eldar-Finkelman, 2006; Rechfeld et al., 2011; B. Zhang et al., 2011).

PROTEIN-PROTEIN INTERACTIONS AS THERAPEUTIC TARGETS

The fundamental functions of living cell such as cell growth, morphology, nutrient uptake, gene expression, signaling network, and apoptosis are accomplished by macromolecular interactions; among them, protein-protein interactions (PPIs) have critical role in various biological and pathological conditions (Ryan & Matthews, 2005). It is reported that a single cell of a human can have around 130,000 to 650,000 PPIs (Stumpf et al., 2008; Venkatesan et al., 2009). Thus, understanding the importance of PPIs in various pathological conditions has wide application in rational drug design and protein engineering. However, previously it was assumed that PPI sites are "undruggable (or) intractable", because of their flat and featureless large hydrophobic region that couldn't accommodate a small molecule inhibitor (Wells & McClendon, 2007). This misconception was assumed due to lack of understanding of these sites, and also, there weren't many known small molecule inhibitors that could bind to PPI sites as compared to the traditional drug targets such as enzymes, receptors and ion channels. Further, lack of experimental assay methods, and small molecule libraries that were suitable for identification of PPI inhibitors were not available in yesteryears. However, recent developments in molecular biology, structural biology, computational chemistry and associated fields have helped in gaining better knowledge of PPIs (Cierpicki & Grembecka, 2015; Kuenemann et al., 2015; Villoutreix et al., 2014; Watkins & Arora, 2015).

Traditional drug targets such as receptors, enzymes and ion channels have served as attractive drug targets for many of the pathological conditions, including cancer. At the same time, toxicity issues and drug resistance development was concurrently observed for many of these drug targets (Barouch-Bentov & Sauer, 2011; Liu & Kurzrock, 2014; Wallace, 2015) Consequent to this, quest for new targets for therapeutic intervention turned the focus on PPIs, because of their importance in various pathologies (Ryan & Matthews, 2005; Wells & McClendon, 2007). Thus, design of small molecule inhibitors that directly block PPIs has become an attractive and an alternative strategy to develop small molecule inhibitors for various diseases. However, structural analysis of PPIs revealed that as compared to traditional drug targets, PPI sites are complex and flexible; they lack cavities that are essential for small molecule binding (Cierpicki & Grembecka, 2015). The interfaces of PPIs are considerably flat, and contain featureless large hydrophobic surface area (around 1200 to 3000 Å^2) (Cierpicki & Grembecka, 2015; Lo Conte, Chothia, & Janin, 1999). Further, the intrinsic properties of PPIs such as flexibility and dynamic nature is different in solution as compared to crystal binding mode, and it is very difficult to design small molecule inhibitors for PPIs when structure based drug design methods are applied.

In spite of these challenges, the current body of literature evidences suggest that only few of the clustered residues of PPIs sites contributes to majority of the binding free energy (called as 'hot spots'), and these hot spot residues are critical for protein-protein structure stability and function (Bogan & Thorn, 1998; Clackson & Wells, 1995; Kuttner & Engel, 2012; Moreira, Fernandes, & Ramos, 2007). Thus, targeting PPIs hot spots residues provide an attractive opportunity to explore these drug targets for therapeutic intervention (Cukuroglu, Engin, Gursoy, & Keskin, 2014; Guo, Wisniewski, & Ji, 2014). Further, structural analysis revealed that PPIs sites could be subdivided into three to four sub-pockets (Fuller, Burgoyne, & Jackson, 2009; Jubb, Blundell, & Ascher, 2015) and the binding to these sub-pockets could be explored using natural products and peptidomimetics like compounds. Recently, few small molecule and peptide inhibitors of PPIs are discovered using the above information (Arkin, Tang, & Wells, 2014). The interface residues of PPIs are conserved and are less prone to the development of drug resistance (Kozakov et al., 2011; X. Li, Keskin, Ma, Nussinov, & Liang, 2004; Ma, Elkayam, Wolfson, & Nussinov, 2003). Moreover, growing evidences indicate that PPIs are challenging but tractable drug targets for various diseases. As a proof of concept, few small molecule inhibitors have already entered into clinical trials and a number of small molecule inhibitors are in pre-clinical development (Arkin et al., 2014; Bogoyevitch, Barr, & Ketterman, 2005; Cierpicki & Grembecka, 2015). In the following section, few important PPI drug targets (IAP family proteins, bromodomains, BCL family proteins, MDM2-p53 complex interface, and HIV integrase) and their inhibitors that have progressed towards clinical trials are described in brief. Interested readers may also refer to other review articles for additional information (Arkin et al., 2014; Ivanov, Khuri, & Fu, 2013; Jin et al., 2014; Kuenemann et al., 2015; Laraia, McKenzie, Spring, Venkitaraman, & Huggins, 2015; Sheng, Dong, Miao, Zhang, & Wang, 2015).

IAP Inhibitors

The Inhibitors of Apoptosis Proteins (IAPs), including XIAP, cIAP1, cIAP2, Survivin, and Levivin, play vital role in various biological function including cell division, duplication, cell death, DNA damage, autophagy and immunity (Budhidarmo & Day, 2015; Damgaard & Gyrd-Hansen, 2011). On the other hand, it has been demonstrated that these proteins are over expressed in many of the human solid tumors. Consequently, these proteins have become attractive drug targets for cancer therapy (Dubrez, Berthelet, & Glorian, 2013). These proteins contain at least one conserved BIR domain, and involves in PPIs. IAPs BIR domain directly binds to N-terminal tetra-peptide sequence of caspase-3, caspase-7 and caspase-9 during their proteolytic activation; as a result it inhibits their pro-apoptotic mechanism. However, an endogenous peptide inhibitor of IAPs called Smac (Second mitochondrial activators of caspases) binds to their BIR domain, and activates the apoptosis mechanism. Therefore, developing a small molecule inhibitor of IAPs that mimics the Smac is an attractive approach to inhibit the IAPs protein-protein interaction with caspases. Several Smac mimetics have been developed by many scientific groups; among them, seven compounds have reached clinical trials and five compounds including LCL-161, AT-406, TL-32711, GDC-0152, and GDC-0917 are actively pursued in clinical development for the treatment of cancer (Arkin et al., 2014).

Bromodomain (BRD) Inhibitors

The bromodomain family proteins (BET, BRD2, BRD3, BRD4 and BRDT) are 'epigenetic readers', which have significant role in targeting multiple enzymes, protein scaffolds and DNA binding proteins to their specific sites. Usually, these proteins recognize the acetylated lysines (Kacs) on histone tail, and thereby govern chromatin remodeling and gene transcription (Sanchez & Zhou, 2009). However, dysfunctions of these proteins have been reported in various cancers, and targeting these proteins interaction with acetylated lysine peptide has become an attractive approach for the design of small molecule inhibitors for cancer (Muller, Filippakopoulos, & Knapp, 2011). Consequently, the hydrophobic pocket of bromodomains that recognizes acetyl-lysine peptide has been used to design and develop the Kac mimetic inhibitors. Currently, few compounds including I-BET762, CPI-0610, Ten-010 and OTX15 are in clinical development for cancer (Arkin et al., 2014).

BCL Family Protein Inhibitors

BCL family proteins include pro- and anti-apoptotic proteins which regulates cell death mechanism in various biological events (Hardwick & Soane, 2013). The pro-apoptotic proteins BAD and BAK are tightly regulated or inhibited by anti-apoptotic proteins including BCL-2, BCL-xl and MCL1. Over expression or deregulation of anti-apoptotic proteins have been observed in various cancers (Besbes, Mirshahi, Pocard, & Billard, 2015). The anti-apoptotic proteins are helical proteins which forms an open cavity that binds to BH3 domain (single helix) of pro-apoptotic proteins. Hence, disrupting these PPIs using small molecule inhibitors has become an attractive approach for cancer therapy. In the past, several inhibitors were reported to modulate these PPIs; currently, ABT-199 and ABT-263 are in clinical development for cancer (Arkin et al., 2014).

MDM2-p53 PPI Inhibitors

p53 is the principal tumor suppressor protein, also called as "guardian of genome" because of its central role in protecting and stabilizing the human genome from various mutations. It tightly regulates cell-fate decisions including cell cycle arrest, senescence and apoptosis (Surget, Khoury, & Bourdon, 2013). Mouse Double Minute 2 homolog (MDM2), also known as E3 ubiquitin-protein ligase is one of the major regulator of p53 (Momand, Wu, & Dasgupta, 2000). MDM2, constantly ubiquinates p53 leading to degradation by nuclear and cytoplasmic proteasome enzymes, thereby limiting its tumor suppressor function in unstressed cells. However, mutations are reported in p53 and its regulator protein, which results in loss of p53 tumor suppressor function. Thus, disruption of MDM2-p53 PPIs results in p53 activation in tumor cells. The inhibitor nutlin is the first breakthrough that was discovered by large high-throughput screening for disruption of MDM2-p53 PPIs (Shangary & Wang, 2009). Subsequently, several small molecule inhibitors were discovered by many scientific groups. Currently, RG-7112 (Ro5045337), RG-7388 (Ro5503781) and MI-888 are in clinical development for the treatment of cancer (Arkin et al., 2014; Y. Zhao, Aguilar, Bernard, & Wang, 2015).

HIV Integrase Inhibitors

HIV integrase (IN) allows integration of viral genome into host genome by catalyzing 3′ processing and stand transfer. In this process, the human protein - Lens Endothelial Growth Factor (LEDGF) acts as IN cofactor to activate the integration process, and also to protect from proteolytic degradation. Thus, targeting these host-pathogen PPIs (LEDGF-IN) is an alternative and attractive approach for the design of small molecule inhibitors for treating AIDS. Recently, a second class of IN inhibitors that mimics the LEDGF peptide has entered into clinical trials (Engelman, Kessl, & Kvaratskhelia, 2013). Currently, the inhibitor *tert*–Butoxy-(4phenyl-quinolin-3-ly-acetic acid (*t*BPQA) derivative and BI-224436 are in clinical development for the treatment of AIDS (Arkin et al., 2014).

STRUCTURE-BASED DRUG DESIGN OF PPI INHIBITORS USING *IN SILICO* APPROACHES

The arena of structure based drug design (SBDD) has rapidly grown because of the advancements in genomics, proteomics, structural biology and computational chemistry. SBDD has an important role in drug discovery, and many successful lead molecules have been discovered using this approach (Anderson, 2003; Lionta, Spyrou, Vassilatis, & Cournia, 2014). Once the 3D structure of a particular disease target is resolved using experimental (X-ray or NMR) or computational methods (homology modeling and threading), then it could be used for virtual screening or for de-novo design to identify lead molecule inhibitors. In this chapter, the authors describe in a step-wise manner how SBDD methods could be applied to discover small molecule inhibitors of PPIs. Figure 2 summarizes the SBDD approach for the identification of PPIs inhibitors.

Modeling of Protein and PPI Complex Structures

For the rational design of PPI inhibitors using structure-based drug design (SBDD), it is essential to have 3D structures of PPI complexes involved in the particular disease state. However, available 3D structures of protein-protein complexes are very few in Protein Data Bank (PDB), as compared to the traditional drug targets such as enzyme and receptor structures. Resolving the 3D structures of PPIs using X-ray and NMR methods is much more challenging (Widmer & Jahnke, 2004). Due to the experimental limitations, *in silico* methods including comparative modeling (homology modeling), threading and fold recognition and Ab initio methods can be used to build the protein structure. Knowledge of resolved homologue protein structures and physical principles of protein such as that mimic the protein folding or stochastic (global optimization of energy function) methods are used to search the possible solutions. The *in silico* tools or modules that are available for protein 3D structure predictions are listed in Table 1. Among them, MODELLER (Eswar et al., 2006), SWISS-MODEL (Biasini et al., 2014) (for comparative modeling), I-TASSER (Yang et al., 2015) (for threading), ROBETTA server (Kim, Chivian, &

Figure 2. Structure-based drug design (SBDD) approach for the identification of PPI inhibitors

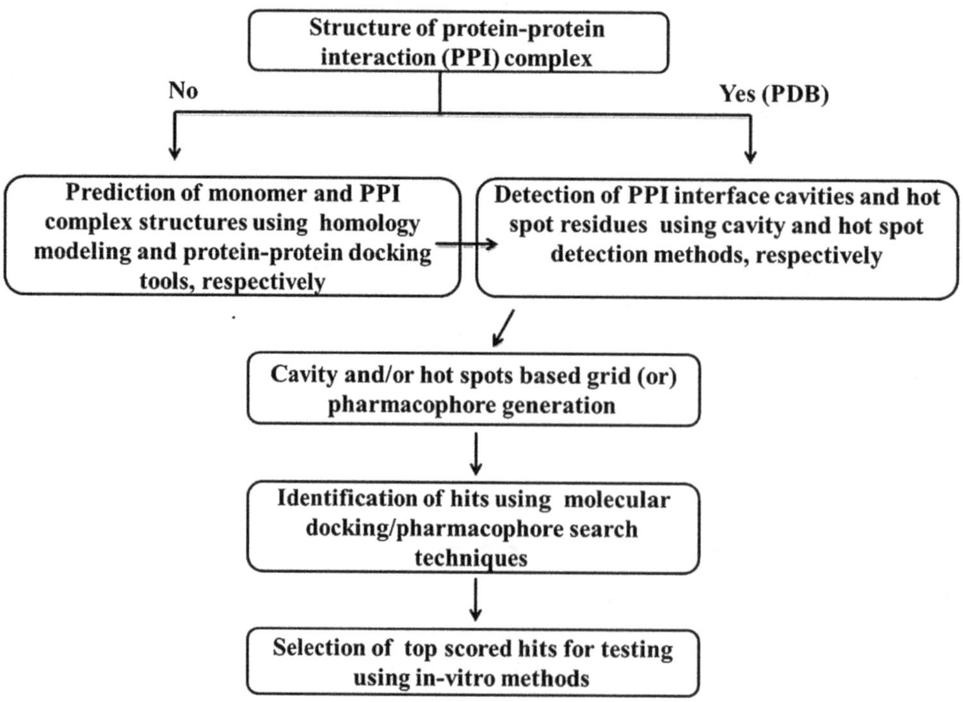

Table 1. List of available protein 3D structure prediction tools and web servers

Sl.No	Name of the Tool	Description and Web-Links
Homology Based or Comparative Methods		
1	Modeller	Downloadable (http://salilab.org/modeller/) and webserver (https://modbase.compbio.ucsf.edu/modweb/)
2	SWISS-MODEL	Webserver (http://swissmodel.expasy.org/)
3	PHYRE2	Webserver (http://www.sbg.bio.ic.ac.uk/phyre2/html/page.cgi?id=index)
4	ESyPred3D	Webserver (http://www.unamur.be/sciences/biologie/urbm/bioinfo/esypred/)
5	AS2TS system	Webserver (http://as2ts.llnl.gov/)
6	3D-JIGSAW	Webserver (http://bmm.cancerresearchuk.org/~3djigsaw/)
7	Yasara	Webserver (http://www.yasara.org/)
8	Biskit	Downloadable (http://biskit.pasteur.fr/)
9	FoldX	Downloadable (http://foldx.crg.es/)
10	Prime	Downloadable (http://www.schrodinger.com/Prime)
11	CPHModel	Webserver (http://www.cbs.dtu.dk/services/CPHmodels/)
12	HHpred	Downloadable (ftp://ftp.tuebingen.mpg.de/pub/protevo/HHsearch/) and webserver (http://toolkit.tuebingen.mpg.de/hhpred)
Threading and Fold Recognition Based Methods		
13	I-TASSER	Downloadable (http://zhanglab.ccmb.med.umich.edu/I-TASSER/download/) and webserver (http://zhanglab.ccmb.med.umich.edu/I-TASSER/)
14	RaptorX	Downloadable (http://raptorx.uchicago.edu/) and Webserver (http://raptorx.uchicago.edu/)
15	LOOPP	Webserver (http://cbsuapps.tc.cornell.edu/loopp.aspx)
16	MUSTER	Webserver (http://zhanglab.ccmb.med.umich.edu/MUSTER/)
17	SPARKS-X	http://sparks-lab.org/yueyang/server/SPARKS-X/_
Ab Initio and Combination of Threading and Ab Initio Methods		
18	ROBETTA	Webserver (http://www.robetta.org/)
19	Bhageerath	Webserver (http://www.scfbio-iitd.res.in/bhageerath/index.jsp)
20	EVfold	Webserver (http://evfold.org/evfold-web/evfold.do)
21	QUARK	Webserver (http://zhanglab.ccmb.med.umich.edu/QUARK/)
22	PEP-FOLD	Webserver (http://bioserv.rpbs.univ-paris-diderot.fr/services/PEP-FOLD/)

Baker, 2004) (combination of comparative and Ab initio methods) are well known modules for protein 3D structure prediction.

Further, in the absence of experimental PPIs complex structures, the available monomer structures (from PDB or modeling) of two individual protein partners could be used to build the PPI complexes using protein-protein docking (Vakser, 2014). The ultimate goal of protein-protein docking is to predict the 3D structure of the macromolecular complexes of interest as they would occur in a living organism. Thus, execution of protein-protein docking of two interested proteins would produces number of plausible candidate complex structures, and these candidates must be ranked using scoring functions to identify structures that are most likely to occur in the native state. There are a number of algorithms and scoring functions developed to model PPIs complexes with reasonable accuracy (Sable & Jois, 2015). The most commonly used protein-protein docking tools and their algorithms are listed in Table 2.

Table 2. List of available protein-protein docking tools and web servers

Sl.No	Name of the Tool	Applied Methodology	Description and Web Links
1	ZODOCK	It uses Fast Fourier transformation (FFT), shape complementarity, desolvation energy and electrostatic interactions	It is available as a webserver (http://zdock.umassmed.edu/) and also as a standalone tool; implemented in Accelrys software suit (discovery studio).
2	HADDOCK	Data-driven bimolecular docking	It is an information-driven flexible docking program used to model bimolecular complexes (http://haddocking.org/)
3	GRAMM-X	Fast Fourier Transformation	It employs smoothed potentials, refinement and knowledge-based scoring function to find the possible solution. (http://vakser.compbio.ku.edu/resources/gramm/grammx/)
4	ClusPro	Rigid-body docking using the FFT correlation approach and RMSD clustering	Webserver (http://nrc.bu.edu/cluster)
5	Patch dock	Global search	It uses shape complementarity and atomic desolvation energy to search possible solution (http://bioinfo3d.cs.tau.ac.il/PatchDock/)
6	RosettaDock	Rigid body docking (energy based method)	It optimizes the rigid body docked poses and side chains to finds lowest energy conformation (http://rosettadock.graylab.jhu.edu/)
7	GPU.proton. DOCK	Fast Fourier Transform bottleneck and electrostatic fields computation	It is a state of the art server for prediction of protein–protein interactions via rigorous and ultrafast docking code (http://gpudock.orgchm.bas.bg/)
8	HexServer	Fast Fourier transformation	(http://hexserver.loria.fr/)
9	3D-Garden	Marching cubes algorithm	http://www.sbg.bio.ic.ac.uk/~3dgarden/_
10	SymmDock	Symmetry by Geometry Based Docking	http://bioinfo3d.cs.tau.ac.il/SymmDock/_
11	SwarmDock	SwarmDock algorithm	This algorithm includes docking using a hybrid particle swarm optimization / local search, and minimization, re-ranking and clustering of docked poses http://bmm.cancerresearchuk.org/~SwarmDock/
12	PIPER	FFT-based docking with pairwise potentials	It is implemented in Schrodinger software suit http://www.schrodinger.com/PIPER/
13	DOT 2.0	FFT based global rigid search	Downloadable (http://www.sdsc.edu/CCMS/DOT/)
14	ICM	Monte Carlo based global rigid search	Downloadable (http://www.molsoft.com/docking.html)
15	MolFit	FFT and shape complementarity based global search	Downloadable http://www.weizmann.ac.il/Chemical_Research_Support/molfit/
Refinement or post docking refinement modules			
16	RDOCK	CHARMM force field	It is implemented in discovery studio (Accelrys software suit). Refines the ZODOCK poses using CHARMM force field
17	FireDock	Fast interaction refinement	It is an efficient method for refinement and re-scoring of rigid-body protein-protein docking poses (http://bioinfo3d.cs.tau.ac.il/FireDock/)
18	HADDOCK-Refinement interface	MD simulated annealing refinement	Refines other protein-protein docking poses http://haddock.science.uu.nl/services/HADDOCK2.2/haddockserver-refinement.html

Identification of Ligand Binding Cavities and Hot Spot Residues at the PPI Interface

The next step for the identification of inhibitors of PPIs is to identify the cavities and ligand binding hot spot residues in PPIs interface region. It has been demonstrated that PPIs interface are flat and contains featureless large hydrophobic surface region (Cierpicki & Grembecka, 2015). Nevertheless, the large hydrophobic surface area is separated into three to four superficial sub-pockets which could be of 50 Å^3 size (Fuller et al., 2009). Moreover, it is also demonstrated that only few clustered residues of each partners of PPIs contributes to the majority of binding free energy (affinity) towards the complex formation. The clustered residues which are mainly contributing to the binding energy are called 'hot spots' (Bogan & Thorn, 1998; Clackson & Wells, 1995). Further, these hot spots of PPIs are conserved throughout evolution and show high propensity for ligand binding. Thus, identification of superficial cavities and ligand binding hot spots are two primary steps that are very critical for the SBDD of PPIs inhibitors. It should also be noted that sometimes it is not possible to get both these structural features at the same site. The ligand binding cavity may exists at one site and the hot spot residues exist at other site or the hot spot residues may be projected into the ligand binding cavity. Therefore, the integrated knowledge that is derived from both the cavity and hot spots detection can be used for SBDD of PPI inhibitors (Kuenemann et al., 2015; Laraia et al., 2015; Watkins & Arora, 2015).

Ligand binding pockets or cavities on the protein surface can be identified using various *in silico* tools which are mentioned in the virtual ligand screening website (http://www.vls3d.com/links/bioinformatics/binding-pockets). Generally, these tools are classified into three categories: (1) protein geometry based methods; (2) energy based methods and (3) combination of above two methods (H. Li, Kasam, Tautermann, Seeliger, & Vaidehi, 2014; Zheng, Gan, Wang, & Wang, 2013). Most of the tools using these methods are also suitable for detecting the cavities in traditional drug targets. More specifically, few energy based methods such as FTMap server (Ngan et al., 2012) and FindBindSite (H. Li et al., 2014) were specifically developed to detect the ligand binding cavities at PPIs interface. These methods use molecular docking approach to map the whole protein surface using small probes or fragments. Consequently, mapping small molecule fragment on protein surface by docking effectively identifies the superficial cavities at PPIs.

The PPIs hot spots can be identified using either experimental or computational methods. Among these methods, experimental alanine scanning mutagenesis is well known and is widely employed in the literature to study the structural and functional aspects of the protein (Lefevre, Remy, & Masson, 1997). Moreover, it exactly identifies the hot spots by substituting the selected residue with alanine. Upon substitution (mutation) it estimates binding free energy by means of thermodynamic properties. After alanine scanning mutagenesis, if the total binding free energy of mutant protein drops more than 1-2 kcal/mol, as compared to the native protein then the mutated amino acid is considered as hot spot residue. However, identification of hot spots using experimental alanine scanning mutagenesis is very expensive and time consuming. On the other hand, various *in silico* methods such as Robetta server (Kim et al., 2004), DrugScorePPI server (Kruger & Gohlke, 2010), HotRegion database (Cukuroglu, Gursoy, & Keskin, 2012), KFC server (Darnell, LeGault, & Mitchell, 2008) and PocketQuery server (Koes & Camacho, 2012a) could be used to detect the hot spot residues in PPIs interface region. These methods use different algorithms, energy functions and parameters to optimize and to correlate with the experimental results. These methods are classified into three categories: (1) simple free energy function based methods, (2) molecular dynamics simulation based methods and (3) machine learning based methods. In fact, all the developed methods have their own advantages and limitation and are widely used by various scientific groups in their research. Moreover, these are very fast, reliable and also correlate with experimental results. Thus, along with the superficial cavities that are identified by cavity detection methods, hot spots knowledge could be used to define the ligand binding sites or to derive the pharmacophore features and used for screening the small molecule databases. Some of the hot spot and ligand binding cavity detection methods and tools that are specially developed for PPIs are listed in Table 3.

Virtual Screening for the Identification of PPI Inhibitors

After the identification of ligand binding sites of PPIs using the cavity detection methods and the detection of hot spots, small molecule inhibitors or initial hits for PPIs can be identified using virtual screening (VS) methods (Reddy, Pati, Kumar, Pradeep, & Sastry, 2007). It is a common practice in the drug discovery project to search the libraries of small molecules to identify molecules that are most likely to bind to a drug target. VS can be carried out using either molecular docking or pharmacophore based methods.

Molecular docking (Ferreira, Dos Santos, Oliva, & Andricopulo, 2015) is a computational technique, used in drug design projects to predict the binding orientation of small molecule inhibitors in their target proteins. Thus, the knowledge derived from binding orientation could be used to predict the binding affinity or activity of the small molecules using scoring functions. Hence, docking has become a powerful *in silico* tool in various pharmaceutical industries. Some of the well-known docking programs that are commonly used by scientific communities are: AutoDock (Morris et al., 2009), GOLD (Jones, Willett,

Table 3. List of PPIs hot spots and ligand binding cavity detection tools and webservers

Sl.No	Name of the Tool	Web Links
Hot Spot Detection Server and Tools		
1	Robetta server	http://www.robetta.org/alascansubmit.jsp_
2	KFC2 server	http://kfc.mitchell-lab.org/_
3	HotRegion	http://prism.ccbb.ku.edu.tr/hotregion/_
4	DrugScorePPI	http://cpclab.uni-duesseldorf.de/dsppi/main.php_
5	HotPoint	http://prism.ccbb.ku.edu.tr/hotpoint/_
6	PredHS	http://nar.oxfordjournals.org/content/early/2014/05/22/nar.gku437_
7	HotSpot Wizard	http://loschmidt.chemi.muni.cz/hotspotwizard/_
8	PCRPi-DB	http://www.bioinsilico.org/cgi-bin/PCRPIDB/htmlPCRPI/home_
9	PCRPi	http://www.bioinsilico.org/PCRPi/_
10	FoldX	http://foldx.crg.es/_
11	Hot Spot Prediction	Standalone tool, can be downloaded from the link http://sfb.kaust.edu.sa/pages/software.aspx
Ligand Binding Cavity Detection Servers and Databases		
12	FTmap	http://ftmap.bu.edu/login.php_
13	FindBindSite	PPIs binding site prediction methods. Not yet developed as a server
14	Pocketquery	http://pocketquery.csb.pitt.edu/_
15	MAPPIS	http://bioinfo3d.cs.tau.ac.il/MAPPIS/_
16	ANCHOR	http://structure.pitt.edu/anchor/_

Glen, Leach, & Taylor, 1997), DOCK (Ewing, Makino, Skillman, & Kuntz, 2001), Glide-XP (Friesner et al., 2006), LibDock (Diller & Merz, 2001), CDOCKER (G. Wu, Robertson, Brooks, & Vieth, 2003), FlexX (Rarey, Kramer, Lengauer, & Klebe, 1996), LigandFit (Venkatachalam, Jiang, Oldfield, & Waldman, 2003), etc. All these programs use different algorithms and parameters, and these could be classified broadly as shape based (e.g. LibDock) and energy based (e.g. Glide XP) methods.

Virtual screening using structure based pharmacophore (SBP) is an important and an alternative method to docking (Pirhadi, Shiri, & Ghasemi, 2013). SBP method uses a set of essential features (steric and electrostatic interactions) that are necessary to maintain the optimal intermolecular interactions with a specific biological target to activate or block its biological response. Generally, pharmacophore features include hydrogen bond acceptor, hydrogen bond donor, hydrophobic, aromatic, positive and negative ionizable groups. Earlier, only ligand based pharmacophore hypothesis were usually employed to screen the small molecule databases in the absence of the protein structure. But, advancement of structural biology and computational chemistry greatly helped to derive the pharmacophore features from protein or protein-ligand complex structures. Thus, several tools are developed to derive the pharmacophore features using protein or protein-ligand structural information. Among them, Catalyst (Discovery Studio 3.1), Phase and E-pharmacophore (Schrödinger Suite 9.2 software package), Ligand and Structure-Based Query Editor (MOE), LiganScout (Wolber & Langer, 2005), and ZINCPharmer (Koes & Camacho, 2012b) are few routinely employed tools in drug discovery projects. Using these tools, the pharmacophore features could be derived from the known protein-ligand crystal structure or exclusively from the identified binding sites or from the identified hot spot residues of the given target PPIs. Moreover, the pharmacophore features can be derived from docking the fragments at the defined binding site of protein using E-pharmacophore module (Schrödinger software suit), which takes the advantages of both ligand and structure based methodologies to generate an energetically optimized pharmacophore hypothesis that can be used to screen the small molecule databases. Interested readers may refer to Clik2Drug website (http://www.click2drug.org/directory_Docking.html) for an exhaustive list of tools available for SBDD.

The PPIs inhibitors identified by VS could be further evaluated using various computational methods. One approach is, assessing the binding stability of the compound with the target protein using molecular dynamics simulation (Mortier et al., 2015). Using this, the dynamic behavior of the compounds with the target protein in the presence solvent molecules could be studied. The second approach is to assess the drug-like properties of the identified hits, using Lipinski's rule of five (Lipinski, Lombardo, Dominy, & Feeney, 2001). Based on the favorable results from the above *in silico* investigations, the identified inhibitors could be subjected to *in vitro* testing and further optimization to identify PPIs inhibitors.

An *In Silico* Approach for the Identification of PPI Inhibitors: Aurora B-INCENP PPI as a Case Study

Aurora kinases are a subclass of protein kinase that plays a crucial role in mitosis through phosphorylation of serine/threonine residues of the substrate protein. There are three isoforms of Aurora kinases (Aurora-A, B and C) which share a conserved ATP binding site. Among these three isoforms, Aurora A and B are over expressed in many of the human solid tumors (Fu, Bian, Jiang, & Zhang, 2007). Consequently, scientific communities are developing Aurora kinase inhibitors for cancer therapy. Presently, over a dozen Aurora kinase inhibitors that target the ATP binding site are in various phases of clinical development (Figure 3) (Cheung, Coumar, Chang, & Hsieh, 2011; Cheung, Coumar, Hsieh, & Chang,

Figure 3. Chemical structures of Aurora kinase inhibitors in clinical development

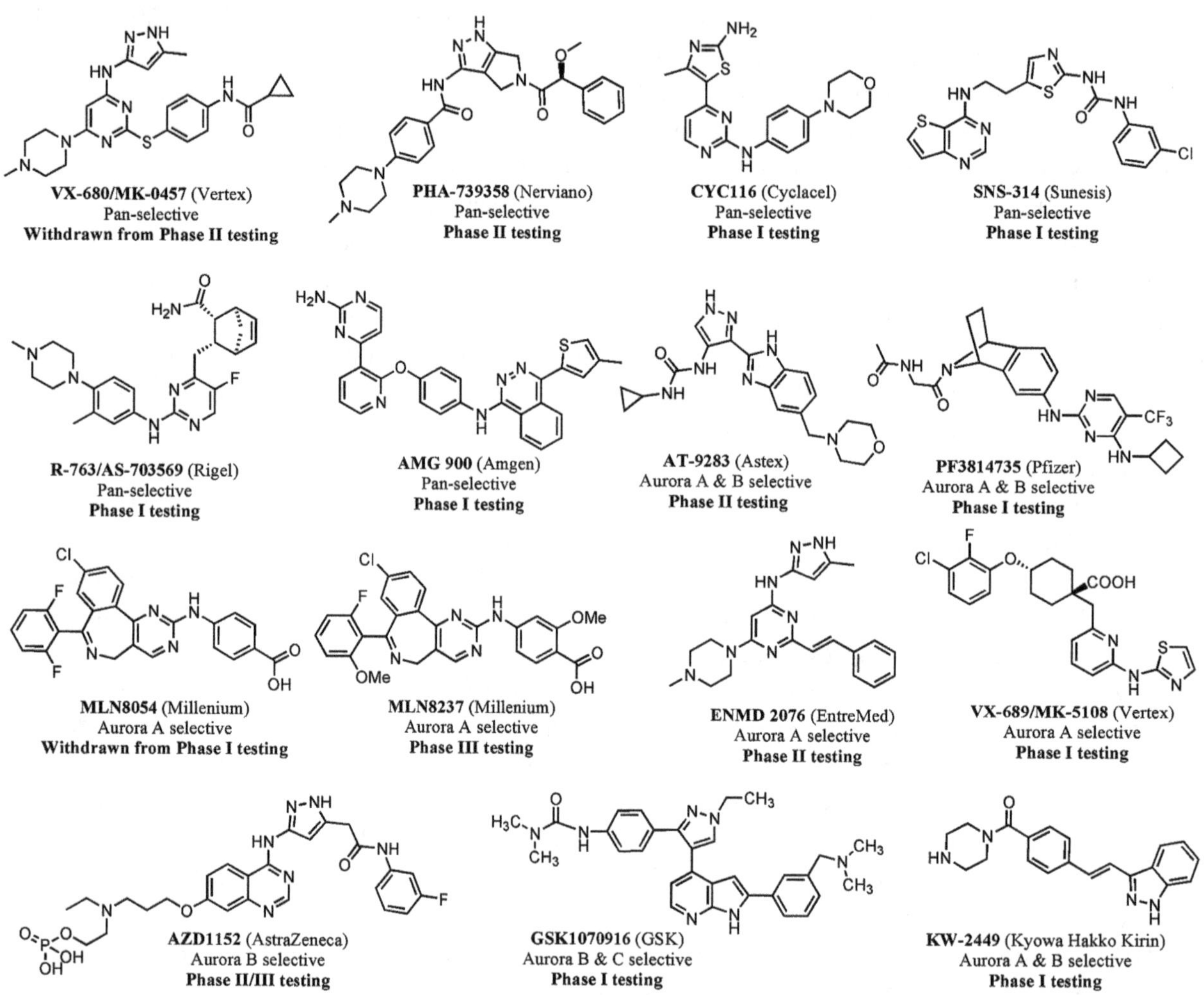

2009; Cheung, Sarvagalla, Lee, Huang, & Coumar, 2014). Most of the developed Aurora kinase inhibitors are pan-selective (Aurora A, B and C selective), and few of them are isoform selective (Aurora A or B selective); but none of them have been approved for clinical use so far. The major reasons for the failure is due to on- and off-target toxicity problems associated with the administration of non-selective Aurora kinase inhibitors (Sarvagalla & Coumar, 2015).

As Aurora kinase family shares a conserved ATP binding site, developing isoform selective inhibitors is very difficult. Hence, most of the developed Aurora kinase inhibitors are pan-selective, and evaluation of these inhibitors in the clinical trials reported on- and off-target toxicity (Kollareddy et al., 2012; Sarvagalla & Coumar, 2015). The major on-target toxicities that are reported in clinics are: neutropenia, leukopenia, myelosuppression, septicemia, pneumonia and thrombocytopenia. In addition to this, several off-target problems including mucositis, proctalgia, hypertension, somnolence, stomatitis, grade 3 increases in aspartate aminotransferase and grade 2 ventricular dysfunctions have been reported. Drug resistance, is another major problem that need to be addressed for Aurora kinase inhibitors (Girdler et al.,

2008). Because of the development of drug resistance, drug efficacy and affinity is greatly affected. It is also reported that various types of cancer cell lines develop resistance to many Aurora kinase inhibitors. Another drawback of Aurora kinase inhibitors is the lack of chemical diversity, because of the conserved ATP binding site, resulting in difficulties in IP protection (Backes et al., 2008).

The above reported problems could be overcome by targeting the PPI sites of Aurora kinases. Aurora kinase A interacts with TPX2 (Bayliss, Sardon, Vernos, & Conti, 2003); whereas Aurora B interacts with INCENP protein (Sessa et al., 2005). The interactions of TPX2 and INCENP with their respective Aurora kinase isoform are very specific and essential for the full activation of the kinase. Thus, inhibiting these PPIs would be very specific and serves as an alternative strategy to target the Aurora kinase for cancer treatment with reduced toxicity profile. Recent evidence also indicates that the design of isoform selective Aurora kinase inhibitor has advantage over the pan-selective Aurora inhibitors (Sarvagalla & Coumar, 2015). This could be achieved by targeting PPI sites of Aurora enzymes (Gohard, St-Cyr, Tyers, & Earnshaw, 2014). To this end, to design isoform selective small molecule inhibitors of Aurora kinase B, here the authors have applied the cavity and hot spot detection techniques discussed above to identify ligand binding PPIs sites in the interface of Aurora B-INCENP. Further, molecular docking was carried out to identify PPI inhibitors by screening approved drugs (http://www.drugbank.ca/). The methodology applied for the identification of Aurora B-INCENP PPIs inhibitors is depicted in Figure 4.

For this process, first the Aurora B-INCENP protein crystal structure (PDB ID: 4AF3) was downloaded from PDB, and was processed for docking as reported earlier (Sarvagalla et al., 2015). Then, the prepared protein was used to identify the ligand binding cavities using two different methods: The first one is shape based method (Define Site- discovery studio module) and the second one is based on the energy based method (FTMap server). These two methods identified a common ligand binding

Figure 4. Schematic representation of the workflow used to identify PPI inhibitors of Aurora B-INCENP complex

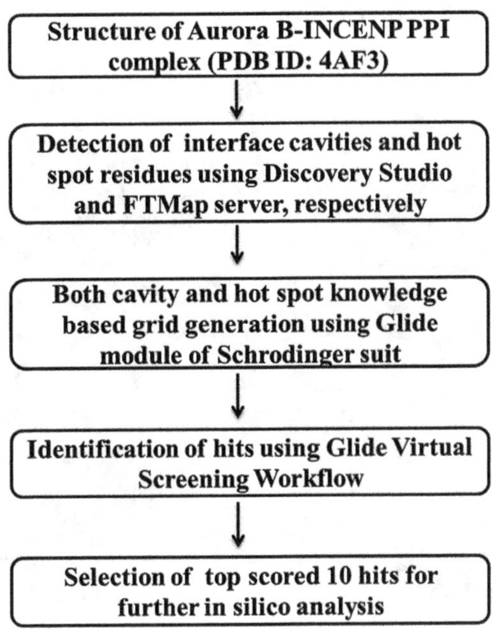

cavity at the interface of Aurora B-INCENP complex. Further, comparison of the results from both the method revealed that most of the residues identified in the cavity are common. The residues involved in the cavity are shown in Table 4, and the location of the cavity that was identified by discovery studio and FTMap are shown in Figure 5.

Further, the hot spot residues of both the proteins (Aurora B and INCENP) were identified by mutating the interface residue to alanine using discovery studio module and Robetta web-server. Results from both the methods are shown in Table 5. Analysis of the results revealed that four hot spots residues (Phe101, Ile102, Tyr141 and Leu210) from Aurora B and six hot spot residues (Ile855, Gln858, Tyr859, Leu865, Leu868 and Phe869) from INCENP protein are projected into the identified cavity. Next, the integrated knowledge derived from the cavity and hot spot detection was used to define the ligand binding site

Table 4. Results from cavity detection using Discovery Studio and FTMap server for Aurora B-INCENP complex

Residues Identified in the Cavity of Aurora B-INCENP Complex	
Discovery Studio Results	**FTMap Server Results**
Aurora B: Leu93, Phe101, Ile102, Val103, Ala104, Leu103, His133, His134, Pro135, Asn136, Ile137, Leu138, Agr139, Leu140, Tyr141, Ile153, Leu154, Glu155, Tyr156, Ala157, Pro158, Agr159, Leu207, Leu208, Leu210, Lys211, Glu213, Leu214, Lys215, Leu335, Pro336, Pro337 and Ser338.	**Aurora B:** Phe101, Ile102, His134, Pro135, Asn136, Ile137, Leu138, Agr139, Tyr141, Glu155, Tyr156, Ala157, Pro158, Leu207, Leu208, Gly209, Leu210, Lys211, Glu213, Leu214, Lys215, Pro336, Pro337 and Ser338.
INCENP: Ala854, Ile855, Ile856, His857, Gln858, Tyr859, Tyr860, His861, Pro862, Pro863, Leu865, Leu868, Phe869	**INCENP:** Ile855, Gln858, Tyr859, His861, Pro862, Pro863 and Leu865

Figure 5. Location of the ligand binding cavity in the Aurora B-INCENP interface region. (a) Identified by Discovery Studio and (b) identified by FTMap server.

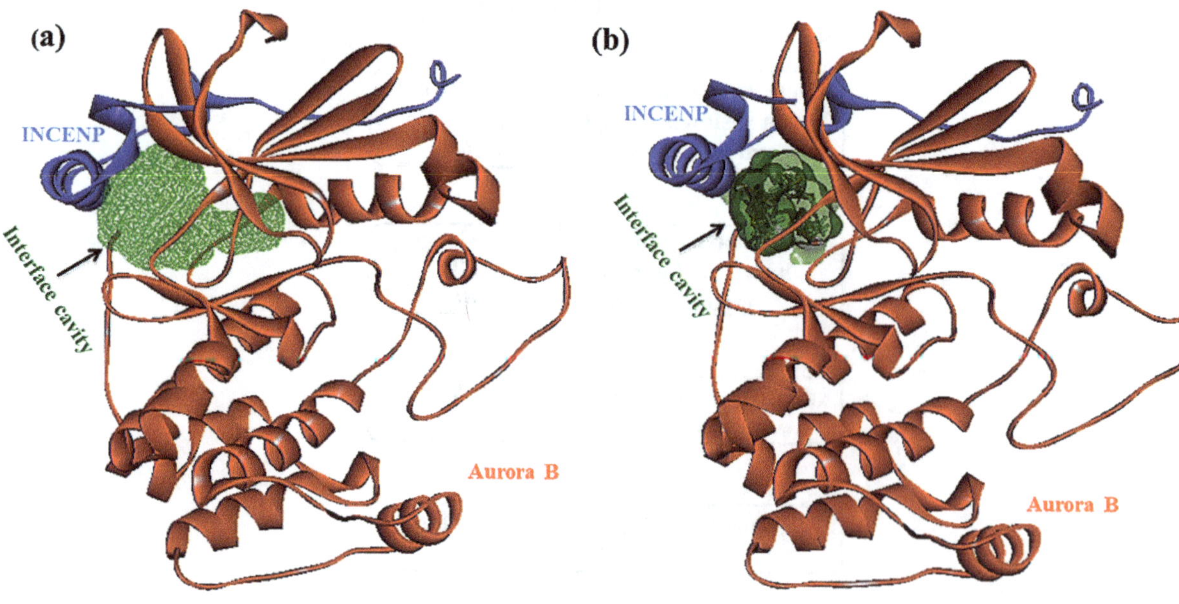

Table 5. List of identified hot spot residues of Aurora B- INCENP complex using Robetta server and Discovery Studio

Sl.No	Residue	Robetta Server	Discovery Studio	Sl.No	Residue	Robetta Server	Discovery Studio
Aurora B Interface Residues				INCENP Interface Residues			
1	Phe72	H (1.20)	H (1.26)	17	Trp845	H (4.62)	H (4.54)
2	Glu78	H (1.00)	NH (0.28)	18	Leu851	H (1.05)	H (1.26)
3	Leu93	H (1.33)	H (1.10)	19	Ile855	H (1.83)	H (1.27)
4	Arg95	H (1.33)	H (1.73)	20	Gln858	H (1.83)	H (1.36)
5	Phe101	H (2.31)	H (1.47)	21	Tyr859	H (3.58)	H (2.89)
6	Ile102	H (1.65)	H (1.25)	22	Leu865	H (1.05)	NH (0.34)
7	Lys110	H (1.46)	H (1.97)	23	Leu868	H (1.05)	H (0.96)
8	Leu122	NH (0.0)	H (0.57)	24	Phe869	H (3.65)	H (2.73)
9	Ile126	NH (0.0)	H (0.69)	25	Ile872	H (2.26)	H (2.21)
10	Tyr141	H (1.40)	H (1.72)	26	Leu875	H (2.32)	H (1.96)
11	Asn142	NH (0.0)	H (1.21)	27	Leu877	H (1.89)	H (1.91)
12	Tyr143	H (2.88)	H (2.48)	28	Ile880	H (1.35)	H (1.30)
13	Tyr145	H (1.87)	H (1.40)	28	Phe881	H (2.99)	H (2.76)
14	Ile150	NH (0.0)	H (1.03)	30	Lys882	NH (0.0)	H (0.94)
15	Tyr156	NH (0.0)	H (0.73)				
16	Leu210	H (1.89)	H (0.95)				

using 'Receptor Grid Generation' option of glide module; this was followed by Virtual Screening of approved drug bank compounds. For VS, Glide module 'Virtual Screening Workflow' was employed; this module provides three stages of screening: first HTVS (high-throughput virtual screening) mode, then SP (standard precision) mode and the final XP (extra precision) mode. The VS workflow, enriches the data in every step by eliminating false positives by applying more extensive sampling and scoring methods. After screening, top scored 10 compounds were selected as initial hits for Aurora B-INCENP PPI inhibition. The selected 10 compounds, Glide XP docking scores and the hot spot residues involved in non-bonded interaction are listed in Table 6. Analysis of these hits in the PPI binding site shows that few hot spot residues of Aurora B (Phe101, Tyr141 and Leu210) and INCENP (Ile855, Gln858, Tys859 and Leu865) are consistently making hydrogen bonds as well as hydrophobic interactions with the identified hits. Thus, the selected hits might disrupt the PPI of Aurora B-INCENP complex by interacting with the important hot spot residues. For illustration, top scored two compounds 3D and 2D binding mode in the Aurora B-INCENP complex is shown in Figure 6. As a next step, the identified hits need to be tested in *in vitro* assay, to confirm their ability to interfere in Aurora B-INCENP PPI.

CONCLUSION AND FUTURE DIRECTIONS

Most of the kinase inhibitors in the market or those under development act by interacting with the ATP binding site of the kinase. Due to the conserved nature of the ATP binding site of the kinase, existing

Table 6. Top 10 hits identified by virtual screening of DrugBank database as Aurora B-INCENP PPI inhibitors

SI.No	Drug Name	Glide XP Docking Score	Hot Spot Residues Involved in Non-Bonded Interaction with the Drugs	
			Aurora B	INCENP
1	Masoprocol	-10.065	Phe101, Tyr141 and Leu210	Ile855, Gln858, Tys859 and Leu865
2	Trifluridine	-9.342	Phe101, Tys141 and Leu210	Ile855, Gln858, Tys859 and Leu865
3	Fenoldopam	-8.948	Phe101 and Tys141	Tys859 and Leu865
4	Telbivudine	-8.942	Phe101, Tys141 and Leu210	Ile855, Gln858, Tys859 and Leu865
5	Dexrazoxane	-8.729	Phe101, Ile102, Tys141 and Leu210	Ile855, Gln858, Tys859 and Leu865
6	Dobutamine	-8.300	Phe101, Tys141 and Leu210	Ile855, Gln858, Tys859 and Leu865
7	Pioglitazone	-8.206	No	No
8	Miglustat	-8.116	Phe101, Tys141 and Leu210	Ile855, Gln858, Tys859 and Leu865
9	Idoxuridine	-8.079	Phe101, Tys141 and Leu210	Ile855, Gln858, Tys859 and Leu865
10	Mitoxantrone	-8.070	Leu210	Leu865

Figure 6. Binding mode of the top two hits in Aurora B (green)-INCENP (red) complex interface region. (a) 3D binding mode of Masoprocol (b) 2D interaction map of Masoprocol (c) 3D binding mode of Trifluridine and (d) 2D interaction map of Trifluridine.

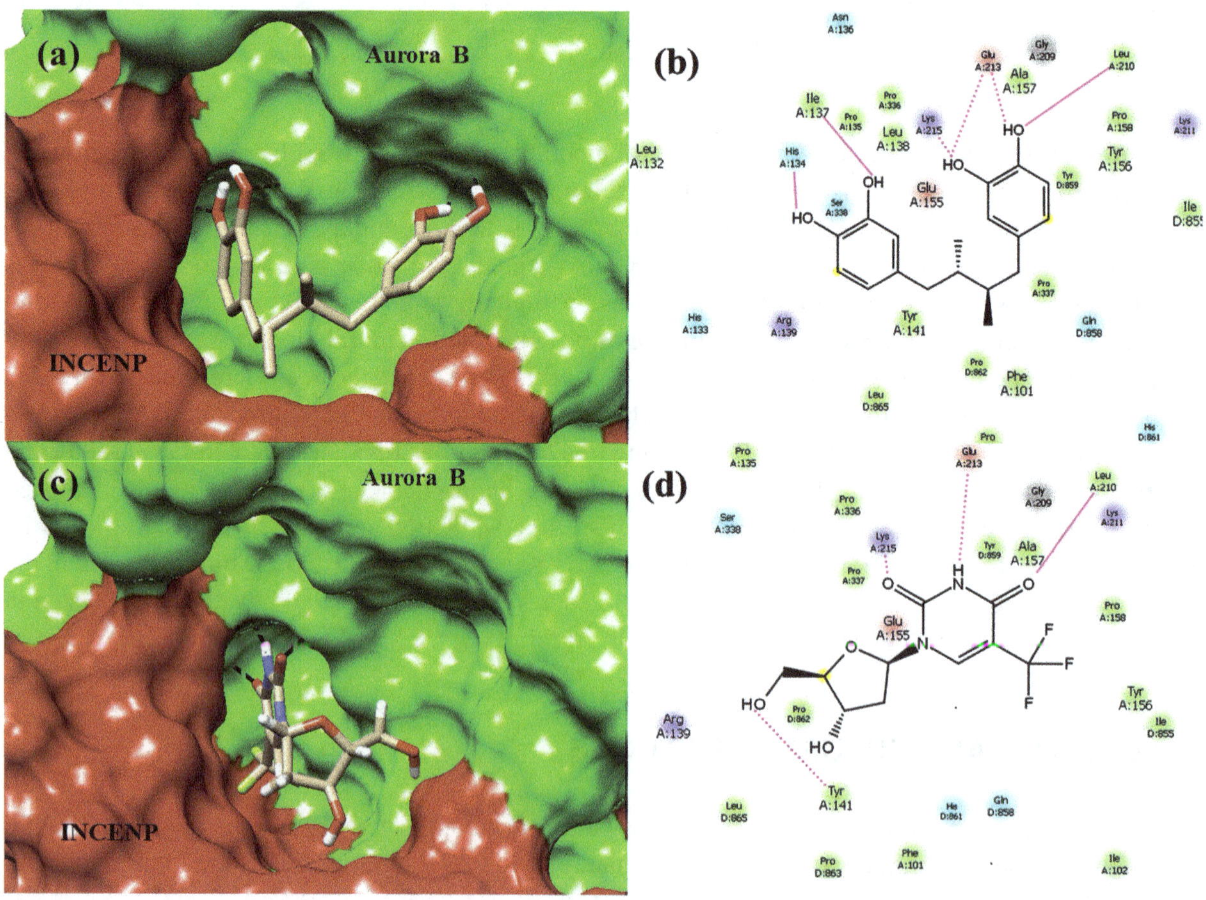

kinase inhibitors are burdened with drawbacks including off- and on-target toxicity and development of drug resistance. These drawbacks could be overcome by targeting the protein-protein interactions (PPIs) of kinase, instead of targeting the ATP binding site. Though targeting the PPIs is a difficult task and success in the form of a marketed drug is yet to come, it is worthwhile exploring these targets, as they are very selective in nature and provides distinct advantages. Moreover, if the PPIs are considered as potential drug targets, then the number of targets would exceed the druggable human kinome; because large numbers of PPIs exist in a living cell as compared to human kinome, and many of them are involved in disease conditions.

Towards this end, here the authors have provided an *in silico* framework for the identification of kinase PPI inhibitors. By utilizing the structural knowledge (cavities and hot spots) of kinase PPI interfaces, structure-based drug design could be carried out. As a case study, Aurora B-INCENP PPI inhibitor identification using the *in silico* framework is discussed. In the future, availability of 3D protein-protein interaction complex, advancement in protein-protein docking algorithms and reliable *in silico* tools for the detection of hot spots, could greatly enhance the speed at which PPI inhibitors could be reliably identified.

ACKNOWLEDGMENT

Financial support from the Science and Engineering Research Board (SR/FT/LS-64/2011 for MSC) and Department of Biotechnology (DBT-JRF/2012-13/80 for SS), Govt. of India, are gratefully acknowledged.

REFERENCES

Abeyrathna, P., & Su, Y. (2015). The critical role of Akt in cardiovascular function. *Vascular Pharmacology*, *74*, 38–48. doi:10.1016/j.vph.2015.05.008 PMID:26025205

Adams, J. A. (2001). Kinetic and catalytic mechanisms of protein kinases. *Chemical Reviews*, *101*(8), 2271–2290. doi:10.1021/cr000230w PMID:11749373

Anderson, A. C. (2003). The process of structure-based drug design. *Chemistry & Biology*, *10*(9), 787–797. doi:10.1016/j.chembiol.2003.09.002 PMID:14522049

Arkin, M. R., Tang, Y., & Wells, J. A. (2014). Small-molecule inhibitors of protein-protein interactions: Progressing toward the reality. *Chemistry & Biology*, *21*(9), 1102–1114. doi:10.1016/j.chembiol.2014.09.001 PMID:25237857

Backes, A., Zech, B., Felber, B., Klebl, B., & Muller, G. (2008). Small-molecule inhibitors binding to protein kinases. Part I: Exceptions from the traditional pharmacophore approach of type I inhibition. *Expert Opin Drug Discov*, *3*(12), 1409–1425. doi:10.1517/17460440802579975 PMID:23506106

Banks, A. S., McAllister, F. E., Camporez, J. P., Zushin, P. J., Jurczak, M. J., Laznik-Bogoslavski, D., & Spiegelman, B. M. et al. (2015). An ERK/Cdk5 axis controls the diabetogenic actions of PPARgamma. *Nature*, *517*(7534), 391–395. doi:10.1038/nature13887 PMID:25409143

Barnes, P. J. (2013). New anti-inflammatory targets for chronic obstructive pulmonary disease. *Nature Reviews. Drug Discovery*, *12*(7), 543–559. doi:10.1038/nrd4025 PMID:23977698

Barouch-Bentov, R., & Sauer, K. (2011). Mechanisms of drug resistance in kinases. *Expert Opinion on Investigational Drugs*, *20*(2), 153–208. doi:10.1517/13543784.2011.546344 PMID:21235428

Bayliss, R., Sardon, T., Vernos, I., & Conti, E. (2003). Structural basis of Aurora-A activation by TPX2 at the mitotic spindle. *Molecular Cell*, *12*(4), 851–862. doi:10.1016/S1097-2765(03)00392-7 PMID:14580337

Besbes, S., Mirshahi, M., Pocard, M., & Billard, C. (2015). New dimension in therapeutic targeting of BCL-2 family proteins. *Oncotarget*, *6*(15), 12862–12871. doi:10.18632/oncotarget.3868 PMID:25970783

Biasini, M., Bienert, S., Waterhouse, A., Arnold, K., Studer, G., Schmidt, T., . . . Schwede, T. (2014). SWISS-MODEL: modelling protein tertiary and quaternary structure using evolutionary information. *Nucleic Acids Res, 42*, W252-258. doi: 10.1093/nar/gku340

Bogan, A. A., & Thorn, K. S. (1998). Anatomy of hot spots in protein interfaces. *Journal of Molecular Biology*, *280*(1), 1–9. doi:10.1006/jmbi.1998.1843 PMID:9653027

Bogoyevitch, M. A., Barr, R. K., & Ketterman, A. J. (2005). Peptide inhibitors of protein kinases-discovery, characterisation and use. *Biochimica et Biophysica Acta*, *1754*(1-2), 79–99. doi:10.1016/j.bbapap.2005.07.025 PMID:16182621

Budhidarmo, R., & Day, C. L. (2015). IAPs: Modular regulators of cell signalling. *Seminars in Cell & Developmental Biology*, *39*, 80–90. doi:10.1016/j.semcdb.2014.12.002 PMID:25542341

Burtness, B. (2014). Targeted agents: Management of dermatologic toxicities. *Journal of the National Comprehensive Cancer Network*, *12*(5Suppl), 793–796. PMID:24853219

Cheung, C. H., Coumar, M. S., Chang, J. Y., & Hsieh, H. P. (2011). Aurora kinase inhibitor patents and agents in clinical testing: An update (2009-10). *Expert Opin Ther Pat*, *21*(6), 857–884. doi:10.1517/13543776.2011.574614 PMID:21591849

Cheung, C. H., Coumar, M. S., Hsieh, H. P., & Chang, J. Y. (2009). Aurora kinase inhibitors in preclinical and clinical testing. *Expert Opinion on Investigational Drugs*, *18*(4), 379–398. doi:10.1517/13543780902806392 PMID:19335272

Cheung, C. H., Sarvagalla, S., Lee, J. Y., Huang, Y. C., & Coumar, M. S. (2014). Aurora kinase inhibitor patents and agents in clinical testing: An update (2011 - 2013). *Expert Opin Ther Pat*, *24*(9), 1021–1038. doi:10.1517/13543776.2014.931374 PMID:24965505

Chico, L. K., Van Eldik, L. J., & Watterson, D. M. (2009). Targeting protein kinases in central nervous system disorders. *Nature Reviews. Drug Discovery*, *8*(11), 892–909. doi:10.1038/nrd2999 PMID:19876042

Cierpicki, T., & Grembecka, J. (2015). Targeting protein-protein interactions in hematologic malignancies: Still a challenge or a great opportunity for future therapies? *Immunological Reviews*, *263*(1), 279–301. doi:10.1111/imr.12244 PMID:25510283

Clackson, T., & Wells, J. A. (1995). A hot spot of binding energy in a hormone-receptor interface. *Science*, *267*(5196), 383–386. doi:10.1126/science.7529940 PMID:7529940

Corbel, C., Wang, Q., Bousserouel, H., Hamdi, A., Zhang, B., Lozach, O., & Bach, S. et al. (2011). First BRET-based screening assay performed in budding yeast leads to the discovery of CDK5/p25 interaction inhibitors. *Biotechnology Journal*, *6*(7), 860–870. doi:10.1002/biot.201100138 PMID:21681968

Cukuroglu, E., Engin, H. B., Gursoy, A., & Keskin, O. (2014). Hot spots in protein-protein interfaces: Towards drug discovery. *Progress in Biophysics and Molecular Biology*, *116*(2-3), 165–173. doi:10.1016/j.pbiomolbio.2014.06.003 PMID:24997383

Cukuroglu, E., Gursoy, A., & Keskin, O. (2012). HotRegion: A database of predicted hot spot clusters. *Nucleic Acids Research*, *40*(Database issue), D829–D833. doi:10.1093/nar/gkr929 PMID:22080558

D'Cruz, O. J., & Uckun, F. M. (2013). Novel Bruton's tyrosine kinase inhibitors currently in development. *Onco Targets Ther*, *6*, 161–176. doi:10.2147/OTT.S33732 PMID:23493945

Damgaard, R. B., & Gyrd-Hansen, M. (2011). Inhibitor of apoptosis (IAP) proteins in regulation of inflammation and innate immunity. *Discovery Medicine*, *11*(58), 221–231. PMID:21447281

Darnell, S. J., LeGault, L., & Mitchell, J. C. (2008). KFC Server: interactive forecasting of protein interaction hot spots. *Nucleic Acids Res, 36*, W265-269. doi: 10.1093/nar/gkn346

Dickson, H. M., Wilbur, A., Reinke, A. A., Young, M. A., & Vojtek, A. B. (2015). Targeted inhibition of the Shroom3-Rho kinase protein-protein interaction circumvents Nogo66 to promote axon outgrowth. *BMC Neuroscience*, *16*(1), 34. doi:10.1186/s12868-015-0171-5 PMID:26077244

Diller, D. J., & Merz, K. M. Jr. (2001). High throughput docking for library design and library prioritization. *Proteins*, *43*(2), 113–124. doi:10.1002/1097-0134(20010501)43:2<113::AID-PROT1023>3.0.CO;2-T PMID:11276081

Dubrez, L., Berthelet, J., & Glorian, V. (2013). IAP proteins as targets for drug development in oncology. *Onco Targets Ther*, *9*, 1285–1304. doi:10.2147/OTT.S33375 PMID:24092992

Engelman, A., Kessl, J. J., & Kvaratskhelia, M. (2013). Allosteric inhibition of HIV-1 integrase activity. *Current Opinion in Chemical Biology*, *17*(3), 339–345. doi:10.1016/j.cbpa.2013.04.010 PMID:23647983

Eswar, N., Webb, B., Marti-Renom, M. A., Madhusudhan, M. S., Eramian, D., Shen, M. Y., . . . Sali, A. (2006). Comparative protein structure modeling using Modeller. Curr Protoc Bioinformatics. doi:10.1002/0471250953.bi0506s15

Eswaran, J., & Knapp, S. (2010). Insights into protein kinase regulation and inhibition by large scale structural comparison. *Biochimica et Biophysica Acta*, *1804*(3), 429–432. doi:10.1016/j.bbapap.2009.10.013 PMID:19854302

Ewing, T. J., Makino, S., Skillman, A. G., & Kuntz, I. D. (2001). DOCK 4.0: Search strategies for automated molecular docking of flexible molecule databases. *Journal of Computer-Aided Molecular Design*, *15*(5), 411–428. doi:10.1023/A:1011115820450 PMID:11394736

Fabbro, D. (2015). 25 years of small molecular weight kinase inhibitors: Potentials and limitations. *Molecular Pharmacology*, *87*(5), 766–775. doi:10.1124/mol.114.095489 PMID:25549667

Fabbro, D., Cowan-Jacob, S. W., Mobitz, H., & Martiny-Baron, G. (2012). Targeting cancer with small-molecular-weight kinase inhibitors. *Methods in Molecular Biology (Clifton, N.J.)*, *795*, 1–34. doi:10.1007/978-1-61779-337-0_1 PMID:21960212

Ferreira, L. G., Dos Santos, R. N., Oliva, G., & Andricopulo, A. D. (2015). Molecular Docking and Structure-Based Drug Design Strategies. *Molecules (Basel, Switzerland)*, *20*(7), 13384–13421. doi:10.3390/molecules200713384 PMID:26205061

Friesner, R. A., Murphy, R. B., Repasky, M. P., Frye, L. L., Greenwood, J. R., Halgren, T. A., & Mainz, D. T. et al. (2006). Extra precision glide: Docking and scoring incorporating a model of hydrophobic enclosure for protein-ligand complexes. *Journal of Medicinal Chemistry*, *49*(21), 6177–6196. doi:10.1021/jm051256o PMID:17034125

Fu, J., Bian, M., Jiang, Q., & Zhang, C. (2007). Roles of Aurora kinases in mitosis and tumorigenesis. *Molecular Cancer Research*, *5*(1), 1–10. doi:10.1158/1541-7786.MCR-06-0208 PMID:17259342

Fuller, J. C., Burgoyne, N. J., & Jackson, R. M. (2009). Predicting druggable binding sites at the protein-protein interface. *Drug Discovery Today*, *14*(3-4), 155–161. doi:10.1016/j.drudis.2008.10.009 PMID:19041415

Gambacorti-Passerini, C., & Piazza, R. (2015). Imatinib-A New Tyrosine Kinase Inhibitor for First-Line Treatment of Chronic Myeloid Leukemia in 2015. *JAMA Oncol*, *1*(2), 143–144. doi:10.1001/jamaoncol.2015.50 PMID:26181011

Garuti, L., Roberti, M., & Bottegoni, G. (2010). Non-ATP competitive protein kinase inhibitors. *Current Medicinal Chemistry*, *17*(25), 2804–2821. doi:10.2174/092986710791859333 PMID:20586715

Garuti, L., Roberti, M., & Bottegoni, G. (2011). Irreversible protein kinase inhibitors. *Current Medicinal Chemistry*, *18*(20), 2981–2994. doi:10.2174/092986711796391705 PMID:21651479

Girdler, F., Sessa, F., Patercoli, S., Villa, F., Musacchio, A., & Taylor, S. (2008). Molecular basis of drug resistance in aurora kinases. *Chemistry & Biology*, *15*(6), 552–562. doi:10.1016/j.chembiol.2008.04.013 PMID:18559266

Godin-Heymann, N., Ulkus, L., Brannigan, B. W., McDermott, U., Lamb, J., Maheswaran, S., & Haber, D. A. et al. (2008). The T790M "gatekeeper" mutation in EGFR mediates resistance to low concentrations of an irreversible EGFR inhibitor. *Molecular Cancer Therapeutics*, *7*(4), 874–879. doi:10.1158/1535-7163.MCT-07-2387 PMID:18413800

Gohard, F. H., St-Cyr, D. J., Tyers, M., & Earnshaw, W. C. (2014). Targeting the INCENP IN-box-Aurora B interaction to inhibit CPC activity in vivo. *Open Biology*, *4*(11), 140–163. doi:10.1098/rsob.140163 PMID:25392451

Guo, W., Wisniewski, J. A., & Ji, H. (2014). Hot spot-based design of small-molecule inhibitors for protein-protein interactions. *Bioorganic & Medicinal Chemistry Letters*, *24*(11), 2546–2554. doi:10.1016/j.bmcl.2014.03.095 PMID:24751445

Han, K. C., Kim, S. Y., & Yang, E. G. (2012). Recent advances in designing substrate-competitive protein kinase inhibitors. *Current Pharmaceutical Design*, *18*(20), 2875–2882. doi:10.2174/138161212800672697 PMID:22571656

Hardwick, J. M., & Soane, L. (2013). Multiple functions of BCL-2 family proteins. *Cold Spring Harbor Perspectives in Biology*, *5*(2), a008722. doi:10.1101/cshperspect.a008722 PMID:23378584

Ho, H. K., Yeo, A. H., Kang, T. S., & Chua, B. T. (2014). Current strategies for inhibiting FGFR activities in clinical applications: Opportunities, challenges and toxicological considerations. *Drug Discovery Today*, *19*(1), 51–62. doi:10.1016/j.drudis.2013.07.021 PMID:23932951

Ivanov, A. A., Khuri, F. R., & Fu, H. (2013). Targeting protein-protein interactions as an anticancer strategy. *Trends in Pharmacological Sciences*, *34*(7), 393–400. doi:10.1016/j.tips.2013.04.007 PMID:23725674

Jin, L., Wang, W., & Fang, G. (2014). Targeting protein-protein interaction by small molecules. *Annual Review of Pharmacology and Toxicology*, *54*(1), 435–456. doi:10.1146/annurev-pharmtox-011613-140028 PMID:24160698

Jones, G., Willett, P., Glen, R. C., Leach, A. R., & Taylor, R. (1997). Development and validation of a genetic algorithm for flexible docking. *Journal of Molecular Biology*, *267*(3), 727–748. doi:10.1006/jmbi.1996.0897 PMID:9126849

Jubb, H., Blundell, T. L., & Ascher, D. B. (2015). Flexibility and small pockets at protein-protein interfaces: New insights into druggability. *Progress in Biophysics and Molecular Biology*, *119*(1), 2–9. doi:10.1016/j.pbiomolbio.2015.01.009 PMID:25662442

Kaidanovich-Beilin, O., & Eldar-Finkelman, H. (2006). Peptides targeting protein kinases: Strategies and implications. *Physiology (Bethesda, MD)*, *21*(6), 411–418. doi:10.1152/physiol.00022.2006 PMID:17119154

Kikuchi, R., Nakamura, K., MacLauchlan, S., Ngo, D. T., Shimizu, I., Fuster, J. J., & Walsh, K. et al. (2014). An antiangiogenic isoform of VEGF-A contributes to impaired vascularization in peripheral artery disease. *Nature Medicine*, *20*(12), 1464–1471. doi:10.1038/nm.3703 PMID:25362254

Kim, D. E., Chivian, D., & Baker, D. (2004). Protein structure prediction and analysis using the Robetta server. *Nucleic Acids Res, 32*, W526-531. doi: 10.1093/nar/gkh468

Koes, D. R., & Camacho, C. J. (2012a). PocketQuery: protein-protein interaction inhibitor starting points from protein-protein interaction structure. *Nucleic Acids Res, 40*, W387-392. doi: 10.1093/nar/gks336

Koes, D. R., & Camacho, C. J. (2012b). ZINCPharmer: pharmacophore search of the ZINC database. *Nucleic Acids Res, 40*, W409-414. doi: 10.1093/nar/gks378

Kollareddy, M., Zheleva, D., Dzubak, P., Brahmkshatriya, P. S., Lepsik, M., & Hajduch, M. (2012). Aurora kinase inhibitors: Progress towards the clinic. *Investigational New Drugs*, *30*(6), 2411–2432. doi:10.1007/s10637-012-9798-6 PMID:22350019

Kozakov, D., Hall, D. R., Chuang, G. Y., Cencic, R., Brenke, R., Grove, L. E., & Vajda, S. et al. (2011). Structural conservation of druggable hot spots in protein-protein interfaces. *Proceedings of the National Academy of Sciences of the United States of America, 108*(33), 13528–13533. doi:10.1073/pnas.1101835108 PMID:21808046

Kruger, D. M., & Gohlke, H. (2010). DrugScorePPI webserver: fast and accurate in silico alanine scanning for scoring protein-protein interactions. *Nucleic Acids Res, 38*, W480-486. doi: 10.1093/nar/gkq471

Kuenemann, M. A., Sperandio, O., Labbe, C. M., Lagorce, D., Miteva, M. A., & Villoutreix, B. O. (2015). In silico design of low molecular weight protein-protein interaction inhibitors: Overall concept and recent advances. *Progress in Biophysics and Molecular Biology, 119*(1), 20–32. doi:10.1016/j.pbiomolbio.2015.02.006 PMID:25748546

Kuttner, Y. Y., & Engel, S. (2012). Protein hot spots: The islands of stability. *Journal of Molecular Biology, 415*(2), 419–428. doi:10.1016/j.jmb.2011.11.009 PMID:22100447

Laraia, L., McKenzie, G., Spring, D. R., Venkitaraman, A. R., & Huggins, D. J. (2015). Overcoming Chemical, Biological, and Computational Challenges in the Development of Inhibitors Targeting Protein-Protein Interactions. *Chemistry & Biology, 22*(6), 689–703. doi:10.1016/j.chembiol.2015.04.019 PMID:26091166

Lefevre, F., Remy, M. H., & Masson, J. M. (1997). Alanine-stretch scanning mutagenesis: A simple and efficient method to probe protein structure and function. *Nucleic Acids Research, 25*(2), 447–448. doi:10.1093/nar/25.2.447 PMID:9016577

Li, H., Kasam, V., Tautermann, C. S., Seeliger, D., & Vaidehi, N. (2014). Computational method to identify druggable binding sites that target protein-protein interactions. *Journal of Chemical Information and Modeling, 54*(5), 1391–1400. doi:10.1021/ci400750x PMID:24762202

Li, X., Keskin, O., Ma, B., Nussinov, R., & Liang, J. (2004). Protein-protein interactions: hot spots and structurally conserved residues often locate in complemented pockets that pre-organized in the unbound states: implications for docking. *Journal of Molecular Biology, 344*(3), 781–795. doi:10.1016/j.jmb.2004.09.051 PMID:15533445

Lionta, E., Spyrou, G., Vassilatis, D. K., & Cournia, Z. (2014). Structure-based virtual screening for drug discovery: Principles, applications and recent advances. *Current Topics in Medicinal Chemistry, 14*(16), 1923–1938. doi:10.2174/1568026614666140929124445 PMID:25262799

Lipinski, C. A., Lombardo, F., Dominy, B. W., & Feeney, P. J. (2001). Experimental and computational approaches to estimate solubility and permeability in drug discovery and development settings. *Advanced Drug Delivery Reviews, 46*(1-3), 3–26. doi:10.1016/S0169-409X(00)00129-0 PMID:11259830

Liu, S., & Kurzrock, R. (2014). Toxicity of targeted therapy: Implications for response and impact of genetic polymorphisms. *Cancer Treatment Reviews, 40*(7), 883–891. doi:10.1016/j.ctrv.2014.05.003 PMID:24867380

Lo Conte, L., Chothia, C., & Janin, J. (1999). The atomic structure of protein-protein recognition sites. *Journal of Molecular Biology, 285*(5), 2177–2198. doi:10.1006/jmbi.1998.2439 PMID:9925793

Ma, B., Elkayam, T., Wolfson, H., & Nussinov, R. (2003). Protein-protein interactions: Structurally conserved residues distinguish between binding sites and exposed protein surfaces. *Proceedings of the National Academy of Sciences of the United States of America*, *100*(10), 5772–5777. doi:10.1073/pnas.1030237100 PMID:12730379

Manning, G., Whyte, D. B., Martinez, R., Hunter, T., & Sudarsanam, S. (2002). The protein kinase complement of the human genome. *Science*, *298*(5600), 1912–1934. doi:10.1126/science.1075762 PMID:12471243

Momand, J., Wu, H. H., & Dasgupta, G. (2000). MDM2--master regulator of the p53 tumor suppressor protein. *Gene*, *242*(1-2), 15–29. doi:10.1016/S0378-1119(99)00487-4 PMID:10721693

Moreira, I. S., Fernandes, P. A., & Ramos, M. J. (2007). Hot spots--a review of the protein-protein interface determinant amino-acid residues. *Proteins*, *68*(4), 803–8012. doi:10.1002/prot.21396 PMID:17546660

Morris, G. M., Huey, R., Lindstrom, W., Sanner, M. F., Belew, R. K., Goodsell, D. S., & Olson, A. J. (2009). AutoDock4 and AutoDockTools4: Automated docking with selective receptor flexibility. *Journal of Computational Chemistry*, *30*(16), 2785–2791. doi:10.1002/jcc.21256 PMID:19399780

Mortier, J., Rakers, C., Bermudez, M., Murgueitio, M. S., Riniker, S., & Wolber, G. (2015). The impact of molecular dynamics on drug design: Applications for the characterization of ligand-macromolecule complexes. *Drug Discovery Today*, *20*(6), 686–702. doi:10.1016/j.drudis.2015.01.003 PMID:25615716

Muller, S., Filippakopoulos, P., & Knapp, S. (2011). Bromodomains as therapeutic targets. *Expert Reviews in Molecular Medicine*, *13*, e29. doi:10.1017/S1462399411001992 PMID:21933453

Murray, B. W., & Miller, N. (2015). Durability of Kinase-Directed Therapies-A Network Perspective on Response and Resistance. *Molecular Cancer Therapeutics*, *14*(9), 1975–1984. doi:10.1158/1535-7163.MCT-15-0088 PMID:26264276

Ngan, C. H., Bohnuud, T., Mottarella, S. E., Beglov, D., Villar, E. A., Hall, D. R., . . . Vajda, S. (2012). FTMAP: extended protein mapping with user-selected probe molecules. *Nucleic Acids Res, 40*(Web Server issue), W271-275. doi: 10.1093/nar/gks441

O'Brien, Z., & Fallah Moghaddam, M. (2013). Small molecule kinase inhibitors approved by the FDA from 2000 to 2011: A systematic review of preclinical ADME data. *Expert Opinion on Drug Metabolism & Toxicology*, *9*(12), 1597–1612. doi:10.1517/17425255.2013.834046 PMID:23992134

O'Hare, T., Eide, C. A., & Deininger, M. W. (2007). Bcr-Abl kinase domain mutations, drug resistance, and the road to a cure for chronic myeloid leukemia. *Blood*, *110*(7), 2242–2249. doi:10.1182/blood-2007-03-066936 PMID:17496200

Pearson, G., Robinson, F., Beers Gibson, T., Xu, B. E., Karandikar, M., Berman, K., & Cobb, M. H. (2001). Mitogen-activated protein (MAP) kinase pathways: Regulation and physiological functions. *Endocrine Reviews*, *22*(2), 153–183. doi:10.1210/edrv.22.2.0428 PMID:11294822

Pirhadi, S., Shiri, F., & Ghasemi, J. B. (2013). Methods and applications of structure based pharmacophores in drug discovery. *Current Topics in Medicinal Chemistry*, *13*(9), 1036–1047. doi:10.2174/1568026611313090006 PMID:23651482

Rarey, M., Kramer, B., Lengauer, T., & Klebe, G. (1996). A fast flexible docking method using an incremental construction algorithm. *Journal of Molecular Biology*, *261*(3), 470–489. doi:10.1006/jmbi.1996.0477 PMID:8780787

Rask-Andersen, M., Zhang, J., Fabbro, D., & Schioth, H. B. (2014). Advances in kinase targeting: Current clinical use and clinical trials. *Trends in Pharmacological Sciences*, *35*(11), 604–620. doi:10.1016/j.tips.2014.09.007 PMID:25312588

Rechfeld, F., Gruber, P., Hofmann, J., & Kirchmair, J. (2011). Modulators of protein-protein interactions: Novel approaches in targeting protein kinases and other pharmaceutically relevant biomolecules. *Current Topics in Medicinal Chemistry*, *11*(11), 1305–1319. doi:10.2174/156802611795589610 PMID:21513500

Reddy, A. S., Pati, S. P., Kumar, P. P., Pradeep, H. N., & Sastry, G. N. (2007). Virtual screening in drug discovery -- a computational perspective. *Current Protein & Peptide Science*, *8*(4), 329–351. doi:10.2174/138920307781369427 PMID:17696867

Rommel, C. (2010). Taking PI3Kdelta and PI3Kgamma one step ahead: Dual active PI3Kdelta/gamma inhibitors for the treatment of immune-mediated inflammatory diseases. *Current Topics in Microbiology and Immunology*, *346*, 279–299. doi:10.1007/82_2010_79 PMID:20593311

Rudolph, J. (2007). Inhibiting transient protein-protein interactions: Lessons from the Cdc25 protein tyrosine phosphatases. *Nature Reviews. Cancer*, *7*(3), 202–211. doi:10.1038/nrc2087 PMID:17287826

Ryan, D. P., & Matthews, J. M. (2005). Protein-protein interactions in human disease. *Current Opinion in Structural Biology*, *15*(4), 441–446. doi:10.1016/j.sbi.2005.06.001 PMID:15993577

Sable, R., & Jois, S. (2015). Surfing the Protein-Protein Interaction Surface Using Docking Methods: Application to the Design of PPI Inhibitors. *Molecules (Basel, Switzerland)*, *20*(6), 11569–11603. doi:10.3390/molecules200611569 PMID:26111183

Sanchez, R., & Zhou, M. M. (2009). The role of human bromodomains in chromatin biology and gene transcription. *Current Opinion in Drug Discovery & Development*, *12*(5), 659–665. PMID:19736624

Sarvagalla, S., & Coumar, M. S. (2015). Structural Biology Insight for the Design of Sub-type Selective Aurora Kinase Inhibitors. *Current Cancer Drug Targets*, *15*(5), 375–393. doi:10.2174/1568009615666150421110401 PMID:25895501

Sarvagalla, S., Singh, V. K., Ke, Y. Y., Shiao, H. Y., Lin, W. H., Hsieh, H. P., & Coumar, M. S. et al. (2015). Identification of ligand efficient, fragment-like hits from an HTS library: Structure-based virtual screening and docking investigations of 2H- and 3H-pyrazolo tautomers for Aurora kinase A selectivity. *Journal of Computer-Aided Molecular Design*, *29*(1), 89–100. doi:10.1007/s10822-014-9807-2 PMID:25344840

Sessa, F., Mapelli, M., Ciferri, C., Tarricone, C., Areces, L. B., Schneider, T. R., & Musacchio, A. et al. (2005). Mechanism of Aurora B activation by INCENP and inhibition by hesperadin. *Molecular Cell*, *18*(3), 379–391. doi:10.1016/j.molcel.2005.03.031 PMID:15866179

Shangary, S., & Wang, S. (2009). Small-molecule inhibitors of the MDM2-p53 protein-protein interaction to reactivate p53 function: A novel approach for cancer therapy. *Annual Review of Pharmacology and Toxicology*, *49*(1), 223–241. doi:10.1146/annurev.pharmtox.48.113006.094723 PMID:18834305

Sheng, C., Dong, G., Miao, Z., Zhang, W., & Wang, W. (2015). State-of-the-art strategies for targeting protein-protein interactions by small-molecule inhibitors. *Chemical Society Reviews*. doi:10.1039/c5cs00252d

Stumpf, M. P., Thorne, T., de Silva, E., Stewart, R., An, H. J., Lappe, M., & Wiuf, C. (2008). Estimating the size of the human interactome. *Proceedings of the National Academy of Sciences of the United States of America*, *105*(19), 6959–6964. doi:10.1073/pnas.0708078105 PMID:18474861

Surget, S., Khoury, M. P., & Bourdon, J. C. (2013). Uncovering the role of p53 splice variants in human malignancy: A clinical perspective. *Onco Targets Ther*, *7*, 57–68. doi:10.2147/OTT.S53876 PMID:24379683

Thaimattam, R., Banerjee, R., Miglani, R., & Iqbal, J. (2007). Protein kinase inhibitors: Structural insights into selectivity. *Current Pharmaceutical Design*, *13*(27), 2751–2765. doi:10.2174/138161207781757042 PMID:17897021

Vakser, I. A. (2014). Protein-protein docking: From interaction to interactome. *Biophysical Journal*, *107*(8), 1785–1793. doi:10.1016/j.bpj.2014.08.033 PMID:25418159

Venkatachalam, C. M., Jiang, X., Oldfield, T., & Waldman, M. (2003). LigandFit: A novel method for the shape-directed rapid docking of ligands to protein active sites. *Journal of Molecular Graphics & Modelling*, *21*(4), 289–307. doi:10.1016/S1093-3263(02)00164-X PMID:12479928

Venkatesan, K., Rual, J. F., Vazquez, A., Stelzl, U., Lemmens, I., Hirozane-Kishikawa, T., & Vidal, M. et al. (2009). An empirical framework for binary interactome mapping. *Nature Methods*, *6*(1), 83–90. doi:10.1038/nmeth.1280 PMID:19060904

Villoutreix, B. O., Kuenemann, M. A., Poyet, J. L., Bruzzoni-Giovanelli, H., Labbe, C., Lagorce, D., & Miteva, M. A. et al. (2014). Drug-Like Protein-Protein Interaction Modulators: Challenges and Opportunities for Drug Discovery and Chemical Biology. *Mol Inform*, *33*(6-7), 414–437. doi:10.1002/minf.201400040 PMID:25254076

Wallace, K. B. (2015). Multiple Targets for Drug-induced Mitochondrial Toxicity. *Current Medicinal Chemistry*, *22*(20), 2488–2492. doi:10.2174/0929867322666150514095424 PMID:25973981

Watkins, A. M., & Arora, P. S. (2015). Structure-based inhibition of protein-protein interactions. *European Journal of Medicinal Chemistry*, *94*, 480–488. doi:10.1016/j.ejmech.2014.09.047 PMID:25253637

Wells, J. A., & McClendon, C. L. (2007). Reaching for high-hanging fruit in drug discovery at protein-protein interfaces. *Nature*, *450*(7172), 1001–1009. doi:10.1038/nature06526 PMID:18075579

White, A. W., Westwell, A. D., & Brahemi, G. (2008). Protein-protein interactions as targets for small-molecule therapeutics in cancer. *Expert Reviews in Molecular Medicine*, *10*, e8. doi:10.1017/S1462399408000641 PMID:18353193

Widmer, H., & Jahnke, W. (2004). Protein NMR in biomedical research. *Cellular and Molecular Life Sciences*, *61*(5), 580–599. doi:10.1007/s00018-003-3382-3 PMID:15004697

Wisniewski, D., Lambek, C. L., Liu, C., Strife, A., Veach, D. R., Nagar, B., & Clarkson, B. et al. (2002). Characterization of potent inhibitors of the Bcr-Abl and the c-kit receptor tyrosine kinases. *Cancer Research*, *62*(15), 4244–4255. PMID:12154026

Wolber, G., & Langer, T. (2005). LigandScout: 3-D pharmacophores derived from protein-bound ligands and their use as virtual screening filters. *Journal of Chemical Information and Modeling*, *45*(1), 160–169. doi:10.1021/ci049885e PMID:15667141

Wu, G., Robertson, D. H., Brooks, C. L. III, & Vieth, M. (2003). Detailed analysis of grid-based molecular docking: A case study of CDOCKER-A CHARMm-based MD docking algorithm. *Journal of Computational Chemistry*, *24*(13), 1549–1562. doi:10.1002/jcc.10306 PMID:12925999

Wu, P., Nielsen, T. E., & Clausen, M. H. (2015). FDA-approved small-molecule kinase inhibitors. *Trends in Pharmacological Sciences*, *36*(7), 422–439. doi:10.1016/j.tips.2015.04.005 PMID:25975227

Wu, Y., & Chakrabarti, S. (2015). ERK5 Mediated Signalling in Diabetic Retinopathy. *Med Hypothesis Discov Innov Ophthalmol*, *4*(1), 17–26. PMID:25861671

Yang, J., Yan, R., Roy, A., Xu, D., Poisson, J., & Zhang, Y. (2015). The I-TASSER Suite: Protein structure and function prediction. *Nature Methods*, *12*(1), 7–8. doi:10.1038/nmeth.3213 PMID:25549265

Zhang, B., Corbel, C., Gueritte, F., Couturier, C., Bach, S., & Tan, V. B. (2011). An in silico approach for the discovery of CDK5/p25 interaction inhibitors. *Biotechnology Journal*, *6*(7), 871–881. doi:10.1002/biot.201100139 PMID:21681969

Zhang, J., Yang, P. L., & Gray, N. S. (2009). Targeting cancer with small molecule kinase inhibitors. *Nature Reviews. Cancer*, *9*(1), 28–39. doi:10.1038/nrc2559 PMID:19104514

Zhao, Y., Aguilar, A., Bernard, D., & Wang, S. (2015). Small-molecule inhibitors of the MDM2-p53 protein-protein interaction (MDM2 Inhibitors) in clinical trials for cancer treatment. *Journal of Medicinal Chemistry*, *58*(3), 1038–1052. doi:10.1021/jm501092z PMID:25396320

Zhao, Z., Wu, H., Wang, L., Liu, Y., Knapp, S., Liu, Q., & Gray, N. S. (2014). Exploration of type II binding mode: A privileged approach for kinase inhibitor focused drug discovery? *ACS Chemical Biology*, *9*(6), 1230–1241. doi:10.1021/cb500129t PMID:24730530

Zheng, X., Gan, L., Wang, E., & Wang, J. (2013). Pocket-based drug design: Exploring pocket space. *The AAPS Journal*, *15*(1), 228–241. doi:10.1208/s12248-012-9426-6 PMID:23180158

KEY TERMS AND DEFINITIONS

Aurora Kinases: Human genome encodes three members of Auora kinases: Aurora A, Aurora B, and Auora C. These members belong to serine/threonine kinase families that are essential for cell division and duplication.

Docking: Docking is a computation tool which predicts predominant binding mode of biomolecule interactions including protein-ligand, protein-protein, protein-DNA, protein-RNA and protein-carbohydrate.

Drug Target: Drug target is a biomolecule within the living organism to which an endogenous ligand or drug molecule binds to produce the therapeutic effect.

Hot Spots: In the protein-protein interaction interface region, a residue or cluster of residues that contributes to the majority of binding free energy for the complex formation is called hot spots.

INCENP: The human gene *INCENP* encodes a protein known as - Inner Centromere Protein, which is essential for the full activation of Aurora B kinase.

Kinase: Kinase is an enzyme which catalyzes the transfer of an ATP terminal phosphate group to a specific substrate protein; the process is called phosphorylation.

Pharmacophore: Pharmacophore is a 3D hypothetical model of molecular features which are essential for the molecular recognition of a ligand by a drug target.

Protein-Protein Interactions: Protein-protein interactions (PPIs) refer to a physical contact that is formed between two or more proteins. As a result of PPIs, a wide array of biological functions including cell division, cell growth, signal transductions, metabolic reactions, and transportation of molecules from one location to another will occur within the living organism.

Chapter 11
Applications of Molecular Docking:
Its Impact and Importance outside the Purview of Drug Discovery

Josephine Anthony
Earth System Science Organization-National Institute of Ocean Technology (ESSO-NIOT), Ministry of Earth Sciences, Govt. of India, India

Kumar T. Shivasankarasubbiah
Earth System Science Organization-National Institute of Ocean Technology (ESSO-NIOT), Ministry of Earth Sciences, Govt. of India, India

Vijaya Raghavan Rangamaran
Earth System Science Organization-National Institute of Ocean Technology (ESSO-NIOT), Ministry of Earth Sciences, Govt. of India, India

Dharani Gopal
Earth System Science Organization-National Institute of Ocean Technology (ESSO-NIOT), Ministry of Earth Sciences, Govt. of India, India

Kirubagaran Ramalingam
Earth System Science Organization-National Institute of Ocean Technology (ESSO-NIOT), Ministry of Earth Sciences, Govt. of India, India

ABSTRACT

Computational tools have extended their reach into different realms of scientific research. Often coupled with molecular dynamics simulation, docking provides comprehensive insight into molecular mechanisms of biological processes. Influence of molecular docking is highly experienced in the field of structure based drug discovery, wherein docking is vital in validating novel lead compounds. The significance of molecular docking is also understood in several environmental and industrial research, in order to untangle the interactions among macromolecules of non-medical interest. Various processes such as bioremediation (REMEDIDOCK), nanomaterial interactions (NANODOCK), nutraceutical interactions (NUTRADOCK), fatty acid biosynthesis (FADOCK), and antifoulers interactions (FOULDOCK) find the application of molecular docking. This chapter emphasizes the involvement of computational techniques in the aforementioned fields to expand our knowledge on macromolecular interacting mechanisms.

DOI: 10.4018/978-1-5225-0362-0.ch011

INTRODUCTION

Interactions between biomolecules are fundamental to all biological processes. Using these interactions, living organisms maintain complex regulatory and metabolic interaction networks that together constitute the processes of life (Cai, Li, Wang, & Chen, 2004). Understanding of the complexity of biological pathways and the interactions between the macromolecules involved has become imperative and molecular docking has catered to this need in a comprehensive manner. Molecular docking may be defined as a concept of computational chemistry which provides solutions to unravel the mechanism behind substrate-ligand interactions. It facilitates the interacting molecule to fit together based on their topography (Figure 1). The main objective of docking is to determine the best possible conformation of protein-ligand, protein-protein and/or other type of interactions with minimal energy. Docking is also seen as a vital technique in modelling the protein-ligand, protein-protein interactions and thereby involved in studies related to deciphering the molecular function of various complexes. Such studies pave way to explore novel products that could be very specific for a particular target (DesJarlais, Sheridan, Dixon, Kuntz, & Venkataraghavan, 1986; Goodford, 1984)

Molecular docking approach is a powerful mode to model the protein and small molecule interaction at the atomic level. It greatly helps in characterizing the behaviour of small molecules at their binding sites. The two major steps involved in docking are: (i) obtaining the stable ligand conformation and (ii) assessing its binding affinity, and in majority of cases, binding sites are predicted before performing docking. Binding sites are generally obtained by comparing the target of interest with the other proteins from same family bearing similar function (Meng, Zhang, Mezei, & Cui, 2011). The mechanism of molecular docking has evolved to a greater extent, starting from the "lock and key" model, which is the most primitive of all (Kuntz, Blaney, Oatley, Langridge, & Ferrin, 1982). "Induced-fit" model was seen as a logical extension of the basic lock and key model, where the active site changes its conformation based on the binding ligand (Koshland, 1963).

The applications are widespread, the information obtained are profound and docking has thrown the doors wide open to stride into new arena of research focussing on lead molecule designing, target discovery and analysing application potential of the compounds. Given the wide scope in studying the binding mode and affinities between various molecules, docking has become an indispensable approach that serves as a precursor to several wet lab researches. Concept of docking and its application has been

Figure 1. Mechanism of molecular docking depicted through the interaction between the ligand, acetyl coA and the receptor, carboxyl transferase domain of acetyl coA carboxylase

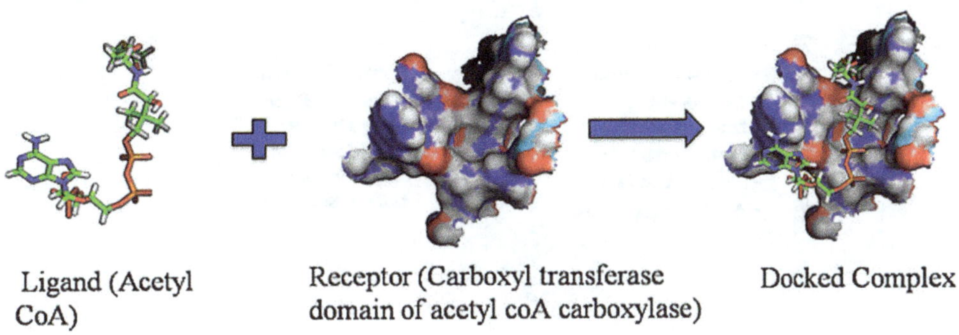

exhaustively conceived by the researcher fraternity and the results obtained have been vital in various industries. Though the primary application of docking is deep into drug discovery and validation, docking has also realized its potential in understanding the basic molecular interactions in various biological processes. This review is an attempt to summarize the diverse applications of molecular docking in various fields apart from drug discovery.

BACKGROUND

Molecular docking has become the most widely employed computational technique. Despite the fact that docking is mostly involved in structure-based screening of novel lead compounds in medical and to some extent non-medical research, docking has also been closely associated as a antecedent to wet lab experiments due to multidimensional confirmatory needs. Molecular docking has found its wide spread application in various avenues of research viz. exploring protein-protein interactions, computer aided drug discovery, high throughput screening, unravelling small molecule interactions with large macromolecules, etc. as highlighted in Figure 2.

1. STRUCTURE BASED DRUG DESIGN

The need of identifying new lead compounds and thrust for novel drugs is ever-increasing and therefore, the role of molecular docking has become critical in the process of drug discovery. The compounds

Figure 2. Schematic diagram featuring the wide range of applications of molecular docking in terms of non-medical and medical fields

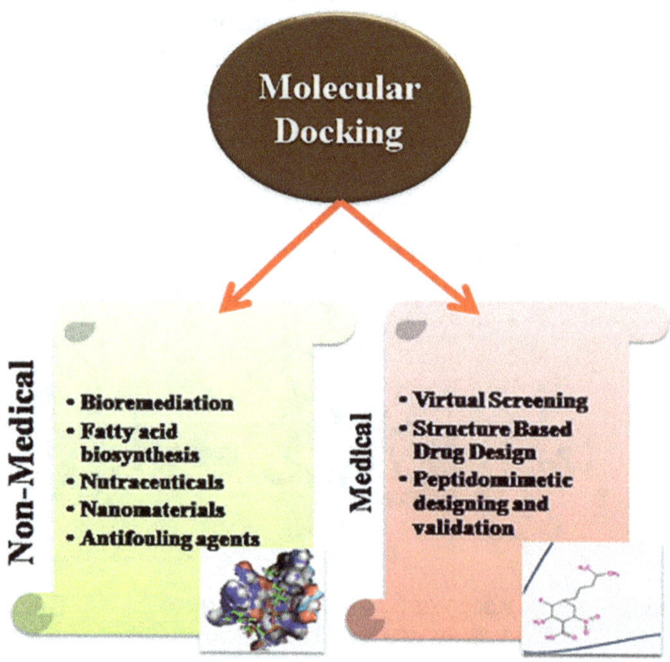

that have been identified through virtual screening can be validated by verifying its interaction with the intended target. Identification of lead compounds for several biological applications has been facilitated through structure based drug design. The entire process has been eased through various optimized docking algorithms over the years (Cai, et al., 2004; Shoichet, Kuntz, & Bodian, 1992; Walls & Sternberg, 1992).The challenging part of carrying out drug discovery has always been target identification, and with the intrusion of docking techniques, identification of specific sites of the lead compound and their binding affinities has been made simpler (Bohm & Schneider, 2000). Molecular docking is of great help in generating various conformations of a small molecule and finding various possible binding positions, orientations and conformations. The binding of these molecules are evaluated based on their docking score, binding energies, etc. The molecule with highest binding score is considered as the candidate molecule for that particular target. With the advent of virtual screening, a huge set of compounds can be designed, docked and validated. The facilitation of testing thousands of compounds through virtual screening at a single stretch has become a boon for computational chemists in discovering potentially newer drugs and drug targets in pharmaceutical and biomedical application (Figure 3).

A classic example of structure based drug design is the identification of inhibitors to repress the anti-apoptotic role of B-cell lymphoma 2 (Bcl2). Bcl2 belongs to a family of proteins that are associated with apoptosis. The levels of Bcl2 were observed to be high in case of breast cancer, prostate cancer, B-cell carcinoma, etc., thereby making it a potential target for anti-cancer therapy. The crystal structure of

Figure 3. The sequential process involved in virtual screening. The initial library of compounds (lead molecules) is filtered based on molecular descriptors and authenticated for their activity at their specified targets based on binding affinity. Subsequently, upon successful validation, these compounds are sent for clinical trials, thereby becoming candidate drugs.

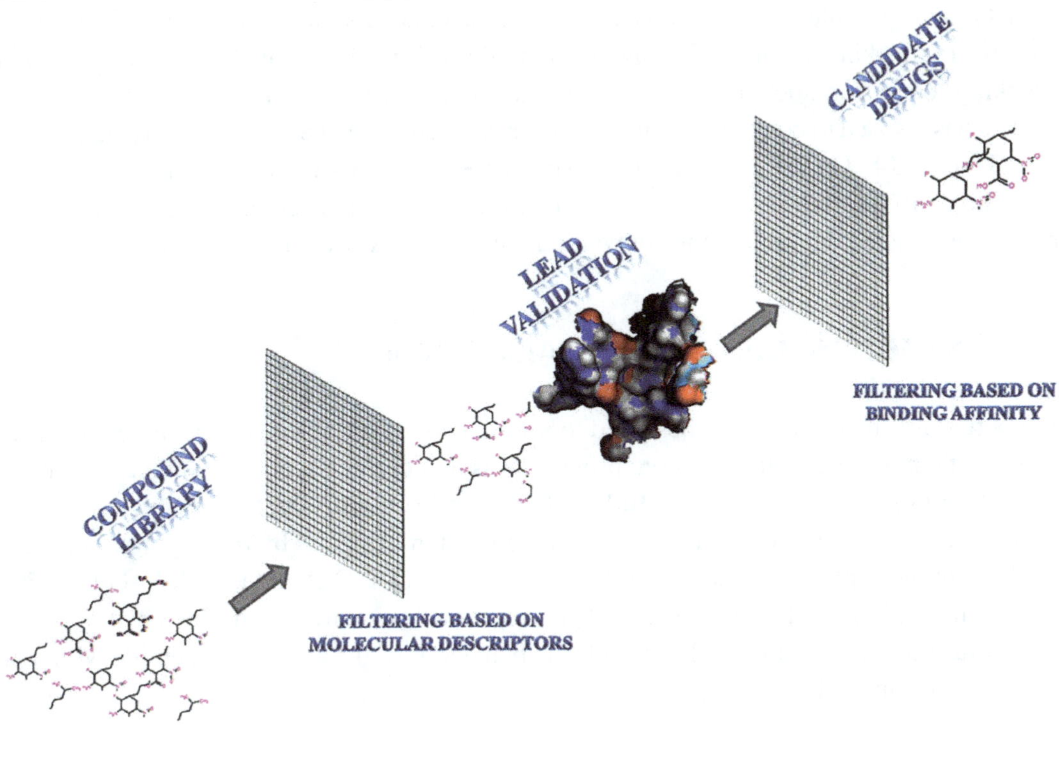

Bcl2-extra large- Bcl2 homologous antagonist killer (BAK)-Bcl2 homology domains (BH3) [Bcl-xL-Bak BH3] peptide complex was considered as the template for homology modelling. It was also found that Bak-BH3 domain binds to the BH1, BH2, BH3 domains of Bcl2. These domains are the active sites for its anti-apoptotic function. Therefore, designing an inhibitor molecule that bind against the BH3 domain and induce the apoptotic functionality would be an interesting approach. Small molecule search was carried out against the 3D database of National Cancer Institute, wherein 2,06,876 small compounds were docked to Bcl2 using DOCK program. The top 500 compounds were chosen based on their affinity and energy scores. Non-peptide small molecules were selected for further screening, and in-vitro assays were carried out to validate their potency of inhibition. The interactions of these small molecules were validated through molecular dynamics simulation and the conformational flexibility of Bcl2 was studied. These small molecules potentially inhibited the anti-apoptotic role of Bcl2. This further highlights that computational screening could be used to design and develop novel inhibitors to various cancer related molecular targets (Evers & Klabunde, 2005).

The computational techniques associated with structure based drug design and High Throughput Screening (HTS) has evolved a lot in due course of time. Some of the most recently evolved approaches are *Fragment based drug designing (FBDD) and Multiple Ligand Simultaneous Docking (MLSD).* Fragment based drug designing has found its importance in designing inhibitors and analogue molecules. In case of FBDD, the optimization of the lead compound is validated through the interaction energies of its individual fragments. This method is adopted as an improvisation step in lead molecule design and optimization (Hajduk, 2006). Multiple Ligand Simultaneous Docking (MLSD) is a novel algorithm that has musterd much importance in the recent past. MLSD helps in docking more than one ligand at the same time, which correlates with the in-vivo condition, wherein more often than not, more than one molecule interacts with a particular receptor. The overall binding efficiency is arrived at based on the contribution from all the ligands (Li & Li, 2010).

Another typical example for the advancement of molecular docking approaches is the concept of "reverse or inverse docking". Unlike docking, where multiple ligands are targeted to a single drug target, reverse docking focuses on docking a single ligand molecule to an array of macromolecular targets. reverse docking has also led to conceptualising a new term called as "Reverse virtual screening" (Kharkar, Warrier, & Gaud, 2014). One of the widely used web-server for such reverse docking strategy is "PharmMapper", which aims in identifying targets for small molecules. This server proves to be beneficial in performing reverse screening unlike the conventional methods (X. Liu et al., 2010).

2. PEPTIDOMIMETICS: DESIGN AND VALIDATION

Peptidomimetics are small molecules designed to mimic native peptides in their core elements (Figure 4). These mimetic molecules retain the core properties as that of the peptide, such that the function and biological activity of the peptide is unperturbed. Designing peptidomimetics as a potential substitute to peptides is on the rising trend. A mimetic molecule is often designed by modifying a selected set of amino acids from the peptide at certain specific sites (Hruby, Qui, Okayama, & Soloshonok, 2002). In case of designing a mimic molecule, the central pharmacophore features are defined by the non-peptide variants and thus the resulting molecule upholds the behaviour of the aboriginal peptide with respect to target specificity (Lam et al., 1991).

Figure 4. Conversion of a small peptide to its equivalent peptidomimetic molecule that is designed based on the basic physical properties of the native peptide

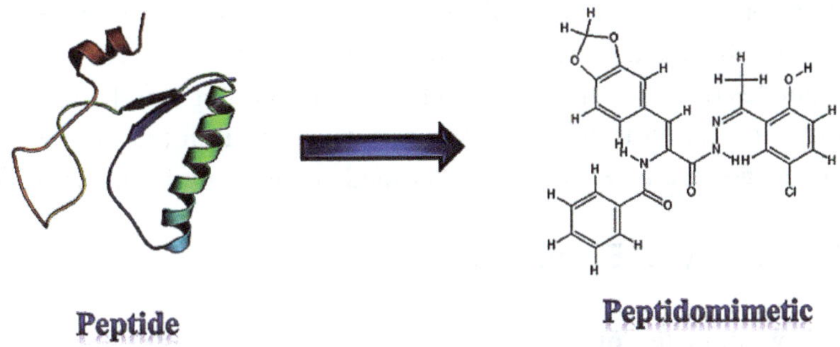

There is a giant leap in studies on genomes, transcriptomes and proteomes, which has interestingly increased the need for looking alternate molecules derived from their natural counterparts. Hence, several approaches are attempted to study the intricacies of these interactions of mimic molecules, which has led to various research works viz. designing agonists, antagonists, etc. that can perform the desired action. Designing of peptidomimetic molecule that target a particular receptor has been on the rise ever since its inception.

The computational chemistry tools have been put into great use in examining the mimetic molecule through docking to the specific target followed by simulation to ascertain their biological function (Hruby, 1982). With the steady increase in industrial requirements in pursuit of novel compounds with potent activity, peptidomimetics is already identified as an indispensable entity in computational chemistry. One typical example is the designing of novel peptidomimetic molecule to trigger Follicle Stimulating Hormone Receptor (FSHR). A novel molecule was designed based on the residues involved in the complex formation of FSH and its receptor. The obtained mimic molecule was then docked to FSHR and its stability was validated through molecular dynamics simulation. It was observed that the mimic molecule-FSHR complex was rigid and the intermolecular interactions were retained throughout the simulation process. The peptidomimetic molecule thus acquired can be synthesized in the laboratory and used as a trigger to activate FSHR (Sonawani, Niazi, & Idicula-Thomas, 2013).

3. APPLICATIONS OF MOLECULAR DOCKING OTHER THAN DRUG DISCOVERY

Importance of molecular docking in drug design and identification of new targets have been very well studied and documented. A plethora of drugs have been discovered and validated using molecular docking as their base, and indeed some of them have successfully reached the clinical trials. More often than not, molecular docking is viewed as a technique confined to drug discovery processes. On the contrary, molecular docking is a powerful strategy that extends its ambit by many leap and helps in understanding the basic mechanisms of several biological processes as well. Role of computational techniques in deciphering the intricacies involved in several molecular functions has witnessed a meteoric rise in the recent past. Researchers have sought the support of molecular docking techniques in order to find novel answers to various yesteryear questions. The solutions provided by such tools and algorithms have proved to be comprehensive and convincing. This advantage has been exploited and many industries are now keen in performing this computational analysis as a precursor to their wet lab research.

Major representative applications of molecular docking in certain research avenues, especially, where the product discovery is not the goal, rather to construe the basics of various processes has been brought into limelight. The examples dealt are only representative subsets of research areas involving molecular docking and still there are a lot more to look upon as well. Based on the momentous applications of docking in specific fields, a suitable term has been coined accordingly in this review (Figure 5).

3.1 REMEDIDOCK: Study of Bioremediation Process

Revamping of places contaminated on account of pollution due to several factors is on the rise and bioremediation have offered more viable solutions to fix these issues. Regardless of the fact that micro-biological interventions are highly effective in terms of decontaminating the environmental niches, they

Figure 5. MOLINDOCK- Understanding Molecular Interactions using Docking. The applications of molecular docking in fields outside drug discovery along with their specific terminologies were coined accordingly, as REMEDIDCOK, FADOCK, NUTRADOCK, NANODOCK and FOULDOCK. This encompasses a representative set of disciplines, where computational chemistry techniques are implemented to study the basic interacting mechanism.

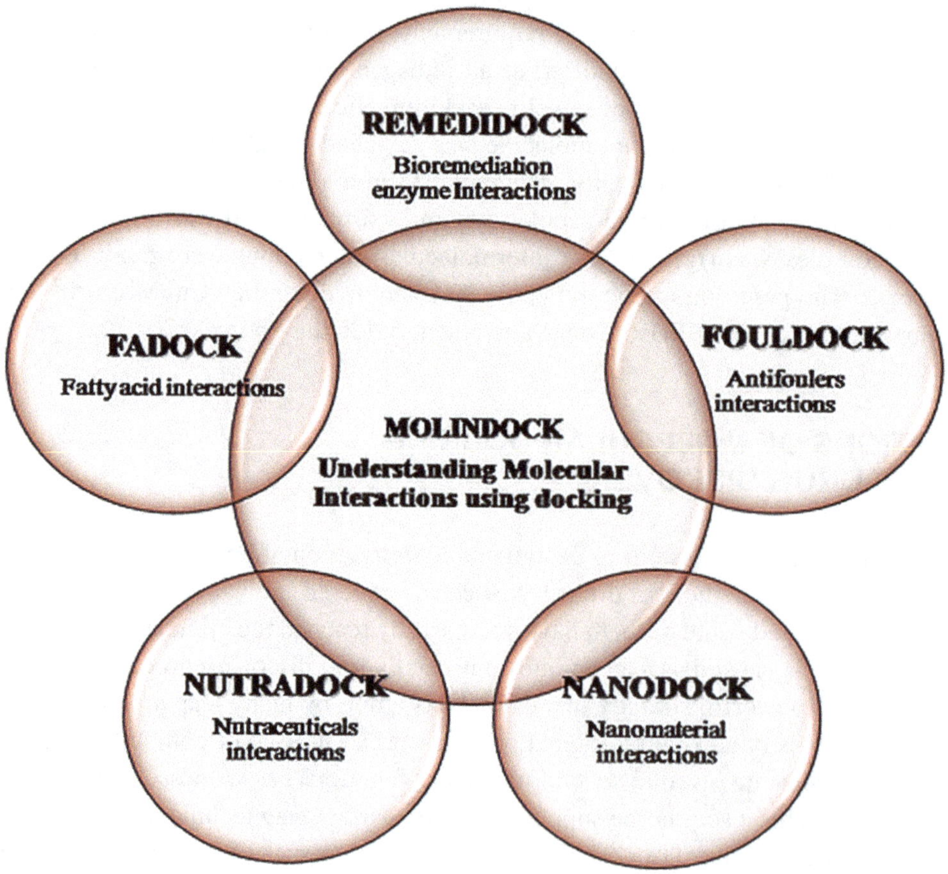

are cost effective as well (Khan, Pal, Vikram, & Cameotra, 2013; Lovley, 2003; Paul, Pandey, Pandey, & Jain, 2005). It has been well documented that certain bacteria have the ability to catabolize organic hydrocarbons and other chemical substances in the environment (Parales, Bruce, Schmid, & Wackett, 2002). Composting is another naturally offered mechanism, which hastens the natural biodegradation by converting organic wastes to valuable resources. Yet again, composting is not applicable to all wastes (Fulekar, 2009). Treatment of waste water using microalgal deployment is on the rise as well, which nevertheless, cannot be extended for solid wastes. Thus, the search to unearth a potential mode for bioremediation is still on the long roads.

Bioinformatics approaches have a crucial role to play in bioremediation, as accumulation of huge datasets on various chemical structures and their sequences with activity information is obligatory. Furthermore, repository of comparative genomics and proteomics data are essential as well (Alves, Chaleil, & Sternberg, 2002; Fraser, Hirsh, Steinmetz, Scharfe, & Feldman, 2002). Unfortunately, the number of available resources in this perspective is very scarce. The University of Minnesota Biocatalysts/Biodegradation Database (UMBBD) is one such quintessential repository which holds information on 740 chemical compounds, 820 reactions, 502 enzymes and 253 organisms (Ellis, Hou, Kang, & Wackett, 2003). An interactive interface, "MetaRouter" has been designed and implemented to query the UMBBD database and mine more information (Pazos, Guijas, Valencia, & De Lorenzo, 2005).

Molecular docking has become critical in studying the role of various chemical compounds involved in the bioremediation process. It helps in understanding their binding to the target and the subsequent mechanisms. Application of docking lies in obtaining robust information on the specific activity of the compounds in coupling with 3D-QSAR and molecular dynamics simulation techniques (Khan, Sajid, & Cameotra, 2013). The list of enzymes from microorganisms and other sources involved in bioremediation and the applications of molecular docking in the above process are provided in Table 1.

Following are the two sample reports describing the role of computational chemistry techniques in evaluating the bioremediation processes.

3.1.1 Biodegradation of Herbicide Diuron by Laccase-1

Diuron is one of the widely used herbicides belonging to phenylamide family finds a great application in the field of agriculture. This herbicide, at higher concentration can have a devastating effect on the photosystem-II of plants, thereby affecting its photosynthetic ability (Vieira et al., 2015). Diuron is a chlorophenol compound and it can be dechlorinated by laccase enzyme. The laccase enzyme is produced by the fungus, *Ceriporiopsis subvermispora* and the interaction of this enzyme with the ligand, diuron was studied through molecular docking. It was observed that the best docked pose of diuron to the laccase enzyme was also the catalytic site for oxidation catalysed by the residue, His 457. This is further mediated by the presence of copper ions that causes the protonation of this residue to be moved from Nδ1 to the Nδ2 and helps in hydrogen bond formation with the dimethylurea oxygen of diuron. This suggests that the biodegradation of diuron does not involve the conventional demethylation process, wherein 3-dichloroaniline (DCA) is formed and is even more toxic than diuron, thereby suggesting that the current methodology employed facilitates more eco-friendly mode of bioremediation (Vieira, et al., 2015). Molecular docking found itself a vital role in this study, where the mechanism of degradation was inferred by studying the *in silico* interaction of diuron and the enzyme, laccase. This is a typical example of applying computational techniques in conceiving the underlying mechanism of bioremediation.

Table 1. List of various enzymes involved in bioremediation process and implementation of docking and other computational techniques

Enzymes/ Biochemicals	Mode of Action	Computational Tools Employed	References
Laccase	Degrade polycyclic aromatic hydrocarbons like anthracene and benzopyrene	Docking	(Suresh, Kumar, Kumar, & Singh, 2008)
Naphthalene dioxygenase	Degrade a broad range of pollutants	Molecular dynamic simulation	(Librando & Forte, 2005)
Laccase	Degrade herbicides like Diuron	Molecular docking	(Vieira, et al., 2015)
Laccase and Peroxidase	Degrade triphenylmethane dyes and textile red dyes	Molecular docking	(Weisburger, 2002)
Aniline dioxygenase	Simultaneous deamination and oxygenation of aniline and α-toluidine to produce catechol and 3-methylcatechol, which acts as a biocatalyst for bioremediation	Docking	(Lui, 2007)
Carbamoylase	Degrade a wide range of toxic pollutants comprising of aromatic hydrocarbons	Homology modelling	(Dalal & Malik, 2013)
Catechol 2,3-dioxygenase	Degrade aromatic compounds	Docking and homology modelling	(Pennacchio, 2009)
Azoreductase	Degrade synthetic textile azo dyes	Docking	(Sridhar & Chandra, 2014)
Nitroreductase	Degrade nitroaromatic and nitroheterocyclic explosives, scavenges hydrogen peroxide through NADH oxidase activity, degrade NAD+ to free nicotinamide through an oxidative catalysis	Docking studies	(Cortial et al., 2010)
Hydroxamate siderophore from marine cyanobacteia	Complex with Uranium, a radionuclide and sequester it	Docking- an *In silico* approach	(Rashmi et al., 2013)

3.1.2 Bioremediation of Triphenylmethane Dyes: Docking Approach

Synthetic dyes such as azo, anthroquinone, triphenylmethane, etc. are being widely used in the current day trend. Alarming rate of release of the coloured compounds into the environment poses a serious threat to the ecological balance. In addition to the damage caused to the photosystem of plants by these coloured compounds, there is a deleterious impact on the marine fauna, as these dyes act as chemical mutagens (Weisburger, 2002). There are reports that document the role of laccase and peroxidase enzymes obtained from fungi in decolourising these toxic dyes (Wessels & Van der Veen, 1956). More recently, it was observed that, a strain of *Citrobacter sp.* was capable of producing extracellular metabolites, with wide range of activity in decolourising these toxic dyes. The enzyme, triphenylmethane reductase (TMR) was found to be more active in reducing malachite green and other related dyes. Molecular docking was employed to investigate the decolourising mechanism of the enzyme TMR. Hence, a crystal structure of TMR was elucidated with NADPH cofactor. This enzyme structure was docked with the dye, malachite green and it was discerned that this enzyme profoundly enabled the breakdown of the dye, as it binds to the enzyme's active site. The binding pocket was seemed to be characterized by Leu-142, Phe-146, Phe-147, and Ile-251residues. This study can also be expanded to other dyes such as Congo red, Magenta fuschin etc. thereby facilitating better understanding of the enzyme behaviour (Kim et al., 2008; Tekere, Mswaka, Zvauya, & Read, 2001)

Thus computational tool can be widely used in designing bioremediation systems depending upon the nature of dyes involved. Characterising the mechanism of such enzymes screened from various bacteria or fungi in degrading environmental toxins, through molecular docking and molecular dynamic simulation approaches, could prove beneficial in unravelling novel methods of bioremediation with more specificity towards the toxins that demand immediate attention and action.

3.2 FADOCK: Study of Fatty Acid Biosynthesis

Molecular docking has also found diverse applications in fatty acid biosynthesis in drug designing as well in identification of potent herbicides and algal lipid production. These wide aspects are listed in Table 2.

Following are the two examples that depict the association of molecular docking studies in unravelling the interactions of fatty acid precursors with the corresponding enzymes.

3.2.1 Docking in Fatty Acid Biosynthesis: A Case Study

The use of biofuels has shown a sharp growth in the recent past, primarily riding on the global policies towards mitigating the detrimental effects of greenhouse gas emissions (International Energy Agency, 2007). The change in climatic conditions and the predictions that have been made for the future have questioned the sustainability of fossil fuel usage, given the fact that, fossil fuel is a major contributor of greenhouse gases. In order to address this issue on greenhouse gas emissions and climatic changes, biofuel has emerged as a worthwhile alternative. Speaking about biofuels, microalgae are being considered as an extremely potential source, considering the amount of lipid it could accumulate in its cells. An avalanche of research methodologies have been developed and are being implemented in extracting biofuel from microalgae, thus making it available for commercial use eventually.

The biochemical pathway of lipid metabolism in algae has been exhaustively studied and in fact, mutational studies are tried to validate whether or not they increase the efficiency of lipid synthesis

Table 2. Application of molecular docking in studying enzymes associated with fatty acid biosynthesis

Enzymes / Inhibitors	Mode of Action	Computational Tools Employed	References
Fatty acid synthase (FASN) / cerulenin, triclosan and orlistat	FASN leads to retinoblastoma tumour invasiveness and differentiation. The inhibitors listed targets various FASN domains and act as anti-cancer drug candidates in the cast of retinoblastoma	Molecular modelling and docking simulations	(Deepa et al., 2010)
Acetyl co A carboxylase / herbicides and insecticides	Apart from fatty acid biosynthesis, its catalytic and regulatory subunits made it as a effective target for herbicides and insecticides, which inhibits the activity of this enzyme Enhances the fatty acid biosynthesis by converting acetyl coA to malonyl coA	Molecular docking and homology modelling Molecular docking and ligand interaction studies	(Chandrakar et al., 2013) (Anthony, et al., 2015)
Acyl carrier protein (ACP)-Thioesterase TE)	Involved in fatty acid biosynthesis and functionally establishes the length and identity of the fatty acid end product	Molecular docking and homology modelling	(Blatti, 2012)
β-ketoacyl-ACP synthase III (FabH)	FabH is a key factor involved in elongating the type II fatty acid chains to find use in biolipid production. Another enzyme associated with called "acyl carrier protein (ACP)", catalyses the transfer of fatty acid intermediates to the respective active sites.	Homology modelling, docking followed by molecular dynamics simulation	(Misra, et al., 2013)

in algae (International Energy Agency, 2007). Such studies and research works are complemented by computational tools and techniques. The role of molecular modelling, molecular docking and molecular dynamics simulation has been well conceived and implemented by the scientists and researchers. Such computational methods are of great help in deciphering the process behind lipid metabolism and provide greater assistance in carrying out site directed mutagenesis studies. The critical role of molecular docking and other computational strategies that were implemented in studying the biofuel production process has been cited below.

An attempt was made in our laboratory to improve the production of fatty acids from marine microalgae, *Chlorella vulgaris,* which eventually contributed to enhanced biodiesel production (Anthony et al., 2015). Among various genetic engineering approaches documented in the vicinity of biodiesel production, triggering fatty acid biosynthesis through random mutagenesis is one successful method. Random mutagenesis by physical mutagen (ultraviolet light, exposed for 60 s) and chemical mutagen (5'-Fluorodeoxyuridine at 0.25 mM concentration) augmented lipid production as well as improved the fatty acid saturation-desaturation ratio, favouring more of saturated fatty acids. In an attempt to unearth the mechanism behind, the key enzyme, acetyl coA carboxylase involved in fatty acid biosynthesis was taken into consideration. The mutant strains that improved the lipid content have been sequenced and the point mutations were identified. Based on the sequences obtained with the mutant strains, three dimensional structure of carboxyl transferase β subunit of acetyl coA carboxylase was modelled using homology modelling methods.

The obtained structures were then docked with the ligand, acetyl coA, in order to study the variations caused by the mutagens with respect to the wild type. This docking study revealed few variations in the pattern of amino acid interactions between the wild type and the mutant strains (Figure 6).

It was observed that the leucine residue at the sixth position (Leu6) has been replaced by a serine residue (Ser6) in the mutant strain. This Ser6 residue was found to introduce an additional hydrogen bonding with the carboxyl transferase domain, thereby leading to more rigid binding of the ligand. This change could be the underlying mechanism behind the enhancement of the lipid content observed. This

Figure 6. Comparison of interaction between acetyl coA and acetyl coA carboxylase in wild type and UV mutated strain of C. vulgaris. The mutation, L6S is highlighted.

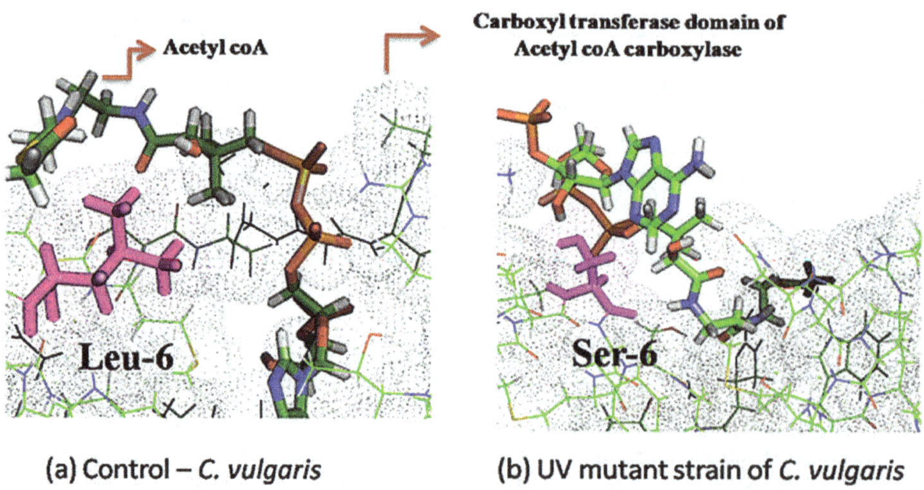

study has led to interesting correlations to the wet lab experiments, further raising various research questions that are being worked out.

3.2.2 Modelling and Docking of FabH Enzyme in *Chlorella variabilis*

The microalga, *C. variabilis* is also one of the widely studied species for its potential to produce biolipids, which are the precursors to biofuel. The enzyme FabH is a key factor involved in elongating the type II fatty acid chains. Another enzyme, acyl carrier protein (ACP), catalyses the transfer of fatty acid intermediates to the respective active sites. Molecular modelling followed by molecular docking was carried out to study the interaction mechanism between FabH and ACP. Docking studies revealed strong hydrogen bonding interactions between FabH and ACP. A hydrogen bond between the active site residues of both the enzymes (Arg91 of FabH and Glu53 & Asp56 of ACP) was observed, this suggesting a strong affinity between both the enzymes. This observation was further ascertained by molecular dynamic simulations, which revealed that this interaction was upheld throughout the course of simulation process. Hence, the computation chemistry approaches can be implemented in designing mutation studies that could be vital in increasing the synthesis of biolipids in algae (Misra, Patra, Panda, Sukla, & Mishra, 2013).

3.3 NUTRADOCK: Study of Nutraceuticals Interactions

The role of digestive enzymes in breaking macromolecules present in food into smaller units has been elaborately studied and documented. The intervention of these enzymes is pivotal in facilitating the adsorption of food particles, which in turn mediates digestion, thereby offering the required nutrition for the proper functioning of the human system. Digestive enzymes are located across the gastrointestinal tract, i.e. right from the salivary glands till the intestines. Major enzymes involved in the breakdown of food molecules are amylases, proteases and lipases, which act on starch, proteins and fats respectively and break them into simpler blocks. Apart from the aforementioned enzymes, there exists array of other co-enzymes and co-factors, which are critical in the digestion process. Understanding these processes is ideal and it offers potential assistance in designing novel nutraceuticals, which would act as a viable food supplements (Table 3).

Following are few classic examples for the intervention of molecular docking in the disentanglement of interaction between digestive enzymes and the food particles, which in turn paves way to design new formulae for nutraceuticals and validating them.

3.3.1 Docking and Simulation of Catechins Binding to Trypsin

Polyphenols are chemical compounds that contain multiple phenol units and are widely known for their benefits in health promotion. Catechins are a major representative group of polyphenols that has persuasive activity in reducing the risk of several chronic diseases (Mendilaharsu, De Stefani, Deneo-Pellegrini, Carzoglio, & Ronco, 1998; Wang, Yu, Wu, & Zhang, 2012). (-)-Epigallocatechin gallate (EGCG), (-)-epigallocatechin, (EGC), (-)-epicatechingallate (ECG) and (-)-epicatechin (EC) are the various subtypes of catechins that possess proven activities against wide range of cancers (Ju et al., 2005; Suganuma, Ohkura, Okabe, & Fujiki, 2001). It has been found that enhanced activity of trypsin, a major digestive enzyme, causes proliferation and metastasis of tumors by degrading the extracellular matrix (Nyberg, Ylipalosaari, Sorsa, & Salo, 2006; Soreide, Janssen, Körner, & Baak, 2006). Hence, it

Table 3. List of nutraceuticals and their mode of action

Nutraceuticals /Enzymes	Mode of Action	Computational Tools Employed	References
Binding of Nutraceuticals with Farnesyl Pyrophosphate Synthase	Addition of a farnesyl or geranyl moiety to the CAAX motif of the protein prevents the formation of mutant lamin A, thereby preventing the progerin accumulation, which further prevents the Hutchinson Gilford Progeria syndrome	Molecular docking	(Jeyam, Sushma, Sharanya, & Poornima, 2011)
a. Coagulant protein (MO2.1) of *Moringa oleifera* b. Limonoid from neem tree called azadirachtin c. Plant alkaloid called 'palmatine'	It enhances the oligomerization property, which is correlated to interfere with its medicinal properties. This nutraceutical disrupt the detoxification process by inhibiting glutathione S-transferase of *P. mirabilis*, thereby acts as an effective drug candidate for urinary tract infections. Palmatine binds with varius polynucleotides that were sequence dependent.	*In silico* modelling	(Kayathri, 2013)
Inhibition of matrix metalloproteinase by difluorinated benzylidene, novel analog of curcumin	Compared to curcumin, the analog was found to inhibit the invasion and metastasis, thus warranting its further evaluation as an effective anticancer agent	Molecular docking	(Ahmad et al., 2015)
Inhibition of serine protease by compounds from *Murraya koenigii* for Dengue fever	The viral polyproteins is cleaved at various regions in order to produce structural and non-structural viral proteins by serine protease	*In silico* docking and virtual screening	(Sing Yong, Bing Choi, & A Wahab, 2015)
Inhibition of trypsin by catechins	Trypsin forms soluble complexes, which further aggregate and lead to precipitation. The strong binding of catechins alters the natural conformation of proteins, and thus disturbs the stability and bioactivity of proteins.	Molecular docking, molecular dynamic simulations and binding free energy calculations	(Cui, et al., 2015)
Inhibition of Urokinase plasminogen activator (uPA) by benzamidine, p-benzamidine, and amiloride	uPA binds to its specific GPI-anchored receptor on the cell surface. This in turn lead to augmenting the rate of plasminogen activation and directing the uPA proteolytic activity at migrating cells. This factor is actually responsible for the invasive ability of tumour cells	Molecular docking	(Conese & Blasi, 1995);(Rabbani, et al., 1995)

becomes imperative to mediate the balanced regulation of trypsin in order to evade the detrimental effects and this is achieved by inhibiting trypsin with catechins. Catechins also tend to alter the secondary structure of trypsin, thereby altering its function (Huang & Zhao, 2008; X. Wu et al., 2013).

Function of trypsin, which is to breakdown the protein backbone is mediated and catalysed by the residues, His57, Asp102 and Ser195, which are present in its active sites (also known as S1 pocket). This motif is also observed as a common pattern in serine protease family (Graf et al., 1988; Ma, Tang, & Lai, 2005). Binding of catechins to trypsin and its inhibitory activities were studied through combined molecular docking, molecular dynamic simulation and binding free energy calculation. All the four types of catechins were subjected to docking with trypsin, followed by molecular dynamic simulations. It was construed that catechins were binding to the S1 pocket, which were stabilized by hydrogen bonds and hydrophobic interactions. Ser190, Gln192, Ser195, Val213, Ser214 and Trp215 were involved in hydrogen bond formation with all the four subtypes of catechins, and established that the catechins-trypsin complex was more stable than free trypsin through the changes in root mean square deviation (RMSD) values. Binding free energy calculations revealed that Asp189, located adjacent to the S1 pocket, contributed significantly for the electrostatic interaction and hence proven as the most important residue required for catechins binding. Apart from Asp189, residues such as Gln192, Trp215 and Gly216 also played significant roles in upholding catechins binding.

Moreover, it was also inferred that EGCG had greater binding affinity than its other associate catechins by possessing higher binding free energy (Cui, Yang, & Li, 2015). These results propose the action of catechins in inhibiting trypsin, thereby providing a suitable roadmap for understanding and interpreting the biochemical mechanisms, which in turn leads to designing novel nutraceuticals. These novel compounds could act as health supplements and open doors for prospective research works in this avenue.

3.3.2 Nutraceutical in Prostate Cancer Treatment: Docking Involvement

Prostate cancer has become more prevalent disease, particularly amongst elderly men. There are wide ranges of treatments available such as radiotherapy, radical prostatectomy, cryosurgical ablation, etc. Prolonged treatments results in metabolic failures in patients and the major limitation is that all these treatments can carry morbidity. Role of nutraceuticals in inhibiting the factors causing prostate cancer has been studied and it has been established that curcumin and EGCG are found to possess beneficial activity against cancer (Achbarou et al., 1994). Urokinase plasminogen activator (uPA), an enzyme belonging to the serine protease family is a major restraining factor in prostate cancer formation. The activation of cancer proteolytic machinery as well as the hydrolysis of basement membrane and connective tissue proteins is the major underlying mechanisms behind the cancer promoting effect of uPA. When this enzyme is overexpressed in prostate cancer cells, it exhibited a profound increase in metastasis. While, on the other hand, inhibition of uPA using inhibitors such as p-aminobenzamidine and Amiloride, led to the tremendous reduction in the metastatic potential (Conese & Blasi, 1995; Rabbani, Harakidas, Davidson, Henkin, & Mazar, 1995). The list of nutraceuticals considered for validating their activity were antipain, leupeptin, folic acid, rosamarinic acid, lavendustinA, fisetin, myricetin, tolfenamic acid, pterins, etc. From the aforementioned list, only those nutraceuticals that were commercially available were taken forward for further analyses. A library of isomers of each of these nutraceuticals was generated and their inhibitory efficiency against uPA was validated through molecular docking technique.

It was observed that the nutraceuticals bound to the active site residues 187-197 and 212-229 of the uPA, which is the actual site for inhibition. The binding of these small molecules at the active site prevents uPA from binding to plasmin, hence repressing its metastatic potential. Nutraceuticals such as benzamidine, p-benzamidine, and amiloride bound onto the active site, whereas bigger molecules such as folic acid, etc. were half fit into the active site. The remaining half were bound around the catalytic site, thereby causing steric hindrance to uPA's activity. These results paved way in figuring out the nutraceuticals that could be added to the regular diet, which could act as a prophylactic strategy for prostate cancer treatment.

3.4 NANODOCK: Study of Nanomaterial Interactions

Nanotechnology has emerged as an interdisciplinary approach and has found its importance in a variety of fields such as drug delivery, therapeutics, medical diagnosis and cancer biology. Nanotechnology is also being implemented as an important strategy in designing and validation of biosensors. A variety of nanoparticles are available viz. silica, gold, silver, iron, cerium oxide, carbon nanotubes, etc. The deduction of orientation of protein on nanoscale surfaces finds vital insinuations in the field of integrating proteins onto micro and nanofabricated devices (Shrivastava, Nuffer, Siegel, & Dordick, 2012).

The nature of interactions between the nanomaterials and the corresponding protein targets are often mediated by covalent attachment or physical adsorption, in order to deliver the required function (Gagner, Lopez, Dordick, & Siegel, 2011). The properties of the protein interacting with the nanoparticles have strong influence on the cellular uptake and the absorption of those materials. The study investigating the interactions of nanoparticles take into account, the change in size of nanoparticle, amount of protein adsorbed and the structural changes that occur (Calzolai, Laera, Ceccone, & Rossi, 2012).

The structure, stability and chemical nature of the nanomaterial are the vital factors that define its ideal conjugate functionality. Although, nanoparticles have been successfully implemented by the scientific community in terms of structure, stability and activity of proteins attached on the surface of nanomaterial, understanding the mode of action and the binding of these nanoparticles to their specified target and their cytotoxicity is still subjected to heated discussions, which is greatly answered by the computational tools (Table 4). Few examples from the above mentioned studies are discussed below.

3.4.1 Silica Nanoparticles and Enzymes Interactions

The interdisciplinary approach "Nano-biotechnology" has embossed its application in a variety of research works involving physics, chemistry, biology and medicine. Nanoparticle that possesses several attractive properties including biocompatibility, huge surface to volume ratio with robust functionality have

Table 4. Use of docking in studying nanomaterial interactions

Nanomaterials	Mode of Action	Computational Tools Employed	References
Interactions of pyrazinamide drug with functionalized single-wall carbon nanotubes	Pyrazinamide is only used in combination with other drugs such as isoniazid and rifampicin in the treatment of tuberculosis. Carbon nanotubes helps in delivering the drug exactly at the targeted sites.	Docking and molecular simulations	(Saikia, Rajkhowa, & Deka, 2013)
Binding of ubiquitin to gold nanoparticles	Ubiquitin – It is involved in protein degradation and upon binding with gold nanaoparticles, it elucidated the role of nanoparticle surfactants such as citrate in the association process.	Molecular docking	(Brancolini, Kokh, Calzolai, Wade, & Corni, 2012)
Interaction of proteins such as taq DNA polymerase, tubulin and ubiquitin with two fullerene (fullerol and fullerene)-based nanoparticles	Taq DNA polymerase- Involved in DNA replication Tubulin – Provides the mechanical strength for the cells to maintain their size and shape Ubiquitin –It is involved in protein degradation	Docking and molecular simulations	(Nedumpully Govindan, 2013)
Nanomaterial interactions	It explore the hazardous and toxic effects of nanomateials	Docking and Nano-QSAR	(Leszczynski & Puzyn, 2012)
Interactions between silica nanoparticles and various enzymes such as cytochrome C, Robonuclease A and Lysozyme	Cytochrome c acts as a mobile electron carrier between complexes III and IV of the electron transport chain. Ribonuclease A cleaves single-stranded RNA. It is a key enzyme in studying protein structure, folding and unfolding pathways and enzyme catalysis. Lysozyme is involved in immune system.	Molecular dynamic simulations	(Sun, et al., 2014)
Coagulant protein interaction with precursors	It exerts coagulation and antimicrobial properties. Upon binding with suitable precursors, it determines the function of nanomaterials, thereby develops an interface hybrid structures for various applications	Docking and *in silico* modelling	(Okoli, et al., 2013)

found a range of applications in biology. Silicon nanoparticle is one such material that possess diverse functions and has created great interest amongst researchers (Calzolai, et al., 2012; Gagner, et al., 2011; He, Fan, & Lee, 2010; Shrivastava, et al., 2012; Sun, Feng, Zhang, Hou, & Li, 2014). The orientation and adsorption between various enzymes such as cytochrome-C, ribonuclease-A (RNAse-A), lysozyme and the silica nanoparticles has been studied (He, et al., 2010). Cytochrome-C is an important enzyme present in the inner membrane of the mitochondria and is involved in the electron transport chain (J.-F. Wu, Xu, & Zhao, 2010). RNAse has a vital role in studying protein folding, unfolding and enzyme catalysis (Wlodawer, Svensson, Sjoelin, & Gilliland, 1988), while lysozymes are part of the immune system (Gill, Scanlon, Osipovitch, Madden, & Griswold, 2011).

The changes in the active site of the aforementioned enzymes were compared before and after their interaction with the silica nanoparticles through molecular docking. Molecular dynamic simulations were employed to investigate the adsorption of the enzymes to the nanoparticles of size 4 nm and 11 nm. The RMSD of these enzymes while adsorbing to the 4 nm nanoparticle was found to be ~4.56 Å, ~2.66 Å and ~4.15 Å for cytochrome-C, RNAse-A and lysozyme, respectively, whereas the RMSD in case of 11 nm nanoparticle were ~6.29 Å, ~3.97 Å and ~5.37 Å, respectively. It was found that the enzymes had shown less significant structural changes while being adsorbed onto the 4 nm silica nanoparticle when compared to the 11 nm silica nanoparticles (Sun, et al., 2014).

It was also observed that the binding of cytochrome-C and RNAse to the silica nanoparticles causes structural changes in their active sites. More interestingly, it was seen that the active site of cytochrome-C was not available to the ligands after its adsorption to the 11 nm nanoparticle. This study also analysed the impact of various functional groups such as $-CH_3$, -COOH, $-NH_2$ and –OH upon coating them to the nanoparticles. It was inferred that the active site of cytochrome-C was far away from the silica nanoparticles and that the active site displayed least structural variation upon the presence of –COOH group (Sun, et al., 2014).

This documentation provides a quintessential representation of the interaction of silica nanoparticles with the enzymes and describes the structural alterations that occur in due course. The impact of various functional groups on the binding of enzymes paves way for designing novel nanoparticles with desired functionality based on the target enzyme.

3.4.2 Coagulant Protein Interaction with Precursors

Nanomaterials have been successfully synthesised and applied in various disciplines, paving way for integrating synthetic chemistry with biology (De, Ghosh, & Rotello, 2008). Core metals such as Fe, Ni, Co and metal oxides such as Fe_3O_4 and γFe_2O_3 have been evaluated for nanoparticle production. The physical properties of these metals make them lucrative for nanoparticle synthesis thus making them viable for diverse applications (Burda, Chen, Narayanan, & El-Sayed, 2005; Faraji, Yamini, & Rezaee, 2010). The surface properties of the nanoparticles are defined by their stability upon binding to their targets and their rate of adsorption. The chemical moieties that are attached to these nanoparticles play a significant role in determining their function (Rana, Yeh, & Rotello, 2010). In order to achieve a particular functionality through nanoparticles, more often than not, it requires to be conjugated to a protein target. The nature of functional groups and their associated charge greatly influences the nanoparticle binding in terms of electrostatic interactions, which in turn defines the structural stability of the proteins (Aubin-Tam & Hamad-Schifferli, 2005). Functionalized nanoparticles find its application in drug delivery at the targeted sites, waste water treatment, etc.(Satishkumar & Vertegel, 2008).

An example involving *in silico* modelling behind the coagulant protein interactions with precursors was reported (Okoli et al., 2013). A hybrid system containing a protein-functionalised nanoparticle for a protein from *Moringa oleifera*, which possesses anti-coagulant and anti-microbial properties has been prepared. The binding site information of this protein is scarce and hence theoretical modelling and molecular docking studies was employed to unravel them. The nanoparticle containing iron oxide core with outer shell constituting amino, hydroxyl, silanol or carboxyl groups was used. These superparamagnetic nanoparticles (SPION) were then coated with various linkers such as trisodium citrate (TSC), tetraethoxysilane (TEOS) and 3-aminopropyltriethoxysilane (APTES). The interaction of this precursor coated SPION to the anti-coagulant protein was studied .

Molecular docking studies revealed that the protein had two binding sites. The core binding site, where TEOS and APTES were bound and a secondary binding region in the side chain, where TSC was bound. The study construed that the entire three precursors linked SPION were bound to the protein through strong electrostatic and Vanderwaal's interactions, which stabilises the complex and specifically the TSC-SPION combination showed a two fold increase in electrostatic interaction. It was observed that the –COO moiety in TSC-SPION acted like a clip and bound strongly to the Arg48 and Arg52 residues. Molecular dynamics simulation analyses also substantially confirmed the stability of these complexes under ambient conditions such as temperature and pressure, which can be a pointer for the laboratory experiments. The TSC coated nanoparticle was found to possess one protein-one ligand interaction, since it bound to the one unit of the homodimeric protein, making it more suitable for wet lab synthesis (Okoli, et al., 2013).

The coagulation activity of TSC-coated, TEOS-coated and APTES-coated nanoparticles was also validated through wet lab experiments and it was intriguing to study that TSC-coated nanoparticles increased the coagulation activity. It showed ~80% activity as compared to 60% exhibited by TEOS and 65% by APTES. Thus, the *in silico* docking and dynamics studies proved as a potential precursor to wet lab experiments and they acted as a pivotal tool in studying the nature of target protein and designing a suitable nanoparticle, accordingly (Okoli, et al., 2013). The importance of *in-silico* approaches in designing and validating a nanoparticle has been re-emphasized through the above outcome, thereby serving as an antecedent to wet lab experiments.

3.5 FOULDOCK: Study of Antifoulers Interactions

Accumulation of living organisms such as bacteria, fungus, algae, etc. onto any wet surface is termed as biofouling. Biofouling is a major concern for the marine industry and its implications are indeed huge. Biofouling is capable of creating severe damages to underwater structures, desalination plants, ship hulls, fishing cages, nets and bridge piles. The amassing biomass on the ships increases its weight, leading to reduced velocity and thereby more fuel consumption. Similarly, corrosion on steel, iron, copper and aluminium hulls of the ship is again one of the major impediments faced by the shipping industry, and therefore industries use different types of paints in order to prevent corrosion and thereby to regain the economic loss, which again is associated with environmental obstacles (Yebra, Kiil, & Dam-Johansen, 2004). Biofouling can be classified into two broad groups: microfouling and macrofouling. Microfouling is often caused by bacteria and macrofouling involves algae and barnacles. Some of the other major contributors to macrofouling are mussels, polychaete worms and sea weeds (Olsen, Kiil, Dam-Johansen, Pedersen, & Hermann, 2009).

The process of biofouling happens in a sequential manner. Initial biofilm formation is facilitated through polysaccharides and proteins on the substrate, and this organic layer provides the suitable scaffold for other fouling agents to adhere. The biofilm paves way for adhesion of bacteria and algae. This adhesion is irreversible and is characterized by secretion of extracellular polymeric substances (EPS). Once this biofilm is formed, macrofoulers come and bind to them, thereby resulting in complex biological community (Fletcher & Loeb, 1979; Walt, Smulow, Turesky, & Hill, 1985).

The design of anti-fouling agents has been made by targeting this cellular adhesion of foulers such as bacteria, etc. to prevent biofilm formation. Several research activities are going on across the world to address this global issue, since biofouling causes financial loss as well. Antifouling has been attempted through various strategies in obtaining a sustainable solution. It includes chemicals-based antifouling method involving antifouling paints (Omae, 2003), physical methods mediated through electrolysis, radiation and by altering the zeta potential (Dittrich & Sibler, 2005; Matsunaga & Lim, 2000) and biological methods through enzymes that targets adhesion (Aldred, Phang, Conlan, Clare, & Vancso, 2008) and biofilm matrix (Xavier, Picioreanu, Rani, van Loosdrecht, & Stewart, 2005).The understanding of the mode of action of these antifouling agents is imperative and better knowledge on their targets helps in designing novel antifouling compounds. The gigantic use of molecular docking in the process of biofilm inhibition is depicted in Table 5.

The two examples cited below focus on the role of computational techniques such as molecular modelling, docking and molecular dynamics simulation in interpreting interaction of antifouling agents.

Table 5. Molecular docking in exploring the mechanism of antifoulers

Antifoulers	Mode of Action	Computational Tools Employed	References
Bioactive compounds such as benzyl (6Z,9Z,12Z)-6,9,12-octadecatrienoate, 3-benzyloxy-1-nitro-butan-2-ol and 1,3-cyclohexane dicarbohydrazide	*Salvadora persica* contains bioactive anti-biofilm agents with dual functionalities of growth inhibition and Quorum sensing regulator interaction.	Molecular docking	(Al-Sohaibani & Murugan, 2012)
Antifouling compounds produced by *Streptococcus mutans*	*S. mutans* forms biofilm on dental tissues. Initially, it interacts with proteins/glycoproteins of salivary and microbial origin adsorbed on the acquired pellicle that coats the tooth enamel. It then aggregates enriched with extracellular polysaccharides and leads to the maturation of biofilm.	*In silico* modelling	(da Silva et al., 2014)
Butenolide -1	In *B.amphitrite,* butenolide 1 was found to bind with acetyl-coenzyme A acetyltransferase 1 (ACAT1), which plays a significant role in ketone body synthesis and is essential during the development stages of cyprid larvae of *B.amphitrite*. In *B. Neritina*, butenolide 1 was found to bind with three proteins, such as acyl-coenzyme A dehydrogenase (ACAD), actin and glutathione S-transferase. In *Vibrio sp.,* butenolide 1 binds with succinyl-CoA synthetase (SCS) subunit beta and DNA-directed RNA polymerase subunit alpha.	Molecular docking	(Zhang, et al., 2012)
5-substituted, 3,4-dihalo-5H-furan-2-one compounds	It acts as a quorum sensing inhibitors that targets both LasR and LuxS receptors simultaneously, which are the key receptors where the bacteria, *P. aeruginosa* binds and induces biofilm formation. Thus it acts as an antifoulers by blocking the target receptors.	*In silico* modelling	(G. Y. Liu, et al., 2012)

3.5.1 Antifouling Activity of Butenolide: Role of Docking

The most common practice associated with antifouling strategies is the usage of coatings on the surfaces of water structures. As mentioned earlier, there exists a very diverse group of fouling agents that includes bacteria, algae, mussels, etc. which makes the designing of a specific antifouling compound little challenging. In most cases, the antifouling agents are observed to be toxic, causing ecological problems (Omae, 2003; Xavier, et al., 2005). The chemical compound, 5-octylfuran-2(5H)-one commonly referred as butenolide 1 has been found as a potent antifouling agent with good market value. Butenolide 1, derived from natural products, is also found to be highly specific to the biofoulers with effective antifouling action (Xu et al., 2010; Zhang et al., 2011).

Though the antifouling activity of butenolide 1 is validated, there are no information about its mode of action and its molecular targets. It becomes essential to delve a bit deeper in this aspect, in order to make a stern comparisons with other antifouling agents and understand its pharmacology. For instance, the activity of butenolide 1 against two macrofoulers, *Balanus amphitrite* and *Bugula neritina* and one microfouler, *Vibrio sp. UST020129-010* was evaluated to explore the molecular targets and binding sites of butenolide 1(Zhang et al., 2012).

In case of *B.amphitrite*, it was found that butenolide 1 was bound to acetyl-coenzyme A acetyltransferase 1 (ACAT1), which plays a significant role in ketone body synthesis and is essential during the development stages of cyprid larvae of *B.amphitrite*. Molecular docking studies revealed that the active site residues involved in the interactions were Cys111, Asn338, His370, and Gly400 and this interaction was found to possess least free energy of binding. This revealed the fact that butenolide 1 affects the cyprids development by inhibiting the ketone body metabolism through downregulating ACAT1. Binding of butenolide 1 to *B. neritina* involved three different types of proteins, which were similar to acyl-coenzyme A dehydrogenase (ACAD), actin and glutathione S-transferase (GST). These proteins were involved in lipid metabolism, muscle contractions and detoxification mechanism, respectively and butenolide 1 inhibits their activities. Docking of butenolide 1 to these proteins revealed that Lys284, Glu325, and Glu393 were the important residues involved in butenolide 1 binding. The inhibition of ACAD, actin and GST was found to cause lack of energy, affect the motility and transport of cellular compounds in *B.neritina*. In case of *Vibrio sp. UST020129-010* strain, it was observed that Butenolide 1 possesses bacteriocidal effects. The proteins that were found to be involved were succinyl-CoA synthetase (SCS) subunit beta and DNA-directed RNA polymerase subunit alpha. These proteins are involved in citric acid cycle and also supplies succinyl coA required for ketone body activation. Butenolide 1 was found to inhibit these proteins, thereby resulting in growth depletion of the *Vibrio sp. UST020129-010* strain (Zhang, et al., 2012).

The above demonstration has provided exhaustive insights into investigating the binding sites of antifouling agents, the associated proteins and their mechanism of action. These reports could serve as a base data for designing novel antifouling agents targeting specific proteins and prevent the growth of the fouling agents.

3.5.2 Docking in Antifouling of *Pseudomonas aeruginosa*

The example above discloses a superficial view on the activities of the antifouling agent against bacteria which is a microfouler. The following report provides a thorough information on the mechanism of antifouling compounds against the biofouling activity of *Pseudomonas aeruginosa*. *P.aeruginosa* are commonly observed in most chronic wounds, where they colonize in the fashion of biofilm on the surface by communicating with each other through quorum-sensing, which in turn lead to serious antibiotics resistance and dysfunction of immune system (Bjarnsholt et al., 2008; James et al., 2008; Walters, Roe, Bugnicourt, Franklin, & Stewart, 2003). Current strategies are focussed on designing novel inhibitors that can block the quorum sensing activity, which is triggered by the autoinducer molecule n-acyl homoserine lactones (AHL), through binding to the LasR receptors (Musk, Dinty, & Hergenrother, 2006). Another key enzyme, LuxS, is another autoinducer molecule involved in biomass production in both gram-positive and gram-negative bacteria. It is reported that modifications to these autoinducers increase the quorum sensing inhibitory effect (Geske, Mattmann, & Blackwell, 2008).

Targeting LasR and LuxS enzymes are vital in designing novel inhibitors for quorum sensing activity. The interaction of the inhibitors with the LasR receptor was carried out through molecular docking approach. Modified furanone compounds such as 3,4-Dichloro-2(5H)-furanone and 3,4-Dibromo-2(5H)-furanone were docked to LasR receptor. It was found that all the furanone derivatives docked perfectly well into the active site of the LasR receptor. The interaction was characterised by hydrogen bonding and lipophilic electron-donating aromatic ring. The stabilisation in binding is brought out by Arg61, Leu110 residues found in the binding pocket. This acts as an inhibitor to autoinducer molecule binding, which in turn prevents the quorum sensing activity, resulting in the blocking of biofilm formation by the bacteria, *P.aeruginosa* (G. Y. Liu et al., 2012).

The indulgence of computational docking technique provides insights into the binding mode and action of quorum sensing inhibitors in the microfouler bacteria. This is a representative study, which can be applied to a larger group of microfoulers, thereby developing novel compounds by investigating their interactions through *in silico* methods.

CONCLUSION

With the advent of molecular docking, the field of drug discovery has made giant leaps in the last decade. This computational chemistry technique has introduced robust methodologies to investigate the nature of interactions and their impact between any particular target receptor and its corresponding ligand molecule. New algorithms are being developed every now and then to increase the authenticity of this technique. The magnitude of development attained by the drug designing industry through this method has provoked other industries to use molecular docking as a significant tool to examine the biological processes. Though the prevalence of molecular docking is more dominant in the field of drug discovery, other disciplines of science have started using it for various classical research purposes. And in most cases, these *in silico* studies have been precursors to wet lab experiments, thereby acting as an investigative interface.

The involvement of computational methodologies have been explored and exploited in various research avenues. It was also learnt that these molecular docking studies have often provided conclusive results and have divulged the facts on the molecular mechanisms of the interacting macromolecules. For instance, the case study focussing on the interaction between acetyl coA and its enzyme, acetyl coA carboxylase has led to a striking conclusion on the type of interactions that are altered upon mutation in the fatty acid biosynthetic pathway. This has further paved way to identify the crucial amino acid residues involved in the approach of enhancing lipid production, and thereby biodiesel. Likewise, the role of various secondary metabolites from bacteria and their interactions with the hydrocarbons in order to degrade them as a part of bioremediation are extensively studied through molecular docking and molecular dynamics simulation methods as discussed earlier in this chapter. Computational modelling, docking and simulation studies have bridged the gap between theoretical inferences and experimental reality. The application and reliance on these techniques have always been on the exponential rise and the results are evident. Adding a further note that the disciplines represented and evaluated here are just a random picks from a myriad of scientific researches that are currently being pursued. This chapter is a pointer of such eclectic applications of molecular docking to decipher the basic processes over product development.

ACKNOWLEDGMENT

The authors gratefully acknowledge the Earth System Science Organization (ESSO), Ministry of Earth sciences, Govt. of India, for financial support. The authors are thankful to Dr. M. A. Atmanand, Director, ESSO-National Institute of Ocean Technology, for his constant support and encouragement.

REFERENCES

Achbarou, A., Kaiser, S., Tremblay, G., Ste-Marie, L.-G., Brodt, P., Goltzman, D., & Rabbani, S. A. (1994). Urokinase overproduction results in increased skeletal metastasis by prostate cancer cells in vivo. *Cancer Research*, *54*(9), 2372–2377. PMID:8162583

Ahmad, A., Sayed, A., Ginnebaugh, K. R., Sharma, V., Suri, A., Saraph, A., . . . Sarkar, F. H. (2015). Molecular docking and inhibition of matrix metalloproteinase-2 by novel difluorinatedbenzylidene curcumin analog. *American Journal of Translational Research, 7*(2), 298.

Al-Sohaibani, S., & Murugan, K. (2012). Anti-biofilm activity of Salvadora persica on cariogenic isolates of Streptococcus mutans: In vitro and molecular docking studies. *Biofouling*, *28*(1), 29–38. doi:10.1080/08927014.2011.647308 PMID:22235758

Aldred, N., Phang, I. Y., Conlan, S. L., Clare, A. S., & Vancso, G. J. (2008). The effects of a serine protease, Alcalase®, on the adhesives of barnacle cyprids (Balanus amphitrite). *Biofouling*, *24*(2), 97–107. doi:10.1080/08927010801885908 PMID:18231899

Alves, R., Chaleil, R. A., & Sternberg, M. J. (2002). Evolution of enzymes in metabolism: A network perspective. *Journal of Molecular Biology*, *320*(4), 751–770. doi:10.1016/S0022-2836(02)00546-6 PMID:12095253

Anthony, J., Rangamaran, V. R., Gopal, D., Shivasankarasubbiah, K. T., Thilagam, M. L. J., Dhassiah, M. P., & Dakshinamurthy, S. et al. (2015). Ultraviolet and 5′ Fluorodeoxyuridine Induced Random Mutagenesis in Chlorella vulgaris and Its Impact on Fatty Acid Profile: A New Insight on Lipid-Metabolizing Genes and Structural Characterization of Related Proteins. *Marine Biotechnology (New York, N.Y.)*, *17*(1), 66–80. doi:10.1007/s10126-014-9597-5 PMID:25189135

Aubin-Tam, M.-E., & Hamad-Schifferli, K. (2005). Gold nanoparticle-cytochrome c complexes: The effect of nanoparticle ligand charge on protein structure. *Langmuir*, *21*(26), 12080–12084. doi:10.1021/la052102e PMID:16342975

Bjarnsholt, T., Kirketerp-Møller, K., Jensen, P. Ø., Madsen, K. G., Phipps, R., Krogfelt, K., & Givskov, M. et al. (2008). Why chronic wounds will not heal: A novel hypothesis. *Wound Repair and Regeneration*, *16*(1), 2–10. doi:10.1111/j.1524-475X.2007.00283.x PMID:18211573

Blatti, J. L. (2012). *Investigating fatty acid biosynthesis within the algal chloroplast using Chlamydomonas reinhardtii as a model*. Academic Press.

Brancolini, G., Kokh, D. B., Calzolai, L., Wade, R. C., & Corni, S. (2012). Docking of ubiquitin to gold nanoparticles. *ACS Nano*, *6*(11), 9863–9878. doi:10.1021/nn303444b PMID:23033917

Burda, C., Chen, X., Narayanan, R., & El-Sayed, M. A. (2005). Chemistry and properties of nanocrystals of different shapes. *Chemical Reviews*, *105*(4), 1025–1102. doi:10.1021/cr030063a PMID:15826010

Cai, C. Z., Li, Z. R., Wang, W. L., & Chen, Y. Z. (2004). Advances in modeling of biomolecular interactions. *Acta Pharmacologica Sinica*, *25*, 1–8. PMID:14704115

Calzolai, L., Laera, S., Ceccone, G., & Rossi, F. (2012). *Structure and Stability of Proteins Interacting with Nanoparticles. In Proteins at Interfaces III State of the Art* (Vol. 1120, pp. 839–855). ACS Publications.

Chandrakar, B., Jain, A., Roy, S., Gutlapalli, V. R., Saraf, S., Suppahia, A., & Nayarisseri, A. et al. (2013). Molecular modeling of Acetyl-CoA carboxylase (ACC) from Jatropha curcas and virtual screening for identification of inhibitors. *Journal of Pharmacy Research*, *6*(9), 913–918. doi:10.1016/j.jopr.2013.07.032

Conese, M., & Blasi, F. (1995). 6 The urokinase/urokinase-receptor system and cancer invasion. *Bailliere's Clinical Haematology*, *8*(2), 365–389. doi:10.1016/S0950-3536(05)80273-2 PMID:7549068

Cortial, S., Chaignon, P., Iorga, B. I., Aymerich, S., Truan, G., Gueguen-Chaignon, V., & Ouazzani, J. et al. (2010). NADH oxidase activity of Bacillus subtilis nitroreductase NfrA1: Insight into its biological role. *FEBS Letters*, *584*(18), 3916–3922. doi:10.1016/j.febslet.2010.08.019 PMID:20727352

Cui, F., Yang, K., & Li, Y. (2015). *Investigate the Binding of Catechins to Trypsin Using Docking and Molecular Dynamics Simulation*. Academic Press.

da Silva, A. C. B., da Silva, D. R., de Macêdo Ferreira, S. A., Agripino, G. G., Albuquerque, A. R., & do Rêgo, T. G. (2014). *In Silico Approach for the Identification of Potential Targets and Specific Antimicrobials for Streptococcus mutans*. Advances in Bioscience and Biotechnology.

De, M., Ghosh, P. S., & Rotello, V. M. (2008). Applications of nanoparticles in biology. *Advanced Materials*, *20*(22), 4225–4241. doi:10.1002/adma.200703183

Deepa, P., Vandhana, S., Muthukumaran, S., Umashankar, V., Jayanthi, U., & Krishnakumar, S. (2010). Chemical inhibition of fatty acid synthase: molecular docking analysis and biochemical validation in ocular cancer cells. *Journal of Ocular Biology, Diseases, and Informatics*, *3*(4), 117-128.

DesJarlais, R. L., Sheridan, R. P., Dixon, J. S., Kuntz, I. D., & Venkataraghavan, R. (1986). Docking flexible ligands to macromolecular receptors by molecular shape. *Journal of Medicinal Chemistry, 29*(11), 2149–2153. doi:10.1021/jm00161a004 PMID:3783576

Dittrich, M., & Sibler, S. (2005). Cell surface groups of two picocyanobacteria strains studied by zeta potential investigations, potentiometric titration, and infrared spectroscopy. *Journal of Colloid and Interface Science, 286*(2), 487–495. doi:10.1016/j.jcis.2005.01.029 PMID:15897062

Ellis, L. B., Hou, B. K., Kang, W., & Wackett, L. P. (2003). The University of Minnesota biocatalysis/biodegradation database: Post-genomic data mining. *Nucleic Acids Research, 31*(1), 262–265. doi:10.1093/nar/gkg048 PMID:12519997

Evers, A., & Klabunde, T. (2005). Structure-based drug discovery using GPCR homology modeling: Successful virtual screening for antagonists of the alpha1A adrenergic receptor. *Journal of Medicinal Chemistry, 48*(4), 1088–1097. doi:10.1021/jm0491804 PMID:15715476

Faraji, M., Yamini, Y., & Rezaee, M. (2010). Magnetic nanoparticles: Synthesis, stabilization, functionalization, characterization, and applications. *Journal of the Indian Chemical Society, 7*(1), 1–37.

Fletcher, M., & Loeb, G. (1979). Influence of substratum characteristics on the attachment of a marine pseudomonad to solid surfaces. *Applied and Environmental Microbiology, 37*(1), 67–72. PMID:16345338

Fraser, H. B., Hirsh, A. E., Steinmetz, L. M., Scharfe, C., & Feldman, M. W. (2002). Evolutionary rate in the protein interaction network. *Science, 296*(5568), 750–752. doi:10.1126/science.1068696 PMID:11976460

Fulekar, M. (2009). *Environmental Cleanup Approach Using Bioinformatics in Bioremediation. In Bioinformatics: Applications in Life and Environmental Sciences* (pp. 173–199). Springer. doi:10.1007/978-1-4020-8880-3_10

Gagner, J. E., Lopez, M. D., Dordick, J. S., & Siegel, R. W. (2011). Effect of gold nanoparticle morphology on adsorbed protein structure and function. *Biomaterials, 32*(29), 7241–7252. doi:10.1016/j.biomaterials.2011.05.091 PMID:21705074

Geske, G. D., Mattmann, M. E., & Blackwell, H. E. (2008). Evaluation of a focused library of N-aryl L-homoserine lactones reveals a new set of potent quorum sensing modulators. *Bioorganic & Medicinal Chemistry Letters, 18*(22), 5978–5981. doi:10.1016/j.bmcl.2008.07.089 PMID:18760602

Gill, A., Scanlon, T. C., Osipovitch, D. C., Madden, D. R., & Griswold, K. E. (2011). Crystal structure of a charge engineered human lysozyme having enhanced bactericidal activity. *PLoS ONE, 6*(3), e16788. doi:10.1371/journal.pone.0016788 PMID:21408218

Goodford, P. J. (1984). Drug design by the method of receptor fit. *Journal of Medicinal Chemistry, 27*(5), 557–564. doi:10.1021/jm00371a001 PMID:6716396

Graf, L., Jancso, A., Szilágyi, L., Hegyi, G., Pintér, K., Náray-Szabó, G., & Rutter, W. J. et al. (1988). Electrostatic complementarity within the substrate-binding pocket of trypsin. *Proceedings of the National Academy of Sciences of the United States of America, 85*(14), 4961–4965. doi:10.1073/pnas.85.14.4961 PMID:3134655

Hajduk, P. J. (2006). Fragment-based drug design: How big is too big? *Journal of Medicinal Chemistry*, *49*(24), 6972–6976. doi:10.1021/jm060511h PMID:17125250

He, Y., Fan, C., & Lee, S.-T. (2010). Silicon nanostructures for bioapplications. *Nano Today*, *5*(4), 282–295. doi:10.1016/j.nantod.2010.06.008

Hruby, V. J. (1982). Conformational restrictions of biologically active peptides via amino acid side chain groups. *Life Sciences*, *31*(3), 189–199. doi:10.1016/0024-3205(82)90578-1 PMID:6126794

Hruby, V. J., Qui, W., Okayama, T., & Soloshonok, V. A. (2002). Design of nonpeptides from peptide ligands for peptide receptors. *Methods in Enzymology*, *343*, 91–123. doi:10.1016/S0076-6879(02)43129-1 PMID:11665597

Huang, H., & Zhao, M. (2008). Changes of trypsin in activity and secondary structure induced by complex with trypsin inhibitors and tea polyphenol. *European Food Research and Technology*, *227*(2), 361–365. doi:10.1007/s00217-007-0729-2

James, G. A., Swogger, E., Wolcott, R., Secor, P., Sestrich, J., Costerton, J. W., & Stewart, P. S. (2008). Biofilms in chronic wounds. *Wound Repair and Regeneration*, *16*(1), 37–44. doi:10.1111/j.1524-475X.2007.00321.x PMID:18086294

Jeyam, M., Sushma, R., Sharanya, M., & Poornima, V. (2011). Validating nutraceuticals to alleviate progeria using molecular docking studies. *Journal of Pharmaceutical and Biomedical Sciences*, *13*(13).

Ju, J., Hong, J., Zhou, J.-, Pan, Z., Bose, M., Liao, J., & Lin, Y. et al. (2005). Inhibition of intestinal tumorigenesis in Apcmin/+ mice by (−)-epigallocatechin-3-gallate, the major catechin in green tea. *Cancer Research*, *65*(22), 10623–10631. doi:10.1158/0008-5472.CAN-05-1949 PMID:16288056

Khan, F., Pal, D., Vikram, S., & Cameotra, S. S. (2013). Metabolism of 2-chloro-4-nitroaniline via novel aerobic degradation pathway by Rhodococcus sp. strain MB-P1. *PLoS ONE*, *8*(4), e62178. doi:10.1371/journal.pone.0062178 PMID:23614030

Khan, F., Sajid, M., & Cameotra, S. S. (2013). In Silico Approach for the Bioremediation of Toxic Pollutants. *J Pet Environ Biotechnol*, *4*(161), 2.

Kharkar, P. S., Warrier, S., & Gaud, R. S. (2014). Reverse docking: a powerful tool for drug repositioning and drug rescue. *Future Medicinal Chemistry, 6*(3), 333-342.

Kim, M. H., Kim, Y., Park, H.-J., Lee, J. S., Kwak, S.-N., Jung, W.-H., & Oh, T.-K. et al. (2008). Structural insight into bioremediation of triphenylmethane dyes by Citrobacter sp. triphenylmethane reductase. *The Journal of Biological Chemistry*, *283*(46), 31981–31990. doi:10.1074/jbc.M804092200 PMID:18782772

Koshland, D. (1963). Correlation of structure an function in enzyme action. *Science*, *142*(3599), 1533–1541. doi:10.1126/science.142.3599.1533 PMID:14075684

Kuntz, I. D., Blaney, J. M., Oatley, S. J., Langridge, R., & Ferrin, T. E. (1982). A geometric approach to macromolecule-ligand interactions. *Journal of Molecular Biology*, *161*(2), 269–288. doi:10.1016/0022-2836(82)90153-X PMID:7154081

Lam, K. S., Salmon, S. E., Hersh, E. M., Hruby, V. J., Kazmierski, W. M., & Knapp, R. J. (1991). A new type of synthetic peptide library for identifying ligand-binding activity. *Nature*, *354*(6348), 82–84. doi:10.1038/354082a0 PMID:1944576

Leszczynski, J., & Puzyn, T. (2012). *Towards efficient designing of safe nanomaterials: Innovative merge of computational approaches and experimental techniques*. Royal Society of Chemistry. doi:10.1039/9781849735476

Li, H., & Li, C. (2010). Multiple ligand simultaneous docking: Orchestrated dancing of ligands in binding sites of protein. *Journal of Computational Chemistry*, *31*(10), 2014–2022. PMID:20166125

Librando, V., & Forte, S. (2005). Computer evaluation of protein segments removal effects from naphthalene 1, 2-dioxygenase enzyme on polycyclic aromatic hydrocarbons interaction. *Biochemical Engineering Journal*, *27*(2), 161–166. doi:10.1016/j.bej.2005.08.011

Liu, G. Y., Guo, B. Q., Chen, W. N., Cheng, C., Zhang, Q. L., Dai, M. B., & Chen, W. M. et al. (2012). Synthesis, Molecular Docking, and Biofilm Formation Inhibitory Activity of 5-Substituted 3, 4-Dihalo-5H-furan-2-one Derivatives on Pseudomonas aeruginosa. *Chemical Biology & Drug Design*, *79*(5), 628–638. doi:10.1111/j.1747-0285.2012.01342.x PMID:22268453

Liu, X., Ouyang, S., Yu, B., Liu, Y., Huang, K., Gong, J., & Jiang, H. et al. (2010). PharmMapper server: A web server for potential drug target identification using pharmacophore mapping approach. *Nucleic Acids Research*, *38*(Web Serversuppl 2), W609–W614. doi:10.1093/nar/gkq300 PMID:20430828

Lovley, D. R. (2003). Cleaning up with genomics: Applying molecular biology to bioremediation. *Nature Reviews. Microbiology*, *1*(1), 35–44. doi:10.1038/nrmicro731 PMID:15040178

Lui, A. E. (2007). *Engineering of aniline dioxygenase for bioremediation and industrial applications*. Academic Press.

Ma, W., Tang, C., & Lai, L. (2005). Specificity of trypsin and chymotrypsin: Loop-motion-controlled dynamic correlation as a determinant. *Biophysical Journal*, *89*(2), 1183–1193. doi:10.1529/biophysj.104.057158 PMID:15923233

Matsunaga, T., & Lim, T.-k. (2000). Electrochemical prevention of biofouling. *Electrochemistry*, *68*(11), 847–852.

Mendilaharsu, M., De Stefani, E., Deneo-Pellegrini, H., Carzoglio, J. C., & Ronco, A. (1998). Consumption of tea and coffee and the risk of lung cancer in cigarette-smoking men: A case-control study in Uruguay. *Lung Cancer (Amsterdam, Netherlands)*, *19*(2), 101–107. doi:10.1016/S0169-5002(97)00075-5 PMID:9567246

Meng, X.-Y., Zhang, H.-X., Mezei, M., & Cui, M. (2011). Molecular docking: A powerful approach for structure-based drug discovery. *Current Computer-aided Drug Design*, *7*(2), 146–157. doi:10.2174/157340911795677602 PMID:21534921

Misra, N., Patra, M. C., Panda, P. K., Sukla, L. B., & Mishra, B. K. (2013). Homology modeling and docking studies of FabH (β-ketoacyl-ACP synthase III) enzyme involved in type II fatty acid biosynthesis of Chlorella variabilis: A potential algal feedstock for biofuel production. *Journal of Biomolecular Structure & Dynamics*, *31*(3), 241–257. doi:10.1080/07391102.2012.698247 PMID:22830394

Musk, J., Dinty, J., & Hergenrother, P. J. (2006). Chemical countermeasures for the control of bacterial biofilms: Effective compounds and promising targets. *Current Medicinal Chemistry*, *13*(18), 2163–2177. doi:10.2174/092986706777935212 PMID:16918346

Nedumpully Govindan, P. (2013). *Computational Modeling Investigating the Interactions of Proteins with Fullerene-based Nanoparticles*. Academic Press.

Nyberg, P., Ylipalosaari, M., Sorsa, T., & Salo, T. (2006). Trypsins and their role in carcinoma growth. *Experimental Cell Research*, *312*(8), 1219–1228. doi:10.1016/j.yexcr.2005.12.024 PMID:16457812

Okoli, C., Sengottaiyan, S., Arul Murugan, N., Pavankumar, A. R., Ågren, H., & Kuttuva Rajarao, G. (2013). In silico modeling and experimental evidence of coagulant protein interaction with precursors for nanoparticle functionalization. *Journal of Biomolecular Structure & Dynamics*, *31*(10), 1182–1190. doi:10.1080/07391102.2012.726534 PMID:23163338

Olsen, S. M., Kiil, S., Dam-Johansen, K., Pedersen, L. T., & Hermann, M. H. (2009). *Controlled release of environmentally friendly antifouling agents from marine coatings*. Hempel A/SHempel A/S.

Omae, I. (2003). Organotin antifouling paints and their alternatives. *Applied Organometallic Chemistry*, *17*(2), 81–105. doi:10.1002/aoc.396

Parales, R., Bruce, N., Schmid, A., & Wackett, L. (2002). Biodegradation, biotransformation, and biocatalysis (B3). *Applied and Environmental Microbiology*, *68*(10), 4699–4709. doi:10.1128/AEM.68.10.4699-4709.2002 PMID:12324310

Paul, D., Pandey, G., Pandey, J., & Jain, R. K. (2005). Accessing microbial diversity for bioremediation and environmental restoration. *Trends in Biotechnology*, *23*(3), 135–142. doi:10.1016/j.tibtech.2005.01.001 PMID:15734556

Pazos, F., Guijas, D., Valencia, A., & De Lorenzo, V. (2005). MetaRouter: Bioinformatics for bioremediation. *Nucleic Acids Research*, *33*(suppl 1), D588–D592. doi:10.1093/nar/gki068 PMID:15608267

Pennacchio, F. (2009). *Strategies for using microorganisms and enzymatic systems in bioremediation*. Università degli Studi di Napoli Federico II.

Rabbani, S., Harakidas, P., Davidson, D. J., Henkin, J., & Mazar, A. P. (1995). Prevention of prostate-cancer metastasis in vivo by a novel synthetic inhibitor of urokinase-type plasminogen activator (uPA). *International Journal of Cancer*, *63*(6), 840–845. doi:10.1002/ijc.2910630615 PMID:8847143

Rana, S., Yeh, Y.-C., & Rotello, V. M. (2010). Engineering the nanoparticle–protein interface: Applications and possibilities. *Current Opinion in Chemical Biology*, *14*(6), 828–834. doi:10.1016/j.cbpa.2010.10.001 PMID:21035376

Rashmi, V., ShylajaNaciyar, M., Rajalakshmi, R., D'Souza, S. F., Prabaharan, D., & Uma, L. (2013). Siderophore mediated uranium sequestration by marine cyanobacterium Synechococcus elongatus BDU 130911. *Bioresource Technology*, *130*, 204–210. doi:10.1016/j.biortech.2012.12.016 PMID:23306130

Saikia, N., Rajkhowa, S., & Deka, R. C. (2013). Density functional and molecular docking studies towards investigating the role of single-wall carbon nanotubes as nanocarrier for loading and delivery of pyrazinamide antitubercular drug onto pncA protein. *Journal of Computer-Aided Molecular Design*, *27*(3), 257–276. doi:10.1007/s10822-013-9638-6 PMID:23413106

Satishkumar, R., & Vertegel, A. (2008). Charge-directed targeting of antimicrobial protein-nanoparticle conjugates. *Biotechnology and Bioengineering*, *100*(3), 403–412. doi:10.1002/bit.21782 PMID:18183633

Shoichet, B. K., Kuntz, I. D., & Bodian, D. L. (1992). Molecular docking using shape descriptors. *Journal of Computational Chemistry*, *13*(3), 380–397. doi:10.1002/jcc.540130311

Shrivastava, S., Nuffer, J. H., Siegel, R. W., & Dordick, J. S. (2012). Position-specific chemical modification and quantitative proteomics disclose protein orientation adsorbed on silica nanoparticles. *Nano Letters*, *12*(3), 1583–1587. doi:10.1021/nl2044524 PMID:22296027

Sing Yong, K., Bing Choi, S., & Wahab, A. H. (2015). *Docking of Dengue NS2B-NS3 Protease with Murraya koenigii.* Paper presented at the 3rd International Conference on Computation for Science and Technology (ICCST-3).

Sonawani, A., Niazi, S., & Idicula-Thomas, S. (2013). In Silico Study on Binding Specificity of Gonadotropins and Their Receptors: Design of a Novel and Selective Peptidomimetic for Human Follicle Stimulating Hormone Receptor. *PLoS ONE*, *8*(5), e64475. doi:10.1371/journal.pone.0064475 PMID:23700481

Soreide, K., Janssen, E., Körner, H., & Baak, J. (2006). Trypsin in colorectal cancer: Molecular biological mechanisms of proliferation, invasion, and metastasis. *The Journal of Pathology*, *209*(2), 147–156. doi:10.1002/path.1999 PMID:16691544

Sridhar, S., & Chandra, J. H. (2014). Involvement of Computational tools towards In Silico remediation-Synthetic textile dyes interacting with Azoreductase. *International Journal of Chemical Technology Research*, *6*(9), 4412–4441.

Suganuma, M., Ohkura, Y., Okabe, S., & Fujiki, H. (2001). Combination cancer chemoprevention with green tea extract and sulindac shown in intestinal tumor formation in Min mice. *Journal of Cancer Research and Clinical Oncology*, *127*(1), 69–72. doi:10.1007/s004320000189 PMID:11206275

Sun, X., Feng, Z., Zhang, L., Hou, T., & Li, Y. (2014). *The Selective Interaction between Silica Nanoparticles and Enzymes from Molecular Dynamics Simulations*. Academic Press.

Suresh, P. S., Kumar, A., Kumar, R., & Singh, V. P. (2008). An Insilco approach to bioremediation: Laccase as a case study. *Journal of Molecular Graphics & Modelling*, *26*(5), 845–849. doi:10.1016/j.jmgm.2007.05.005 PMID:17606396

Tekere, M., Mswaka, A., Zvauya, R., & Read, J. (2001). Growth, dye degradation and ligninolytic activity studies on Zimbabwean white rot fungi. *Enzyme and Microbial Technology*, *28*(4), 420–426. doi:10.1016/S0141-0229(00)00343-4 PMID:11240201

Vieira, A. C., Marschalk, C., Biavatti, D. C., Lorscheider, C. A., Peralta, R. M., & Seixas, F. A. V. (2015). Modeling Based Structural Insights into Biodegradation of the Herbicide Diuron by Laccase-1 from Ceriporiopsis subvermispora. *Bioinformation*, *11*, 224–228. doi:PMID:26124565 doi:10.6026/97320630011224

Walls, P. H., & Sternberg, M. J. (1992). New algorithm to model protein-protein recognition based on surface complementarity: Applications to antibody-antigen docking. *Journal of Molecular Biology*, *228*(1), 277–297. doi:10.1016/0022-2836(92)90506-F PMID:1280302

Walt, D. R., Smulow, J. B., Turesky, S. S., & Hill, R. G. (1985). The effect of gravity on initial microbial adhesion. *Journal of Colloid and Interface Science*, *107*(2), 334–336. doi:10.1016/0021-9797(85)90185-7

Walters, M. C., Roe, F., Bugnicourt, A., Franklin, M. J., & Stewart, P. S. (2003). Contributions of antibiotic penetration, oxygen limitation, and low metabolic activity to tolerance of Pseudomonas aeruginosa biofilms to ciprofloxacin and tobramycin. *Antimicrobial Agents and Chemotherapy*, *47*(1), 317–323. doi:10.1128/AAC.47.1.317-323.2003 PMID:12499208

Wang, Y., Yu, X., Wu, Y., & Zhang, D. (2012). Coffee and tea consumption and risk of lung cancer: A dose–response analysis of observational studies. *Lung Cancer (Amsterdam, Netherlands)*, *78*(2), 169–170. doi:10.1016/j.lungcan.2012.08.009 PMID:22964413

Weisburger, J. H. (2002). Comments on the history and importance of aromatic and heterocyclic amines in public health. *Mutation Research/Fundamental and Molecular Mechanisms of Mutagenesis, 506*, 9-20.

Wessels, J., & Van der Veen, R. (1956). The action of some derivatives of phenylurethan and of 3-phenyl-1, 1-dimethylurea on the Hill reaction. *Biochimica et Biophysica Acta*, *19*, 548–549. doi:10.1016/0006-3002(56)90481-4 PMID:13315318

Wlodawer, A., Svensson, L. A., Sjoelin, L., & Gilliland, G. L. (1988). Structure of phosphate-free ribonuclease A refined at 1.26. ANG. *Biochemistry*, *27*(8), 2705–2717. doi:10.1021/bi00408a010 PMID:3401445

Wu, J.-F., Xu, M.-Q., & Zhao, G.-C. (2010). Graphene-based modified electrode for the direct electron transfer of cytochrome c and biosensing. *Electrochemistry Communications*, *12*(1), 175–177. doi:10.1016/j.elecom.2009.11.020

Wu, X., He, W., Wang, W., Luo, X., Cao, H., Lin, L., & Liu, Z. et al. (2013). Investigation of the interaction between (−)-epigallocatechin-3-gallate with trypsin and α-chymotrypsin. *International Journal of Food Science & Technology*, *48*(11), 2340–2347.

Xavier, J. B., Picioreanu, C., Rani, S. A., van Loosdrecht, M. C., & Stewart, P. S. (2005). Biofilm-control strategies based on enzymic disruption of the extracellular polymeric substance matrix–a modelling study. *Microbiology*, *151*(12), 3817–3832. doi:10.1099/mic.0.28165-0 PMID:16339929

Xu, Y., He, H., Schulz, S., Liu, X., Fusetani, N., Xiong, H., & Qian, P.-Y. et al. (2010). Potent antifouling compounds produced by marine Streptomyces. *Bioresource Technology*, *101*(4), 1331–1336. doi:10.1016/j.biortech.2009.09.046 PMID:19818601

Yebra, D. M., Kiil, S., & Dam-Johansen, K. (2004). Antifouling technology—past, present and future steps towards efficient and environmentally friendly antifouling coatings. *Progress in Organic Coatings, 50*(2), 75–104. doi:10.1016/j.porgcoat.2003.06.001

Zhang, Y.-F., Xiao, K., Chandramouli, K. H., Xu, Y., Pan, K., Wang, W.-X., & Qian, P.-Y. (2011). *Acute toxicity of the antifouling compound butenolide in non-target organisms*. Academic Press.

Zhang, Y.-F., Zhang, H., He, L., Liu, C., Xu, Y., & Qian, P.-Y. (2012). Butenolide inhibits marine fouling by altering the primary metabolism of three target organisms. *ACS Chemical Biology*, *7*(6), 1049–1058. doi:10.1021/cb200545s PMID:22458453

Compilation of References

Abagyan, R., Totrov, M., & Kuznetsov, D. (1994). Docking and Structure Prediction from the Distorted Native Conformation. *Journal of Computational Chemistry*, *15*(5), 488–506. doi:10.1002/jcc.540150503

Abbasi, M., Ramezani, F., Elyasi, M., Sadeghi-Aliabadi, H., & Amanlou, M. (2015). A study on quantitative structure-activity relationship and molecular docking of metalloproteinase inhibitors based on L-tyrosine scaffold. *DARU Journal of Pharmaceutical Sciences*, *23*(1), 29. doi:10.1186/s40199-015-0111-z PMID:25925871

Abeyrathna, P., & Su, Y. (2015). The critical role of Akt in cardiovascular function. *Vascular Pharmacology*, *74*, 38–48. doi:10.1016/j.vph.2015.05.008 PMID:26025205

Abrol, R., Bray, J. K., & Goddard, W. A. III. (2011). Bihelix: Towards de novo structure prediction of an ensemble of G-protein coupled receptor conformations. *Proteins*, *80*(2), 505–518. doi:10.1002/prot.23216 PMID:22173949

Abrol, R., Griffith, A. R., Bray, J. K., & Goddard, W. A. III. (2012). Structure prediction of G protein-coupled receptors and their ensemble of functionally important conformations. *Methods in Molecular Biology (Clifton, N.J.)*, *914*, 237–254. PMID:22976032

Abrol, R., Trzaskowski, B., Goddard, W. A., Nesterovb, A., Olave, I., & Irons, C. (2014). Ligand- and mutation-induced conformational selection in the CCR5 chemokine G protein-coupled receptor. *Proceedings of the National Academy of Sciences of the United States of America*, *111*(36), 13040–13045. doi:10.1073/pnas.1413216111 PMID:25157173

Achbarou, A., Kaiser, S., Tremblay, G., Ste-Marie, L.-G., Brodt, P., Goltzman, D., & Rabbani, S. A. (1994). Urokinase overproduction results in increased skeletal metastasis by prostate cancer cells in vivo. *Cancer Research*, *54*(9), 2372–2377. PMID:8162583

Adams, J. A. (2001). Kinetic and catalytic mechanisms of protein kinases. *Chemical Reviews*, *101*(8), 2271–2290. doi:10.1021/cr000230w PMID:11749373

Ahmad, A., Sayed, A., Ginnebaugh, K. R., Sharma, V., Suri, A., Saraph, A., . . . Sarkar, F. H. (2015). Molecular docking and inhibition of matrix metalloproteinase-2 by novel difluorinatedbenzylidene curcumin analog. *American Journal of Translational Research*, *7*(2), 298.

Ahn, K. H., Scott, C. E., Abrol, R., Goddard, W. A. III, & Kendall, D. A. (2013). Computationally-predicted CB1 cannabinoid receptor mutants show distinct patterns of salt-bridges that correlate with their level of constitutive activity reflected in G protein coupling levels, thermal stability, and ligand binding. *Proteinsx*, *81*(8), 1304–1317. doi:10.1002/prot.24264 PMID:23408552

Akaike, H. (1974). A new look at the statistical model identification. *IEEE Transactions on Automatic Control*, *19*(6), 716–723. doi:10.1109/TAC.1974.1100705

Aldred, N., Phang, I. Y., Conlan, S. L., Clare, A. S., & Vancso, G. J. (2008). The effects of a serine protease, Alcalase®, on the adhesives of barnacle cyprids (Balanus amphitrite). *Biofouling*, *24*(2), 97–107. doi:10.1080/08927010801885908 PMID:18231899

Allen, W. J., & Rizzo, R. C. (2014). Implementation of the Hungarian algorithm to account for ligand symmetry and similarity in structure-based design. *Journal of Chemical Information and Modeling*, *54*(2), 518–529. doi:10.1021/ci400534h PMID:24410429

Alonso, H., Bliznyuk, A. A., & Gready, J. E. (2006). Combining docking and molecular dynamic simulations in drug design. *Medicinal Research Reviews*, *26*(5), 531–568. doi:10.1002/med.20067 PMID:16758486

Al-Sohaibani, S., & Murugan, K. (2012). Anti-biofilm activity of Salvadora persica on cariogenic isolates of Streptococcus mutans: In vitro and molecular docking studies. *Biofouling*, *28*(1), 29–38. doi:10.1080/08927014.2011.647308 PMID:22235758

Altschul, S. F., Gish, W., Miller, W., Myers, E. W., & Lipman, D. J. (1990). Basic local alignment search tool. *Journal of Molecular Biology*, *25*(3), 403–410. doi:10.1016/S0022-2836(05)80360-2 PMID:2231712

Alves, R., Chaleil, R. A., & Sternberg, M. J. (2002). Evolution of enzymes in metabolism: A network perspective. *Journal of Molecular Biology*, *320*(4), 751–770. doi:10.1016/S0022-2836(02)00546-6 PMID:12095253

Ambure, P. S., Gangwal, R. P., & Sangamwar, A. T. (2012). 3D-QSAR and molecular docking analysis of biphenyl amide derivatives as p38α mitogen-activated protein kinase inhibitors. *Molecular Diversity*, *16*(2), 377–388. doi:10.1007/s11030-011-9353-y PMID:22228035

Amico, M., Finelli, M., Rossi, I., Zauli, A., Elofsson, A., Viklund, H., & Casadio, R. et al. (2006). PONGO: A web server for multiple predictions of all-alpha transmembrane proteins. *Nucleic Acids Research*, *34*(suppl 2), W169–W172. doi:10.1093/nar/gkl208 PMID:16844984

Analysis Tool Web Services from the EMBL-EBI. (2013). *Nucleic Acids Research, 41*, 597-600.

Anand, P., Singh, B., & Singh, N. (2012). A review on coumarins as acetylcholinesterase inhibitors for alzheimer's disease. *Bioorganic & Medicinal Chemistry*, *20*(3), 1175–1180. doi:10.1016/j.bmc.2011.12.042 PMID:22257528

Anderson, A. C. (2003). The process of structure-based drug design. *Chemistry & Biology*, *10*(9), 787–797. doi:10.1016/j.chembiol.2003.09.002 PMID:14522049

Anderson, M. J., & Legendre, P. (1999). An empirical comparison of permutation methods for test of partial regression coefficients in a linear model. *Journal of Statistical Computation and Simulation*, *62*(3), 271–303. doi:10.1080/00949659908811936

Andersson, K., & Lidholm, J. (2003). Characteristics and immunobiology of grass pollen allergens. *International Archives of Allergy and Immunology*, *130*(2), 87–107. doi:10.1159/000069013 PMID:12673063

Andrusier, N., Nussinov, R., & Wolfson, H. J. (2007). FireDock: Fast Interaction Refinement in Molecular Docking. *Proteins*, *69*(1), 139–159. doi:10.1002/prot.21495 PMID:17598144

Angshuman, B. (2015). Structural Characterizations of Metal Ion Binding Transcriptional Regulator CueR from Opportunistic Pathogen *Pseudomonasaeruginosa* to Identify Its Possible Involvements in Virulence. *Applied Biochemistry and Biotechnology*, *175*(2), 649–656. doi:10.1007/s12010-014-1304-5 PMID:25342258

Anthony, J., Rangamaran, V. R., Gopal, D., Shivasankarasubbiah, K. T., Thilagam, M. L. J., Dhassiah, M. P., & Dakshinamurthy, S. et al. (2015). Ultraviolet and 5′ Fluorodeoxyuridine Induced Random Mutagenesis in Chlorella vulgaris and Its Impact on Fatty Acid Profile: A New Insight on Lipid-Metabolizing Genes and Structural Characterization of Related Proteins. *Marine Biotechnology (New York, N.Y.)*, *17*(1), 66–80. doi:10.1007/s10126-014-9597-5 PMID:25189135

Antonov, V. K., Ginodman, L. M., Kapitannikov, Y. V., Barshevskaya, T. N., Gurova, A. G., & Rumsh, L. D. (1978). Mechanism of pepsin catalysis: General base catalysis by the active-site carboxylate ion. *FEBS Letters*, *88*(1), 87–90. doi:10.1016/0014-5793(78)80613-9 PMID:346376

Arai, Y., Endo, S., Miyagi, N., Abe, N., Miura, T., Nishinaka, T., & El-Kabbani, O. et al. (2015). Structure-activity relationship of flavonoids as potent inhibitors of carbonyl reductase 1 (CBR1). *Fitoterapia*, *101*, 51–56. doi:10.1016/j.fitote.2014.12.010 PMID:25549925

Ardakani, A., & Ghasemi, J. B. (2013). Identification of novel inhibitors of HIV-1 integrase using pharmacophore-based virtual screening combined with molecular docking strategies. *Medicinal Chemistry Research*, *22*(11), 5545–5556. doi:10.1007/s00044-013-0545-0

Arinaminpathy, Y., Khurana, E., Engelman, D. M., & Gerstein, M. B. (2009). Computational analysis of membrane proteins: The largest class of drug targets. *Drug Discovery Today*, *14*(23), 1130–1135. doi:10.1016/j.drudis.2009.08.006 PMID:19733256

Arkin, M. R., Tang, Y., & Wells, J. A. (2014). Small-molecule inhibitors of protein-protein interactions: Progressing toward the reality. *Chemistry & Biology*, *21*(9), 1102–1114. doi:10.1016/j.chembiol.2014.09.001 PMID:25237857

Ash, W. L., Zlomislic, M. R., Oloo, E. O., & Tieleman, D. P. (2004). Computer simulations of membrane proteins. Biochimica et Biophysica Acta (BBA)-. *Biomembranes*, *1666*(1), 158–189. doi:10.1016/j.bbamem.2004.04.012

Astles, P. C., Baker, S. R., Boot, J. R., Broad, L. M., Dell, C. P., & Keenan, M. (2002). Recent progress in the development of subtype selective nicotinic acetylcholine receptor ligands. *Current Drug Targets. CNS and Neurological Disorders*, *1*(4), 337–348. doi:10.2174/1568007023339256 PMID:12769608

Atkins, P., De Paula, J., & Walters, V. (2006). *Physical Chemistry*: Macmillan. *Higher Education.*

Aubin-Tam, M.-E., & Hamad-Schifferli, K. (2005). Gold nanoparticle-cytochrome c complexes: The effect of nanoparticle ligand charge on protein structure. *Langmuir*, *21*(26), 12080–12084. doi:10.1021/la052102e PMID:16342975

Backes, A., Zech, B., Felber, B., Klebl, B., & Muller, G. (2008). Small-molecule inhibitors binding to protein kinases. Part I: Exceptions from the traditional pharmacophore approach of type I inhibition. *Expert Opin Drug Discov*, *3*(12), 1409–1425. doi:10.1517/17460440802579975 PMID:23506106

Bader, A., Tuccinardi, T., Granchi, C., Martinelli, A., Macchia, M., Minutolo, F., & Braca, A. et al. (2015). Phenylpropanoids and flavonoids from Phlomis kurdica as inhibitors of human lactate dehydrogenase. *Phytochemistry*, *116*, 262–268. doi:10.1016/j.phytochem.2015.03.007 PMID:25890391

Balaji, M., Jizhen, W., Haiqing, J., & Lijun, R. (2005). Comprehensive Analysis of Ebola Virus GP1 in Viral Entry. *Journal of Virology*, *79*(8), 4793–4805. doi:10.1128/JVI.79.8.4793-4805.2005 PMID:15795265

Baldwin, R. L. (1996). How Hofmeister ion interactions affect protein stability. *Biophysical Journal*, *71*(4), 2056–2063. doi:10.1016/S0006-3495(96)79404-3 PMID:8889180

Ballesteros, J. A., & Weinstein, H. (1995). Integrated methods for the construction of three-dimensional models and computational probing of structure-function relations in G protein coupled receptors. *Methods Neurosci*, *25*, 366–428. doi:10.1016/S1043-9471(05)80049-7

Ball, H. A., van Scott, M. R., & Robinson, C. B. (2004). Sense and antisense: Therapeutic potential of oligonucleotides and interference RNA in asthma and allergic disorders. *Clinical Reviews in Allergy & Immunology*, *27*(3), 207–217. doi:10.1385/CRIAI:27:3:207 PMID:15630157

Ball, T., Edstrom, W., Mauch, L., Schmitt, J., Leistler, B., Fiebig, H., & Valenta, R. et al. (2005). Gain of structure and IgE epitopes by eukaryotic expression of the major Timothy grass pollen allergen, Phl p 1. *The FEBS Journal*, *272*(1), 217–227. doi:10.1111/j.1432-1033.2004.04403.x PMID:15634345

Banks, A. S., McAllister, F. E., Camporez, J. P., Zushin, P. J., Jurczak, M. J., Laznik-Bogoslavski, D., & Spiegelman, B. M. et al. (2015). An ERK/Cdk5 axis controls the diabetogenic actions of PPARgamma. *Nature*, *517*(7534), 391–395. doi:10.1038/nature13887 PMID:25409143

Bantounas, I., Phylactou, L. A., & Uney, J. B. (2004). RNA interference and the use of small interfering RNA to study gene function in mammalian systems. *Journal of Molecular Endocrinology*, *33*(3), 545–557. doi:10.1677/jme.1.01582 PMID:15591019

Baraldi, P. G., Bovero, A., Fruttarolo, F., Preti, D., Tabrizi, M. A., Pavani, M. G., & Romagnol, R. (2004). DNA minor groove binders as potential antitumor and antimicrobial agents. *Medicinal Research Reviews*, *24*(4), 475–528. doi:10.1002/med.20000 PMID:15170593

Barnes, P. J. (2013). New anti-inflammatory targets for chronic obstructive pulmonary disease. *Nature Reviews. Drug Discovery*, *12*(7), 543–559. doi:10.1038/nrd4025 PMID:23977698

Barouch-Bentov, R., & Sauer, K. (2011). Mechanisms of drug resistance in kinases. *Expert Opinion on Investigational Drugs*, *20*(2), 153–208. doi:10.1517/13543784.2011.546344 PMID:21235428

Bartus, R. T., Dean, R. L., Beer, B., & Lippa, A. S. (1982). The cholinergic hypothesis of geriatric memory dysfunction. *Science*, *217*(4558), 408–414. doi:10.1126/science.7046051 PMID:7046051

Basu, A., & Sarkar, A. (2013). Emphasizing the of role of proteins in construction of the developmental genetic toolkit in plants. In Y. Pan & A. Y. Zomaya (Eds.), *Algorithmic and AI Methods for Protein Bioinformatics* (pp. 3–40). John Wiley and Sons Inc. doi:10.1002/9781118567869.ch1

Basyn, F., Spies, B., Bouffioux, O., Thomas, A., & Brasseur, R. (2003). Insertion of X-ray structures of proteins in membranes. *Journal of Molecular Graphics & Modelling*, *22*(1), 11–21. doi:10.1016/S1093-3263(03)00122-0 PMID:12798387

Bayliss, R., Sardon, T., Vernos, I., & Conti, E. (2003). Structural basis of Aurora-A activation by TPX2 at the mitotic spindle. *Molecular Cell*, *12*(4), 851–862. doi:10.1016/S1097-2765(03)00392-7 PMID:14580337

Berman, H. M., Olson, W. K., Beveridge, D. L., Westbrook, J., Gelbin, A., Demeny, T., & Schneider, B. et al. (1992). The Nucleic Acid Database: A comprehensive relational database of three-dimensional structures of nucleic acids. *Biophysical Journal*, *63*(3), 751–759. doi:10.1016/S0006-3495(92)81649-1 PMID:1384741

Berman, H. M., Westbrook, J., Feng, Z., Gilliland, G., Bhat, T. N., Weissig, H., & Bourne, P. E. et al. (2000). The Protein Data Bank. *Nucleic Acids Research*, *28*(1), 235–242. doi:10.1093/nar/28.1.235 PMID:10592235

Besbes, S., Mirshahi, M., Pocard, M., & Billard, C. (2015). New dimension in therapeutic targeting of BCL-2 family proteins. *Oncotarget*, *6*(15), 12862–12871. doi:10.18632/oncotarget.3868 PMID:25970783

Bhattacharya, S., Das, A., Ghosh, S., Dasgupta, R., & Bagchi, A. (2014). Hypoglycosylation of dystroglycan due to T192M mutation: A molecular insight behind the fact. *Gene*, *537*, 108–114. PMID:24361964

Biasini, M., Bienert, S., Waterhouse, A., Arnold, K., Studer, G., Schmidt, T., . . . Schwede, T. (2014). SWISS-MODEL: modelling protein tertiary and quaternary structure using evolutionary information. *Nucleic Acids Res, 42*, W252-258. doi: 10.1093/nar/gku340

Bidon-Chanal, A., Martí, M. A., Crespo, A., Milani, M., Orozco, M., Bolognesi, M., & Estrin, D. A. et al. (2006). Ligand-induced dynamical regulation of NO conversion in Mycobacterium tuberculosis truncated hemoglobin-N. *Proteins: Structure, Function, and Bioinformatics*, *64*(2), 457–464. doi:10.1002/prot.21004 PMID:16688782

Bikadi, Z., & Hazai, E. (2009). Application of the PM6 semi-empirical method to modeling proteins enhances docking accuracy of AutoDock. *Journal of Cheminformatics*, *1*(15). doi:10.1186/1758-2946-1-15 PMID:20150996

Bi, S. Y., Ding, L., Tian, Y., Song, D. Q., Zhou, X., Liu, X., & Zhang, H. Q. (2004). Investigation of the interaction between flavonoids and human serum albumin. *Journal of Molecular Structure*, *703*(1-3), 37–45. doi:10.1016/j.molstruc.2004.05.026

Bjarnsholt, T., Kirketerp-Møller, K., Jensen, P. Ø., Madsen, K. G., Phipps, R., Krogfelt, K., & Givskov, M. et al. (2008). Why chronic wounds will not heal: A novel hypothesis. *Wound Repair and Regeneration*, *16*(1), 2–10. doi:10.1111/j.1524-475X.2007.00283.x PMID:18211573

Blatti, J. L. (2012). *Investigating fatty acid biosynthesis within the algal chloroplast using Chlamydomonas reinhardtii as a model*. Academic Press.

Blot, W. J., McLaughlin, J. K., & Chow, W. H. (1997). Cancer rates among drinkers of black tea. *Critical Reviews in Food Science and Nutrition*, *37*(8), 739–760. doi:10.1080/10408399709527800 PMID:9447273

Blundell, T. L., Jhoti, H., & Abell, C. (2002). High-throughput crystallography for lead discovery in drug design. *Nature Reviews. Drug Discovery*, *1*(1), 45–54. doi:10.1038/nrd706 PMID:12119609

Bogan, A. A., & Thorn, K. S. (1998). Anatomy of hot spots in protein interfaces. *Journal of Molecular Biology*, *280*(1), 1–9. doi:10.1006/jmbi.1998.1843 PMID:9653027

Bogoyevitch, M. A., Barr, R. K., & Ketterman, A. J. (2005). Peptide inhibitors of protein kinases-discovery, characterisation and use. *Biochimica et Biophysica Acta*, *1754*(1-2), 79–99. doi:10.1016/j.bbapap.2005.07.025 PMID:16182621

Bohm, H. J., & Stahl, M. (1999). Part 1--Docking and Scoring: Methods Development-rapid empirical scoring functions in virtual screening applications. *Medicinal Chemistry Research*, *9*(7-8), 445–462.

Böhm, H.-J. (1992a). The computer program LUDI: A new method for the de novo design of enzyme inhibitors. *Journal of Computer-Aided Molecular Design*, *6*(1), 61–78. doi:10.1007/BF00124387 PMID:1583540

Böhm, H.-J. (1992b). LUDI: Rule-based automatic design of new substituents for enzyme inhibitor leads. *Journal of Computer-Aided Molecular Design*, *6*(6), 593–606. doi:10.1007/BF00126217 PMID:1291628

Bon, R. S., & Waldmann, H. (2010). Bioactivity-guided navigation of chemical space. *Accounts of Chemical Research*, *43*(8), 1103–1114. doi:10.1021/ar100014h PMID:20481515

Bordner, A. J. (2009). Predicting protein-protein binding sites in membrane proteins. *BMC Bioinformatics*, *10*(1), 312. doi:10.1186/1471-2105-10-312 PMID:19778442

Borrelli, K. W., Vitalis, A., Alcantara, R., & Guallar, V. (2005). PELE: Protein energy landscape exploration. A novel Monte Carlo based technique. *Journal of Chemical Theory and Computation*, *1*(6), 1304–1311. doi:10.1021/ct0501811 PMID:26631674

Brady, G. P. Jr, & Stouten, P. F. (2000). Fast prediction and visualization of protein binding pockets with PASS. *Journal of Computer-Aided Molecular Design*, *14*(4), 383–401. doi:10.1023/A:1008124202956 PMID:10815774

Brana, M. F., Cacho, M., Gradillas, A., de Pascual-Teresa, B., & Ramos, A. (2001). Intercalators as anticancer drugs. *Current Pharmaceutical Design*, *7*(17), 1745–1780. doi:10.2174/1381612013397113 PMID:11562309

Brancolini, G., Kokh, D. B., Calzolai, L., Wade, R. C., & Corni, S. (2012). Docking of ubiquitin to gold nanoparticles. *ACS Nano*, *6*(11), 9863–9878. doi:10.1021/nn303444b PMID:23033917

Bray, J. K., Abrol, R., Goddard, W. A. III, Trzaskowski, B., & Scott, C. E. (2014). SuperBiHelix method for predicting the pleiotropic ensemble of G-protein-coupled receptor conformations. *Proceedings of the National Academy of Sciences of the United States of America*, *111*(1), E72–E78. doi:10.1073/pnas.1321233111 PMID:24344284

Brooijmans, N., & Kuntz, I. D. (2003). Molecular recognition and docking algorithms. *Annual Review of Biophysics and Biomolecular Structure*, *32*(1), 335–373. doi:10.1146/annurev.biophys.32.110601.142532 PMID:12574069

Budhidarmo, R., & Day, C. L. (2015). IAPs: Modular regulators of cell signalling. *Seminars in Cell & Developmental Biology*, *39*, 80–90. doi:10.1016/j.semcdb.2014.12.002 PMID:25542341

Buolamwini, J. K., & Assefa, H. (2002). CoMFA and CoMSIA 3D QSAR and docking studies on conformationally-restrained cinnamoyl HIV-1 integrase inhibitors: Exploration of a binding mode at the active site. *Journal of Medicinal Chemistry*, *45*(4), 841–852. doi:10.1021/jm010399h PMID:11831895

Burda, C., Chen, X., Narayanan, R., & El-Sayed, M. A. (2005). Chemistry and properties of nanocrystals of different shapes. *Chemical Reviews*, *105*(4), 1025–1102. doi:10.1021/cr030063a PMID:15826010

Burgen, A., Roberts, G., & Feeney, J. (1975). *Binding of flexible ligands to macromolecules*. Academic Press.

Burtness, B. (2014). Targeted agents: Management of dermatologic toxicities. *Journal of the National Comprehensive Cancer Network*, *12*(5Suppl), 793–796. PMID:24853219

Cai, C. Z., Li, Z. R., Wang, W. L., & Chen, Y. Z. (2004). Advances in modeling of biomolecular interactions. *Acta Pharmacologica Sinica*, *25*, 1–8. PMID:14704115

Calzolai, L., Laera, S., Ceccone, G., & Rossi, F. (2012). *Structure and Stability of Proteins Interacting with Nanoparticles. In Proteins at Interfaces III State of the Art* (Vol. 1120, pp. 839–855). ACS Publications.

Cameron, E., Pauling, L., & Leibovitz, B. (1979). Ascordic acid and cancer: A review. *Cancer Research*, *39*, 663–681. PMID:371790

Campos, L. A., & Sancho, J. (2003). The active site of pepsin is formed in the intermediate conformation dominant at mildly acidic pH. *FEBS Letters*, *538*(1-3), 89–95. doi:10.1016/S0014-5793(03)00152-2 PMID:12633859

Carletti, E., Colletier, J. P., Dupeux, F., Trovaslet, M., Masson, P., & Nachon, F. (2010). Structural evidence that human acetylcholinesterase inhibited by tabun ages through O-dealkylation. *Journal of Medicinal Chemistry*, *53*(10), 4002–4008. doi:10.1021/jm901853b PMID:20408548

Carlson, H. A., Masukawa, K. M., & McCammon, J. A. (1999). Method for including the dynamic fluctuations of a protein in computer-aided drug design. *The Journal of Physical Chemistry A*, *103*(49), 10213–10219. doi:10.1021/jp991997z

Carrillo, O., & Orozco, M. (2008). GRID-MD—A tool for massive simulation of protein channels. *Proteins: Structure, Function, and Bioinformatics*, *70*(3), 892–899. doi:10.1002/prot.21592 PMID:17803238

Cartmill, M., & Smith, F. H. (2011). *The Human Lineage*. John Wiley & Sons.

Case, D. A., Babin, V., Berryman, J. T., Betz, R. M., Cai, Q., Cerutti, D. S. … Kollman, P. A. (2014). AMBER 14. San Francisco, CA: University of California.

Chandrakar, B., Jain, A., Roy, S., Gutlapalli, V. R., Saraf, S., Suppahia, A., & Nayarisseri, A. et al. (2013). Molecular modeling of Acetyl-CoA carboxylase (ACC) from Jatropha curcas and virtual screening for identification of inhibitors. *Journal of Pharmacy Research*, *6*(9), 913–918. doi:10.1016/j.jopr.2013.07.032

Chang, C., Ray, A., & Swaan, P. (2005). In silico strategies for modeling membrane transporter function. *Drug Discovery Today*, *10*(9), 663–671. doi:10.1016/S1359-6446(05)03429-X PMID:15894231

Chang, M. W., Ayeni, C., Breuer, S., & Torbett, B. E. (2010). Virtual screening for HIV protease inhibitors: A comparison of AutoDock 4 and Vina. *PLoS ONE*, *5*(8), e11955. doi:10.1371/journal.pone.0011955 PMID:20694138

Chen, G. F., Ma, X., Meng, F. B., & Li, G. X. (2008). The electron transfer reactivity of kaempferol and its interaction with amino acid residues. *Bioelectrochemistry (Amsterdam, Netherlands)*, *72*(2), 169–173. doi:10.1016/j.bioelechem.2008.02.001 PMID:18321794

Chen, H. M., Liu, B. F., Huang, H. L., Hwang, S. F., & Ho, S. Y. (2007). SODOCK: Swarm optimization for highly flexible protein–ligand docking. *Journal of Computational Chemistry*, *28*(2), 612–623. doi:10.1002/jcc.20542 PMID:17186483

Chen, Q., Shafer, R. H., & Kuntz, I. D. (1997). Structure-based discovery of ligands targeted to the RNA double helix. *Biochemistry*, *36*(38), 11402–11407. doi:10.1021/bi970756j PMID:9298959

Chen, R., Li, L., & Weng, Z. (2003). ZDOCK: An initial-stage protein docking algorithms. *Proteins*, *51*, 82–87.

Chen, Y. C. (2015). Beware of docking! *Trends in Pharmacological Sciences*, *36*(2), 78–95. doi:10.1016/j.tips.2014.12.001 PMID:25543280

Cherezov, V., Rosenbaum, D. M., Hanson, M. A., Rasmussen, S. G. F., Thian, F. S., Kobilka, T. S., & Stevens, R. C. et al. (2007). High-resolution crystal structure of an engineered human beta2-adrenergic G protein-coupled receptor. *Sci.*, *318*(5854), 1258–1265. doi:10.1126/science.1150577 PMID:17962520

Cherkasov, A., Ban, F., Li, Y., Fallahi, M., & Hammond, G. L. (2006). Progressive docking: A hybrid QSAR/docking approach for accelerating in silico high throughput screening. *Journal of Medicinal Chemistry*, *49*(25), 7466–7478. doi:10.1021/jm060961+ PMID:17149875

Cheung, C. H., Coumar, M. S., Chang, J. Y., & Hsieh, H. P. (2011). Aurora kinase inhibitor patents and agents in clinical testing: An update (2009-10). *Expert Opin Ther Pat*, *21*(6), 857–884. doi:10.1517/13543776.2011.574614 PMID:21591849

Cheung, C. H., Coumar, M. S., Hsieh, H. P., & Chang, J. Y. (2009). Aurora kinase inhibitors in preclinical and clinical testing. *Expert Opinion on Investigational Drugs*, *18*(4), 379–398. doi:10.1517/13543780902806392 PMID:19335272

Cheung, C. H., Sarvagalla, S., Lee, J. Y., Huang, Y. C., & Coumar, M. S. (2014). Aurora kinase inhibitor patents and agents in clinical testing: An update (2011 - 2013). *Expert Opin Ther Pat*, *24*(9), 1021–1038. doi:10.1517/13543776.2014.931374 PMID:24965505

Chico, L. K., Van Eldik, L. J., & Watterson, D. M. (2009). Targeting protein kinases in central nervous system disorders. *Nature Reviews. Drug Discovery*, *8*(11), 892–909. doi:10.1038/nrd2999 PMID:19876042

Chien, E. Y., Liu, W., Zhao, Q., Katritch, V., Han, G. W., Hanson, M. A., & Stevens, R. C. et al. (2010). Structure of the human dopamine D3 receptor in complex with a D2/D3 selective antagonist. *Science*, *330*(6007), 1091–1095. doi:10.1126/science.1197410 PMID:21097933

China Pharmacopoeia Committee. (2010). *Pharmacopoeia of the People's Republic of China*. Beijing: Chemical Industry Press.

Chi, Z. X., Liu, R. T., & Zhang, H. (2010). Noncovalent Interaction of Oxytetracycline with the Enzyme Trypsin. *Biomacromolecules*, *11*(9), 2454–2459. doi:10.1021/bm100633h PMID:20681619

Chou, P. Y., & Fasman, G. D. (1974). Prediction of protein conformation. *Biochemistry*, *13*(2), 222–245. doi:10.1021/bi00699a002 PMID:4358940

Chou, P. Y., & Fasman, G. D. (1978). Empirical predictions of protein conformation. *Annual Review of Biochemistry*, *47*(1), 251–276. doi:10.1146/annurev.bi.47.070178.001343 PMID:354496

Cierpicki, T., & Grembecka, J. (2015). Targeting protein-protein interactions in hematologic malignancies: Still a challenge or a great opportunity for future therapies? *Immunological Reviews*, *263*(1), 279–301. doi:10.1111/imr.12244 PMID:25510283

Ciolkowski, M., Rozanek, M., Bryszewska, M., & Klajnert, B. (2013). The influence of PAMAM dendrimers surface groups on their interaction with porcine pepsin. *Biochimica et Biophysica Acta*, *1834*(10), 1982–1987. doi:10.1016/j.bbapap.2013.06.020 PMID:23851144

Clackson, T., & Wells, J. A. (1995). A hot spot of binding energy in a hormone-receptor interface. *Science*, *267*(5196), 383–386. doi:10.1126/science.7529940 PMID:7529940

Clark, K. P., & Ajay, . (1995). Flexible ligand docking without parameter adjustment across four ligand–receptor complexes. *Journal of Computational Chemistry*, *16*(10), 1210–1226. doi:10.1002/jcc.540161004

Clore, G. M., & Gronenborn, A. M. (1991). Structures of larger proteins in solution: Three-and four-dimensional heteronuclear NMR spectroscopy. *Science*, *252*(5011), 1390–1399. doi:10.1126/science.2047852 PMID:2047852

Collis, A. V., Brouwer, A. P., & Martin, A. C. (2003). Analysis of the antigen combining site: Correlations between length and sequence composition of the hypervariable loops and the nature of the antigen. *Journal of Molecular Biology*, *325*(2), 337–354. doi:10.1016/S0022-2836(02)01222-6 PMID:12488099

Comeau, S. R., Gatchel, D. W., Vajda, S., & Camacho, C. J. (2004). ClusPro: An automated docking and discrimination method for the prediction of protein complexes. *Bioinformatics (Oxford, England)*, *20*(1), 45–50. doi:10.1093/bioinformatics/btg371 PMID:14693807

Conese, M., & Blasi, F. (1995). 6 The urokinase/urokinase-receptor system and cancer invasion. *Bailliere's Clinical Haematology*, *8*(2), 365–389. doi:10.1016/S0950-3536(05)80273-2 PMID:7549068

Conn, P. J., Christopoulos, A., & Lindsley, C. W. (2009). Allosteric modulators of GPCRs: A novel approach for the treatment of CNS disorders. *Nature Reviews. Drug Discovery*, *8*(1), 41–54. doi:10.1038/nrd2760 PMID:19116626

Consortium, T. U. (2015). UniProt: A hub for protein information. *Nucleic Acids Research*, *43*(D1), D204–D212. doi:10.1093/nar/gku989 PMID:25348405

Contreras-Romo, M. C., Martínez-Archundia, M., Deeb, O., Ślusarz, M. J., Ramírez-Salinas, G., Garduño-Juárez, R., & Correa-Basurto, J. et al. (2014). Exploring the ligand recognition properties of the human vasopressin v1a receptor using QSAR and molecular modeling studies. *Chemical Biology & Drug Design*, *83*(2), 207–223. doi:10.1111/cbdd.12229 PMID:24010681

Cooper, G. M. (2000). *The Cell: A Molecular Approach*. Sunderland: Sinauer Associates.

Corbeil, C. R., Englebienne, P., & Moitessier, N. (2007). Docking ligands into flexible and solvated macromolecules. 1. Development and validation of FITTED 1.0. *Journal of Chemical Information and Modeling*, *47*(2), 435–449. doi:10.1021/ci6002637 PMID:17305329

Corbel, C., Wang, Q., Bousserouel, H., Hamdi, A., Zhang, B., Lozach, O., & Bach, S. et al. (2011). First BRET-based screening assay performed in budding yeast leads to the discovery of CDK5/p25 interaction inhibitors. *Biotechnology Journal*, *6*(7), 860–870. doi:10.1002/biot.201100138 PMID:21681968

Cornell, W. D., Cieplak, P., Bayly, C. I., Gould, I. R., Merz, K. M., Ferguson, D. M., & Kollman, P. A. et al. (1995). A Second Generation Force Field for the Simulation of Proteins, Nucleic Acids, and Organic Molecules. *Journal of the American Chemical Society*, *117*(19), 5179–5197. doi:10.1021/ja00124a002

Cortial, S., Chaignon, P., Iorga, B. I., Aymerich, S., Truan, G., Gueguen-Chaignon, V., & Ouazzani, J. et al. (2010). NADH oxidase activity of Bacillus subtilis nitroreductase NfrA1: Insight into its biological role. *FEBS Letters*, *584*(18), 3916–3922. doi:10.1016/j.febslet.2010.08.019 PMID:20727352

Coupez, B., & Lewis, R. A. (2006). Docking and scoring--theoretically easy, practically impossible? *Current Medicinal Chemistry*, *13*(25), 2995–3003. doi:10.2174/092986706778521797 PMID:17073642

Cramer, R. D., Patterson, D. E., & Bunce, J. D. (1988). Comparative molecular field analysis (CoMFA). 1. Effect of shape on binding of steroids to carrier proteins. *Journal of the American Chemical Society*, *110*(18), 5959–5967. doi:10.1021/ja00226a005 PMID:22148765

Cui, F., Yang, K., & Li, Y. (2015). *Investigate the Binding of Catechins to Trypsin Using Docking and Molecular Dynamics Simulation*. Academic Press.

Cui, F. C., Yang, K. C., & Li, Y. Q. (2015). Investigate the binding of catechins to trypsin using docking and molecular dynamics simulation. *PLoS ONE*, *10*(5), e0125848. doi:10.1371/journal.pone.0125848 PMID:25938485

Cukuroglu, E., Engin, H. B., Gursoy, A., & Keskin, O. (2014). Hot spots in protein-protein interfaces: Towards drug discovery. *Progress in Biophysics and Molecular Biology*, *116*(2-3), 165–173. doi:10.1016/j.pbiomolbio.2014.06.003 PMID:24997383

Cukuroglu, E., Gursoy, A., & Keskin, O. (2012). HotRegion: A database of predicted hot spot clusters. *Nucleic Acids Research*, *40*(Database issue), D829–D833. doi:10.1093/nar/gkr929 PMID:22080558

Cushman, D. W., Ondetti, M. A., Gordon, E. M., Natarajan, S., Karanewsky, D. S., Krapcho, J., & Petrillo, E. W. Jr. (1987). Rational design and biochemical utility of specific inhibitors of angiotensin-converting enzyme. *Journal of Cardiovascular Pharmacology*, *10*, 17–30. doi:10.1097/00005344-198706107-00004 PMID:2485059

da Silva, A. C. B., da Silva, D. R., de Macêdo Ferreira, S. A., Agripino, G. G., Albuquerque, A. R., & do Rêgo, T. G. (2014). *In Silico Approach for the Identification of Potential Targets and Specific Antimicrobials for Streptococcus mutans*. Advances in Bioscience and Biotechnology.

Da, C., Mooberry, S. L., Gupton, J. T., & Kellogg, G. E. (2013). How to deal with low-resolution target structures: Using SAR, ensemble docking, hydropathic analysis, and 3D-QSAR to definitively map the αβ-tubulin colchicine site. *Journal of Medicinal Chemistry*, *56*(18), 7382–7395. doi:10.1021/jm400954h PMID:23961916

Dahl, S. G., Sylte, I., & Ravna, A. W. (2004). Structures and models of transporter proteins. *The Journal of Pharmacology and Experimental Therapeutics*, *309*(3), 853–860. doi:10.1124/jpet.103.059972 PMID:14988415

Dailey, M. M., Hait, C., Holt, P. A., Maguire, J. M., Meier, J. B., Miller, M. C., & Trent, J. O. et al. (2009). Structure-based drug design: From nucleic acid to membrane protein targets. *Experimental and Molecular Pathology*, *86*(3), 141–150. doi:10.1016/j.yexmp.2009.01.011 PMID:19454265

Damgaard, R. B., & Gyrd-Hansen, M. (2011). Inhibitor of apoptosis (IAP) proteins in regulation of inflammation and innate immunity. *Discovery Medicine*, *11*(58), 221–231. PMID:21447281

Dangles, O., & Dufore, C. (2005). Flavonoid-protein interactions. In Ø. M. Andersen & K. R. Markham (Eds.), *Flavonoids: chemistry, biochemistry and applications* (pp. 443–469). CRC Press. doi:10.1201/9781420039443.ch9

Danser, A. H., & Deinum, J. (2005). Renin, prorenin and the putative (pro) renin receptor. *Hypertension*, *46*(5), 1069–1076. doi:10.1161/01.HYP.0000186329.92187.2e PMID:16186442

Dantzig, A. H., de Alwis, D. P., & Burgess, M. (2003). Considerations in the design and development of transport inhibitors as adjuncts to drug therapy. *Advanced Drug Delivery Reviews*, *55*(1), 133–150. doi:10.1016/S0169-409X(02)00175-8 PMID:12535578

Darnell, S. J., LeGault, L., & Mitchell, J. C. (2008). KFC Server: interactive forecasting of protein interaction hot spots. *Nucleic Acids Res*, *36*, W265-269. doi: 10.1093/nar/gkn346

D'Cruz, O. J., & Uckun, F. M. (2013). Novel Bruton's tyrosine kinase inhibitors currently in development. *Onco Targets Ther*, *6*, 161–176. doi:10.2147/OTT.S33732 PMID:23493945

de Groot, H., & Rauen, U. (1998). Tissue injury by reactive oxygen species and the protective effects of flavonoids. *Fundamental & Clinical Pharmacology*, *12*(3), 249–255. doi:10.1111/j.1472-8206.1998.tb00951.x PMID:9646056

Deb, P. K., Sharma, A., Piplani, P., & Akkinepally, R. R. (2012). Molecular docking and receptor-specific 3D-QSAR studies of acetylcholinesterase inhibitors. *Molecular Diversity*, *16*(4), 803–823. doi:10.1007/s11030-012-9394-x PMID:22996404

Deeb, O., da Cunha, E. F., Cormanich, R. A., Ramalho, T. C., & Freitas, M. P. (2012). Computer-assisted assessment of potentially useful non-peptide HIV-1 protease inhibitors. *Chemometrics and Intelligent Laboratory Systems*, *116*, 123–127. doi:10.1016/j.chemolab.2012.05.003

Deeb, O., Rosales-Hernández, M. C., Gómez-Castro, C., Garduño-Juárez, R., & Correa-Basurto, J. (2010). Exploration of human serum albumin binding sites by docking and molecular dynamics flexible ligand–protein interactions. *Biopolymers*, *93*(2), 161–170. doi:10.1002/bip.21314 PMID:19785033

Deepa, P., Vandhana, S., Muthukumaran, S., Umashankar, V., Jayanthi, U., & Krishnakumar, S. (2010). Chemical inhibition of fatty acid synthase: molecular docking analysis and biochemical validation in ocular cancer cells. *Journal of Ocular Biology, Diseases, and Informatics*, *3*(4), 117-128.

Deepshikha, D. (2002). *Protein-Ligand Interactions: Docking, Design and Protein Conformational Change*. (Thesis). California Institute of Technology, Pasadena, CA.

DeLano, W. L. (2002). *The PyMOL Molecular Graphics System DeLano Scientific*. San Carlos, CA, USA; doi:10.1093/nar/gki408

Deligkaris, C., Ascone, A. T., Sweeney, K. J., & Greene, A. J. Q. (2014). Validation of a computational docking methodology to identify the non-covalent binding site of ligands to DNA. *Molecular BioSystems*, *10*(8), 2106–2125. doi:10.1039/C4MB00239C PMID:24853173

De, M., Ghosh, P. S., & Rotello, V. M. (2008). Applications of nanoparticles in biology. *Advanced Materials*, *20*(22), 4225–4241. doi:10.1002/adma.200703183

DesJarlais, R. L., Sheridan, R. P., Dixon, J. S., Kuntz, I. D., & Venkataraghavan, R. (1986). Docking flexible ligands to macromolecular receptors by molecular shape. *Journal of Medicinal Chemistry*, *29*(11), 2149–2153. doi:10.1021/jm00161a004 PMID:3783576

Detering, C., & Varani, G. (2004). Validation of automated docking programs for docking and database screening against RNA drug targets. *Journal of Medicinal Chemistry*, *47*(17), 4188–4201. doi:10.1021/jm030650o PMID:15293991

Devillers, J., & Balaban, A. T. (1999). *Topological Indices and Related Descriptors in QSAR and QSPR* (p. 811). Amsterdam: Gordon Breach Scientific Publishers.

Dhanik, A., McMurray, J. S., & Kavraki, L. E. (2013). DINC: A new AutoDock-based protocol for docking large ligands. *BMC Structural Biology*, *13*(Suppl 1), S11. doi:10.1186/1472-6807-13-S1-S11 PMID:24564952

Dickson, H. M., Wilbur, A., Reinke, A. A., Young, M. A., & Vojtek, A. B. (2015). Targeted inhibition of the Shroom3-Rho kinase protein-protein interaction circumvents Nogo66 to promote axon outgrowth. *BMC Neuroscience*, *16*(1), 34. doi:10.1186/s12868-015-0171-5 PMID:26077244

Diller, D. J., & Merz, K. M. Jr. (2001). High throughput docking for library design and library prioritization. *Proteins*, *43*(2), 113–124. doi:10.1002/1097-0134(20010501)43:2<113::AID-PROT1023>3.0.CO;2-T PMID:11276081

Dittrich, M., & Sibler, S. (2005). Cell surface groups of two picocyanobacteria strains studied by zeta potential investigations, potentiometric titration, and infrared spectroscopy. *Journal of Colloid and Interface Science*, *286*(2), 487–495. doi:10.1016/j.jcis.2005.01.029 PMID:15897062

Dominguez, C., Boelens, R., & Bonvin, A. M. (2003). HADDOCK: A protein-protein docking approach based on biochemical or biophysical information. *Journal of the American Chemical Society*, *125*(7), 1731–1737. doi:10.1021/ja026939x PMID:12580598

Doré, A. S., Robertson, N., Errey, J. C., Ng, I., Hollenstein, K., Tehan, B., & Marshall, F. H. et al. (2011). Structure of the adenosine A(2A) receptor in complex with ZM241385 and the xanthines XAC and caffeine. *Structure (London, England)*, *19*(9), 1283–1293. doi:10.1016/j.str.2011.06.014 PMID:21885291

Dor, O., & Zhou, Y. (2007). Achieving 80% ten-fold cross-validated accuracy for secondary structure prediction by large-scale training. Proteins. *Structure, Function, and Bioinformatics*, *66*(4), 838–845. doi:10.1002/prot.21298 PMID:17177203

Douglas, E. V. Pires, David, B. Ascher, & Tom, L. Blundell. (2013). mCSM: Predicting the effects of mutations in proteins using graph-based signatures. *Bioinformatics (Oxford, England)*. doi:10.1093/bioinformatics/btt691

Dounin, V., Constantinof, A., Schulze, H., Bachmann, T. T., & Kerman, K. (2011). Electrochemical detection of interaction between thioflavin T and acetylcholinesterase. *Analyst (London)*, *136*(6), 1234–1238. doi:10.1039/c0an00743a PMID:21264414

Dubrez, L., Berthelet, J., & Glorian, V. (2013). IAP proteins as targets for drug development in oncology. *Onco Targets Ther*, *9*, 1285–1304. doi:10.2147/OTT.S33375 PMID:24092992

Duffy, B. C., Liu, S., Martin, G. S., Wang, R., Hsia, M. M., Zhao, H., & Kitchen, D. B. et al. (2015). Discovery of a new chemical series of BRD4(1) inhibitors using protein-ligand docking and structure-guided design. *Bioorganic & Medicinal Chemistry Letters*, *25*(14), 2818–2823. doi:10.1016/j.bmcl.2015.04.107 PMID:26022843

Duhovny, D., Nussinov, R., & Wolfson, H. J. (2002) Efficient Unbound Docking of Rigid Molecules. In *Proceedings of the 2nd Workshop on Algorithms in Bioinformatics(WABI)* (LNCS), (vol. 2452, pp. 185-200). Springer Verlag.

Dunnett, S. B., & Fibiger, H. C. (1993). Role of forebrain cholinergic systems in learning and memory: Relevance to the cognitive deficits of aging and Alzheimer's dementia. *Progress in Brain Research*, *98*, 413–420. doi:10.1016/S0079-6123(08)62425-5 PMID:8248529

Duprat, A. F., Huynh, T., & Dreyfus, G. (1998). Toward a principled methodology for neural network design and performance evaluation in QSAR. Application to the prediction of log P. *Journal of Chemical Information and Computer Sciences*, *38*(4), 586–594. doi:10.1021/ci980042v PMID:9691473

Duthie, E. S., & Chain, E. A. (1939). A mucolytic enzyme in tests extract. *Nature*, *144*(3658), 977–978. doi:10.1038/144977b0

Duthie, G. G., Duthie, S. J., & Kyle, J. A. M. (2000). Plant polyphenols in cancer and heart disease: Implications as nutritional antioxidants. *Nutrition Research Reviews*, *13*(01), 79–106. doi:10.1079/095442200108729016 PMID:19087434

Eisenberg, D., Luthy, R., & Bowie, J. U. (1997). VERIFY3D: Assessment of protein models with three-dimensional profiles. *Methods in Enzymology*, *277*, 396–404. doi:10.1016/S0076-6879(97)77022-8 PMID:9379925

Eldridge, M. D., Murray, C. W., Auton, T. R., Paolini, G. V., & Mee, R. P. (1997). Empirical scoring functions: I. The development of a fast empirical scoring function to estimate the binding affinity of ligands in receptor complexes. *Journal of Computer-Aided Molecular Design*, *11*(5), 425–445. doi:10.1023/A:1007996124545 PMID:9385547

Ellis, C. (2004). The state of GPCR research in 2004. *Nature Reviews. Drug Discovery*, *3*(7), 577–626. doi:10.1038/nrd1458 PMID:15272499

Ellis, L. B., Hou, B. K., Kang, W., & Wackett, L. P. (2003). The University of Minnesota biocatalysis/biodegradation database: Post-genomic data mining. *Nucleic Acids Research*, *31*(1), 262–265. doi:10.1093/nar/gkg048 PMID:12519997

Elofsson, A., & Heijne, G. V. (2007). Membrane protein structure: Prediction versus reality. *Annual Review of Biochemistry*, *76*(1), 125–140. doi:10.1146/annurev.biochem.76.052705.163539 PMID:17579561

Engelman, A., Kessl, J. J., & Kvaratskhelia, M. (2013). Allosteric inhibition of HIV-1 integrase activity. *Current Opinion in Chemical Biology*, *17*(3), 339–345. doi:10.1016/j.cbpa.2013.04.010 PMID:23647983

Erickson, J. A., Jalaie, M., Robertson, D. H., Lewis, R. A., & Vieth, M. (2004). Lessons in molecular recognition: The effects of ligand and protein flexibility on molecular docking accuracy. *Journal of Medicinal Chemistry*, *47*(1), 45–55. doi:10.1021/jm030209y PMID:14695819

Esbenshade, T. A., Krueger, K. M., Miller, T. R., Kang, C. H., Denny, L. I., Witte, D. G., & Hancock, A. A. et al. (2003). Two novel and selective nonimidazole histamine H3 receptor antagonists A-304121 and A-317920: I. In vitro pharmacological effects. *The Journal of Pharmacology and Experimental Therapeutics*, *305*(3), 887–896. doi:10.1124/jpet.102.047183 PMID:12606603

Escalante, T., Franceschi, A., Rucavado, A., & Gutierrez, J.M. (2000). *Effectiveness of batimastat, a synthetic inhibitor of matrix metalloproteinases, in neutralizing local tissue damage induced by BaP1, a hemorrhagic metalloproteinase from the venom of the snake* Bothrops asper. Academic Press.

Eswar, N., Webb, B., Marti-Renom, M. A., Madhusudhan, M. S., Eramian, D., Shen, M. Y., . . . Sali, A. (2006). Comparative protein structure modeling using Modeller. Curr Protoc Bioinformatics. doi:10.1002/0471250953.bi0506s15

Eswaran, J., & Knapp, S. (2010). Insights into protein kinase regulation and inhibition by large scale structural comparison. *Biochimica et Biophysica Acta*, *1804*(3), 429–432. doi:10.1016/j.bbapap.2009.10.013 PMID:19854302

Eswar, N., Webb, B., Marti-Renom, M. A., Madhusudhan, M. S., Eramian, D., Shen, M. Y., & Sali, A. et al. (2006). Comparative protein structure modeling using Modeller. *Current Protocols in Bioinformatics*, 5–6. PMID:18428767

Evans, D. A., & Neidle, S. (2006). Virtual screening of DNA minor groove bonders. *Journal of Medicinal Chemistry*, *49*(14), 4232–4238. doi:10.1021/jm0601957 PMID:16821782

Evers, A., & Klabunde, T. (2005). Structure-based drug discovery using GPCR homology modeling: Successful virtual screening for antagonists of the alpha1A adrenergic receptor. *Journal of Medicinal Chemistry*, *48*(4), 1088–1097. doi:10.1021/jm0491804 PMID:15715476

Ewing, T. J. A., & Kuntz, I. D. (1997). Critical evaluation of search algorithms for automated molecular docking and database screening. *Journal of Computational Chemistry*, *18*(9), 1175–1189. doi:10.1002/(SICI)1096-987X(19970715)18:9<1175::AID-JCC6>3.0.CO;2-O

Ewing, T. J., Makino, S., Skillman, A. G., & Kuntz, I. D. (2001). DOCK 4.0: Search strategies for automated molecular docking of flexible molecule databases. *Journal of Computer-Aided Molecular Design*, *15*(5), 411–428. doi:10.1023/A:1011115820450 PMID:11394736

Fabbro, D. (2015). 25 years of small molecular weight kinase inhibitors: Potentials and limitations. *Molecular Pharmacology*, *87*(5), 766–775. doi:10.1124/mol.114.095489 PMID:25549667

Fabbro, D., Cowan-Jacob, S. W., Mobitz, H., & Martiny-Baron, G. (2012). Targeting cancer with small-molecular-weight kinase inhibitors. *Methods in Molecular Biology (Clifton, N.J.)*, *795*, 1–34. doi:10.1007/978-1-61779-337-0_1 PMID:21960212

Faraji, M., Yamini, Y., & Rezaee, M. (2010). Magnetic nanoparticles: Synthesis, stabilization, functionalization, characterization, and applications. *Journal of the Indian Chemical Society*, *7*(1), 1–37.

Fatemi, M. H., Heidari, A., & Gharagani, S. (2015). QSAR prediction of HIV-1 protease inhibitory activities using docking derived molecular descriptors. *Journal of Theoretical Biology*, *369*, 13–22. doi:10.1016/j.jtbi.2015.01.008 PMID:25600056

Fernández-Recio, J., Totrov, M., & Abagyan, R. (2002). Soft protein–protein docking in internal coordinates. *Protein Science*, *11*(2), 280–291. doi:10.1110/ps.19202 PMID:11790838

Ferreira, L. G., dos Santos, R. N., Oliva, G., & Andricopulo, A. D. (2015). Molecular docking and structure-based drug design strategies. *Molecules (Basel, Switzerland)*, *20*(7), 13384–13421. doi:10.3390/molecules200713384 PMID:26205061

Fischer, E. (1894). Einfluss der Configuration auf die Wirkung der Enzyme, Ber. dtsch. Chem. Ges. 27, 2985–2993.

Fiser, A., & Sali, A. (2003). ModLoop: Automated modeling of loops in protein structures. *Bioinformatics (Oxford, England)*, *19*(18), 2500–2501. doi:10.1093/bioinformatics/btg362 PMID:14668246

Fletcher, M., & Loeb, G. (1979). Influence of substratum characteristics on the attachment of a marine pseudomonad to solid surfaces. *Applied and Environmental Microbiology*, *37*(1), 67–72. PMID:16345338

Flora, K., Hahn, M., Rosen, H., & Benner, K. (1998). Milk thistle (Silybum marianum) for the therapy of liver disease. *The American Journal of Gastroenterology*, *93*(2), 139–143. doi:10.1111/j.1572-0241.1998.00139.x PMID:9468229

Forbes, I. T., Ham, P., Booth, D. H., Martin, R. T., Thompson, M., Baxter, G. S., & Wood, M. D. et al. (1995). 5-Methyl-1-(3-pyridiiylcarbamoyl)-1,2,3,5-tetrahydropyrrolo[2,3-f]indole: A novel 5-HT2C/5-HT2B receptor antagonist with improved affinity, selectivity, and oral activity. *Journal of Medicinal Chemistry*, *38*(14), 2524–2530. doi:10.1021/jm00014a004 PMID:7629791

Forli, S., & Olson, A. J. (2012). A Force Field with Discrete Displaceable Waters and Desolvation Entropy for Hydrated Ligand Docking. *Journal of Medicinal Chemistry*, *55*(2), 623–638. doi:10.1021/jm2005145 PMID:22148468

Fournie-Zaluski, M. C., Coric, P., Thery, V., Gonzalez, W., Meudal, H., Turcaud, S., & Roques, B. P. et al. (1996). Design of orally active dual inhibitors of neutral endopeptidase and angiotensin-converting enzyme with long duration of action. *Journal of Medicinal Chemistry*, *39*(13), 2594–2608. doi:10.1021/jm950783c PMID:8691458

Fradera, X., Kaur, J., & Mestres, J. (2004). Unsupervised guided docking of covalently bound ligands. *Journal of Computer-Aided Molecular Design*, *18*(10), 635–650. doi:10.1007/s10822-004-5291-4 PMID:15849994

Fradera, X., Knegtel, R., & Mestres, J. (2000). Similarity-driven flexible ligand docking. *Proteins: Structure, Function, and Bioinformatics*, *40*(4), 623–636. doi:10.1002/1097-0134(20000901)40:4<623::AID-PROT70>3.0.CO;2-I PMID:10899786

Fraser, H. B., Hirsh, A. E., Steinmetz, L. M., Scharfe, C., & Feldman, M. W. (2002). Evolutionary rate in the protein interaction network. *Science*, *296*(5568), 750–752. doi:10.1126/science.1068696 PMID:11976460

Fredriksson, R., Lagerstrom, M. C., Lundin, L. G., & Schioth, H. B. (2003). The G-protein-coupled receptors in the human genome form five main families. Phylogenetic analysis, paralogon groups, and fingerprints. *Molecular Pharmacology*, *63*(6), 1256–1272. doi:10.1124/mol.63.6.1256 PMID:12761335

Friesner, R. A., Banks, J. L., Murphy, R. B., Halgren, T. A., Klicic, J. J., Mainz, D. T., & Perry, J. K. et al. (2004). Glide: A new approach for rapid, accurate docking and scoring. 1. Method and assessment of docking accuracy. *Journal of Medicinal Chemistry*, *47*(7), 1739–1749. doi:10.1021/jm0306430 PMID:15027865

Friesner, R. A., Murphy, R. B., Repasky, M. P., Frye, L. L., Greenwood, J. R., Halgren, T. A., & Mainz, D. T. et al. (2006). Extra precision glide: Docking and scoring incorporating a model of hydrophobic enclosure for protein-ligand complexes. *Journal of Medicinal Chemistry*, *49*(21), 6177–6196. doi:10.1021/jm051256o PMID:17034125

Frishman, D., & Mewes, H. W. (1997). Protein structural classes in five complete genomes. *Nature Structural Biology*, *4*(8), 626–628. doi:10.1038/nsb0897-626 PMID:9253410

Fu, J., Bian, M., Jiang, Q., & Zhang, C. (2007). Roles of Aurora kinases in mitosis and tumorigenesis. *Molecular Cancer Research*, *5*(1), 1–10. doi:10.1158/1541-7786.MCR-06-0208 PMID:17259342

Fujita, T. (1990). The extrathermodynamic approach to drug design. In Comprehensive Medicinal Chemistry (vol. 4). Pergamon.

Fulekar, M. (2009). *Environmental Cleanup Approach Using Bioinformatics in Bioremediation. In Bioinformatics: Applications in Life and Environmental Sciences* (pp. 173–199). Springer. doi:10.1007/978-1-4020-8880-3_10

Fuller, J. C., Burgoyne, N. J., & Jackson, R. M. (2009). Predicting druggable binding sites at the protein-protein interface. *Drug Discovery Today*, *14*(3-4), 155–161. doi:10.1016/j.drudis.2008.10.009 PMID:19041415

Furusawa, M., Narita, Y., Iwai, K., Fukunaga, T., & Nakagiri, O. (2011). Inhibitory effect of a hot water extract of coffee 'silverskin' on hyaluronidase. *Bioscience, Biotechnology, and Biochemistry*, *75*(6), 1205–1207. doi:10.1271/bbb.110106 PMID:21670509

Gabb, H. A., Jackson, R. M., & Sternberg, M. J. (1997). Modelling protein docking using shape complementarity, electrostatics and biochemical information. *Journal of Molecular Biology*, *272*(1), 106–120. doi:10.1006/jmbi.1997.1203 PMID:9299341

Gagner, J. E., Lopez, M. D., Dordick, J. S., & Siegel, R. W. (2011). Effect of gold nanoparticle morphology on adsorbed protein structure and function. *Biomaterials*, *32*(29), 7241–7252. doi:10.1016/j.biomaterials.2011.05.091 PMID:21705074

Galvez, J., Coelho, G., Crespo, M. E., Cruz, T., Rodriguez-Cabezas, M. E., Concha, A., & Zarzuelo, A. et al. (2001). Intestinal anti-inflammatory activity of morin on chronic experimental colitis in the rat. *Alimentary Pharmacology & Therapeutics*, *15*(12), 2027–2039. doi:10.1046/j.1365-2036.2001.01133.x PMID:11736735

Gambacorti-Passerini, C., & Piazza, R. (2015). Imatinib-A New Tyrosine Kinase Inhibitor for First-Line Treatment of Chronic Myeloid Leukemia in 2015. *JAMA Oncol*, *1*(2), 143–144. doi:10.1001/jamaoncol.2015.50 PMID:26181011

Garnier, J., Gibrat, J., & Robson, B. (1996). GOR method for predicting protein secondary structure from amino acid sequence. *Methods in Enzymology Computer Methods for Macromolecular Sequence Analysis*, 540-553.

Garuti, L., Roberti, M., & Bottegoni, G. (2010). Non-ATP competitive protein kinase inhibitors. *Current Medicinal Chemistry*, *17*(25), 2804–2821. doi:10.2174/092986710791859333 PMID:20586715

Garuti, L., Roberti, M., & Bottegoni, G. (2011). Irreversible protein kinase inhibitors. *Current Medicinal Chemistry*, *18*(20), 2981–2994. doi:10.2174/092986711796391705 PMID:21651479

Gasteiger, J., & Marsili, M. (1980). Iterative partial equalization of orbital electronegativity—a rapid access to atomic charges. *Tetrahedron*, *36*(22), 3219–3228. doi:10.1016/0040-4020(80)80168-2

Gautam, R., & Jachak, S. M. (2009). Recent developments in anti-inflammatory natural products. *Medicinal Research Reviews*, *29*(5), 767–820. doi:10.1002/med.20156 PMID:19378317

Gazak, R., Purchartova, K., Marhol, P., Zivna, L., Sedmera, P., Valentova, K., & Kren, V. et al. (2010). Antioxidant and antiviral activities of silybin fatty acid conjugates. *European Iournal of Medicinal Chemistry*, *45*(3), 1059–1067. doi:10.1016/j.ejmech.2009.11.056 PMID:20036447

Gazák, R., Walterová, D., & Kren, V. (2007). Silybin and silymarin—new and emerging applications in medicine. *Current Medicinal Chemistry*, *14*(3), 315–338. doi:10.2174/092986707779941159 PMID:17305535

Gentili, P. L., Ortica, F., & Favaro, G. (2008). Static and dynamic interaction of a naturally occurring photochromic molecule with bovine serum albumin studied by UV-visible absorption and fluorescence spectroscopy. *The Journal of Physical Chemistry B*, *112*(51), 16793–16801. doi:10.1021/jp805922g PMID:19367911

George, R. A., & Heringa, J. (2002). An analysis of protein domain linkers: Their classification and role in protein folding. *Protein Engineering*, *15*(11), 871–879. doi:10.1093/protein/15.11.871 PMID:12538906

Geske, G. D., Mattmann, M. E., & Blackwell, H. E. (2008). Evaluation of a focused library of N-aryl L-homoserine lactones reveals a new set of potent quorum sensing modulators. *Bioorganic & Medicinal Chemistry Letters*, *18*(22), 5978–5981. doi:10.1016/j.bmcl.2008.07.089 PMID:18760602

Ghalia, S., & Noura, B. (2015). Docking studies of flavonoid compounds as inhibitors of beta-ketoacyl acyl carrier protein synthase I (Kas I) of Escherichia coli. *Journal of Molecular Graphics & Modelling*, *61*, 214–223. doi:10.1016/j.jmgm.2015.07.005 PMID:26292066

Ghasemi, J. B., & Hooshmand, S. (2013). 3D-QSAR, docking and molecular dynamics for factor Xa inhibitors as anticoagulant agents. *Molecular Simulation*, *39*(6), 453–471. doi:10.1080/08927022.2012.741235

Ghasemi, J. B., & Nouri, M. (2013). Docking and CoMFA study on novel human CCR5 receptor antagonists. *Medicinal Chemistry Research*, *22*(3), 1356–1364. doi:10.1007/s00044-012-0118-7

Ghasemi, J. B., Safavi-Sohi, R., & Barbosa, E. G. (2012). 4D-LQTA-QSAR and docking study on potent gram-negative specific LpxC inhibitors: A comparison to CoMFA modeling. *Molecular Diversity*, *16*(1), 203–213. doi:10.1007/s11030-011-9340-3 PMID:22127637

Giacomini, K. M., Huang, S. M., Tweedie, D. J., Benet, L. Z., Brouwer, K. L., Chu, X., & Zhang, L. et al. (2010). Membrane transporters in drug development. *Nature Reviews. Drug Discovery*, *9*(3), 215–236. doi:10.1038/nrd3028 PMID:20190787

Gilad, Y., & Senderowitz, H. (2014). Docking studies on DNA intercalators. *Journal of Chemical Information and Modeling*, *54*(1), 96–107. doi:10.1021/ci400352t PMID:24303988

Gill, A., Scanlon, T. C., Osipovitch, D. C., Madden, D. R., & Griswold, K. E. (2011). Crystal structure of a charge engineered human lysozyme having enhanced bactericidal activity. *PLoS ONE*, *6*(3), e16788. doi:10.1371/journal.pone.0016788 PMID:21408218

Girdler, F., Sessa, F., Patercoli, S., Villa, F., Musacchio, A., & Taylor, S. (2008). Molecular basis of drug resistance in aurora kinases. *Chemistry & Biology*, *15*(6), 552–562. doi:10.1016/j.chembiol.2008.04.013 PMID:18559266

Glaser, F., Morris, R. J., Najmanovich, R. J., Laskowski, R. A., & Thornton, J. M. (2006). A method for localizing ligand binding pockets in protein structures. *Proteins*, *62*(2), 479–488. doi:10.1002/prot.20769 PMID:16304646

Goddard, W. A. III, & Abrol, R. (2007). 3-Dimensional structures of G protein-coupled receptors and binding sites of agonists and antagonists. *The Journal of Nutrition*, *137*, 1528S–1538S. PMID:17513420

Godin-Heymann, N., Ulkus, L., Brannigan, B. W., McDermott, U., Lamb, J., Maheswaran, S., & Haber, D. A. et al. (2008). The T790M "gatekeeper" mutation in EGFR mediates resistance to low concentrations of an irreversible EGFR inhibitor. *Molecular Cancer Therapeutics*, *7*(4), 874–879. doi:10.1158/1535-7163.MCT-07-2387 PMID:18413800

Gohard, F. H., St-Cyr, D. J., Tyers, M., & Earnshaw, W. C. (2014). Targeting the INCENP IN-box-Aurora B interaction to inhibit CPC activity in vivo. *Open Biology*, *4*(11), 140–163. doi:10.1098/rsob.140163 PMID:25392451

Gohlke, H., & Klebe, G. (2001). Statistical potentials and scoring functions applied to protein-ligand binding. *Current Opinion in Structural Biology*, *11*(2), 231–235. doi:10.1016/S0959-440X(00)00195-0 PMID:11297933

Goodford, P. J. (1984). Drug design by the method of receptor fit. *Journal of Medicinal Chemistry*, *27*(5), 557–564. doi:10.1021/jm00371a001 PMID:6716396

Goodford, P. J. (1985). A computational procedure for determining energetically favorable binding sites on biologically important macromolecules. *Journal of Medicinal Chemistry*, *28*(7), 849–857. doi:10.1021/jm00145a002 PMID:3892003

Gozalbes, R., Simon, L., Froloff, N., Sartori, E., Monteils, C., & Baudelle, R. (2008). Development and experimental validation of a docking strategy for the generation of kinase-targeted libraries. *Journal of Medicinal Chemistry*, *51*(11), 3124–3132. doi:10.1021/jm701367r PMID:18479119

Graf, B. A., Mibury, P. E., & Blumberg, J. B. (2005). Flavonols, flavones, flavanones, and human health: Epidemiological evidence. *Journal of Medicinal Food*, *8*(3), 281–290. doi:10.1089/jmf.2005.8.281 PMID:16176136

Graf, L., Jancso, A., Szilágyi, L., Hegyi, G., Pintér, K., Náray-Szabó, G., & Rutter, W. J. et al. (1988). Electrostatic complementarity within the substrate-binding pocket of trypsin. *Proceedings of the National Academy of Sciences of the United States of America*, *85*(14), 4961–4965. doi:10.1073/pnas.85.14.4961 PMID:3134655

Granier, S., Manglik, A., Kruse, A. C., Kobilka, T. S., Thian, F. S., Weis, W. I., & Kobilka, B. K. (2012). Structure of the delta-opioid receptor bound to naltrindole. *Nature*, *485*(7398), 400–404. doi:10.1038/nature11111 PMID:22596164

Granseth, E., Viklund, H., & Elofsson, A. (2006). ZPRED: Predicting the distance to the membrane center for residues in α-helical membrane proteins. *Bioinformatics (Oxford, England)*, *22*(14), e191–e196. doi:10.1093/bioinformatics/btl206 PMID:16873471

Greenblatt, H. M., Kryger, G., Lewis, T., Silman, I., & Sussman, J. L. (1999). Structure of acetylcholinesterase complexedwith (-)-galanthamine at 2.3 Å resolution. *FEBS Letters*, *463*(3), 321–326. doi:10.1016/S0014-5793(99)01637-3 PMID:10606746

Groves, C. P. (2001). *Primate Taxonomy*. Smithsonian Institution Press.

Gruber, A. R., Lorenz, R., Bernhart, S. H., Neuböck, R., & Hofacker, I. L. (2008). The Vienna RNA Websuite. *Nucleic Acids Research*, *36*, W70–W74. doi:10.1093/nar/gkn188 PMID:18424795

Guilbert, C., & James, T. L. (2008). Docking to RNA via Root-Mean-Square-Deviation-Driven Energy Minimization with Flexible Ligands and Flexible Targets. *Journal of Chemical Information and Modeling*, *48*(6), 1257–1268. doi:10.1021/ci8000327 PMID:18510306

Guo, W., Wisniewski, J. A., & Ji, H. (2014). Hot spot-based design of small-molecule inhibitors for protein-protein interactions. *Bioorganic & Medicinal Chemistry Letters*, *24*(11), 2546–2554. doi:10.1016/j.bmcl.2014.03.095 PMID:24751445

Guzzo, A. V. (1965). The influence of amino acid sequence on protein structure. *Biophysical Journal*, *5*(6), 809–822. doi:10.1016/S0006-3495(65)86753-4 PMID:5884309

Haga, K., Kruse, A. C., Asada, H., Yurugi-Kobayashi, T., Shiroishi, M., Zhang, C., & Kobayashi, T. et al. (2012). Structure of the human M2 muscarinic acetylcholine receptor bound to an antagonist. *Nature*, *482*(7386), 547–551. doi:10.1038/nature10753 PMID:22278061

Hajduk, P. J. (2006). Fragment-based drug design: How big is too big? *Journal of Medicinal Chemistry*, *49*(24), 6972–6976. doi:10.1021/jm060511h PMID:17125250

Hajduk, P. J., & Greer, J. (2007). A decade of fragment-based drug design: Strategic advances and lessons learned. *Nature Reviews. Drug Discovery*, *6*(3), 211–219. doi:10.1038/nrd2220 PMID:17290284

Halgren, T. A. (1996). Merck molecular force field. I. Basis, form, scope, parameterization, and performance of MMFF94. *Journal of Computational Chemistry*, *17*(5-6), 490–519. doi:10.1002/(SICI)1096-987X(199604)17:5/6<490::AID-JCC1>3.0.CO;2-P

Halperin, I., Ma, B., Wolfson, H., & Nussinov, R. (2002). Principles of docking: An overview of search algorithms and a guide to scoring functions. *Proteins*, *47*(4), 409–443. doi:10.1002/prot.10115 PMID:12001221

Han, K. C., Kim, S. Y., & Yang, E. G. (2012). Recent advances in designing substrate-competitive protein kinase inhibitors. *Current Pharmaceutical Design*, *18*(20), 2875–2882. doi:10.2174/138161212800672697 PMID:22571656

Hansch, C. (1969). A quantitative approach to biochemical structure-activity relationships. *Accounts of Chemical Research*, *2*(8), 232–239. doi:10.1021/ar50020a002

Hanson, M. A., Roth, C. B., Jo, E., Griffith, M. T., Scott, F. L., Reinhart, G., & Stevens, R. C. et al. (2012). Crystal structure of a lipid G protein-coupled receptor. *Science*, *335*(6070), 851–855. doi:10.1126/science.1215904 PMID:22344443

Hansson, G. K., & Edfeldt, K. (2005). Toll to be paid at the gateway to the vessel wall. *Arteriosclerosis, Thrombosis, and Vascular Biology*, *25*(6), 1085–1087. doi:10.1161/01.ATV.0000168894.43759.47 PMID:15923538

Hardwick, J. M., & Soane, L. (2013). Multiple functions of BCL-2 family proteins. *Cold Spring Harbor Perspectives in Biology*, *5*(2), a008722. doi:10.1101/cshperspect.a008722 PMID:23378584

Harel, M., Schalk, I., Ehret-sabatier, L., Bouet, F., Goeldner, M., Hirth, C., & Sussman, J. L. et al. (1993). Quaternary ligand binding to aromatic residues in the active-site gorge of acetylcholinesterase. *Proceedings of the National Academy of Sciences of the United States of America*, *90*(19), 9031–9035. doi:10.1073/pnas.90.19.9031 PMID:8415649

Heim, K. E., Tagliaferro, A. R., & Bobilya, D. J. (2002). Flavonoid antioxidants: Chemistry, metabolism and structure-activity relationships. *The Journal of Nutritional Biochemistry*, *13*(10), 572–584. doi:10.1016/S0955-2863(02)00208-5 PMID:12550068

Herman, T., & Tor, Y. (2005) RNA as a target for small-molecule therapeutics. *Expert Opinion on Therapeutic Patents*, *15*(1), 49-62. doi:10.1517/13543776.15.1.49

Hertog, M. G. L., Feskens, E. J. M., Hollman, P. C. H., Katan, M. B., & Kromhout, D. (1993). Dietary antioxidant flavonoids and risk of coronary heart disease: The Zutphen elderly study. *Lancet*, *342*(8878), 1007–1011. doi:10.1016/0140-6736(93)92876-U PMID:8105262

He, Y., Fan, C., & Lee, S.-T. (2010). Silicon nanostructures for bioapplications. *Nano Today*, *5*(4), 282–295. doi:10.1016/j.nantod.2010.06.008

Hill, J. R., & Deane, C. M. (2013). MP-T: Improving membrane protein alignment for structure prediction. *Bioinformatics (Oxford, England)*, *29*(1), 54–61. doi:10.1093/bioinformatics/bts640 PMID:23110969

Ho, H. K., Yeo, A. H., Kang, T. S., & Chua, B. T. (2014). Current strategies for inhibiting FGFR activities in clinical applications: Opportunities, challenges and toxicological considerations. *Drug Discovery Today*, *19*(1), 51–62. doi:10.1016/j.drudis.2013.07.021 PMID:23932951

Holt, P. A., Buscaglia, R., Trent, J. O., & Chaires, J. B. (2011). A discovery funnel for nucleic acid binding drug candidates. *Drug Development Research*, *72*(2), 178–186. doi:10.1002/ddr.20414 PMID:21566705

Holt, P. A., Chaires, J. B., & Trent, J. O. (2008). Molecular docking of intercalators and groove-binders to nucleic acids using Autodock and Surflex. *Journal of Chemical Information and Modeling*, *48*(8), 1602–1615. doi:10.1021/ci800063v PMID:18642866

Hooft, R. W., Vriend, G., Sander, C., & Abola, E. E. (1996). Errors in protein structures. *Nature*, *381*(6580), 272. doi:10.1038/381272a0 PMID:8692262

Hopkins, A. L., & Groom, C. R. (2002). The druggable genome. *Nature Reviews. Drug Discovery*, *1*(9), 727–730. doi:10.1038/nrd892 PMID:12209152

Hou, X., Du, J., Zhang, J., Du, L., Fang, H., & Li, M. (2013). How to improve docking accuracy of AutoDock4. 2: A case study using different electrostatic potentials. *Journal of Chemical Information and Modeling*, *53*(1), 188–200. doi:10.1021/ci300417y PMID:23244516

Hruby, V. J. (1982). Conformational restrictions of biologically active peptides via amino acid side chain groups. *Life Sciences*, *31*(3), 189–199. doi:10.1016/0024-3205(82)90578-1 PMID:6126794

Hruby, V. J., Qui, W., Okayama, T., & Soloshonok, V. A. (2002). Design of nonpeptides from peptide ligands for peptide receptors. *Methods in Enzymology*, *343*, 91–123. doi:10.1016/S0076-6879(02)43129-1 PMID:11665597

Huang, H., & Zhao, M. (2008). Changes of trypsin in activity and secondary structure induced by complex with trypsin inhibitors and tea polyphenol. *European Food Research and Technology*, *227*(2), 361–365. doi:10.1007/s00217-007-0729-2

Huang, L., Yuan, X., Yu, D., Lee, K. H., & Chen, C. H. (2005). Mechanism of action and resistant profile of anti-HIV-1 coumarin derivatives. *Virology*, *332*(2), 623–628. doi:10.1016/j.virol.2004.11.033 PMID:15680427

Huang, N., & Shoichet, B. K. (2008). Exploiting ordered waters in molecular docking. *Journal of Medicinal Chemistry*, *51*(16), 4862–4865. doi:10.1021/jm8006239 PMID:18680357

Huang, S. Y., Grinter, S. Z., & Zou, X. (2010). Scoring functions and their evaluation methods for protein-ligand docking: Recent advances and future directions. *Physical Chemistry Chemical Physics*, *12*(40), 12899–12908. doi:10.1039/c0cp00151a PMID:20730182

Huang, X. Y., Shan, Z. J., Zhai, H. L., Su, L., & Zhang, X. Y. (2011). Study on the anticancer activity of coumarin derivatives by molecular modeling. *Chemical Biology & Drug Design*, *78*(4), 651–658. doi:10.1111/j.1747-0285.2011.01195.x PMID:21791009

Huang, Y. B., Yan, J., Liu, B. Z., Yu, Z., Gao, X. Y., Tang, Y. C., & Zi, Y. Q. (2010). Investigation on interaction of prulifloxacin with pepsin: A spectroscopic analysis. *Spectrochimica Acta. Part A: Molecular and Biomolecular Spectroscopy*, *75*(3), 1024–1029. doi:10.1016/j.saa.2009.12.044 PMID:20045662

Huey, R., Morris, G. M., Olson, A. J., & Goodsell, D. S. (2007). A semiempirical free energy force field with charge-based desolvation. *Journal of Computational Chemistry*, *28*(6), 1145–1152. doi:10.1002/jcc.20634 PMID:17274016

Hui, C., Quan, L., Jian, S., Jian, B. X., & Ming, X. (2008). Comparing the affinities of flavonoid isomers with protein by fluorescence spectroscopy. *Analytical Letters*, *41*(4), 521–532. doi:10.1080/00032710801910486

Humphrey, W., Dalke, A., & Schulten, K. (1996). VMD - Visual Molecular Dynamics. *Journal of Molecular Graphics*, *14*(1), 33–38. doi:10.1016/0263-7855(96)00018-5 PMID:8744570

Isberg, V., Balle, T., Sander, T., Jørgensen, F. S., & Gloriam, D. E. (2011). G protein- and agonist-bound serotonin 5-HT2A receptor model activated by steered molecular dynamics simulations. *Journal of Chemical Information and Modeling*, *51*(2), 315–325. doi:10.1021/ci100402f PMID:21261291

Ivanov, A. A., Khuri, F. R., & Fu, H. (2013). Targeting protein-protein interactions as an anticancer strategy. *Trends in Pharmacological Sciences*, *34*(7), 393–400. doi:10.1016/j.tips.2013.04.007 PMID:23725674

Jaakola, V. P., Griffith, M. T., Hanson, M. A., Cherezov, V., Chien, E. Y., Lane, J. R., & Stevens, R. C. et al. (2008). The 2.6 angstrom crystal structure of a human A2A adenosine receptor bound to an antagonist. *Sci.*, *322*(5905), 1211–1217. doi:10.1126/science.1164772 PMID:18832607

Jacobson, K. A., Ji, X., Li, A. H., Melman, N., Siddiqui, M. A., Shin, K. J., & Ravi, R. G. et al. (2000). Methanocarba analogues of purine nucleosides as potent and selective adenosine receptor agonists. *Journal of Medicinal Chemistry*, *43*(11), 2196–2203. doi:10.1021/jm9905965 PMID:10841798

Jacobson, K. A., Klutz, A. M., Tosh, D. K., Ivanov, A. A., Preti, D., & Baraldi, P. G. (2009). Medicinal chemistry of the A3 adenosine receptor: Agonists, antagonists, and receptor engineering. *Handbook of Experimental Pharmacology*, *193*, 123–159. doi:10.1007/978-3-540-89615-9_5 PMID:19639281

Jain, A. N. (2003). Surflex: Fully automatic flexible molecular docking using a molecular similarity-based search engine. *Journal of Medicinal Chemistry*, *46*(4), 499–511. doi:10.1021/jm020406h PMID:12570372

Jain, A. N. (2007). Surflex-Dock 2.1: Robust performance from ligand energetic modeling, ring flexibility, and knowledge-based search. *Journal of Computer-Aided Molecular Design*, *21*(5), 281–306. doi:10.1007/s10822-007-9114-2 PMID:17387436

James, G. A., Swogger, E., Wolcott, R., Secor, P., Sestrich, J., Costerton, J. W., & Stewart, P. S. (2008). Biofilms in chronic wounds. *Wound Repair and Regeneration*, *16*(1), 37–44. doi:10.1111/j.1524-475X.2007.00321.x PMID:18086294

Jayaraman, J., Jesudoss, V. A. S., Menon, V. P., & Namasivayam, N. (2012). Anti-inflammatory role of naringenin in rats with ethanol induced liver injury. *Toxicology Mechanisms and Methods*, *22*(7), 568–576. doi:10.3109/15376516.2012.707255 PMID:22900548

Jeffers, S. A., Sanders, D. A., & Sanchez, A. (2002). Covalent modifications of the Ebola virus glycoprotein. *Journal of Virology*, *76*(24), 12463–12472. doi:10.1128/JVI.76.24.12463-12472.2002 PMID:12438572

Jeong, L. S., Choe, S. A., Gunaga, P., Kim, H. O., Lee, H. W., Lee, S. K., & Moon, H. R. et al. (2007). Discovery of a New Nucleoside Template for Human A3 Adenosine Receptor Ligands: D-4'-Thioadenosine Derivatives without 4'-Hydroxymethyl Group as Highly Potent and Selective Antagonists. *Journal of Medicinal Chemistry*, *50*(14), 3159–3162. doi:10.1021/jm070259t PMID:17555308

Jeon, I. H., Kim, H. S., Kang, H. J., Lee, H. S., Jeong, S. I., Kim, S. J., & Jang, S. I. (2014). Anti-inflammatory and antipruritic effects of luteolin from Perilla (*P. frutescens* L.) leaves. *Molecules (Basel, Switzerland)*, *19*(6), 6941–6951. doi:10.3390/molecules19066941 PMID:24871572

Jeyam, M., Sushma, R., Sharanya, M., & Poornima, V. (2011). Validating nutraceuticals to alleviate progeria using molecular docking studies. *Journal of Pharmaceutical and Biomedical Sciences*, *13*(13).

Jiang, F., & Kim, S. H. (1991). "Soft docking": Matching of molecular surface cubes. *Journal of Molecular Biology*, *219*(1), 79–102. doi:10.1016/0022-2836(91)90859-5 PMID:2023263

Ji, G. Y., Yang, Q. H., Hao, J., Guo, L. N., Chen, X., Hu, J. P., & Jiang, Z. Q. et al. (2011). Anti-inflammatory effect of genistein on non-alcoholic steatohepatitis rats induced by high fat diet and its potential mechanisms. *International Immunopharmacology*, *11*(6), 762–768. doi:10.1016/j.intimp.2011.01.036 PMID:21320636

Jimenez-Botello, L. C., Soriano-Ursúa, M. A., Deeb, O., Segura-Cabrera, A., Martínez-Archundia, M., & Correa-Basurto, J. (2013). Homology modeling, molecular dynamics and docking simulations of rat A2A receptor: A three-dimensional model validation under QSAR studies. *Current Trends in Medicinal Chemistry*, *7*, 35–52.

Jin, K. S., Rho, Y., Kim, J., Kim, H., Kim, I. J., & Ree, M. (2008). Synchrotron small-angle X-ray scattering studies of the structure of porcine pepsin under various pH conditions. *Journal of Physical Chemistry*, *112*(49), 15821–15827. doi:10.1021/jp805940d PMID:19367902

Jin, L., Wang, W., & Fang, G. (2014). Targeting protein-protein interaction by small molecules. *Annual Review of Pharmacology and Toxicology*, *54*(1), 435–456. doi:10.1146/annurev-pharmtox-011613-140028 PMID:24160698

Jones, S., Stewart, M., Michie, A., Swindells, MB., Orengo, C., & Thornton, J. M., (1998). Domain assignment for protein structures using a consensus approach: characterization and analysis. *Protein Sci,* *7*(2), 233–42. doi:10.1002/pro.5560070202

Jones, G., Willett, P., Glen, R. C., Leach, A. R., & Taylor, R. (1997). Development and validation of a genetic algorithm for flexible docking. *Journal of Molecular Biology*, *267*(3), 727–748. doi:10.1006/jmbi.1996.0897 PMID:9126849

Jorgensen, W. L., & Tirado-Rives, J. (1988). The OPLS [optimized potentials for liquid simulations] potential functions for proteins, energy minimizations for crystals of cyclic peptides and crambin. *Journal of the American Chemical Society*, *110*(6), 1657–1666. doi:10.1021/ja00214a001

Joseph-McCarthy, D., Thomas, B. E., Belmarsh, M., Moustakas, D., & Alvarez, J. C. (2003). Pharmacophore-based molecular docking to account for ligand flexibility. *Proteins: Structure, Function, and Bioinformatics*, *51*(2), 172–188. doi:10.1002/prot.10266 PMID:12660987

Jubb, H., Blundell, T. L., & Ascher, D. B. (2015). Flexibility and small pockets at protein-protein interfaces: New insights into druggability. *Progress in Biophysics and Molecular Biology*, *119*(1), 2–9. doi:10.1016/j.pbiomolbio.2015.01.009 PMID:25662442

Ju, J., Hong, J., Zhou, J.-, Pan, Z., Bose, M., Liao, J., & Lin, Y. et al. (2005). Inhibition of intestinal tumorigenesis in Apcmin/+ mice by (−)-epigallocatechin-3-gallate, the major catechin in green tea. *Cancer Research*, *65*(22), 10623–10631. doi:10.1158/0008-5472.CAN-05-1949 PMID:16288056

Jung, H. J., Kim, S. J., Song, Y. S., Park, E. H., & Lim, C. J. (2010). Evaluation of the antiangiogenic, anti-inflammatory, and antinociceptive activities of morin. *Planta Medica*, *76*(03), 273–275. doi:10.1055/s-0029-1186079 PMID:19774509

Jutel, M., Blaser, K., & Akdis, C. A. (2005). Histamine in chronic allergic responses. *Journal of Investigational Allergology & Clinical Immunology*, *15*, 1–8. PMID:15864876

Kabsch, W., & Sander, C. (1983). Dictionary of protein secondary structure: Pattern recognition of hydrogen-bonded and geometrical features. *Biopolymes*, *22*(12), 2577–2637. doi:10.1002/bip.360221211 PMID:6667333

Kaidanovich-Beilin, O., & Eldar-Finkelman, H. (2006). Peptides targeting protein kinases: Strategies and implications. *Physiology (Bethesda, MD)*, *21*(6), 411–418. doi:10.1152/physiol.00022.2006 PMID:17119154

Kakarla, P., Devireddy, A. R., Inupakutika, M. A., Cheeti, U. R., Floyd, J. T., Mun, M. M., & Varela, M. F. et al. (2014). Molecular modelling, 3d-qsar, and rug docking studies on the role of natural anticoagulant compounds in antithrombotic therapy. *International Journal of Pharmaceutical Sciences and Research*, *5*(10), 4141. PMID:25750914

Kakegawa, H., Matsumoto, H., & Satoh, T. (1999). Inhibitory effects of some natural products on the activation of hyaluronidase and their anti-allergic action. *Chemical & Pharmaceutical Bulletin*, *40*(6), 1439–1442. doi:10.1248/cpb.40.1439 PMID:1382876

Källblad, P., & Dean, P. M. (2003). Efficient conformational sampling of local side-chain flexibility. *Journal of Molecular Biology*, *326*(5), 1651–1665. doi:10.1016/S0022-2836(03)00083-4 PMID:12595271

Kam, V. W. T., & Goddard, W. A. III. (2008). Flat-Bottom Strategy for Improved Accuracy in Protein Side-Chain Placements. *Journal of Chemical Theory and Computation*, *4*(12), 2160–2169. doi:10.1021/ct800196k PMID:26620487

Kandeel M., Kitade Y. (2013) Computational Analysis of siRNA Recognition by the Ago2 PAZ Domain and Identification of the Determinants of RNA-Induced Gene Silencing. *PLoS ONE, 8*(2). doi:10.1371/journal.pone.0057140

Kang, C.-, Liu, D.-, Wang, X.-y., Dai, Y.-, Cheng, J.-, & Ly, Y.-. (2015). 3D-QSAR and docking study on 3-benzimidazol-2-ylhydroquinolin-2-one derivatives as VEGFR-2 tyrosine kinase inhibitors. *Medicinal Chemistry Research*, *24*(3), 934–943. doi:10.1007/s00044-014-1161-3

Kang, X., Shafer, R. H., & Kuntz, I. D. (2004). Calculation of ligand-nucleic acid binding free energies with the generalized-born model in DOCK. *Biopolymers*, *73*(2), 192–204. doi:10.1002/bip.10541 PMID:14755577

Karelson, M. (2000). *Molecular Descriptors in QSAR/QSPR*. New York: Wiley-InterScience.

Karplus, M. (1983). CHARMM: A program for macromolecular energy, minimization, and dynamics calculations. *Journal of Computational Chemistry*, *4*, 187217.

Kastenholz, M. A., Pastor, M., Cruciani, G., Haaksma, E. E., & Fox, T. (2000). GRID/CPCA: A new computational tool to design selective ligands. *Journal of Medicinal Chemistry*, *43*(16), 3033–3044. doi:10.1021/jm000934y PMID:10956211

Katchalski-Katzir, E., Shariv, I., Eisenstein, M., Friesem, A. A., Aflalo, C., & Vakser, I. A. (1992). Molecular surface recognition: Determination of geometric fit between proteins and their ligands by correlation techniques. *Proceedings of the National Academy of Sciences of the United States of America*, *89*(6), 2195–2199. doi:10.1073/pnas.89.6.2195 PMID:1549581

Katoh, K., Kuma, K., Toh, H., & Miyata, T. (2005). MAFFT version 5: Improvement in accuracy of multiple sequence alignment. *Nucleic Acids Research*, *33*(2), 511–518. doi:10.1093/nar/gki198 PMID:15661851

Katritch, V., Kufareva, I., & Abagyan, R. (2011). Structure based prediction of subtype-selectivity for adenosine receptor antagonists. *Neuropharmacology*, *60*(1), 108–115. doi:10.1016/j.neuropharm.2010.07.009 PMID:20637786

Katritzky, A. R., Lobanov, V. S., & Karelson, M. (1994). CODESSA, Reference Manual. Gainesville, FL: University of Florida. Retrieved from http://www.semichem.com/codessa

Kay, A. B. (2008). *Allergy and Allergic Diseases*. Oxford, UK: Blackwell Publishing Ltd. doi:10.1002/9781444300918

Khan, F., Pal, D., Vikram, S., & Cameotra, S. S. (2013). Metabolism of 2-chloro-4-nitroaniline via novel aerobic degradation pathway by Rhodococcus sp. strain MB-P1. *PLoS ONE*, *8*(4), e62178. doi:10.1371/journal.pone.0062178 PMID:23614030

Khan, F., Sajid, M., & Cameotra, S. S. (2013). In Silico Approach for the Bioremediation of Toxic Pollutants. *J Pet Environ Biotechnol*, *4*(161), 2.

Khan, S. N., Islam, B., Yennamalli, R., Sultan, A., Subbarao, N., & Khan, A. U. (2008). Interaction of mitoxantrone with human serum albumin: Spectroscopic and molecular modeling studies. *European Journal of Pharmaceutical Sciences*, *35*(5), 371–382. doi:10.1016/j.ejps.2008.07.010 PMID:18762252

Kharkar, P. S., Warrier, S., & Gaud, R. S. (2014). Reverse docking: a powerful tool for drug repositioning and drug rescue. *Future Medicinal Chemistry, 6*(3), 333-342.

Kikuchi, R., Nakamura, K., MacLauchlan, S., Ngo, D. T., Shimizu, I., Fuster, J. J., & Walsh, K. et al. (2014). An anti-angiogenic isoform of VEGF-A contributes to impaired vascularization in peripheral artery disease. *Nature Medicine*, *20*(12), 1464–1471. doi:10.1038/nm.3703 PMID:25362254

Kim, D. E., Chivian, D., & Baker, D. (2004). Protein structure prediction and analysis using the Robetta server. *Nucleic Acids Res, 32*, W526-531. doi: 10.1093/nar/gkh468

Kim, M. H., Kim, Y., Park, H.-J., Lee, J. S., Kwak, S.-N., Jung, W.-H., & Oh, T.-K. et al. (2008). Structural insight into bioremediation of triphenylmethane dyes by Citrobacter sp. triphenylmethane reductase. *The Journal of Biological Chemistry*, *283*(46), 31981–31990. doi:10.1074/jbc.M804092200 PMID:18782772

Kim, S. K., Fristrup, P., Abrol, R., & Goddard, W. A. III. (2011). Structure-based prediction of subtype selectivity of histamine H3 receptor selective antagonists in clinical trials. *Journal of Chemical Information and Modeling*, *51*(12), 3262–3274. doi:10.1021/ci200435b PMID:22035233

Kim, S. K., Riley, L., Abrol, R., Jacobson, K. A., & Goddard, W. A. III. (2011). Predicted structures of agonist and antagonist bound complexes of adenosine A3 receptor. *Proteins*, *79*(6), 1878–1897. doi:10.1002/prot.23012 PMID:21488099

Kim, S.-K., Gao, Z.-G., Van Rompaey, P., Gross, A. S., Chen, A., Van Calenbergh, S., & Jacobson, K. A. (2003). Modeling the adenosine receptors: Comparison of the binding domains of A2A agonists and antagonists. *Journal of Medicinal Chemistry*, *46*(23), 4847–4859. doi:10.1021/jm0300431 PMID:14584936

Kim, S.-K., Li, Y., Abrol, R., Heo, J., & Goddard, W. A. III. (2011). Predicted structures and dynamics for agonists and antagonists bound to serotonin 5-HT2B and 5-HT2C receptors. *Journal of Chemical Information and Modeling*, *51*(2), 420–433. doi:10.1021/ci100375b PMID:21299232

Kitchen, D. B., Decornez, H., Furr, J. R., & Bajorath, J. (2004). Docking and scoring in virtual screening for drug discovery: Methods and applications. *Nature Reviews. Drug Discovery*, *3*(11), 935–949. doi:10.1038/nrd1549 PMID:15520816

Klco, J. M., Wiegand, C. B., Narzinski, K., & Baranski, T. J. (2005). Essential role for the second extracellular loop in C5a receptor activation. *Nature Structural & Molecular Biology*, *12*(4), 320–326. doi:10.1038/nsmb913 PMID:15768031

Klebe, G. (2006). Virtual ligand screening: Strategies, perspectives and limitations. *Drug Discovery Today*, *11*(13), 580–594. doi:10.1016/j.drudis.2006.05.012 PMID:16793526

Kleywegt, G. J., & Jones, T. A. (1994). Detection, delineation, measurement and display of cavities in macromolecular structures. *Acta Crystallographica. Section D, Biological Crystallography*, *50*(2), 178–185. doi:10.1107/S0907444993011333 PMID:15299456

Knegtel, R. M., Kuntz, I. D., & Oshiro, C. (1997). Molecular docking to ensembles of protein structures. *Journal of Molecular Biology*, *266*(2), 424–440. doi:10.1006/jmbi.1996.0776 PMID:9047373

Knox, R. B., & Suphioglu, C. (1996). Pollen allergens: Development and function. *Sexual Plant Reproduction*, *9*(6), 318–323. doi:10.1007/BF02441949

Koes, D. R., & Camacho, C. J. (2012a). PocketQuery: protein-protein interaction inhibitor starting points from protein-protein interaction structure. *Nucleic Acids Res*, *40*, W387-392. doi: 10.1093/nar/gks336

Koes, D. R., & Camacho, C. J. (2012b). ZINCPharmer: pharmacophore search of the ZINC database. *Nucleic Acids Res*, *40*, W409-414. doi: 10.1093/nar/gks378

Kokh, D. B., Wade, R. C., & Wenzel, W. (2011). Receptor flexibility in small-molecule docking calculations. *Wiley Interdisciplinary Reviews: Computational Molecular Science*, *1*(2), 298–314.

Kollareddy, M., Zheleva, D., Dzubak, P., Brahmkshatriya, P. S., Lepsik, M., & Hajduch, M. (2012). Aurora kinase inhibitors: Progress towards the clinic. *Investigational New Drugs*, *30*(6), 2411–2432. doi:10.1007/s10637-012-9798-6 PMID:22350019

Kontogiorgis, C. A., Xu, Y., Hadjipavlou-Litina, D., & Luo, Y. (2007). Coumarin derivatives protection against ROS production in cellular models of A beta toxicities. *Free Radical Research*, *41*(10), 1168–1180. doi:10.1080/10715760701447884 PMID:17886039

Kontoyianni, M., Madhav, P., Suchanek, E., & Seibel, W. (2008). Theoretical and practical considerations in virtual screening: A beaten field? *Current Medicinal Chemistry*, *15*(2), 107–116. doi:10.2174/092986708783330566 PMID:18220766

Koshland, D. (1963). Correlation of structure an function in enzyme action. *Science*, *142*(3599), 1533–1541. doi:10.1126/science.142.3599.1533 PMID:14075684

Kozakov, D., Beglov, D., Bohnuud, T., Mottarella, S., Xia, B., Hall, D. R., & Vajda, S. (2013). How good is automated protein docking? *Proteins: Structure, Function, and Bioinformatics*, *81*(12), 2159–2166. doi:10.1002/prot.24403 PMID:23996272

Kozakov, D., Brenke, R., Comeau, S. R., & Vajda, S. (2006). PIPER: An FFT-based protein docking program with pairwise potentials. *Proteins: Structure, Function, and Bioinformatics*, *65*(2), 392–406. doi:10.1002/prot.21117 PMID:16933295

Kozakov, D., Hall, D. R., Chuang, G. Y., Cencic, R., Brenke, R., Grove, L. E., & Vajda, S. et al. (2011). Structural conservation of druggable hot spots in protein-protein interfaces. *Proceedings of the National Academy of Sciences of the United States of America*, *108*(33), 13528–13533. doi:10.1073/pnas.1101835108 PMID:21808046

Kramer, B., Rarey, M., & Lengauer, T. (1999). Evaluation of the FLEXX incremental construction algorithm for protein–ligand docking. *Proteins: Structure, Function, and Bioinformatics*, *37*(2), 228–241. doi:10.1002/(SICI)1097-0134(19991101)37:2<228::AID-PROT8>3.0.CO;2-8 PMID:10584068

Krieger, E., Nabuurs, S. B., & Vriend, G. (2003). Homology modeling. In Structural Bioinformatics. Wiley. doi:10.1002/0471721204.ch25

Kruger, D. M., & Gohlke, H. (2010). DrugScorePPI webserver: fast and accurate in silico alanine scanning for scoring protein-protein interactions. *Nucleic Acids Res*, *38*, W480-486. doi: 10.1093/nar/gkq471

Kruse, A. C., Hu, J., Pan, A. C., Arlow, D. H., Rosenbaum, D. M., Rosemond, E., & Kobilka, B. K. et al. (2012). Structure and dynamics of the M3 muscarinic acetylcholine receptor. *Nature*, *482*(7386), 552–556. doi:10.1038/nature10867 PMID:22358844

Kryger, G., Silman, I., & Sussman, J. L. (1999). Structure of acetylcholinesterase complexed with E2020 (Aricept): Implications for the design of new anti-Alzheimer drugs. *Structure (London, England)*, *7*(3), 297–307. doi:10.1016/S0969-2126(99)80040-9 PMID:10368299

Kuang, R., Leslie, C. S., & Yang, A. S. (2004). Protein backbone angle prediction with machine learning approaches. *Bioinformatics (Oxford, England)*, *20*(10), 1612–1621. doi:10.1093/bioinformatics/bth136 PMID:14988121

Kubiny, H. (1994). Variable selection in QSAR studies: 2. A highly efficient combination of systematic search and evolution. *Quantitative Structure-activity Relationships*, *13*, 393–401.

Kubinyi, H. (1997a). QSAR and 3D QSAR in drug design Part 1: Methodology. *Drug Discovery Today*, *2*(11), 457–467. doi:10.1016/S1359-6446(97)01079-9

Kubinyi, H. (1997b). QSAR and 3D QSAR in drug design Part 2: Applications and problems. *Drug Discovery Today*, *2*(12), 538–546. doi:10.1016/S1359-6446(97)01084-2

Kubinyi, H., Folkers, G., & Martin, Y. C. (Eds.). (1998). *3D QSAR in drug design: Ligand-protein interactions and molecular similarity* (Vol. 2). Springer Science & Business Media.

Kubinyi, H., Folkers, G., & Martin, Y. C. (Eds.). (2006). *3D QSAR in drug design: Recent advances* (Vol. 3). Springer Science & Business Media.

Kuenemann, M. A., Sperandio, O., Labbe, C. M., Lagorce, D., Miteva, M. A., & Villoutreix, B. O. (2015). In silico design of low molecular weight protein-protein interaction inhibitors: Overall concept and recent advances. *Progress in Biophysics and Molecular Biology*, *119*(1), 20–32. doi:10.1016/j.pbiomolbio.2015.02.006 PMID:25748546

Kuhn, J. H., Becker, S., Ebihara, H., Geisbert, T. W., Johnson, K. M., Kawaoka, Y., … Ana, I. (2010). Proposal for a revised taxonomy of the family Filoviridae: Classification, names of taxa and viruses, and virus abbreviations. *Archives of Virology*, *155*(12), 2083–103. doi:10.1007/s00705-010-0814-x

Kumar, A., & Zhang, K. Y. (2013). Investigation on the effect of key water molecules on docking performance in CSARdock exercise. *Journal of Chemical Information and Modeling*, *53*(8), 1880–1892. doi:10.1021/ci400052w PMID:23617355

Kuntz, I. D., Blaney, J. M., Oatley, S. J., Langridge, R., & Ferrin, T. E. (1982). A geometric approach to macromolecule-ligand interactions. *Journal of Molecular Biology*, *161*(2), 269–288. doi:10.1016/0022-2836(82)90153-X PMID:7154081

Kuppusamy, U. R., & Das, N. P. (1991). Inhibitory effects of flavonoids on several venom hyaluronidases. *Experientia*, *47*(11-12), 1196–1200. doi:10.1007/BF01918384 PMID:1765130

Kuttner, Y. Y., & Engel, S. (2012). Protein hot spots: The islands of stability. *Journal of Molecular Biology*, *415*(2), 419–428. doi:10.1016/j.jmb.2011.11.009 PMID:22100447

LaBute, M. X., Zhang, X., Lenderman, J., Bennion, B. J., Wong, S. E., & Lightstone, F. C. (2014). Adverse Drug Reaction Prediction Using Scores Produced by Large-Scale Drug-Protein Target Docking on High-Performance Computing Machines. *PLoS ONE*, *9*(9), e106298. doi:10.1371/journal.pone.0106298 PMID:25191698

Lacapere, J. J., Pebay-Peyroula, E., Neumann, J. M., & Etchebest, C. (2007). Determining membrane protein structures: Still a challenge! *Trends in Biochemical Sciences*, *32*(6), 259–270. doi:10.1016/j.tibs.2007.04.001 PMID:17481903

Lakowicz, J. R., & Weber, G. (1973). Quenching of protein fluorescence by oxygen. Detection of structural fluctuations in proteins on the nanosecond time scale. *Biochemistry*, *12*(21), 4171–4179. doi:10.1021/bi00745a021 PMID:4200894

Lam, K. S., Salmon, S. E., Hersh, E. M., Hruby, V. J., Kazmierski, W. M., & Knapp, R. J. (1991). A new type of synthetic peptide library for identifying ligand-binding activity. *Nature*, *354*(6348), 82–84. doi:10.1038/354082a0 PMID:1944576

Lang, P. T., Brozell, S. R., Mukherjee, S., Pettersen, E. F., Meng, E. C., Thomas, V., & Kuntz, I. D. et al. (2009). DOCK6: Combining techniques to model RNA-small molecule complexes. *RNA (New York, N.Y.)*, *15*(6), 1219–1230. doi:10.1261/rna.1563609 PMID:19369428

Laraia, L., McKenzie, G., Spring, D. R., Venkitaraman, A. R., & Huggins, D. J. (2015). Overcoming Chemical, Biological, and Computational Challenges in the Development of Inhibitors Targeting Protein-Protein Interactions. *Chemistry & Biology*, *22*(6), 689–703. doi:10.1016/j.chembiol.2015.04.019 PMID:26091166

Laskowski RA. (1995). SURFNET: a program for visualizing molecular surfaces, cavities, and intermolecular interactions. *J Mol Graph, 13*(5), 307–328.

Laskowski, R. A. (1995). SURFNET: a program for visualizing molecular surfaces, cavities, and intermolecular interactions. *J Mol Graph, 13*(5), 307-328.

Laskowski, R. A., MacArthur, M. W., Moss, D. S., & Thornton, J. M. (1993). PROCHECK: A program to check the stereochemistry of protein structures. *Journal of Applied Crystallography*, *26*(2), 283–291. doi:10.1107/S0021889892009944

Lawrence, M. C., & Davis, P. C. (1992). CLIX: A search algorithm for finding novel ligands capable of binding proteins of known three-dimensional structure. *Proteins: Structure, Function, and Bioinformatics*, *12*(1), 31–41. doi:10.1002/prot.340120105 PMID:1313175

Le Guilloux, V., Schmidtke, P., & Tuffery, P. (2009). Fpocket: An open source platform for ligand pocket detection. *BMC Bioinformatics*, *10*(1), 168. doi:10.1186/1471-2105-10-168 PMID:19486540

Leach, A. R. (1994). Ligand docking to proteins with discrete side-chain flexibility. *Journal of Molecular Biology*, *235*(1), 345–356. doi:10.1016/S0022-2836(05)80038-5 PMID:8289255

Leach, A. R., Shoichet, B. K., & Peishoff, C. E. (2006). Prediction of protein-ligand interactions. Docking and scoring: Successes and gaps. *Journal of Medicinal Chemistry*, *49*(20), 5851–5855. doi:10.1021/jm060999m PMID:17004700

Lebon, G., Warne, T., Edwards, P. C., Bennett, K., Langmead, C. J., Leslie, A. G. W., & Tate, C. G. (2011). Agonist-bound adenosine A2A receptor structures reveal common features of GPCR activation. *Nature*, *474*(7352), 521–525. doi:10.1038/nature10136 PMID:21593763

Lee, J. H., & Kim, G. H. (2010). Evaluation of antioxidant and inhibitory activities for different subclasses flavonoids on enzymes for *Rheumatoid arthritis*. *Journal of Food Science*, *75*(7), H212–H217. doi:10.1111/j.1750-3841.2010.01755.x PMID:21535545

Lee, K. K., & Choi, J. D. (1999). The effects of *Areca catechu* L. extract on anti-inflammation and anti-melanogensis. *International Journal of Cosmetic Science*, *21*(4), 275–284. doi:10.1046/j.1467-2494.1999.196590.x PMID:18503445

Lee, S. J., Lee, U. S., Kim, W. J., & Moon, S. K. (2011). Inhibitory effect of esculetin on migration, invasion and matrix metalloproteinase-9 expression in TNF-alpha-induced vascular smooth muscle cells. *Molecular Medicine Reports*, *4*, 337–341. PMID:21468574

Lefevre, F., Remy, M. H., & Masson, J. M. (1997). Alanine-stretch scanning mutagenesis: A simple and efficient method to probe protein structure and function. *Nucleic Acids Research*, *25*(2), 447–448. doi:10.1093/nar/25.2.447 PMID:9016577

Lefkowitz, R. J., Pierce, K. L., & Luttrell, L. M. (2002). Dancing with different partners: Protein kinase A phosphorylation of seven membrane-spanning receptors regulates their G protein-coupling specificity. *Molecular Pharmacology*, *62*(5), 971–974. doi:10.1124/mol.62.5.971 PMID:12391258

Lengauer, T., & Rarey, M. (1996). Computational methods for biomolecular docking. *Current Opinion in Structural Biology*, *6*(3), 402–406. doi:10.1016/S0959-440X(96)80061-3 PMID:8804827

Leslie, E. M., Deeley, R. G., & Cole, S. P. (2005). Multidrug resistance proteins: Role of P-glycoprotein, MRP1, MRP2, and BCRP (ABCG2) in tissue defense. *Toxicology and Applied Pharmacology*, *204*(3), 216–237. doi:10.1016/j.taap.2004.10.012 PMID:15845415

Leszczynski, J., & Puzyn, T. (2012). *Towards efficient designing of safe nanomaterials: Innovative merge of computational approaches and experimental techniques*. Royal Society of Chemistry. doi:10.1039/9781849735476

Levitt, D. G., & Banaszak, L. J. (1992). POCKET: A computer graphics method for identifying and displaying protein cavities and their surrounding amino acids. *Journal of Molecular Graphics*, *10*(4), 229–234. doi:10.1016/0263-7855(92)80074-N PMID:1476996

Lewis, P. N., Kotelchuck, D., & Scheraga, H. A. (1970). Helix probability profiles of denatured proteins and their correlation with native structures. *Proceedings of the National Academy of Sciences of the United States of America*, *65*(4), 810–815. doi:10.1073/pnas.65.4.810 PMID:5266152

Lewis, R. A., & Leach, A. R. (1994). Current methods for site-directed structure generation. *Journal of Computer-Aided Molecular Design*, *8*(4), 467–475. doi:10.1007/BF00125381 PMID:7815098

Lewis, R. A., Roe, D. C., Huang, C., Ferrin, T. E., Langridge, R., & Kuntz, I. D. (1992). Automated site-directed drug design using molecular lattices. *Journal of Molecular Graphics*, *10*(2), 66–78. doi:10.1016/0263-7855(92)80059-M PMID:1637751

Lexa, K. W., & Carlson, H. A. (2012). Protein flexibility in docking and surface mapping. *Quarterly Reviews of Biophysics*, *45*(03), 301–343. doi:10.1017/S0033583512000066 PMID:22569329

Librando, V., & Forte, S. (2005). Computer evaluation of protein segments removal effects from naphthalene 1, 2-dioxygenase enzyme on polycyclic aromatic hydrocarbons interaction. *Biochemical Engineering Journal*, *27*(2), 161–166. doi:10.1016/j.bej.2005.08.011

Li, C. H., Ma, X. H., Chen, W. Z., & Wang, C. X. (2003). A soft docking algorithm for predicting the structure of antibody-antigen complexes. *Proteins: Structure, Function, and Bioinformatics*, *52*(1), 47–50. doi:10.1002/prot.10382 PMID:12784367

Li, C. H., Ma, X. H., Zu Chen, W., & Wang, C. X. (2003). A protein–protein docking algorithm dependent on the type of complexes. *Protein Engineering*, *16*(4), 265–269. doi:10.1093/proeng/gzg035 PMID:12736369

Li, D. D., Qin, Y. J., Zhang, X., Yin, Y., Zhu, H. L., & Zhao, L. G. (2015). Combined Molecular Docking, 3D-QSAR, and Pharmacophore Model: Design of Novel Tubulin Polymerization Inhibitors by Binding to Colchicine-binding Site. *Chemical Biology & Drug Design*. doi:10.1111/cbdd.12545 PMID:25711282

Li, D. J., Zhang, T., Xu, C., & Ji, B. M. (2011). Effect of pH on the interaction of baicalein with lysozyme by spectroscopic approaches. *Journal of Photochemistry and Photobiology. B, Biology*, *104*(3), 414–424. doi:10.1016/j.jphotobiol.2011.04.009 PMID:21570863

Li, H. J., & Crothers, D. M. (1969). Relaxation studies of the proflavine DNA complex: The kinetics of an intercalation reaction. *Journal of Molecular Biology*, *39*(3), 461–477. doi:10.1016/0022-2836(69)90138-7 PMID:5390557

Li, H., Kasam, V., Tautermann, C. S., Seeliger, D., & Vaidehi, N. (2014). Computational method to identify druggable binding sites that target protein-protein interactions. *Journal of Chemical Information and Modeling*, *54*(5), 1391–1400. doi:10.1021/ci400750x PMID:24762202

Li, H., & Li, C. (2010). Multiple ligand simultaneous docking: Orchestrated dancing of ligands in binding sites of protein. *Journal of Computational Chemistry*, *31*(10), 2014–2022. PMID:20166125

Li, J., Edwards, P. C., Burghammer, M., Villa, C., & Schertler, G. F. X. (2004). Structure of bovine rhodopsin in a trigonal crystal form. *Journal of Molecular Biology*, *343*(5), 1409–1438. doi:10.1016/j.jmb.2004.08.090 PMID:15491621

Lill, M. A. (2007). Multi-dimensional QSAR in drug discovery. *Drug Discovery Today*, *12*(23), 1013–1017. doi:10.1016/j.drudis.2007.08.004 PMID:18061879

Lindahl, E., & Sansom, M. S. (2008). Membrane proteins: Molecular dynamics simulations. *Current Opinion in Structural Biology*, *18*(4), 425–431. doi:10.1016/j.sbi.2008.02.003 PMID:18406600

Lingel, A., Simon, B., Izaurralde, E., & Sattler, M. (2004). Nucleic acid 3'-end recognition by the Argonaute2 PAZ domain. *Nature Structural & Molecular Biology*, *11*(6), 576–577. doi:10.1038/nsmb777 PMID:15156196

Lin, L., Wei, D. Q., Wang, J. F., & Chou, K. C. (2007). Computational studies of the binding mechanism of calmodulin with chrysin. *Biochemical and Biophysical Research Communications*, *358*(4), 1102–1107. doi:10.1016/j.bbrc.2007.05.053 PMID:17521610

Lin, S. Y., Zhang, G. W., Liao, Y. J., Pan, J. H., & Gong, D. M. (2015). Dietary flavonoids as xanthine oxidase inhibitors: Structure-affinity and structure-activity relationships. *Journal of Agricultural and Food Chemistry*, *63*(35), 7784–7794. doi:10.1021/acs.jafc.5b03386 PMID:26285120

Lionta, E., Spyrou, G., Vassilatis, D. K., & Cournia, Z. (2014). Structure-based virtual screening for drug discovery: Principles, applications and recent advances. *Current Topics in Medicinal Chemistry*, *14*(16), 1923–1938. doi:10.2174/1568026614666140929124445 PMID:25262799

Lipinski, C. A. (2004). Lead- and drug-like compounds: The rule-of-five revolution. *Drug Discovery Today. Technologies, 1*(4), 337–341. doi:10.1016/j.ddtec.2004.11.007 PMID:24981612

Lipinski, C. A., Lombardo, F., Dominy, B. W., & Feeney, P. J. (2001). Experimental and computational approaches to estimate solubility and permeability in drug discovery and development settings1. *Advanced Drug Delivery Reviews, 46*(1–3), 3–26. doi:10.1016/S0169-409X(00)00129-0 PMID:11259830

Lisgarten, J. N., Coll, M., Portugal, J., Wright, C. W., & Aymami, J. (2002). The antimalarial and cytotoxic drug cryptolepine intercalates into DNA at cytosine-cytosine sites.[Letter to the editor]. *Nature Structural Biology, 9*(1), 57–60. doi:10.1038/nsb729 PMID:11731803

Liu, G. Y., Guo, B. Q., Chen, W. N., Cheng, C., Zhang, Q. L., Dai, M. B., & Chen, W. M. et al. (2012). Synthesis, Molecular Docking, and Biofilm Formation Inhibitory Activity of 5-Substituted 3, 4-Dihalo-5H-furan-2-one Derivatives on Pseudomonas aeruginosa. *Chemical Biology & Drug Design, 79*(5), 628–638. doi:10.1111/j.1747-0285.2012.01342.x PMID:22268453

Liu, J., Wang, X. R., Cai, Z. W., & Lee, F. S. C. (2008). Effect of tanshinone HA on the noncovalent interaction between warfarin and human serum albumin studied by electrospray ionization mass spectrometry. *Journal of the American Chemical Society, 19*, 1568–1575. PMID:18657993

Liu, M. H., Lin, Y. S., Sheu, S. Y., & Sun, J. S. (2009). Anti-inflammatory effects of daidzein on primary astroglial cell culture. *Nutritional Neuroscience, 12*(3), 123–134. doi:10.1179/147683009X423274 PMID:19356315

Liu, M., Yin, H., Dong, J. J., Xiao, L., Liu, G., Qian, Z. H., & Miao, J. L. (2013). Inhibition and interaction with hyaluronidase by compounds from Hop (*Humulus iupulus* L.) flowers. *Asian Journal of Chemistry, 25*, 10262–10266.

Liu, S., & Kurzrock, R. (2014). Toxicity of targeted therapy: Implications for response and impact of genetic polymorphisms. *Cancer Treatment Reviews, 40*(7), 883–891. doi:10.1016/j.ctrv.2014.05.003 PMID:24867380

Liu, X. H., Xi, P. X., Chen, F. J., Xu, Z. H., & Zeng, Z. Z. (2008). Spectroscopic studies on binding of 1-phenyl-3-(coumarin-6-yl) sulfonylurea to bovine serum albumin. *Journal of Photochemistry and Photobiology. B, Biology, 92*(2), 98–102. doi:10.1016/j.jphotobiol.2008.04.008 PMID:18571426

Liu, X., Ouyang, S., Yu, B., Liu, Y., Huang, K., Gong, J., & Jiang, H. et al. (2010). PharmMapper server: A web server for potential drug target identification using pharmacophore mapping approach. *Nucleic Acids Research, 38*(Web Server-suppl 2), W609–W614. doi:10.1093/nar/gkq300 PMID:20430828

Li, X., Keskin, O., Ma, B., Nussinov, R., & Liang, J. (2004). Protein-protein interactions: hot spots and structurally conserved residues often locate in complemented pockets that pre-organized in the unbound states: implications for docking. *Journal of Molecular Biology, 344*(3), 781–795. doi:10.1016/j.jmb.2004.09.051 PMID:15533445

Li, Y., Shen, J., Sun, X., Li, W., Liu, G., & Tang, Y. (2010). Accuracy assessment of protein-based docking protocols against RNA targets. *Journal of Chemical Information and Modeling, 50*(6), 1134–1146. doi:10.1021/ci9004157 PMID:20481574

Li, Y., Yao, X., Jin, J., Chen, X., & Hu, Z. (2007). Interaction of rhein with human serum albumin investigation by optic spectroscopic technique and modeling studies. *Biochimca et Biophysica Acta-general Subjects, 1774*(1), 51–58. doi:10.1016/j.bbapap.2006.09.020 PMID:17174615

Li, Z., & Lazaridis, T. (2003). Thermodynamic contributions of the ordered water molecule in HIV-1 protease. *Journal of the American Chemical Society, 125*(22), 6636–6637. doi:10.1021/ja0299203 PMID:12769565

Lo Conte, L., Chothia, C., & Janin, J. (1999). The atomic structure of protein-protein recognition sites. *Journal of Molecular Biology, 285*(5), 2177–2198. doi:10.1006/jmbi.1998.2439 PMID:9925793

Lorber, D. M., & Shoichet, B. K. (1998). Flexible ligand docking using conformational ensembles. *Protein Science*, *7*(4), 938–950. doi:10.1002/pro.5560070411 PMID:9568900

Lorenzi, S., Mor, M., Bordi, F., Rivara, S., Rivara, M., Morini, G., & Plazzi, P. V. et al. (2005). Validation of a histamine H3 receptor model through structure-activity relationships for classical H3 antagonists. *Bioorganic & Medicinal Chemistry*, *13*(19), 5647–5657. doi:10.1016/j.bmc.2005.05.072 PMID:16085419

Lovley, D. R. (2003). Cleaning up with genomics: Applying molecular biology to bioremediation. *Nature Reviews. Microbiology*, *1*(1), 35–44. doi:10.1038/nrmicro731 PMID:15040178

Lui, A. E. (2007). *Engineering of aniline dioxygenase for bioremediation and industrial applications.* Academic Press.

Lu, Q. H., Ba, C. D., & Chen, D. Y. (2008). Investigating noncovalent interactions of rutin-serum albumin by capillary electrophoresis-frontal analysis. *Journal of Pharmaceutical and Biomedical Analysis*, *47*(4-5), 888–891. doi:10.1016/j.jpba.2008.03.007 PMID:18436413

Lu, X.-J., & Olson, W. K. (2008). 3DNA: A versatile, integrated software system for the analysis, rebuilding and visualization of three-dimensional nucleic-acid structures. *Nature Protocols*, *3*(7), 1213–1227. doi:10.1038/nprot.2008.104 PMID:18600227

Ma, B., Elkayam, T., Wolfson, H., & Nussinov, R. (2003). Protein-protein interactions: Structurally conserved residues distinguish between binding sites and exposed protein surfaces. *Proceedings of the National Academy of Sciences of the United States of America*, *100*(10), 5772–5777. doi:10.1073/pnas.1030237100 PMID:12730379

MacKerell, A. D. Jr, Bashford, D., Bellott, M., Dunbrack, R. L. Jr, Evanseck, J. D., Field, M. J., & Karplus, M. et al. (1998). All-Atom Empirical Potential for Molecular Modeling and Dynamics Studies of Proteins. *The Journal of Physical Chemistry B*, *102*(18), 3586–3616. doi:10.1021/jp973084f PMID:24889800

Magnani, F., Pappas, C. G., Crook, T., Magafa, V., Cordopatis, P., Ishiguro, S., & Tzakos, A. G. et al. (2014). Electronic sculpting of ligand-GPCR subtype selectivity: The case of angiotensin II. *ACS Chemical Biology*, *9*(7), 1420–1425. doi:10.1021/cb500063y PMID:24787922

Mahajan, A., Gill, N.S., & Arora, R. (2014). *A Review on Molecular Docking*. Academic Press.

Mahesha, H. G., Singh, S. A., Srinivasan, N., & Rao, A. G. A. (2006). A spectroscopic study of the interaction of isoflavones with human serum albumin. *The FEBS Journal*, *273*(3), 451–467. doi:10.1111/j.1742-4658.2005.05071.x PMID:16420470

Ma, J., Yin, Y. M., Liu, H. L., & Xie, M. X. (2011). Interactions of flavonoids with biomacromolecules. *Current Organic Chemistry*, *15*(15), 2627–2640. doi:10.2174/138527211796367345

Manglik, A., Kruse, A. C., Kobilka, T. S., Thian, F. S., Mathiesen, J. M., Sunahara, R. K., & Granier, S. et al. (2012). Crystal structure of the μ-opioid receptor bound to a morphinan antagonist. *Nature*, *485*(7398), 321–326. doi:10.1038/nature10954 PMID:22437502

Mangoni, M., Roccatano, D., & Di Nola, A. (1999). Docking of flexible ligands to flexible receptors in solution by molecular dynamics simulation. *Proteins: Structure, Function, and Bioinformatics*, *35*(2), 153–162. doi:10.1002/(SICI)1097-0134(19990501)35:2<153::AID-PROT2>3.0.CO;2-E PMID:10223288

Manning, G., Whyte, D. B., Martinez, R., Hunter, T., & Sudarsanam, S. (2002). The protein kinase complement of the human genome. *Science*, *298*(5600), 1912–1934. doi:10.1126/science.1075762 PMID:12471243

Marchler-Bauer, A., Derbyshire, M. K., Gonzalez, N. R., Lu, S., Chitsaz, F., Geer, L. Y., & Bryant, S. H. et al. (2015). CDD: NCBI's conserved domain database. *Nucleic Acids Research*, *43*(D), 222–226. doi:10.1093/nar/gku1221 PMID:25414356

Marrink, S. J., De Vries, A. H., & Mark, A. E. (2004). Coarse grained model for semiquantitative lipid simulations. *The Journal of Physical Chemistry B*, *108*(2), 750–760. doi:10.1021/jp036508g

Martínez-Pacheco, H., Espinosa-Raya, J., Picazo-Picazo, O., Viñas-Bravo, O., & Ramírez-Galicia, G. (2015). *Rational design, docking and QSAR studies of a series of histone deacetylase isoforma 2 inhibitor*. In preparation.

Martínez-Pacheco, H. (2012). *Evaluación in silico –Docking y QSAR- de derivados del ácido valproico sobre la HDAC8 con potencial uso antineoplástico (Evaluation in silico –Docking and QSAR- of valproic acid derivatives on HDAC8 as potential anti-neoplastic)*. Universidad del Papaloapan.

Martínez-Pacheco, H. (2015). *Diseño, síntesis y evaluación de efectos bioquímicos y conductuales de posibles inhibidores de la desacetilación de histonas (Desing, synthesis and biochemistry and behavior efects evaluation of inhibitors of histon desacetilation)*. Escuela Superior de Medicina Instituto Politécnico Nacional.

Marti-Renom, M. A., Stuart, A. C., Fiser, A., Sanchez, R., Melo, F., & Sali, A. (2000). Comparative protein structure modeling of genes and genomes. *Annual Review of Biophysics and Biomolecular Structure*, *29*(1), 291–325. doi:10.1146/annurev.biophys.29.1.291 PMID:10940251

Mashiach E., Schneidman-Duhovny D., Andrusier N., Nussinov, R. Wolfson H. J. (2008). FireDock: a web server for fast interaction refinement in molecular docking. *Nucleic Acids Res.*, *36*, W229-32.

Masood, M. M., Pillalamarri, V. K., Irfan, M., Aneja, B., Jairajpuri, M. A., Zafaryab, M., & Abid, M. et al. (2015). Diketo acids and their amino acid/dipeptidic analogues as promising scaffolds for the development of bacterial methionine aminopeptidase inhibitors. *RSC Advances*, *5*(43), 34173–34183. doi:10.1039/C5RA03354C

Mathews, D. H. (2014). Article. *Current Protocols in Bioinformatics, 46*, 12.6.1-12.6.25.

Matsunaga, T., & Lim, T.-k. (2000). Electrochemical prevention of biofouling. *Electrochemistry*, *68*(11), 847–852.

Ma, W., Tang, C., & Lai, L. (2005). Specificity of trypsin and chymotrypsin: Loop-motion-controlled dynamic correlation as a determinant. *Biophysical Journal*, *89*(2), 1183–1193. doi:10.1529/biophysj.104.057158 PMID:15923233

Mayo, S. L., Olafson, B. D., & Goddard, W. A. III. (1990). DREIDING - a generic force field for molecular simulations. *Journal of Physical Chemistry*, *94*(26), 8897–8909. doi:10.1021/j100389a010

McGann, M. R., Almond, H. R., Nicholls, A., Grant, J. A., & Brown, F. K. (2003). Gaussian docking functions. *Biopolymers*, *68*(1), 76–90. doi:10.1002/bip.10207 PMID:12579581

McMartin, C., & Bohacek, R. S. (1997). QXP: Powerful, rapid computer algorithms for structure-based drug design. *Journal of Computer-Aided Molecular Design*, *11*(4), 333–344. doi:10.1023/A:1007907728892 PMID:9334900

McQuown, S., Barrett, R. M., Matheos, D. P., Post, R. J., Rogge, G. A., Alenghat, T., & Wood, M. A. et al. (2011). HDAC3 is a critical negative regulator of long-term memory formation. *The Journal of Neuroscience*, *31*(2), 764–774. doi:10.1523/JNEUROSCI.5052-10.2011 PMID:21228185

McWilliam, H., Li, W., Uludag, M., Squizzato, S., Park, Y.M., Buso, N., Cowley, A.P., & Lopez, R. (2013). Analysis Tool Web Services from the EMBL-EBI. *Nucleic Acids Research, 41*, 597-600. DOI: 10.1093/nar/gkt376

Megumi, F. T., Kei, N., Kenji, T., Tadahiko, M., & Tadashi, K. (2011). Anti-inflammatory activity of structurally related flavonoids, apigenin, luteolin and fisetin. *International Immunopharmacology*, *11*(9), 1150–1159. doi:10.1016/j.intimp.2011.03.012 PMID:21443976

Mei, D., Yin, Y., Wu, F., Cui, J., Zhou, H., Sun, G., & Feng, Y. et al. (2015). Discovery of potent and selective urea-based ROCK inhibitors: Exploring the inhibitor's potency and ROCK2/PKA selectivity by 3D-QSAR, molecular docking and molecular dynamics simulations. *Bioorganic & Medicinal Chemistry*, *23*(10), 2505–2517. doi:10.1016/j.bmc.2015.03.047 PMID:25882521

Meiler, J., & Baker, D. (2006). ROSETTALIGAND: Protein–small molecule docking with full side-chain flexibility. *Proteins: Structure, Function, and Bioinformatics*, *65*(3), 538–548. doi:10.1002/prot.21086 PMID:16972285

Melo-Filho, C. C., Braga, R. C., & Andrade, C. H. (2014). 3D-QSAR approaches in drug design: Perspectives to generate reliable CoMFA models. *Current Computer-aided Drug Design*, *10*(2), 148–159. doi:10.2174/1573409910666140410111043 PMID:24724896

Mendilaharsu, M., De Stefani, E., Deneo-Pellegrini, H., Carzoglio, J. C., & Ronco, A. (1998). Consumption of tea and coffee and the risk of lung cancer in cigarette-smoking men: A case-control study in Uruguay. *Lung Cancer (Amsterdam, Netherlands)*, *19*(2), 101–107. doi:10.1016/S0169-5002(97)00075-5 PMID:9567246

Meng, X.-Y., Zhang, H.-X., Mezei, M., & Cui, M. (2011). Molecular Docking: A powerful approach for structure-based drug discovery. *Curr Comput Aided Drug Des.*, *7*(2), 146–157. doi:10.2174/157340911795677602 PMID:21534921

Mezei, M. (2003). A new method for mapping macromolecular topography. *Journal of Molecular Graphics & Modelling*, *21*(5), 463–472. doi:10.1016/S1093-3263(02)00203-6 PMID:12543141

Mihasan, M. (2012). What *in silico* molecular docking can do for the 'bench-working biologists'. *Journal of Biosciences*, *37*(6), 1089–1095. doi:10.1007/s12038-012-9273-8 PMID:23151798

Miller, R. L., Eppinger, T. M., McConnell, D., Cunningham-Rundles, C., & Rothman, P. (1998). Analysis of cytokine signaling in patients with extrinsic asthma and hyperimmunoglobulin E. *The Journal of Allergy and Clinical Immunology*, *102*(3), 503–511. doi:10.1016/S0091-6749(98)70141-1 PMID:9768594

Miranker, A., & Karplus, M. (1995). An automated method for dynamic ligand design. *Proteins: Structure, Function, and Bioinformatics*, *23*(4), 472–490. doi:10.1002/prot.340230403 PMID:8749844

Miroshnychenko, K. V., & Shestopalova, A. V. (2013). The study of different binding modes of proflavine with DNA and RNA sequences by molecular docking method: evidence for a proflavine as a minor groove binder. In Book of Abstracts of the 5th International Symposium. Kharkiv, Ukraine: Methods and Applications of Computational Chemistry. Retrieved from http://macc.org.ua/MACC-5/MACC5colored.pdf

Miroshnychenko, K. V., & Shestopalova, A. V. (2010). The effect of drug-DNA interactions on the intercalation site formation. *International Journal of Quantum Chemistry*, *110*(1), 161–176. doi:10.1002/qua.22290

Misra, N., Patra, M. C., Panda, P. K., Sukla, L. B., & Mishra, B. K. (2013). Homology modeling and docking studies of FabH (β-ketoacyl-ACP synthase III) enzyme involved in type II fatty acid biosynthesis of Chlorella variabilis: A potential algal feedstock for biofuel production. *Journal of Biomolecular Structure & Dynamics*, *31*(3), 241–257. doi:10.1080/07391102.2012.698247 PMID:22830394

Mohan, V., Gibbs, A. C., Cummings, M. D., Jaeger, E. P., & DesJarlais, R. L. (2005). Docking: Successes and challenges. *Current Pharmaceutical Design*, *11*(3), 323–333. doi:10.2174/1381612053382106 PMID:15723628

Moitessier, N., Englebienne, P., Lee, D., Lawandi, J., & Corbeil, C. R. (2008). Towards the development of universal, fast and highly accurate docking/scoring methods: A long way to go. *British Journal of Pharmacology*, *153*(S1), S7–S26. doi:10.1038/sj.bjp.0707515 PMID:18037925

Molecular Operating Environment (MOE). (2013). Chemical Computing Group Inc.

Momand, J., Wu, H. H., & Dasgupta, G. (2000). MDM2--master regulator of the p53 tumor suppressor protein. *Gene*, *242*(1-2), 15–29. doi:10.1016/S0378-1119(99)00487-4 PMID:10721693

Montañez-Godinez, N., Martínez-Olguin, A. C., Deeb, O., Garduño-Juárez, R., & Ramírez-Galicia, G. (2014). *QSAR/QSPR as an application of artifitial neural network. Artifitial Neuronal Networks* (pp. 319–333). Springer New York.

Montgomery, D. C., & Peck, E. A. (1992). *Introduction to linear regression analysis*. New York: Wiley.

Monto, A. S., Fleming, D. M., Henry, D., De Groot, R., Makela, M., Klein, T., & Man, C. Y. et al. (1999). Efficacy and safety of the neuraminidase inhibitor zanamivir in the treatment of influenza A and B virus infections. *The Journal of Infectious Diseases*, *180*(2), 254–261. doi:10.1086/314904 PMID:10395837

Moon, S. H., Kim, K. T., & Lee, N. K. (2009). Inhibitory effects of naringenin and its novel derivatives on hyaluronidase. *Food Science and Biotechnology*, *18*, 267–270.

Moreira, I. S., Fernandes, P. A., & Ramos, M. J. (2007). Hot spots--a review of the protein-protein interface determinant amino-acid residues. *Proteins*, *68*(4), 803–8012. doi:10.1002/prot.21396 PMID:17546660

Morelli, X., Dolla, A., Czjzek, M., Palma, P. N., Blasco, F., Krippahl, L., Moura, J.J.G., Guerlesquin, F. (2000). Heteronuclear NMR and soft docking: an experimental approach for a structural model of the cytochrome c 553-ferredoxin complex. *Biochemistry, 39*(10), 2530-2537.

Morelli, X. J., Palma, P. N., Guerlesquin, F., & Rigby, A. C. (2001). A novel approach for assesing macromolecular complexes combining soft-docking calculations with NMR data. *Protein Science*, *10*(10), 2131–2137. doi:10.1110/ps.07501 PMID:11567104

Morelli, X., Czjzek, M., Hatchikian, C. E., Bornet, O., Fontecilla-Camps, J. C., Palma, N. P., & Guerlesquin, F. et al. (2000). Structural Model of the Fe-Hydrogenase/Cytochromec 553 Complex Combining Transverse Relaxation-optimized Spectroscopy Experiments and Soft Docking Calculations. *The Journal of Biological Chemistry*, *275*(30), 23204–23210. doi:10.1074/jbc.M909835199 PMID:10748163

Morley, S. D., & Afshar, M. (2004). Validation of an empirical RNA-ligand scoring function for fast flexible docking using RiboDock®. *Journal of Computer-Aided Molecular Design*, *18*(3), 189–208. doi:10.1023/B:JCAM.0000035199.48747.1e PMID:15368919

Morphy, J. R., & Harris, C. J. (2012). *Designing multi-target drugs* (Vol. 21). Royal Society of Chemistry. doi:10.1039/9781849734912

Morris, G. M., Goodsell, D. S., Halliday, R. S., Huey, R., Hart, W. E., Belew, R. K., & Olson, A. J. (1998). Automated docking using a Lamarckian genetic algorithm and an empirical binding free energy function. *Journal of Computational Chemistry*, *19*(14), 1639–1662. doi:10.1002/(SICI)1096-987X(19981115)19:14<1639::AID-JCC10>3.0.CO;2-B

Morris, G. M., Huey, R., Lindstrom, W., Sanner, M. F., Belew, R. K., Goodsell, D. S., & Olson, A. J. (2009). AutoDock4 and AutoDockTools4: Automated Docking with Selective Receptor Flexibility. *Journal of Computational Chemistry*, *30*(16), 2785–2791. doi:10.1002/jcc.21256 PMID:19399780

Mortier, J., Rakers, C., Bermudez, M., Murgueitio, M. S., Riniker, S., & Wolber, G. (2015). The impact of molecular dynamics on drug design: Applications for the characterization of ligand-macromolecule complexes. *Drug Discovery Today*, *20*(6), 686–702. doi:10.1016/j.drudis.2015.01.003 PMID:25615716

Morton, A., & Matthews, B. W. (1995). Specificity of ligand binding in a buried nonpolar cavity of T4 lysozyme: linkage of dynamics and structural plasticity. *Biochemistry, 34*(27), 8576-8588.

Moustakas, D. T., Lang, P. T., Pegg, S., Pettersen, E., Kuntz, I. D., Brooijmans, N., & Rizzo, R. C. (2006). Development and validation of a modular, extensible docking program: DOCK 5. *Journal of Computer-Aided Molecular Design, 20*(10-11), 601–619. doi:10.1007/s10822-006-9060-4 PMID:17149653

Muegge, I., & Martin, Y. C. (1999). A general and fast scoring function for protein-ligand interactions: A simplified potential approach. *Journal of Medicinal Chemistry*, *42*(5), 791–804. doi:10.1021/jm980536j PMID:10072678

Muller, S., Filippakopoulos, P., & Knapp, S. (2011). Bromodomains as therapeutic targets. *Expert Reviews in Molecular Medicine*, *13*, e29. doi:10.1017/S1462399411001992 PMID:21933453

Murray, B. W., & Miller, N. (2015). Durability of Kinase-Directed Therapies-A Network Perspective on Response and Resistance. *Molecular Cancer Therapeutics*, *14*(9), 1975–1984. doi:10.1158/1535-7163.MCT-15-0088 PMID:26264276

Musk, J., Dinty, J., & Hergenrother, P. J. (2006). Chemical countermeasures for the control of bacterial biofilms: Effective compounds and promising targets. *Current Medicinal Chemistry*, *13*(18), 2163–2177. doi:10.2174/092986706777935212 PMID:16918346

Natt. (2011). *BMC Ear, Nose and Throat Disorders*. Retrieved from http://www.biomedcentral.com/1472-6815/11/3

Nedumpully Govindan, P. (2013). *Computational Modelling Investigating the Interactions of Proteins with Fullerene based Nanoparticles*. Academic Press.

Neese, F. (2012). The ORCA program system. *Wiley Interdisciplinary Reviews: Computational Molecular Science*, *2*(1), 73–78. doi:10.1002/wcms.81

Neidle, S. (2001). DNA minor-groove recognition by small molecules. *Natural Product Reports*, *18*(3), 291–309. doi:10.1039/a705982e PMID:11476483

Ngan, C. H., Bohnuud, T., Mottarella, S. E., Beglov, D., Villar, E. A., Hall, D. R., . . . Vajda, S. (2012). FTMAP: extended protein mapping with user-selected probe molecules. *Nucleic Acids Res, 40*(Web Server issue), W271-275. doi: 10.1093/nar/gks441

Ng, P. C., Henikoff, J. G., & Henikoff, S. (2000). PHAT: A transmembrane-specific substitution matrix. *Bioinformatics (Oxford, England)*, *16*(9), 760–766. doi:10.1093/bioinformatics/16.9.760 PMID:11108698

Nguyen, T. L., McGrath, C., Hermone, A. R., Burnett, J. C., Zaharevitz, D. W., Day, B. W., & Gussio, R. et al. (2005). A common pharmacophore for a diverse set of colchicine site inhibitors using a structure-based approach. *Journal of Medicinal Chemistry*, *48*(19), 6107–6116. doi:10.1021/jm050502t PMID:16162011

Nichols, D. E., & Nichols, C. D. (2008). Serotonin receptors. *Chemical Reviews*, *108*(5), 1614–1641. doi:10.1021/cr078224o PMID:18476671

Norgan, A. P., Coffman, P. K., Kocher, J.-P. A., Katzmann, D. J., & Sosa, C. P. (2011). Multilevel parallelization of Autodock 4.2. *Journal of Cheminformatics*, *3*(12). doi:10.1186/1758-2946-3-12 PMID:21527034

Nyberg, P., Ylipalosaari, M., Sorsa, T., & Salo, T. (2006). Trypsins and their role in carcinoma growth. *Experimental Cell Research*, *312*(8), 1219–1228. doi:10.1016/j.yexcr.2005.12.024 PMID:16457812

Nygaard, R., Zou, Y., Dror, R. O., Mildorf, T. J., Arlow, D. H., Manglik, A., & Kobilka, B. K. et al. (2013). The dynamic process of β 2-adrenergic receptor activation. *Cell*, *152*(3), 532–542. doi:10.1016/j.cell.2013.01.008 PMID:23374348

Nymeyer, H., Woolf, T. B., & Garcia, A. E. (2005). Folding is not required for bilayer insertion: Replica exchange simulations of an α-helical peptide with an explicit lipid bilayer. Proteins. *Structure, Function, and Bioinformatics*, *59*(4), 783–790. doi:10.1002/prot.20460

Oberai, A., Ihm, Y., Kim, S., & Bowie, J. U. (2006). A limited universe of membrane protein families and folds. *Protein Science*, *15*(7), 1723–1734. doi:10.1110/ps.062109706 PMID:16815920

O'Boyle, N. M., Banck, M., James, C. A., Morley, C., Vandermeersch, T., & Hutchison, G. R. (2011). Open Babel: An open chemical toolbox. *Journal of Cheminformatics*, *3*(33). doi:10.1186/1758-2946-3-33 PMID:21982300

O'Brien, Z., & Fallah Moghaddam, M. (2013). Small molecule kinase inhibitors approved by the FDA from 2000 to 2011: A systematic review of preclinical ADME data. *Expert Opinion on Drug Metabolism & Toxicology*, *9*(12), 1597–1612. doi:10.1517/17425255.2013.834046 PMID:23992134

O'Hare, T., Eide, C. A., & Deininger, M. W. (2007). Bcr-Abl kinase domain mutations, drug resistance, and the road to a cure for chronic myeloid leukemia. *Blood*, *110*(7), 2242–2249. doi:10.1182/blood-2007-03-066936 PMID:17496200

Okada, T., Fujiyoshi, Y., Silow, M., Navarro, J., Landau, E. M., & Shichida, Y. (2002). Fundamental role of internal water molecules in rhodopsin revealed by Xray crystallography. *Proceedings of the National Academy of Sciences of the United States of America*, *99*(9), 5982–5987. doi:10.1073/pnas.082666399 PMID:11972040

Okada, T., Sugihara, M., Bondar, A. N., Elstner, M., Entel, P., & Buss, V. (2004). The retinal conformation and its environment in rhodopsin in light of a new 2.2 Å crystal structure. *Journal of Molecular Biology*, *342*(2), 571–583. doi:10.1016/j.jmb.2004.07.044 PMID:15327956

Okoli, C., Sengottaiyan, S., Arul Murugan, N., Pavankumar, A. R., Ågren, H., & Kuttuva Rajarao, G. (2013). In silico modeling and experimental evidence of coagulant protein interaction with precursors for nanoparticle functionalization. *Journal of Biomolecular Structure & Dynamics*, *31*(10), 1182–1190. doi:10.1080/07391102.2012.726534 PMID:23163338

Okumura, A., Pitha, P. M., Yoshimura, A., & Harty, R. N. (2010). Interaction between Ebola virus glycoprotein and host toll-like receptor 4 leads to induction of proinflammatory cytokines and SOCS1. *Journal of Virology*, *1984*(1), 27–33. doi:10.1128/JVI.01462-09 PMID:19846529

Olah, M. E., & Stiles, G. L. (1995). Adenosine receptor subtypes: Characterization and therapeutic regulation *Annu. Rev. Phamloeol. Toxieol.*, *35*, 581–606. PMID:7598508

Olano, L. R., & Rick, S. W. (2004). Hydration free energies and entropies for water in protein interiors. *Journal of the American Chemical Society*, *126*(25), 7991–8000. doi:10.1021/ja049701c PMID:15212549

Olsen, S. M., Kiil, S., Dam-Johansen, K., Pedersen, L. T., & Hermann, M. H. (2009). *Controlled release of environmentally friendly antifouling agents from marine coatings*. Hempel A/SHempel A/S.

Omae, I. (2003). Organotin antifouling paints and their alternatives. *Applied Organometallic Chemistry*, *17*(2), 81–105. doi:10.1002/aoc.396

Osborn, G. G., & Saunders, A. V. (2010). Current treatments for patients with Alzheimer disease. *The Journal of the American Osteopathic Association*, *110*, s16–s26. PMID:20926739

Oshiro, C. M., Kuntz, I. D., & Dixon, J. S. (1995). Flexible ligand docking using a genetic algorithm. *Journal of Computer-Aided Molecular Design*, *9*(2), 113–130. doi:10.1007/BF00124402 PMID:7608743

Österberg, F., Morris, G. M., Sanner, M. F., Olson, A. J., & Goodsell, D. S. (2002). Automated docking to multiple target structures: Incorporation of protein mobility and structural water heterogeneity in AutoDock. *Proteins: Structure, Function, and Bioinformatics*, *46*(1), 34–40. doi:10.1002/prot.10028 PMID:11746701

Ota, F., Monika, A., & Roman, J. (2012). Suppression of TLR Signaling by Targeting TIR domain-Containing Proteins. *Current Protein & Peptide Science*, *13*(8), 776–788. doi:10.2174/138920312804871148 PMID:23305364

Palczewski, K., Kumasaka, T., Hori, T., Behnke, C.A., Motoshima, H. (2000). Crystal structure of rhodopsin: a G protein-coupled receptor. *Sci.*, 739-745.

Palma, P. N., Krippahl, L., Wampler, J. E., & Moura, J. J. (2000). BiGGER: A new (soft) docking algorithm for predicting protein interactions. *Proteins: Structure, Function, and Bioinformatics*, *39*(4), 372–384. doi:10.1002/(SICI)1097-0134(20000601)39:4<372::AID-PROT100>3.0.CO;2-Q PMID:10813819

Pande, M., Kumari, N. K. P., Dubey, V. K., Tripathi, P., & Jagannadham, M. V. (2009). Stability and unfolding studies on alkaline denatured state (I_p) of pepsin. *Process Biochemistry*, *44*(8), 906–911. doi:10.1016/j.procbio.2009.04.016

Parales, R., Bruce, N., Schmid, A., & Wackett, L. (2002). Biodegradation, biotransformation, and biocatalysis (B3). *Applied and Environmental Microbiology*, *68*(10), 4699–4709. doi:10.1128/AEM.68.10.4699-4709.2002 PMID:12324310

Pasqualoto, K. F. M., & Ferreira, M. M. C. (2009). Molecular Modeling and Receptor-Dependent (RD) 3D-QSAR Approach to a Set of Antituberculosis Derivatives. *QSAR & Combinatorial Science*, *28*(11-12), 1455–1464. doi:10.1002/qsar.200960035

Paul, D. (1993). Local and nonlocal interactions in globular proteins and mechanisms of alcohol denaturation. *Protein Science*, *2*(12), 2050–2065. doi:10.1002/pro.5560021206 PMID:8298455

Paul, D., Pandey, G., Pandey, J., & Jain, R. K. (2005). Accessing microbial diversity for bioremediation and environmental restoration. *Trends in Biotechnology*, *23*(3), 135–142. doi:10.1016/j.tibtech.2005.01.001 PMID:15734556

Pazos, F., Guijas, D., Valencia, A., & De Lorenzo, V. (2005). MetaRouter: Bioinformatics for bioremediation. *Nucleic Acids Research*, *33*(suppl 1), D588–D592. doi:10.1093/nar/gki068 PMID:15608267

Pearson, G., Robinson, F., Beers Gibson, T., Xu, B. E., Karandikar, M., Berman, K., & Cobb, M. H. (2001). Mitogen-activated protein (MAP) kinase pathways: Regulation and physiological functions. *Endocrine Reviews*, *22*(2), 153–183. doi:10.1210/edrv.22.2.0428 PMID:11294822

Pei, J., Wang, Q., Liu, Z., Li, Q., Yang, K., & Lai, L. (2006). PSI-DOCK: Towards highly efficient and accurate flexible ligand docking. *Proteins: Structure, Function, and Bioinformatics*, *62*(4), 934–946. doi:10.1002/prot.20790 PMID:16395666

Pei, J., Yin, N., Ma, X., & Lai, L. (2014). Systems biology brings new dimensions for structure-based drug design. *Journal of the American Chemical Society*, *136*(33), 11556–11565. doi:10.1021/ja504810z PMID:25061983

Pennacchio, F. (2009). *Strategies for using microorganisms and enzymatic systems in bioremediation*. Università degli Studi di Napoli Federico II.

Peramaiyan, R., Thamaraiselvan, R., Natarajan, N., Rajendran, P., Yutaka, N., & Ikuo, N. (2014). Kaempferol, a potential cytostatic and cure for inflammatory disorders. *European Journal of Medicinal Chemistry*, *86*, 103–112. doi:10.1016/j.ejmech.2014.08.011 PMID:25147152

Pessini, A. C., Takao, T. T., Cavalheiro, E. C., Vichnewski, W., Sampaio, S. V., Giglio, J. R., & Arantes, E. C. (2001). A hyaluronidase from *Tityus serrulatus* scorpion venom: Isolation, characterization and inhibition by flavonoids. *Toxicon*, *39*(10), 1495–1504. doi:10.1016/S0041-0101(01)00122-2 PMID:11478957

Petřek, M., Košinová, P., Koča, J., & Otyepka, M. (2007). MOLE: A Voronoi diagram-based explorer of molecular channels, pores, and tunnels. *Structure (London, England)*, *15*(11), 1357–1363. doi:10.1016/j.str.2007.10.007 PMID:17997961

Pettersen, E. F., Goddard, T. D., Huang, C. C., Couch, G. S., Greenblatt, D. M., Meng, E. C., & Ferrin, T. E. (2004). UCSF Chimera--a visualization system for exploratory research and analysis. *Journal of Computational Chemistry, 25*(13), 1605–1612. doi:10.1002/jcc.20084 PMID:15264254

Pfeffer, P., & Gohlke, H. (2007). DrugScore^{RNA} – knowledge-based scoring function to predict RNA-ligand interactions. *Journal of Chemical Information and Modeling, 47*(5), 1868–1876. doi:10.1021/ci700134p PMID:17705464

Philips, A., Milanowska, K., Lach, G., & Bujnicki, J. M. (2013). LigandRNA: Computational predictor of RNA-ligand interactions. *RNA (New York, N.Y.), 19*(12), 1605–1616. doi:10.1261/rna.039834.113 PMID:24145824

Phillips, J. C., Braun, R., Wang, W., Gumbart, J., Tajkhorshid, E., Villa, E., & Schulten, K. et al. (2005). Scalable molecular dynamics with NAMD. *Journal of Computational Chemistry, 26*(16), 1781–1802. doi:10.1002/jcc.20289 PMID:16222654

Piazzi, L., Cavalli, A., Colizzi, F., Belluti, F., Bartolini, M., Mancini, F., & Rampa, A. et al. (2008). Multi-target-direct coumarin derivatives: hAChE and BACE 1 inhibitors as potential anti-Alzheimer compounds. *Bioorganic & Medicinal Chemistry Letters, 18*(1), 423–426. doi:10.1016/j.bmcl.2007.09.100 PMID:17998161

Pirhadi, S., Shiri, F., & Ghasemi, J. B. (2013). Methods and applications of structure based pharmacophores in drug discovery. *Current Topics in Medicinal Chemistry, 13*(9), 1036–1047. doi:10.2174/1568026611313090006 PMID:23651482

Pirovano, W., Feenstra, K. A., & Heringa, J. (2008). PRALINE™: A strategy for improved multiple alignment of transmembrane proteins. *Bioinformatics (Oxford, England), 24*(4), 492–497. doi:10.1093/bioinformatics/btm636 PMID:18174178

Pollastri, G., & McLysaght, A. (2005). Porter: A new, accurate server for protein secondary structure prediction. *Bioinformatics (Oxford, England), 21*(8), 1719–1720. doi:10.1093/bioinformatics/bti203 PMID:15585524

Pollastri, G., Przybylski, D., Rost, P., & Baldi, B. (2002). Improving the Prediction of Protein Secondary Structure in Three and Eight Classes Using Recurrent Neural Networks and Profiles. *Proteins, 47*(2), 228–235. doi:10.1002/prot.10082 PMID:11933069

Popescu, F. D. (2005). Antisense- and RNA interference-based therapeutic strategies in allergy. *Journal of Cellular and Molecular Medicine, 9*(4), 840–853. doi:10.1111/j.1582-4934.2005.tb00383.x PMID:16364194

Prothero, J. W. (1966). Correlation between the distribution of amino acids and alpha helices. *Biophysical Journal, 6*(3), 367–370. doi:10.1016/S0006-3495(66)86662-6 PMID:5962284

Puiatti, M., Borioni, J. L., Vallejo, M. G., Cabrera, J. L., Agnese, A. M., Ortega, M. G., & Pierini, A. B. (2013). Study of the interaction of *Huperzia saururus* Lycopodium alkaloids with the acetylcholinesterase enzyme. *Journal of Molecular Graphics & Modelling, 44*, 136–144. doi:10.1016/j.jmgm.2013.05.009 PMID:23827878

Punta, M., Coggill, P. C., Eberhardt, R. Y., Mistry, J., Tate, J., Boursnell, C., Pang, N., & Forslun (2011). The Pfam protein families database. *Nucleic Acids Research, 40*(D1), D290–D301.. PMC 3245129. PMID 22127870.10.1093/nar/gkr1065

Qin, C., Xie, M. X., & Liu, Y. (2007). Characterization of the myricetin-human serum albumin complex by spectroscopic and molecular modeling approaches. *Biomacromolecules, 8*(7), 2182–2189. doi:10.1021/bm070319c PMID:17559264

Rabbani, S., Harakidas, P., Davidson, D. J., Henkin, J., & Mazar, A. P. (1995). Prevention of prostate-cancer metastasis in vivo by a novel synthetic inhibitor of urokinase-type plasminogen activator (uPA). *International Journal of Cancer, 63*(6), 840–845. doi:10.1002/ijc.2910630615 PMID:8847143

Ramachandran, G. N., & Sashisekharan, V. (1968). Conformation of polypeptides and proteins. *Advances in Protein Chemistry, 23*, 283–438. doi:10.1016/S0065-3233(08)60402-7 PMID:4882249

Ramachandran, S., Kota, P., Ding, F., & Dokholyan, N. V. (2011). PROTEINS: Structure. *Function and Bioinformatics*, *79*, 261–270. doi:10.1002/prot.22879

Rana, S., Yeh, Y.-C., & Rotello, V. M. (2010). Engineering the nanoparticle–protein interface: Applications and possibilities. *Current Opinion in Chemical Biology*, *14*(6), 828–834. doi:10.1016/j.cbpa.2010.10.001 PMID:21035376

Rarey, M., Kramer, B., Lengauer, T., & Klebe, G. (1996). A fast flexible docking method using an incremental construction algorithm. *Journal of Molecular Biology*, *261*(3), 470–489. doi:10.1006/jmbi.1996.0477 PMID:8780787

Rarey, M., Wefing, S., & Lengauer, T. (1996). Placement of medium-sized molecular fragments into active sites of proteins. *Journal of Computer-Aided Molecular Design*, *10*(1), 41–54. doi:10.1007/BF00124464 PMID:8786414

Rashmi, V., ShylajaNaciyar, M., Rajalakshmi, R., D'Souza, S. F., Prabaharan, D., & Uma, L. (2013). Siderophore mediated uranium sequestration by marine cyanobacterium Synechococcus elongatus BDU 130911. *Bioresource Technology*, *130*, 204–210. doi:10.1016/j.biortech.2012.12.016 PMID:23306130

Rask-Andersen, M., Zhang, J., Fabbro, D., & Schioth, H. B. (2014). Advances in kinase targeting: Current clinical use and clinical trials. *Trends in Pharmacological Sciences*, *35*(11), 604–620. doi:10.1016/j.tips.2014.09.007 PMID:25312588

Razavi, S. F., Khoobi, M., Nadri, H., Sakhteman, A., Moradi, A., Emami, S., & Shafiee, A. et al. (2013). Synthesis and evaluation of 4-substituted coumarins as novel acetylcholinesterase inhibitors. *European Journal of Medicinal Chemistry*, *64*, 252–259. doi:10.1016/j.ejmech.2013.03.021 PMID:23644208

Rechfeld, F., Gruber, P., Hofmann, J., & Kirchmair, J. (2011). Modulators of protein-protein interactions: Novel approaches in targeting protein kinases and other pharmaceutically relevant biomolecules. *Current Topics in Medicinal Chemistry*, *11*(11), 1305–1319. doi:10.2174/156802611795589610 PMID:21513500

Reddy, A. S., Pati, S. P., Kumar, P. P., Pradeep, H. N., & Sastry, G. N. (2007). Virtual screening in drug discovery -- a computational perspective. *Current Protein & Peptide Science*, *8*(4), 329–351. doi:10.2174/138920307781369427 PMID:17696867

Re, F., & Strominger, J. L. (2002). Monomeric recombinant MD-2 binds toll-like receptor 4 tightly and confers lipopolysaccharide responsiveness. *The Journal of Biological Chemistry*, *277*(26), 23427–23432. doi:10.1074/jbc.M202554200 PMID:11976338

Reymond, J.-L., & Awale, M. (2012). Exploring chemical space for drug discovery using the chemical universe database. *ACS Chemical Neuroscience*, *3*(9), 649–657. doi:10.1021/cn3000422 PMID:23019491

Riberson, M. R., & Harrell, L. E. (1997). Cholinergic activity and amyloid precursor protein metabolism. *Brain Research. Brain Research Reviews*, *25*(1), 50–69. doi:10.1016/S0165-0173(97)00016-7 PMID:9370050

Ricci, C. G., & Netz, P. A. (2009). Docking studies on DNA-ligand interactions: Building and application of a protocol to identify the binding mode. *Journal of Chemical Information and Modeling*, *49*(8), 1925–1935. doi:10.1021/ci9001537 PMID:19655805

Richard, D. E., Leo, G., Eitan, E. W., & Chris, P. P. (2003). Comparison of the genomes of human and mouse lays the foundation of genome zoology. *Human Molecular Genetics*, *12*(7), 701–709. doi:10.1093/hmg/ddg078 PMID:12651866

Richard, T., Lefeuvre, D., Descendit, A., Quideau, S., & Monti, J. P. (2006). Recognition characters in peptide-polyphenol complex formation. *Biochimica et Biophysica Acta-general Subjects*, *1760*(6), 951–958. doi:10.1016/j.bbagen.2006.01.005 PMID:16527409

Roberts, B. C., & Mancera, R. L. (2008). Ligand-protein docking with water molecules. *Journal of Chemical Information and Modeling*, *48*(2), 397–408. doi:10.1021/ci700285e PMID:18211049

Rock, F. L., Hardiman, G, Timans, J. C., Kastelein, R. A., Bazan, J. F. (1998). A family of human receptors structurally related to Drosophila Toll. *Proc Natl Acad Sci U S A., 95*(2), 588–93. doi:10.1073/pnas.95.2.588

Rommel, C. (2010). Taking PI3Kdelta and PI3Kgamma one step ahead: Dual active PI3Kdelta/gamma inhibitors for the treatment of immune-mediated inflammatory diseases. *Current Topics in Microbiology and Immunology*, *346*, 279–299. doi:10.1007/82_2010_79 PMID:20593311

Ross, P. D., & Subramanian, S. (1981). Thermodynamics of protein association reactions: Forces contributing to stability. *Biochemistry*, *20*(11), 3096–3102. doi:10.1021/bi00514a017 PMID:7248271

Rother, M., Rother, K., Puton, T., & Bujnicki, J. M. (2011). Mode RNA: A tool for comparative modelling of RNA 3D structure. *Nucleic Acids Research*, *39*(10), 4007–4022. doi:10.1093/nar/gkq1320 PMID:21300639

Rudolph, J. (2007). Inhibiting transient protein-protein interactions: Lessons from the Cdc25 protein tyrosine phosphatases. *Nature Reviews. Cancer*, *7*(3), 202–211. doi:10.1038/nrc2087 PMID:17287826

Ryan, D. P., & Matthews, J. M. (2005). Protein-protein interactions in human disease. *Current Opinion in Structural Biology*, *15*(4), 441–446. doi:10.1016/j.sbi.2005.06.001 PMID:15993577

Rydberg, E. H., Brumshtein, B., Greenblatt, H. M., Wong, D. M., Shaya, D., Williams, L. D., & Sussman, J. L. et al. (2006). Complexes of alkylene-linked tacrine dimers with *Torpedo californica* acetylcholinesterase: Binding of bis(5)-tacrine produces a dramatic rearrangement in the active-site gorge. *Journal of Medicinal Chemistry*, *49*(18), 5491–5500. doi:10.1021/jm060164b PMID:16942022

Sable, R., & Jois, S. (2015). Surfing the Protein-Protein Interaction Surface Using Docking Methods: Application to the Design of PPI Inhibitors. *Molecules (Basel, Switzerland)*, *20*(6), 11569–11603. doi:10.3390/molecules200611569 PMID:26111183

Saier, M. H. (2000). A functional-phylogenetic classification system for transmembrane solute transporters. *Microbiology and Molecular Biology Reviews*, *64*(2), 354–411. doi:10.1128/MMBR.64.2.354-411.2000 PMID:10839820

Saikia, N., Rajkhowa, S., & Deka, R. C. (2013). Density functional and molecular docking studies towards investigating the role of single-wall carbon nanotubes as nanocarrier for loading and delivery of pyrazinamide antitubercular drug onto pncA protein. *Journal of Computer-Aided Molecular Design*, *27*(3), 257–276. doi:10.1007/s10822-013-9638-6 PMID:23413106

Saleh, N. A. (2015). The QSAR and docking calculations of fullerene derivatives as HIV-1 protease inhibitors. *Spectrochimica Acta: Molecular and Biomolecular Spectroscopy*, *136PC*, 1523–1529. doi:10.1016/j.saa.2014.10.045 PMID:25459714

Sánchez, R., Pieper, U., Melo, F., Eswar, N., Martí-Renom, M. A., Madhusudhan, M. S., & Šali, A. et al. (2000). Protein structure modeling for structural genomics. *Nature Structural & Molecular Biology*, *7*, 986–990. doi:10.1038/80776 PMID:11104007

Sanchez, R., & Zhou, M. M. (2009). The role of human bromodomains in chromatin biology and gene transcription. *Current Opinion in Drug Discovery & Development*, *12*(5), 659–665. PMID:19736624

Sander, C., & Schneider, R. (1991). Database of homology-derived protein structures and the structural meaning of sequence alignment. *Proteins*, *9*(1), 56–68. doi:10.1002/prot.340090107 PMID:2017436

Santos-Filho, O. A., & Cherkasov, A. (2008). Using molecular docking, 3D-QSAR, and cluster analysis for screening structurally diverse data sets of pharmacological interest. *Journal of Chemical Information and Modeling*, *48*(10), 2054–2065. doi:10.1021/ci8001952 PMID:18816024

Santos-Filho, O. A., & Hopfinger, A. J. (2002). The 4D-QSAR paradigm: Application to a novel set of non-peptidic HIV protease inhibitors. *Quantitative Structure Activity Relationships*, *21*(4), 369–381. doi:10.1002/1521-3838(200210)21:4<369::AID-QSAR369>3.0.CO;2-1

Sarvagalla, S., & Coumar, M. S. (2015). Structural Biology Insight for the Design of Sub-type Selective Aurora Kinase Inhibitors. *Current Cancer Drug Targets*, *15*(5), 375–393. doi:10.2174/1568009615666150421110401 PMID:25895501

Sarvagalla, S., Singh, V. K., Ke, Y. Y., Shiao, H. Y., Lin, W. H., Hsieh, H. P., & Coumar, M. S. et al. (2015). Identification of ligand efficient, fragment-like hits from an HTS library: Structure-based virtual screening and docking investigations of 2H- and 3H-pyrazolo tautomers for Aurora kinase A selectivity. *Journal of Computer-Aided Molecular Design*, *29*(1), 89–100. doi:10.1007/s10822-014-9807-2 PMID:25344840

Sashidhara, K. V., Kumar, A., Chatterjee, M., Rao, K. B., Singh, S., Verma, A. K., & Palit, G. (2011). Discovery and synthesis of novel 3-phenylcoumarin derivatives as antidepressant agents. *Bioorganic & Medicinal Chemistry Letters*, *21*(7), 1937–1941. doi:10.1016/j.bmcl.2011.02.040 PMID:21377878

Sasikala, W., & Mukherjee, A. (2013). Intercalation and de-intercalation pathway of proflavine through the minor and major grooves of DNA: Roles of water and entropy. *Physical Chemistry Chemical Physics*, *15*(17), 6446–6455. doi:10.1039/c3cp50501d PMID:23525232

Satishkumar, R., & Vertegel, A. (2008). Charge-directed targeting of antimicrobial protein-nanoparticle conjugates. *Biotechnology and Bioengineering*, *100*(3), 403–412. doi:10.1002/bit.21782 PMID:18183633

Schiffer, M., & Edmundson, A. B. (1967). Use of helical wheels to represent the structures of proteins and to identify segments with helical potential. *Biophysical Journal*, *7*(2), 121–135. doi:10.1016/S0006-3495(67)86579-2 PMID:6048867

Schnecke, V., & Kuhn, L. A. (2000). Virtual screening with solvation and ligand-induced complementarity. *Perspectives in Drug Discovery and Design*, *20*(1), 171–190. doi:10.1023/A:1008737207775

Schneidman-Duhovny, D., Inbar, Y., Nussinov, R., & Wolfson, H. J. (2005). PatchDock and SymmDock: Servers for rigid and symmetric docking. *Nucleic Acids Research*, *33*(Web Serversuppl 2), W363–W367. doi:10.1093/nar/gki481 PMID:15980490

Schneidman-Duhovny, D., Inbar, Y., Polak, V., Shatsky, M., Halperin, I., Benyamini, H., & Nussinov, R. et al. (2003). Taking geometry to its edge: Fast unbound rigid (and hinge-bent) docking. *Proteins: Structure, Function, and Bioinformatics*, *52*(1), 107–112. doi:10.1002/prot.10397 PMID:12784375

Schümann, J., Prockl, J., Kiemer, A. K., Vollmar, A. M., Bang, R., & Tiegs, G. (2003). Silibinin protects mice from T cell-dependent liver injury. *Journal of Hepatology*, *39*(3), 333–340. doi:10.1016/S0168-8278(03)00239-3 PMID:12927918

Searle, M. S., & Williams, D. H. (1992). The cost of conformational order: Entropy changes in molecular associations. *Journal of the American Chemical Society*, *114*(27), 10690–10697. doi:10.1021/ja00053a002

Seifert, M. H. (2005). ProPose: Steered virtual screening by simultaneous protein-ligand docking and ligand-ligand alignment. *Journal of Chemical Information and Modeling*, *45*(2), 449–460. doi:10.1021/ci0496393 PMID:15807511

Sepehri, S., Gharagani, S., Saghaie, L., Aghasadeghi, M. R., & Fassigi, A. (2015). QSAR and docking studies of some 1,2,3,4-tetrahydropyrimidines: Evaluation of gp41 as possible target for anti-HIV-1 activity. *Medicinal Chemistry Research*, *24*(4), 1707–1724. doi:10.1007/s00044-014-1246-z

Sessa, F., Mapelli, M., Ciferri, C., Tarricone, C., Areces, L. B., Schneider, T. R., & Musacchio, A. et al. (2005). Mechanism of Aurora B activation by INCENP and inhibition by hesperadin. *Molecular Cell*, *18*(3), 379–391. doi:10.1016/j.molcel.2005.03.031 PMID:15866179

Sgrignani, J., Bonaccini, C., Grazioso, G., Chioccioli, M., Cavalli, A., & Gratteri, P. (2009). Insights into docking and scoring neuronal α4β2 nicotinic receptor agonists using Molecular Dynamics simulations and QM/MM calculations. *Journal of Computational Chemistry*, *30*(15), 2443–2454. doi:10.1002/jcc.21251 PMID:19360794

Sgrignani, J., Novati, B., Colombo, G., & Grazioso, G. (2015). Covalent docking of selected boron-based serine beta-lactamase inhibitors. *Journal of Computer-Aided Molecular Design*, *29*(5), 441–450. doi:10.1007/s10822-015-9834-7 PMID:25676821

Shangary, S., & Wang, S. (2009). Small-molecule inhibitors of the MDM2-p53 protein-protein interaction to reactivate p53 function: A novel approach for cancer therapy. *Annual Review of Pharmacology and Toxicology*, *49*(1), 223–241. doi:10.1146/annurev.pharmtox.48.113006.094723 PMID:18834305

Shanmugaraj, K., Anandakumar, S., & Ilanchelian, M. (2015). Probing the binding interaction of thionine with lysozyme: A spectroscopic and molecular docking investigation. *Dyes and Pigments*, *112*, 210–219. doi:10.1016/j.dyepig.2014.07.003

Shaw, D. E. (2005). A fast, scalable method for the parallel evaluation of distance-limited pairwise particle interactions. *Journal of Computational Chemistry*, *26*(13), 1318–1328. doi:10.1002/jcc.20267 PMID:16013057

Sheng, C., Dong, G., Miao, Z., Zhang, W., & Wang, W. (2015). State-of-the-art strategies for targeting protein-protein interactions by small-molecule inhibitors. *Chemical Society Reviews*. doi:10.1039/c5cs00252d

Shen, L. L., Xu, H., Huang, F. W., Li, Y., Xiao, H. F., Yang, Z., & Li, Y. N. et al. (2015). Investigation on interaction between ligupurpuroside A and pepsin by spectroscopic and docking methods. *Spectrochimica Acta. Part A: Molecular and Biomolecular Spectroscopy*, *135*, 256–263. doi:10.1016/j.saa.2014.06.087 PMID:25078459

Shibata, T., Fujimoto, K., Nagayama, K., Yamaguchi, K., Yamaguchi, K., & Nakamura, T. (2002). Inhibitory activity of brown algal phlorotannins against hyaluronidase. *International Journal of Food Science & Technology*, *37*(6), 703–709. doi:10.1046/j.1365-2621.2002.00603.x

Shi, J. H., Zhu, Y. Y., Wang, J., Chen, J., & Shen, Y. J. (2013). Intermolecular interaction of prednisolone with bovine serum albumin: Spectroscopic and molecular docking methods. *Spectrochimica Acta. Part A: Molecular and Biomolecular Spectroscopy*, *103*, 287–294. doi:10.1016/j.saa.2012.11.034 PMID:23261625

Shi, J., Blundell, T. L., & Mizuguchi, K. (2001). FUGUE: Sequence-structure homology recognition using environment-specific substitution tables and structure-dependent gap penalties. *Journal of Molecular Biology*, *310*(1), 243–257. doi:10.1006/jmbi.2001.4762 PMID:11419950

Shimamura, T., Shiroishi, M., Weyand, S., Tsujimoto, H., Winter, G., Katritch, V., & Iwata, S. et al. (2011). Structure of the human histamine H1 receptor complex with doxepin. *Nature*, *475*(7354), 65–70. doi:10.1038/nature10236 PMID:21697825

Shintani, T., Nomura, K., & Ichishima, E. (1997). Engineering of porcine pepsin alteration of S1 substrate specificity of pepsin to those of fungal aspartic proteinases by site-directed mutagenesis. *The Journal of Biological Chemistry*, *272*(30), 18855–18861. doi:10.1074/jbc.272.30.18855 PMID:9228062

Shoichet, B. K., & Kuntz, I. D. (1993). Matching chemistry and shape in molecular docking. *Protein Engineering*, *6*(7), 723–732. doi:10.1093/protein/6.7.723 PMID:7504257

Shoichet, B. K., Kuntz, I. D., & Bodian, D. L. (1992). Molecular docking using shape descriptors. *Journal of Computational Chemistry*, *13*(3), 380–397. doi:10.1002/jcc.540130311

Shoichet, B. K., Leach, A. R., & Kuntz, I. D. (1999). Ligand solvation in molecular docking. Proteins. *Structure, Function, and Bioinformatics*, *34*(1), 4–16. doi:10.1002/(SICI)1097-0134(19990101)34:1<4::AID-PROT2>3.0.CO;2-6

Shrivastava, S., Nuffer, J. H., Siegel, R. W., & Dordick, J. S. (2012). Position-specific chemical modification and quantitative proteomics disclose protein orientation adsorbed on silica nanoparticles. *Nano Letters*, *12*(3), 1583–1587. doi:10.1021/nl2044524 PMID:22296027

Siegal, G., Van Duynhoven, J., & Baldus, M. (1999). Biomolecular NMR: Recent advances in liquids, solids and screening. *Current Opinion in Chemical Biology*, *3*(5), 530–536. doi:10.1016/S1367-5931(99)00004-6 PMID:10508666

Silman, I., & Sussman, J. L. (2008). Acetylcholinesterase: How is structure related to function. *Chemico-Biological Interactions*, *175*(1-3), 3–10. doi:10.1016/j.cbi.2008.05.035 PMID:18586019

Simmosis, V. A., & Heringa, J. (2005). PRALINE: A multiple sequence alignment toolbox that integrates homology-extended and secondary structure information. *Nucleic Acids Research*, *33*(Web Server), W289–W294. doi:10.1093/nar/gki390 PMID:15980472

Sing Yong, K., Bing Choi, S., & Wahab, A. H. (2015). *Docking of Dengue NS2B-NS3 Protease with Murraya koenigii.* Paper presented at the 3rd International Conference on Computation for Science and Technology (ICCST-3).

Sinko, G., Brglez, J., & Kovarik, Z. (2010). Interactions of pyridinium oximes with acetylcholinesterase. *Chemico-Biological Interactions*, *187*(1-3), 172–176. doi:10.1016/j.cbi.2010.04.017 PMID:20417629

Sinko, W., Lindert, S., & McCammon, J. A. (2013). Accounting for Receptor Flexibility and Enhanced Sampling Methods in Computer-Aided Drug Design. *Chemical Biology & Drug Design*, *81*(1), 41–49. doi:10.1111/cbdd.12051 PMID:23253130

Sliwoski, G., Kothiwale, S., Meiler, J., & Lowe, E. W. (2014). Computational Methods in Drug Discovery. *Pharmacological Reviews*, *66*(1), 334–395. doi:10.1124/pr.112.007336 PMID:24381236

Slynko, I., Scharfe, M., Rumpf, T., Eib, J., Metzger, E., Schule, R., & Sippl, W. et al. (2014). Virtual screening of PRK1 inhibitors: Ensemble docking, rescoring using binding free energy calculation and QSAR model development. *Journal of Chemical Information and Modeling*, *54*(1), 138–150. doi:10.1021/ci400628q PMID:24377786

Smid, P., Coolen, H. K., Keizer, H. G., van Hes, R., de Moes, J. P., den Hartog, A. P., & Kruse, C. G. et al. (2005). Synthesis, Structure-Activity Relationships, and Biological Properties of 1-Heteroaryl-4-[ω-(1 H-indol-3-yl) alkyl] piperazines, Novel Potential Antipsychotics Combining Potent Dopamine D2 Receptor Antagonism with Potent Serotonin Reuptake Inhibition. *Journal of Medicinal Chemistry*, *48*(22), 6855–6869. doi:10.1021/jm050148z PMID:16250644

Sonawani, A., Niazi, S., & Idicula-Thomas, S. (2013). In Silico Study on Binding Specificity of Gonadotropins and Their Receptors: Design of a Novel and Selective Peptidomimetic for Human Follicle Stimulating Hormone Receptor. *PLoS ONE*, *8*(5), e64475. doi:10.1371/journal.pone.0064475 PMID:23700481

Song, J. J., Liu, J., Tolia, N. H., Schneiderman, J., Smith, S. K., Martienssen, R. A., & Joshua-Tor, L. et al. (2003). The crystal structure of the Argonaute2 PAZ domain reveals an RNA binding motif in RNAi effector complexes. *Nature Structural Biology*, *10*(12), 1026–1032. doi:10.1038/nsb1016 PMID:14625589

Song, Z., Antzutkin, O. N., Lee, Y. K., Shekar, S. C., Rupprecht, A., & Levitt, M. H. (1997). Conformational transitions of the phosphodiester backbone in native DNA: Two-dimensional magic-angle-spinning 31P-NMR of DNA fibers. *Biophysical Journal*, *73*(3), 1539–1552. doi:10.1016/S0006-3495(97)78186-4 PMID:9284321

Soreide, K., Janssen, E., Körner, H., & Baak, J. (2006). Trypsin in colorectal cancer: Molecular biological mechanisms of proliferation, invasion, and metastasis. *The Journal of Pathology*, *209*(2), 147–156. doi:10.1002/path.1999 PMID:16691544

Soto-Ortega, D. D., Murphy, B. P., Gonzalez-Velasquez, F. J., Wilson, K. A., Xie, F., Wang, Q., & Moss, M. A. (2011). Inhibition of amyloid-β aggregation by coumarin analogs can be manipulated by functionalization of the aromatic center. *Bioorganic & Medicinal Chemistry*, *19*(8), 2596–2602. doi:10.1016/j.bmc.2011.03.010 PMID:21458277

Sousa, S. F., Fernandes, P. A., & Ramos, M. J. (2006). Protein–ligand docking: Current status and future challenges. *Proteins: Structure, Function, and Bioinformatics*, *65*(1), 15–26. doi:10.1002/prot.21082 PMID:16862531

Spyrakis, F., Faggiano, S., Abbruzzetti, S., Dominici, P., Cacciatori, E., Astegno, A., & Luque, J. et al. (2011). Histidine E7 dynamics modulates ligand exchange between distal pocket and solvent in AHb1 from Arabidopsis thaliana. *The Journal of Physical Chemistry B*, *115*(14), 4138–4146. doi:10.1021/jp110816h PMID:21428382

Sridhar, S., & Chandra, J. H. (2014). Involvement of Computational tools towards In Silico remediation-Synthetic textile dyes interacting with Azoreductase. *International Journal of Chemical Technology Research*, *6*(9), 4412–4441.

Srivastava, H. K., Chourasia, M., Kumar, D., & Sastry, G. N. (2011). Comparison of computational methods to model DNA minor groove binders. *Journal of Chemical Information and Modeling*, *51*(3), 558–571. doi:10.1021/ci100474n PMID:21375336

Stan, D., Matei, I., Mihailescu, C., Savin, M., Matache, M., Hillebrand, M., & Baciu, I. (2009). Spectroscopic investigations of the binding interaction of a new indanedione derivative with human and bovine serum albumins. *Molecules (Basel, Switzerland)*, *14*(4), 1614–1626. doi:10.3390/molecules14041614 PMID:19396020

Stave, J. W., & Lindpaintner, K. (2013). Antibody and Antigen Contact Residues Define Epitope and Paratope Size and Structure. *Journal of Immunology (Baltimore, MD.: 1950)*, *191*(3), 1428–1435. doi:10.4049/jimmunol.1203198 PMID:23797669

Stern, R., Asari, A. A., & Sugahara, K. N. (2006). Hyaluronan fragments: An information-rich system. *European Journal of Cell Biology*, *85*(8), 699–715. doi:10.1016/j.ejcb.2006.05.009 PMID:16822580

Stoffler, D., Sanner, M. F., Morris, G. M., Olson, A. J., & Goodsell, D. S. (2002). Evolutionary analysis of HIV-1 protease inhibitors: Methods for design of inhibitors that evade resistance. Proteins. *Structure, Function, and Bioinformatics*, *48*(1), 63–74. doi:10.1002/prot.10130 PMID:12012338

Stumpf, M. P., Thorne, T., de Silva, E., Stewart, R., An, H. J., Lappe, M., & Wiuf, C. (2008). Estimating the size of the human interactome. *Proceedings of the National Academy of Sciences of the United States of America*, *105*(19), 6959–6964. doi:10.1073/pnas.0708078105 PMID:18474861

Suganuma, M., Ohkura, Y., Okabe, S., & Fujiki, H. (2001). Combination cancer chemoprevention with green tea extract and sulindac shown in intestinal tumor formation in Min mice. *Journal of Cancer Research and Clinical Oncology*, *127*(1), 69–72. doi:10.1007/s004320000189 PMID:11206275

Sun, X., Feng, Z., Zhang, L., Hou, T., & Li, Y. (2014). *The Selective Interaction between Silica Nanoparticles and Enzymes from Molecular Dynamics Simulations*. Academic Press.

Suresh, P. S., Kumar, A., Kumar, R., & Singh, V. P. (2008). An Insilco approach to bioremediation: Laccase as a case study. *Journal of Molecular Graphics & Modelling*, *26*(5), 845–849. doi:10.1016/j.jmgm.2007.05.005 PMID:17606396

Surget, S., Khoury, M. P., & Bourdon, J. C. (2013). Uncovering the role of p53 splice variants in human malignancy: A clinical perspective. *Onco Targets Ther*, *7*, 57–68. doi:10.2147/OTT.S53876 PMID:24379683

Sussman, J. L., Harel, M., Frolow, F., Oefner, C., Goldman, A., Toker, L., & Silman, I. (1991). Atomic structure of acetylcholinesterase from *Torpedo californica*: A prototypic acetylcholine-binding protein. *Science*, *253*(5022), 872–879. doi:10.1126/science.1678899 PMID:1678899

Taha, M. O., Habash, M., Al-Hadidi, Z., Al-Bakri, A., Younis, K., & Sisan, S. (2011). Docking-based comparative intermolecular contacts analysis as new 3-D QSAR concept for validating docking studies and in silico screening: NMT and GP inhibitors as case studies. *Journal of Chemical Information and Modeling*, *51*(3), 647–669. doi:10.1021/ci100368t PMID:21370899

Takada, A., Robison, C., Goto, H., Sanchez, A., Murti, K. G., Whitt, M. A., & Kawaoka, Y. (1997). A system for functional analysis of Ebola virus glycoprotein. *Proceedings of the National Academy of Sciences of the United States of America*, *94*(26), 14764–14769. doi:10.1073/pnas.94.26.14764 PMID:9405687

Taliani, S., Pugliesi, I., Barresi, E., Salerno, S., Marchand, C., Agama, K., & Da Settimo, F. et al. (2013). Phenylpyrazolo[1,5-a]quinazolin-5(4H)-one: A Suitable Scaffold for the Development of Noncamptothecin Topoisomerase I (Top1) Inhibitors. *Journal of Medicinal Chemistry*, *56*(18), 7458–7462. doi:10.1021/jm400932c PMID:23987476

Tame, J. R. (1999). Scoring functions: A view from the bench. *Journal of Computer-Aided Molecular Design*, *13*(2), 99–108. doi:10.1023/A:1008068903544 PMID:10091117

Tang, C. M., & Insel, P. A. (2005). Genetic variation in G-protein-coupled receptorsconsequences for G-protein-coupled receptors as drug targets. *Expert Opinion on Therapeutic Targets*, *9*(6), 1247–1265. doi:10.1517/14728222.9.6.1247 PMID:16300474

Tartaglia, G. G., Pawar, A. P., Campioni, S., Dobson, C. M., Chiti, F., & Vendruscolo, M. (2008). Prediction of aggregation-prone regions in structured proteins. *Journal of Molecular Biology*, *380*(2), 425–436. doi:10.1016/j.jmb.2008.05.013 PMID:18514226

Tatulian, S. A. (2013). Structural characterization of membrane proteins and peptides by FTIR and ATR-FTIR spectroscopy. In *Lipid-Protein Interactions* (pp. 177–218). Humana Press. doi:10.1007/978-1-62703-275-9_9

Taylor, J. S., & Burnett, R. M. (2000). DARWIN: A program for docking flexible molecules. *Proteins: Structure, Function, and Bioinformatics*, *41*(2), 173–191. doi:10.1002/1097-0134(20001101)41:2<173::AID-PROT30>3.0.CO;2-3 PMID:10966571

Teillet, F., Boumendjel, A., Boutonnat, J., & Ronot, X. (2008). Flavonoids as RTK inhibitors and potential anticancer agents. *Medicinal Research Reviews*, *28*(5), 715–745. doi:10.1002/med.20122 PMID:18080331

Tekere, M., Mswaka, A., Zvauya, R., & Read, J. (2001). Growth, dye degradation and ligninolytic activity studies on Zimbabwean white rot fungi. *Enzyme and Microbial Technology*, *28*(4), 420–426. doi:10.1016/S0141-0229(00)00343-4 PMID:11240201

Teller, D. C., Okada, T., Behnke, C. A., Palczewski, K., & Stenkamp, R. E. (2001). Advances in determination of a high-resolution three-dimensional structure of rhodopsin, a model of G-protein-coupled receptors (GPCRs). *Biochemical*, *40*(26), 7761–7762. doi:10.1021/bi0155091 PMID:11425302

Teodoro, M., & Kavraki, L. (2003). Conformational flexibility models for the receptor in structure based drug design. *Current Pharmaceutical Design*, *9*(20), 1635–1648. doi:10.2174/1381612033454595 PMID:12871062

Tetko, I. V., Alessandro, E., Villa, P., & Livingstone, D. J. (1996). Neural network studies. 2. Variable selection. *Journal of Chemical Information and Computer Sciences*, *36*(4), 794–803. doi:10.1021/ci950204c PMID:8768768

Thaimattam, R., Banerjee, R., Miglani, R., & Iqbal, J. (2007). Protein kinase inhibitors: Structural insights into selectivity. *Current Pharmaceutical Design, 13*(27), 2751–2765. doi:10.2174/138161207781757042 PMID:17897021

Thompson, A. A., Liu, W., Chun, E., Katritch, V., Wu, H., Vardy, E., & Stevens, R. C. et al. (2012). Structure of the nociceptin/orphanin FQ receptor in complex with a peptide mimetic. *Nature, 485*(7398), 395–399. doi:10.1038/nature11085 PMID:22596163

Thompson, J. D., Higgins, D. G., & Gibson, T. J. (1994). CLUSTAL W: Improving the sensitivity of progressive multiple sequence alignment through sequence weighting, position-specific gap penalties and weight matrix choice. *Nucleic Acids Research, 22*(22), 4673–4680. doi:10.1093/nar/22.22.4673 PMID:7984417

Thomsen, R., & Christensen, M. H. (2006). MolDock: A new technique for high-accuracy molecular docking. *Journal of Medicinal Chemistry, 49*(11), 3315–3321. doi:10.1021/jm051197e PMID:16722650

Tina, K. G., Bhadra, R., & Srinivasan, N. (2007). PIC: Protein Interactions Calculator. *Nucleic Acids Research, 35*, W473–W476. doi:10.1093/nar/gkm423 PMID:17584791

Todeschini, R., Consonni, V., & Pavan, M. (2001). *DRAGON-Software for the Calculation of Molecular Descriptors. Release 1.12 for Windows*. Retrieved from http://www.disat.unimib.it/chm

Tomasselli, A. G., & Heinrikson, R. L. (2000). Targeting the HIV-protease in AIDS therapy: A current clinical perspective. Biochimica et Biophysica Acta (BBA)-. *Protein Structure and Molecular Enzymology, 1477*(1), 189–214. doi:10.1016/S0167-4838(99)00273-3

Tøndel, K., Anderssen, E., & Drabløs, F. (2006). Protein Alpha Shape (PAS) Dock: A new gaussian-based score function suitable for docking in homology modelled protein structures. *Journal of Computer-Aided Molecular Design, 20*(3), 131–144. doi:10.1007/s10822-006-9041-7 PMID:16652207

Toniolo, C., & Benedetti, E. (1991). Article. *Trends in Biochemical Sciences, 16*, 350–353. doi:10.1016/0968-0004(91)90142-I PMID:1949158

Topham, C. M., Srinivasan, N., & Blundell, T. L. (1997). Prediction of protein mutants based on structural environment-dependent amino acid substitution and propensity tables. *Protein Engineering, 10*(1), 7–21. doi:10.1093/protein/10.1.7 PMID:9051729

Tritsch, D., Zingle, C., Rohmer, M., & Grosdemange-Billiard, C. (2015). Flavonoids: True or promiscuous inhibitors of enzyme? The case of deoxyxylulose phosphate reductoisomerase. *Bioorganic Chemistry, 59*, 140–144. doi:10.1016/j.bioorg.2015.02.008 PMID:25800132

Trott, O., & Olson, A. J. (2010). AutoDock Vina: Improving the speed and accuracy of docking with a new scoring function, efficient optimization, and multithreading. *Journal of Computational Chemistry, 31*(2), 455–461. doi:10.1002/jcc.21334 PMID:19499576

Tsai, S. J., Huang, C. S., Mong, M. C., Kam, W. Y., Huang, H. Y., & Yin, M. C. (2012). Anti-inflammatory and antifibrotic effects of naringenin in diabetic mice. *Journal of Agricultural and Food Chemistry, 60*(1), 514–521. doi:10.1021/jf203259h PMID:22117528

Tuszyńska, I., Magnus, M., Jonak, K., Dawson, W., & Bujnicki, J.M. (2015) NPDock – a web server for protein-nucleic acid docking *Nucleic Acid Research, 43*(W1), W425-30.

Tuszynska, I., & Bujnicki, J. M. (2011). DARS-RNP and QUASI-RNP: New statistical potentials for protein-RNA docking. *BMC Bioinformatics, 12*(1), 348. doi:10.1186/1471-2105-12-348 PMID:21851628

Ul-Haq, Z., Usmani, S., Shamshad, H., Mahmood, U., & Halim, S. A. (2013). A combined 3D-QSAR and docking studies for the *In-silico* prediction of HIV-protease inhibitors. *Chemistry Central Journal, 7*(1), 88. doi:10.1186/1752-153X-7-88 PMID:23683267

Ulmschneider, J. P., Smith, J. C., White, S. H., & Ulmschneider, M. B. (2011). In silico partitioning and transmembrane insertion of hydrophobic peptides under equilibrium conditions. *Journal of the American Chemical Society, 133*(39), 15487–15495. doi:10.1021/ja204042f PMID:21861483

Ulmschneider, J. P., & Ulmschneider, M. B. (2009). Sampling efficiency in explicit and implicit membrane environments studied by peptide folding simulations. Proteins. *Structure, Function, and Bioinformatics, 75*(3), 586–597. doi:10.1002/prot.22270 PMID:19003985

Utpal, C.D., Tanusree, D., Debanjan, S., & Sudhan, D. (2015). Three-dimensional quantitative structure-activity relationships and docking studies of some structurally diverse flavonoids and design of new aldose reductase inhibitors. *Journal of Advanced Pharmaceutical technology and Research, 6*, 13-18.

Vaidehi, N., Floriano, W. B., Trabanino, R., Hall, S. E., Freddolino, P., Choi, E. J., & Goddard, W. A. III. (2002). Structure and Function of GPCRs. *Proceedings of the National Academy of Sciences of the United States of America, 99*, 12622–12627. doi:10.1073/pnas.122357199 PMID:12351677

Vakser, I. A. (1995). Protein docking for low-resolution structures. *Protein Engineering, 8*(4), 371–377. doi:10.1093/protein/8.4.371 PMID:7567922

Vakser, I. A. (2014). Protein-protein docking: From interaction to interactome. *Biophysical Journal, 107*(8), 1785–1793. doi:10.1016/j.bpj.2014.08.033 PMID:25418159

Valenta, R. (2008). Biochemistry of allergens and recombinant allergens. In A. B. Kay (Ed.), *Allergy and Allergic Diseases* (pp. 895–912). Oxford, UK: Blackwell Publishing Ltd. doi:10.1002/9781444300918.ch42

Valenta, R., Campana, R., Marth, K., & van Hage, M. (2012). Allergen-specific immunotherapy: From therapeutic vaccines to prophylactic approaches. *Journal of Internal Medicine, 272*(2), 144–157. doi:10.1111/j.1365-2796.2012.02556.x PMID:22640224

van den Berg, B., Tessari, M., Boelens, R., Dijkman, R., de Hass, G. H., Kaptein, R., & Verheij, H. M. (1995). NMR structures of phospholipase A2 reveal conformational changes during interfacial activation. *Nature Structural & Molecular Biology, 2*(5), 402–406. doi:10.1038/nsb0595-402 PMID:7664098

van der Horst, E., E. Peironcely, J., J. P. van Westen, G., O. van den Hoven, O., R. J. D. Galloway, W., R. Spring, D., & Bender, A. et al. (2011). Chemogenomics approaches for receptor deorphanization and extensions of the chemogenomics concept to phenotypic space. *Current Topics in Medicinal Chemistry, 11*(15), 1964–1977. doi:10.2174/156802611796391230 PMID:21470175

Van Der Spoel, D., Lindahl, E., Hess, B., Groenhof, G., Mark, A. E., & Berendsen, H. J. (2005). GROMACS: Fast, flexible, and free. *Journal of Computational Chemistry, 26*(16), 1701–1718. doi:10.1002/jcc.20291 PMID:16211538

Van Drie, J. H. (2007). Computer-aided drug design: The next 20 years. *Journal of Computer-Aided Molecular Design, 21*(10-11), 591–601. doi:10.1007/s10822-007-9142-y PMID:17989929

Vats, C., Dhanjal, J. K., Goyal, S., Bharadvaja, N., & Grover, A. (2015). Computational design of novel flavonoid analogues as potential AChE inhibitors: Analysis using group-based QSAR, molecular docking and molecular dynamics simulations. *Structural Chemistry, 26*(2), 467–476. doi:10.1007/s11224-014-0494-3

Vedani, A., Dobler, M., & Zbinden, P. (1998). Quasi-atomistic receptor surface models: A bridge between 3-D QSAR and receptor modeling. *Journal of the American Chemical Society*, *120*(18), 4471–4477. doi:10.1021/ja973976t

Venkatachalam, C. M., Jiang, X., Oldfield, T., & Waldman, M. (2003). LigandFit: A novel method for the shape-directed rapid docking of ligands to protein active sites. *Journal of Molecular Graphics & Modelling*, *21*(4), 289–307. doi:10.1016/S1093-3263(02)00164-X PMID:12479928

Venkatesan, K., Rual, J. F., Vazquez, A., Stelzl, U., Lemmens, I., Hirozane-Kishikawa, T., & Vidal, M. et al. (2009). An empirical framework for binary interactome mapping. *Nature Methods*, *6*(1), 83–90. doi:10.1038/nmeth.1280 PMID:19060904

Verdonk, M. L., Chessari, G., Cole, J. C., Hartshorn, M. J., Murray, C. W., Nissink, J. W. M., & Taylor, R. et al. (2005). Modeling water molecules in protein-ligand docking using GOLD. *Journal of Medicinal Chemistry*, *48*(20), 6504–6515. doi:10.1021/jm050543p PMID:16190776

Verdonk, M. L., Cole, J. C., Hartshorn, M. J., Murray, C. W., & Taylor, R. D. (2003). Improved protein-ligand docking using GOLD. *Proteins*, *52*(4), 609–623. doi:10.1002/prot.10465 PMID:12910460

Verdrengh, M., Jonsson, I. M., Holmdahl, R., & Tarkowski, A. (2003). Genistein as an anti-inflammatory agent. *Inflammation Research*, *52*(8), 341–346. doi:10.1007/s00011-003-1182-8 PMID:14504672

Verma, J., Khedkar, V. M., & Coutinho, E. C. (2010). 3D-QSAR in drug design-a review. *Current Topics in Medicinal Chemistry*, *10*(1), 95–115. doi:10.2174/156802610790232260 PMID:19929826

Vieira, A. C., Marschalk, C., Biavatti, D. C., Lorscheider, C. A., Peralta, R. M., & Seixas, F. A. V. (2015). Modeling Based Structural Insights into Biodegradation of the Herbicide Diuron by Laccase-1 from Ceriporiopsis subvermispora. *Bioinformation*, *11*, 224–228. doi:PMID:26124565 doi:10.6026/97320630011224

Vieth, M., Hirst, J. D., Kolinski, A., & Brooks, C. L. (1998). Assessing energy functions for flexible docking. *Journal of Computational Chemistry*, *19*(14), 1612–1622. doi:10.1002/(SICI)1096-987X(19981115)19:14<1612::AID-JCC7>3.0.CO;2-M

Villoutreix, B. O., Kuenemann, M. A., Poyet, J. L., Bruzzoni-Giovanelli, H., Labbe, C., Lagorce, D., & Miteva, M. A. et al. (2014). Drug-Like Protein-Protein Interaction Modulators: Challenges and Opportunities for Drug Discovery and Chemical Biology. *Mol Inform*, *33*(6-7), 414–437. doi:10.1002/minf.201400040 PMID:25254076

Volchkov, V. E., Feldmann, H., Volchkova, V. A., & Klenk, H. D. (1998). Processing of the Ebola virus glycoprotein by the proprotein convertase furin. *Proceedings of the National Academy of Sciences of the United States of America*, *95*(10), 5762–5767. doi:10.1073/pnas.95.10.5762 PMID:9576958

Wacker, D., Wang, C., Katritch, V., Han, G. W., Huang, X. P., Vardy, E., & Stevens, R. C. et al. (2013). Structural features for functional selectivity at serotonin receptors. *Science*, *340*(6132), 615–619. doi:10.1126/science.1232808 PMID:23519215

Wahlström, R. (2009). *Validation of docking performance in the context of a structural water molecule using model system*. Academic Press.

Walford HH, Doherty T A. (2013) STAT6 and lung inflammation. *JAKSTAT, 2*(4).

Wallace, A. C., Laskowski, R. A., & Thornton, J. M. (1995). LIGPLOT: A program to generate schematic diagrams of protein-ligand interactions. *Protein Engineering*, *8*(2), 127–134. doi:10.1093/protein/8.2.127 PMID:7630882

Wallace, K. B. (2015). Multiple Targets for Drug-induced Mitochondrial Toxicity. *Current Medicinal Chemistry, 22*(20), 2488–2492. doi:10.2174/0929867322666150514095424 PMID:25973981

Wallin, E., & Von Heijne, G. (1998). Genome-wide analysis of integral membrane proteins from eubacterial, archaean, and eukaryotic organisms. *Protein Science, 7*(4), 1029.

Walls, P. H., & Sternberg, M. J. (1992). New algorithm to model protein-protein recognition based on surface complementarity: Applications to antibody-antigen docking. *Journal of Molecular Biology, 228*(1), 277–297. doi:10.1016/0022-2836(92)90506-F PMID:1280302

Walt, D. R., Smulow, J. B., Turesky, S. S., & Hill, R. G. (1985). The effect of gravity on initial microbial adhesion. *Journal of Colloid and Interface Science, 107*(2), 334–336. doi:10.1016/0021-9797(85)90185-7

Walters, M. C., Roe, F., Bugnicourt, A., Franklin, M. J., & Stewart, P. S. (2003). Contributions of antibiotic penetration, oxygen limitation, and low metabolic activity to tolerance of Pseudomonas aeruginosa biofilms to ciprofloxacin and tobramycin. *Antimicrobial Agents and Chemotherapy, 47*(1), 317–323. doi:10.1128/AAC.47.1.317-323.2003 PMID:12499208

Wang, C., Jiang, Y., Ma, J., & Xu, H. E. (2013). Structural Basis for Molecular Recognition at Serotonin Receptors. *Science, 340*(6132), 610–614. doi:10.1126/science.1232807 PMID:23519210

Wang, J.-C., Lin, J.-H., Chen, C.-M., Perryman, A. L., & Olson, A. J. (2011). Robust scoring functions for protein-ligand interactions with quantum chemical charge models. *Journal of Chemical Information and Modeling, 51*(10), 2528–2537. doi:10.1021/ci200220v PMID:21932857

Wang, J., Wolf, R. M., Caldwell, J. W., Kollman, P. A., & Case, D. A. (2004). Development and testing of a general amber force field. *Journal of Computational Chemistry, 25*(9), 1157–1174. doi:10.1002/jcc.20035 PMID:15116359

Wang, Y. C., & Huang, K. M. (2013). In vitro anti-inflammatory effect of apigenin in the helicobacter pylori-infected gastric adenocarcinoma cells. *Food and Chemical Toxicology, 53*, 376–383. doi:10.1016/j.fct.2012.12.018 PMID:23266501

Wang, Y., Yu, X., Wu, Y., & Zhang, D. (2012). Coffee and tea consumption and risk of lung cancer: A dose–response analysis of observational studies. *Lung Cancer (Amsterdam, Netherlands), 78*(2), 169–170. doi:10.1016/j.lungcan.2012.08.009 PMID:22964413

Warne, T., Serrano-Vega, M. J., Baker, J. G., Moukhametzianov, R., Edwards, P. C., Henderson, R., & Schertler, G. F. X. et al. (2008). Structure of a β1-adrenergic G-protein coupled receptor. *Nature, 454*(7203), 486–491. doi:10.1038/nature07101 PMID:18594507

Warren, G. L., Andrews, C. W., Capelli, A.-M., Clarke, B., LaLonde, J., Lambert, M. H., & Head, M. S. et al. (2006). A critical assessment of docking programs and scoring functions. *Journal of Medicinal Chemistry, 49*(20), 5912–5931. doi:10.1021/jm050362n PMID:17004707

Warren, G. L., Do, T. D., Kelley, B. P., Nicholls, A., & Warren, S. D. (2012). Essential considerations for using protein–ligand structures in drug discovery. *Drug Discovery Today, 17*(23), 1270–1281. doi:10.1016/j.drudis.2012.06.011 PMID:22728777

Watkins, A. M., & Arora, P. S. (2015). Structure-based inhibition of protein-protein interactions. *European Journal of Medicinal Chemistry, 94*, 480–488. doi:10.1016/j.ejmech.2014.09.047 PMID:25253637

Wei, B. Q., Weaver, L. H., Ferrari, A. M., Matthews, B. W., & Shoichet, B. K. (2004). Testing a flexible-receptor docking algorithm in a model binding site. *Journal of Molecular Biology, 337*(5), 1161–1182. doi:10.1016/j.jmb.2004.02.015 PMID:15046985

Weiderstein, & Sippl. (2007). ProSA-web: Interactive web service for the recognition of errors in three-dimensional structures of proteins. *Nucleic Acid Residues, 35*, W407-W410.

Weill, N., Therrien, E., Campagna-Slater, V., & Moitessier, N. (2014). Methods for docking small molecules to macromolecules: A user's perspective. 1. The theory. *Current Pharmaceutical Design, 20*(20), 3338–3359. doi:10.2174/13816128113199990605 PMID:23947645

Weisburger, J. H. (2002). Comments on the history and importance of aromatic and heterocyclic amines in public health. *Mutation Research/Fundamental and Molecular Mechanisms of Mutagenesis, 506*, 9-20.

Weisssenhorn, W., Carfi, A., Lee, K. H., Skehel, J. J., & Wiley, D. C. (1998). Crystal Structure of the Ebola virus membrane fusion subunit, GP2, from the envelope glycoprotein ectodomain. *Molecular Cell, 2*(5), 605–616. doi:10.1016/S1097-2765(00)80159-8 PMID:9844633

Wells, J. A., & McClendon, C. L. (2007). Reaching for high-hanging fruit in drug discovery at protein-protein interfaces. *Nature, 450*(7172), 1001–1009. doi:10.1038/nature06526 PMID:18075579

Wessels, J., & Van der Veen, R. (1956). The action of some derivatives of phenylurethan and of 3-phenyl-1, 1-dimethylurea on the Hill reaction. *Biochimica et Biophysica Acta, 19*, 548–549. doi:10.1016/0006-3002(56)90481-4 PMID:13315318

White, A. W., Westwell, A. D., & Brahemi, G. (2008). Protein-protein interactions as targets for small-molecule therapeutics in cancer. *Expert Reviews in Molecular Medicine, 10*, e8. doi:10.1017/S1462399408000641 PMID:18353193

Widmer, H., & Jahnke, W. (2004). Protein NMR in biomedical research. *Cellular and Molecular Life Sciences, 61*(5), 580–599. doi:10.1007/s00018-003-3382-3 PMID:15004697

Wiederstein, M. (2004). *Evolutionary methods in biotechnology*. Weinheim: Wiley-VCH.

Wieman, H., Anderssen, E., & Drablos, F. (2004). Homology-based modelling of targets for rational drug design. *Mini Reviews in Medicinal Chemistry, 4*(7), 793–804. PMID:15379646

Wimley, W. C., Creamer, T. P., & White, S. H. (1996). Solvation energies of amino acid side chains and backbone in a family of host-guest pentapeptides. *Biochem, 35*(16), 5109–5124. doi:10.1021/bi9600153 PMID:8611495

Wisniewski, D., Lambek, C. L., Liu, C., Strife, A., Veach, D. R., Nagar, B., & Clarkson, B. et al. (2002). Characterization of potent inhibitors of the Bcr-Abl and the c-kit receptor tyrosine kinases. *Cancer Research, 62*(15), 4244–4255. PMID:12154026

Wlodawer, A., Svensson, L. A., Sjoelin, L., & Gilliland, G. L. (1988). Structure of phosphate-free ribonuclease A refined at 1.26. ANG. *Biochemistry, 27*(8), 2705–2717. doi:10.1021/bi00408a010 PMID:3401445

Wolber, G., & Langer, T. (2005). LigandScout: 3-D pharmacophores derived from protein-bound ligands and their use as virtual screening filters. *Journal of Chemical Information and Modeling, 45*(1), 160–169. doi:10.1021/ci049885e PMID:15667141

Wold, S. (1991). Validation of QSARs. *Quantitative Structure Activity Relationship, 10*(3), 191–193. doi:10.1002/qsar.19910100302

Wong, T. W., Lee, F. Y., Yu, C., Luo, F. R., Oppenheimer, S., Zhang, H., & Vite, G. D. et al. (2006). Preclinical antitumor activity of BMS-599626, a pan-HER kinase inhibitor that inhibits HER1/HER2 homodimer and heterodimer signaling. *Clinical Cancer Research, 12*(20), 6186–6193. doi:10.1158/1078-0432.CCR-06-0642 PMID:17062696

Wool-Lewis, R. J., & Bates, P. (1998). Characterization of Ebola virus entry by using pseudotyped viruses: Identification of receptor-deficient cell lines. *Journal of Virology*, *72*(4), 3155–3160. PMID:9525641

Wu, B., Chien, E. Y., Mol, C. D., Fenalti, G., Liu, W., Katritch, V., & Stevens, R. C. et al. (2010). Structures of the CXCR4 chemokine GPCR with small-molecule and cyclic peptide antagonists. *Sci.*, *330*(6007), 1066–1071. doi:10.1126/science.1194396 PMID:20929726

Wu, C. R., Chang, C. L., Hsieh, P. Y., Lin, L. W., & Ching, H. (2007). Psoralen and isopsoralen, two coumarins of Psoraleae fructus, can alleviate scopolamine-induced amnesia in rats. *Planta Medica*, *73*(3), 275–278. doi:10.1055/s-2007-967127 PMID:17318785

Wu, G., Robertson, D. H., Brooks, C. L., & Vieth, M. (2003). Detailed analysis of grid-based molecular docking: A case study of CDOCKER—A CHARMm-based MD docking algorithm. *Journal of Computational Chemistry*, *24*(13), 1549–1562. doi:10.1002/jcc.10306 PMID:12925999

Wu, G., & Vieth, M. (2004). SDOCKER: A method utilizing existing X-ray structures to improve docking accuracy. *Journal of Medicinal Chemistry*, *47*(12), 3142–3148. doi:10.1021/jm040015y PMID:15163194

Wu, H., Wacker, D., Mileni, M., Katritch, V., Han, G. W., Vardy, E., & Stevens, R. C. et al. (2012). Structure of the human k-opioid receptor in complex with JDTic. *Nature*, *485*(7398), 327–332. doi:10.1038/nature10939 PMID:22437504

Wu, J.-F., Xu, M.-Q., & Zhao, G.-C. (2010). Graphene-based modified electrode for the direct electron transfer of cytochrome c and biosensing. *Electrochemistry Communications*, *12*(1), 175–177. doi:10.1016/j.elecom.2009.11.020

Wu, P., Nielsen, T. E., & Clausen, M. H. (2015). FDA-approved small-molecule kinase inhibitors. *Trends in Pharmacological Sciences*, *36*(7), 422–439. doi:10.1016/j.tips.2015.04.005 PMID:25975227

Wu, X., He, W., Wang, W., Luo, X., Cao, H., Lin, L., & Liu, Z. et al. (2013). Investigation of the interaction between (−)-epigallocatechin-3-gallate with trypsin and α-chymotrypsin. *International Journal of Food Science & Technology*, *48*(11), 2340–2347.

Wu, Y., & Chakrabarti, S. (2015). ERK5 Mediated Signalling in Diabetic Retinopathy. *Med Hypothesis Discov Innov Ophthalmol*, *4*(1), 17–26. PMID:25861671

Xavier, J. B., Picioreanu, C., Rani, S. A., van Loosdrecht, M. C., & Stewart, P. S. (2005). Biofilm-control strategies based on enzymic disruption of the extracellular polymeric substance matrix–a modelling study. *Microbiology*, *151*(12), 3817–3832. doi:10.1099/mic.0.28165-0 PMID:16339929

Xiang, F., Xiang, J., Fang, Y., Zhang, M., & Li, M. (2014). Discovering isozyme-selective inhibitor scaffolds of human carbonic anhydrases using structural alignment and de novo drug design approaches. *Chemical Biology & Drug Design*, *83*(2), 247–258. doi:10.1111/cbdd.12234 PMID:24112770

Xiang, Z. (2006). Advances in Homology Protein Structure Modeling. *Current Protein & Peptide Science*, *7*(3), 217–227. doi:10.2174/138920306777452312 PMID:16787261

Xu, D., & Zhang, Y. (2001). Improving the Physical Realism and Structural Accuracy of Protein Models by a Two-step Atomic-level Energy Minimization. *Biophysical Journal*, *101*(10), 2525–2534. doi:10.1016/j.bpj.2011.10.024 PMID:22098752

Xu, F., Wu, H., Katritch, V., Han, G. W., Jacobson, K. A., Gao, Z. G., & Stevens, R. C. et al. (2011). Structure of an agonist-bound human A2A adenosine receptor. *Science*, *332*(6027), 322–327. doi:10.1126/science.1202793 PMID:21393508

Xu, L., & Zhang, W. J. (2001). Comparison of different methods for variable selection. *Analytica Chimica Acta*, *446*(1-2), 477–483. doi:10.1016/S0003-2670(01)01271-5

Xu, Y., He, H., Schulz, S., Liu, X., Fusetani, N., Xiong, H., & Qian, P.-Y. et al. (2010). Potent antifouling compounds produced by marine Streptomyces. *Bioresource Technology*, *101*(4), 1331–1336. doi:10.1016/j.biortech.2009.09.046 PMID:19818601

Yadava, U., Gupta, H., & Roychoudhury, M. (2015). Stabilization of microtubules by taxane diterpenoids: Insight from docking and MD simulations. *Journal of Biological Physics*, *41*(2), 117–133. doi:10.1007/s10867-014-9369-5 PMID:25542396

Yadava, U., Shukla, B. K., Roychoudhury, M., & Kumar, D. (2015). Pyrazolo[3,4-*d*]pyrimidines as novel inhibitors of O-acetyl-L-serine sulfhydrylase of *Entamoeba histolytica*: An in silico study. *Journal of Molecular Modeling*, *21*(4), 7–13. doi:10.1007/s00894-015-2631-3 PMID:25799964

Yadava, U., Singh, M., & Roychoudhury, M. (2013). Pyrazolo [3,4-*d*] pyrimidines as inhibitor of anti-coagulation and inflammation activities of phospholipase A2: Insight from molecular docking studies. *Journal of Biological Physics*, *39*(3), 419–438. doi:10.1007/s10867-013-9299-7 PMID:23860918

Yaffe, E., Fishelovitch, D., Wolfson, H. J., Halperin, D., & Nussinov, R. (2008). MolAxis: Efficient and accurate identification of channels in macromolecules. *Proteins: Structure, Function, and Bioinformatics*, *73*(1), 72–86. doi:10.1002/prot.22052 PMID:18393395

Yahaya, Y. A., Don, M. M., & Yahaya, A. S. (2014). The effect of culture conditions on the growth of *T. lactinea* and anti-inflammatory activities via in vitro inhibition of hyaluronidase and lipoxygenase enzyme activities. *Journal of the TAIWAN Institute of Chemical Engineers*, *45*(5), 2054–2059. doi:10.1016/j.jtice.2014.04.003

Yamamoto, M, Sato, S, Hemmi, H, Uematsu, S., Hoshino, K., Kaisho, T., Takeuchi, O., Takeda, K. & Akira, S. (2003). TRAM is specifically involved in the Toll-like receptor 4-mediated MyD88-independent signaling pathway. *Nat. Immunol.*, *4*(11), 1144–50. doi: 14556004.10.1038/ni986.PMID

Yang, H., Avila, M. Y., Peterson-Yantorno, K., Coca-Prados, M., Stone, R. A., Jacobson, K. A., & Civan, M. M. (2005). The cross-species A3 adenosine-receptor antagonist MRS1292 inhibits adenosine-triggered human nonpigmented ciliary epithelial cell fluid release and reduces mouse intraocular pressure. *Current Eye Research*, *30*(9), 747–754. doi:10.1080/02713680590953147 PMID:16146920

Yang, J., Yan, R., Roy, A., Xu, D., Poisson, J., & Zhang, Y. (2015). The I-TASSER Suite: Protein structure and function prediction. *Nature Methods*, *12*(1), 7–8. doi:10.1038/nmeth.3213 PMID:25549265

Yang, R., Yu, L. L., Zeng, H. J., Liang, R. L., Chen, X. L., & Qu, L. B. (2012). The interaction of flavonoid-lysozyme and the relationship between molecular structure of flavonoids and their binding activity to lysozyme. *Journal of Fluorescence*, *22*(6), 1449–1459. doi:10.1007/s10895-012-1082-9 PMID:22772804

Yang, Y. M., Hu, Q. L., Fan, Y. L., & Shen, H. S. (2008). Study of the binding of luteolin to bovine serum albumin. *Spectrochimica Acta. Part A: Molecular and Biomolecular Spectroscopy*, *69*(2), 432–436. doi:10.1016/j.saa.2007.04.019 PMID:17719269

Yarov-Yarovoy, V., Schonbrun, J., & Baker, D. (2006). Multipass membrane protein structure prediction using Rosetta. Proteins. *Structure, Function, and Bioinformatics*, *62*(4), 1010–1025. doi:10.1002/prot.20817

Yebra, D. M., Kiil, S., & Dam-Johansen, K. (2004). Antifouling technology—past, present and future steps towards efficient and environmentally friendly antifouling coatings. *Progress in Organic Coatings*, *50*(2), 75–104. doi:10.1016/j.porgcoat.2003.06.001

Yilmaz, M. Z., Guzel, A., Torun, A. C., Okuyucu, A., Salis, O., Karli, R., & Alacam, H. et al. (2014). The therapeutic effects of anti-oxidant and anti-inflammatory drug quercetin on aspiration-induced lung injury in rats. *Journal of Molecular Histology*, *45*(2), 195–203. doi:10.1007/s10735-013-9542-3 PMID:24122261

Ying, M., Huang, F. W., Ye, H. D., Xu, H., Shen, L. L., Huan, T. W., & Zhou, K. et al. (2015). Study on interaction between curcumin and pepsin by spectroscopic and docking methods. *International Journal of Biological Macromolecules*, *79*, 201–208. doi:10.1016/j.ijbiomac.2015.04.057 PMID:25940524

Yin, L., Zheng, L., Xu, L., Dong, D., Han, X., Qi, Y., & Peng, J. et al. (2015). In-silico prediction of drug targets, biological activities, signal pathways and regulating networks of dioscin based on bioinformatics. *BMC Complementary and Alternative Medicine*, *15*(1), 41. doi:10.1186/s12906-015-0579-6 PMID:25879470

Yong-Chen, L., Wen-Chen, Y., & Pamela, S. O. (2008). LPS/TLR4 signal transduction pathway. *Cytokine*, *42*(2), 145–151. doi:10.1016/j.cyto.2008.01.006 PMID:18304834

Yuan, Y., Pei, J., & Lai, L. (2011). LigBuilder 2: A Practical de Novo Drug Design Approach. *Journal of Chemical Information and Modeling*, *51*(5), 1083–1091. doi:10.1021/ci100350u PMID:21513346

Yuedong, Y., & Yaoqi, Z. (2008). Specific interactions for ab initio folding of protein terminal regions with secondary structures. *Proteins*, *72*(2), 793–803. doi:10.1002/prot.21968 PMID:18260109

Yu, H., Fang, Y., Lu, X., Liu, Y., & Zhang, H. (2014). Combined 3D-QSAR, Molecular Docking, Molecular Dynamics Simulation, and Binding Free Energy Calculation Studies on the 5-Hydroxy-2H-Pyridazin-3-One Derivatives as HCV NS5B Polymerase Inhibitors. *Chemical Biology & Drug Design*, *83*(1), 89–105. doi:10.1111/cbdd.12203 PMID:23941500

Yuriev, E., Agostino, M., & Ramsland, P. A. (2011). Challenges and advances in computational docking: 2009 in review. *Journal of Molecular Recognition*, *24*(2), 149–164. doi:10.1002/jmr.1077 PMID:21360606

Yuriev, E., Holien, J., & Ramsland, P. A. (2014). Improvements, trends, and new ideas in molecular docking: 2012-2013 in review. *Journal of Molecular Recognition*, *28*(10), 581–604. doi:10.1002/jmr.2471 PMID:25808539

Yuriev, E., & Ramsland, P. A. (2013). Latest developments in molecular docking: 2010–2011 in review. *Journal of Molecular Recognition*, *26*(5), 215–239. doi:10.1002/jmr.2266 PMID:23526775

Zavodszky, M. I., & Kuhn, L. A. (2005). Side-chain flexibility in protein–ligand binding: The minimal rotation hypothesis. *Protein Science*, *14*(4), 1104–1114. doi:10.1110/ps.041153605 PMID:15772311

Zeng, H. J., Hu, G. Z., You, J., Yang, R., & Qu, L. B. (2015). Spectroscopic and molecular modeling investigation on the interactions between hyaluronidase and baicalein and chrysin. *Process Biochemistry*, *50*(5), 738–744. doi:10.1016/j.procbio.2015.02.007

Zeng, H. J., Ma, J., Yang, R., You, J., & Qu, L. B. (2015). Molecular interactios of flavonoids to hyaluronidase: Insights from spectroscopic and molecular modeling studies. *Journal of Fluorescence*, *25*(4), 941–959. doi:10.1007/s10895-015-1576-3 PMID:26006100

Zeng, H. J., You, J., Liang, H. L., Yang, R., & Qu, L. B. (2014). Investigation on the binding interaction between silybin and pepsin by spectral and molecular docking. *International Journal of Biological Macromolecules*, *67*, 105–111. doi:10.1016/j.ijbiomac.2014.02.051 PMID:24608028

Zhang, Y.-F., Xiao, K., Chandramouli, K. H., Xu, Y., Pan, K., Wang, W.-X., & Qian, P.-Y. (2011). *Acute toxicity of the antifouling compound butenolide in non-target organisms*. Academic Press.

Zhang, B., Corbel, C., Gueritte, F., Couturier, C., Bach, S., & Tan, V. B. (2011). An in silico approach for the discovery of CDK5/p25 interaction inhibitors. *Biotechnology Journal*, *6*(7), 871–881. doi:10.1002/biot.201100139 PMID:21681969

Zhang, J., Yang, P. L., & Gray, N. S. (2009). Targeting cancer with small molecule kinase inhibitors. *Nature Reviews. Cancer*, *9*(1), 28–39. doi:10.1038/nrc2559 PMID:19104514

Zhang, X. X., Wang, G. J., Gurley, E. C., & Zhou, H. P. (2014). Flavonoid apigenin inhibits lipopolysaccharide-induced inflammatory response through multiple mechanisms in macrophages. *PLoS ONE*, *9*, e107072. PMID:25192391

Zhang, X., Wong, S. E., & Lightstone, F. C. (2013). Message passing interface and multithreading hybrid for parallel molecular docking of large databases on petascale high performance computing machines. *Journal of Computational Chemistry*, *34*(11), 915–927. doi:10.1002/jcc.23214 PMID:23345155

Zhang, Y. Z., Zhou, B., Zhang, X. P., Huang, P., Li, C. H., & Liu, Y. (2009). Interaction of malachite green with bovine serum albumin: Determination of the binding mechanism and binding site by xpectroxcopic methods. *Journal of Hazardous Materials*, *163*(2-3), 1345–1352. doi:10.1016/j.jhazmat.2008.07.132 PMID:18786760

Zhang, Y.-F., Zhang, H., He, L., Liu, C., Xu, Y., & Qian, P.-Y. (2012). Butenolide inhibits marine fouling by altering the primary metabolism of three target organisms. *ACS Chemical Biology*, *7*(6), 1049–1058. doi:10.1021/cb200545s PMID:22458453

Zhang, Y., & Skolnick, J. (2005). TM-Align: A protein structure alignment algorithm based on TM-score. *Nucleic Acids Research*, *33*(7), 2302–2309. doi:10.1093/nar/gki524 PMID:15849316

Zhao, Y., Aguilar, A., Bernard, D., & Wang, S. (2015). Small-molecule inhibitors of the MDM2-p53 protein-protein interaction (MDM2 Inhibitors) in clinical trials for cancer treatment. *Journal of Medicinal Chemistry*, *58*(3), 1038–1052. doi:10.1021/jm501092z PMID:25396320

Zhao, Y., & Sanner, M. F. (2007). FLIPDock: Docking flexible ligands into flexible receptors. *Proteins: Structure, Function, and Bioinformatics*, *68*(3), 726–737. doi:10.1002/prot.21423 PMID:17523154

Zhao, Z., Wu, H., Wang, L., Liu, Y., Knapp, S., Liu, Q., & Gray, N. S. (2014). Exploration of type II binding mode: A privileged approach for kinase inhibitor focused drug discovery? *ACS Chemical Biology*, *9*(6), 1230–1241. doi:10.1021/cb500129t PMID:24730530

Zheng, M., Liu, X., Xu, Y., Li, H., Luo, C., & Jiang, H. (2013). Computational methods for drug design and discovery: Focus on China. *Trends in Pharmacological Sciences*, *34*(10), 549–559. doi:10.1016/j.tips.2013.08.004 PMID:24035675

Zheng, X., Gan, L., Wang, E., & Wang, J. (2013). Pocket-based drug design: Exploring pocket space. *The AAPS Journal*, *15*(1), 228–241. doi:10.1208/s12248-012-9426-6 PMID:23180158

Zhou, T., Commodore, L., Huang, W. S., Wang, Y., Thomas, M., Keats, J., & Zhu, X. et al. (2011). Structural mechanism of the Pan-BCR-ABL inhibitor ponatinib (AP24534): Lessons for overcoming kinase inhibitor resistance. *Chemical Biology & Drug Design*, *77*(1), 1–11. doi:10.1111/j.1747-0285.2010.01054.x PMID:21118377

Zhou, X., Wang, X. B., Wang, T., & Kong, L. Y. (2008). Design, synthesis, and acetylcholinesterase inhibitory activity of novel coumarin analogues. *Bioorganic & Medicinal Chemistry*, *16*(17), 8011–8021. doi:10.1016/j.bmc.2008.07.068 PMID:18701305

Zimmermann, O., & Hansmann, U. H. (2006). Support vector machines for prediction of dihedral angle regions. *Bioinformatics (Oxford, England)*, *22*(24), 3009–3015. doi:10.1093/bioinformatics/btl489 PMID:17005536

Zsoldos, Z., Reid, D., Simon, A., Sadjad, B. S., & Peter Johnson, A. (2006). eHiTS: An innovative approach to the docking and scoring function problems. *Current Protein & Peptide Science*, *7*(5), 421–435. doi:10.2174/138920306778559412 PMID:17073694

Zuker, M., & Stiegler, P. (1981). Optimal computer folding of large RNA sequences using thermodynamics and auxiliary information. *Nucleic Acids Research*, *9*(1), 133–148. doi:10.1093/nar/9.1.133 PMID:6163133

About the Contributors

Siavoush Dastmalchi graduated as Doctor of Pharmacy from Tabriz University of Medical Sciences (TUMS). Then he moved to Sydney where he received his PhD from the Faculty of Pharmacy at the University of Sydney in 2002. Since then he has worked as a full academic in the Medicinal Chemistry Department at the School of Pharmacy, TUMS, teaching medicinal chemistry, instrumental drug analysis and bioinformatics to graduate and postgraduate students. He is currently the Director of the Biotechnology Research Centre at TUMS where he leads his research team mainly with interests in molecular modelling, structural biology, and chemo-bioinformatics for their application to drug discovery.

Maryam Hamzeh-Mivehroud is an associate professor who graduated as Doctor of Pharmacy from Tabriz University of Medical Sciences in 2004 and received her PhD in Medicinal Chemistry from this university in 2011. Since then she has worked as a full academic member in the Medicinal Chemistry Department at the School of Pharmacy, and teaches medicinal chemistry at the undergraduate and postgraduate levels. Her main research interests are focused on QSAR and molecular modelling based on various description selection algorithms and machine learning methods. She is also professional in working with modelling and statistical softwares such as GROMACS, GOLD, AMBER, GOLPE, MOE, etc.

Babak Sokouti has over 17 years IT technical management and consulting experience, including managing and maintaining sophisticated network infrastructures. He has obtained Bachelor of Science in Electrical Engineering with a specialization in Control from Isfahan University of Technology, Isfahan, Iran; a Master of Science in Electrical Engineering with a specialization in Electronics (with background of biomedical engineering) from Tabriz Branch, Islamic Azad University, Tabriz, Iran; a Master of Science in Information Security with Distinction from Royal Holloway University of London, London, UK; and obtained PhD in Bioinformatics from Biotechnology Research Center, Tabriz University of Medical Sciences, Tabriz, Iran. In addition, he has obtained IT industry certifications including MCP, MCSA 2003, MCDBA 2000, MCSE 2003, and MCTS 2008. His research interests include cryptographic algorithms, information security, network security and protocols, image processing, protein structure prediction, and hybrid intelligent neural network systems based on genetic algorithms.

* * *

Rahul Agarwal is a 3rd year PhD student under Dr. Ashutosh Singh, in Department of Life Science, School of Natural Sciences, Shiv Nadar University.

Piyali Basak is an Assistant Professor at School of Bioscience & Engineering, Jadavpur University, Kolkata, India. She received her M.Pharm & Ph.D degrees in Biomedical Engineering and Biopolymers from Jadapur University and Indian Institute of Technology- Kharagpur respectively. Her research interests include biopolymers and targeted drug delivery, with focus on nanomaterials in drug delivery system. She has to her credit over 50 national & international publications and several book chapters.

Anamika Basu obtained her M Sc degree in Biochemistry in 2001 and M Tech degree in Biotechnology in 2003. She is a lecturer in Dept. of Biochemistry of Gurudas College in Kolkata from the year 2004. She teaches Immunology, Molecular Biology, Cell Biology and Clinical Biochemistry in graduate level. She is a co-author of two book chapters and seven conference papers. Her research interest includes Computational Biology, Bioinformatics, Molecular Modelling, Molecular Docking, Molecular Biology and Plant Biology.

Mohane Selvaraj Coumar is an Assistant Professor at the Centre for Bioinformatics Pondicherry University. He received his M.Pharm & Ph.D degrees in Pharmaceutical Sciences from Panjab University, Chandigarh, India. His research interests include Computer-aided drug design and Medicinal chemistry, with focus on Cancer drug discovery. He has to his credit over 50 national & international publications, three US and one EU patents.

Omar Deeb received his MSc (Analytical Chemistry) in 1986 and his PhD (Molecular Modelling and Drug Design) in 1993 from the Hebrew University of Jerusalem. This was followed by post-doctoral research at Chemometrics Laboratory, Orleans University, France, on QSAR of different pharmaceutical compounds. He joined Al-Quds University in 1995 and is currently a Professor in the faculty of Pharmacy. He is the group leader of computational chemistry/drug design laboratory. His research interests include QSAR and molecular modelling of pharmaceutical compounds, hydrogen bonding interactions, and proton transfer. His research achievements include QSAR of different groups of pharmaceutical compounds which leads to designing new drugs, application of new techniques such as flip regression for handling the symmetry of different systems, and the use of the orbital nodal angle descriptors in QSAR. Most of his publications including key articles, chapters in books, proceedings and oral presentations at international meetings related to QSAR.

Ramón Garduño-Juárez received his Ph.D. in Biophysics from the State University of New York at Buffalo in 1980. He was a NSF postdoctoral fellow at NASA Ames Research Center. Later on he held Full Professor positions at the Instituto Politécnico Nacional and Instituto Tecnológico de Tijuana. In 1984 he was hired by Universidad Nacional Autónoma de México where he is a Senior Professor up to date. He has written 50 research articles and has an h index of 12 and an i10 index of 15. He is a member of the Mexican Academy of Sciences and the Morelos Academy of Sciences. He has been distinguished with a fellowship from the Ministry of Education of México.

William A. Goddard III is the Charles and Mary Ferkel Professor of Chemistry, Materials Science, and Applied Physics and Director of the Materials and Process Simulation Center at California Institute of Technology (Caltech), where he has been on the faculty since Nov. 1964. Goddard has been a pioneer in developing methods for quantum mechanics (QM), force fields (FF), reactive dynamics (ReaxFF RD), electron dynamics (eFF), molecular dynamics (MD), and Monte Carlo (MC) predictions on chemical, catalytic, and biochemical systems and is actively involved in applying these methods to many materials, including proteins and nuclei acids. He has been developing methodology (GEnSeMBLE) to predict the 3D structures of G-Protein Coupled Receptors by evaluating the energies of a very complete sampling (13 trillion 7-helix bundles) of helix tilts and rotations starting from xray templates and for predicting ligand binding sites and energies using methods that sample complete sets of poses (DarwinDock). Chapter 6 shows validations for these methods.

Soo-Kyung Kim is a director at Biomacromolecular modeling section at California Institute of Technology (Caltech). She has been involved in numerous projects involving molecular modeling of the G protein-coupled receptors (GPCRs) which are important as drug targets; (1) Uroteinsin II, (2) Serotonin 5-HT2B/ 2C, (3) β2 adrenergic, (4) Chemokine CXCR3/ CCR5, (5) Adenosine A3/ A2A, (6) P2Y, (7) rhodopsin, (8) Muscarinic M2, (9) orphan GPR81, (10) Histamine H3, (11) kappa Opioid and (12) sensory (smell and taste) receptors. She has been studying highly subtype selective agonists or antagonists at various GPCR subtypes to reduce undesirable drug side effects. She also performed dynamic simulation to GPCRs in explicit membrane and water to understand the activation mechanism of agonist binding to specific GPCRs.

Heidy Martínez-Pachecho obtained her bachelor´s degree in Chemistry from Universidad del Papaloapan, Oaxaca, México in February 2013 and her Master degree of Science in Pharmacology from Instituto Politécnico Nacional, in March 2015. She is currently a student at the Biomedical Sciences PhD, Universidad Nacional Autónoma de México. Her research is centered on design in silico -Docking and QSAR-, biochemical and behavioral effects of new histone deacetylase inhibitors as a possible therapy of neurodegenerative diseases.

Kateryna V. Miroshnychenko received her MSc in Biophysics from the V. N. Karazin Kharkiv National University in 2004. During 2004-2007 she was Ph.D. student in the O.Ya. Usikov Institute for Radiophysics and Electronics of NAS of Ukraine. Since 2007 she works as a junior researcher in the department of Biological Physics of O.Ya. Usikov Institute for Radiophysics and Electronics of NAS of Ukraine under the supervision of the Sc.D. Anna V. Shestopalova. Her research interests include molecular docking of biologically active substances (BAS) to nucleic acids and proteins, molecular dynamic simulations of nucleic acids and proteins and their complexes with BAS.

Ling-bo Qu is a professor of College of Chemistry and Molecular Engineering and School of Chemistry & Chemical Engineering of Henan University of Technology. His research interests are focused on computational chemistry, pharmaceutical analysis and bioanalytical chemistry.

Guillermo Ramírez-Galicia received his PhD degree in Chemistry (Quantum Chemistry) from the Universidad Nacional Autónoma de México (UNAM) in 2002. His PhD research involved applications of ab initio electronic structure calculations for reaction mechanism. He completed six months postdoctoral fellowship in the Masaryk University, Czech Republic in 2004-2005 and two years postdoctoral fellowship in the Instituto de Ciencias Físicas, UNAM, Mexico in 2005-2007. He is currently Professor-Research in the Universidad del Papaloapan, Mexico since 2007.

Anasua Sarkar has completed her Ph D work on Bioinformatics in Jadavpur University, Kolkata, India in 2014. She was also awarded EMMA-EPU fellowship, 2011 to pursue her research work at LaBRI, University Bordeaux1, France. She is presently an Assistant Professor in Computer Science and Engineering Dept., Jadavpur University, Kolkata, India, from 2016. She has published 11 original research papers in peer-reviewed journals. She also a co-author in 8 book chapters and 15 conference papers. She is also reviewer in journals of Parallel and Distributed Computing (JPDC), Elsevier and IEEE SMCC-C. She is also a Student member, ISCB, 2010-2011, SMIEEE since 2014, member CSTA, ACM Chapter, ACM IGUCCS and IAENG. She has worked with INRIA MAGNOME group for 18 months in France for the expansion of Genolevures database for inclusion of PISO and ARAD species. Her research interests include Proteomics, Phylogenetics, Computational Biology, Pattern Recognition, Data Mining, Bioinformatics, Embedded, Parallel Systems.

Sailu Sarvagalla is a DBT-SRF pursuing his Ph.D program at the Centre for Bioinformatics, Pondicherry University. His interest include bioinformatics and computer-aided drug design in the area of cancer drug discovery. To his credit he has four publications in international journals.

Subhabrata Sen is an associate professor in Shiv Nadar University. He did his PhD from University of MIssouri-Columbia, MO, USA and postdoc from Colorado State Fort Collins with Dr. Albert I Meyers. After about 10 years in the industry he joined academia in 2013. He has several papers and patents to his credit.

Anna V. Shestopalova received high education in Biophysics in Kharkov State University in 1980. During 1980-1983 she was Ph.D. student in the Department of Molecular and Applied Biophysics, Radiophysics Faculty, Kharkiv State University. She received her Ph.D in Biophysics in 1987 from Institute for Physics Georgian Academy of Sciences, Tbilisi, Georgia (Ph.D. Thesis: Caffeine hydration and its interaction with system DNA-water) and D.Sc. degree in Biophysics from the V. N. Karazin Kharkiv National University in 2007 (D.Sc. Thesis: Molecular mechanisms of hydration and complex formation of nucleic acids and biological active compounds). Her research interests include the mechanisms of interactions in molecular systems, namely: biomacromolecules (or their components and derivatives) - water - ions - biologically active agents and mechanisms of molecular recognition in protein - nucleic acid complexes. Anna V. Shestopalova is Head of Biophysical Department in O. Ya. Usikov Institute for Radiophysics and Electronics of NAS of Ukraine. She is the author and co-author of 126 scientific publications, including 3 monographs.

Ashutosh Singh is currently working as Assistant Professor in Shiv Nadar University and currently he is working on Bioinformatics methods for identification of Novel Biomarkers in Cancer and rare diseases. He has worked in Academia,Industry and R&D institutions, has good expertise on drug discovery methods specially Molecular Modelling and Docking.

Ran Yang is associate professor of College of Chemistry and Molecular Engineering, Zhengzhou University. Her research interests are focused on biosensor based on the nanomaterials.

Nanda Kumar Yellapu is a Bioinformatics Scientist at Biomedical Informatics Centre of Vector Control Research Centre (ICMR), Pondicherry, India. He obtained his Masters degree from SVIMS University and PhD from Sri Venkateswara University, Tirupati, India. His research focuses on the protein conformational studies and drug discovery strategies through molecular modelling studies. His current research is to identify the insecticidal and drug resistance mechanisms due to mutations in the malarial vectors and pathogens.

Hua-jin Zeng is an associate Professor of Zhengzhou University in China. He received his PhD degree in Pharmcognosy from China Pharmaceutical University in 2006. His research interests cover pharmaceutical analysis, biological chemistry and molecular docking.

Index

T

V

www.ingramcontent.com/pod-product-compliance
Lightning Source LLC
LaVergne TN
LVHW080535300926
849162LV00007B/1613